GUIDE TO EUROPEAN AUTOMOTIVE ELECTRICAL SYSTEMS

by **ROB SIEGEL**

and the Bentley Publishers Technical Team

[B] BentleyPublishers®
.com

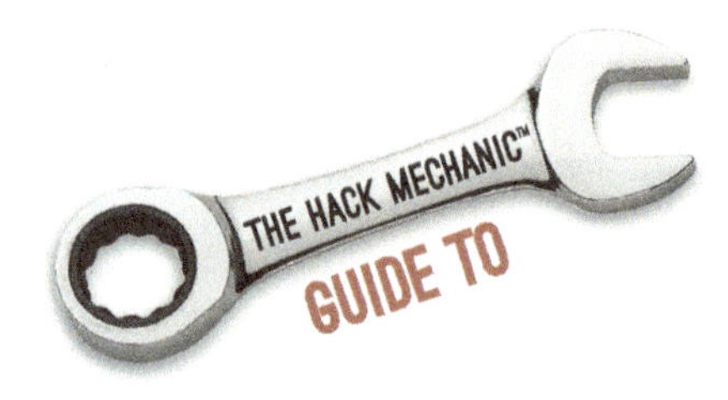

EUROPEAN AUTOMOTIVE ELECTRICAL SYSTEMS

Testing battery voltage
Page 2-4

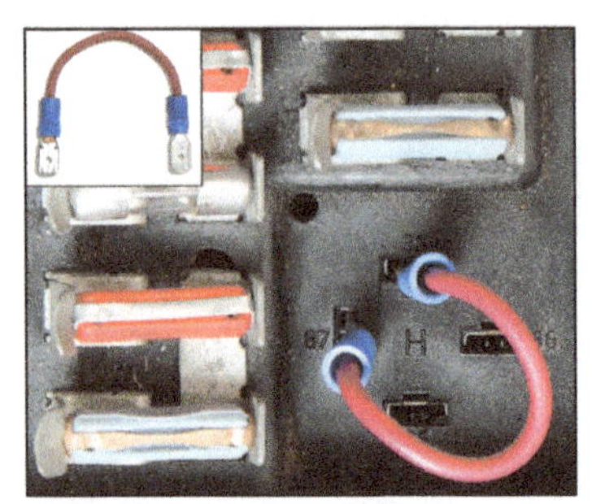

Installing homemade relay jumper
Page 7-13

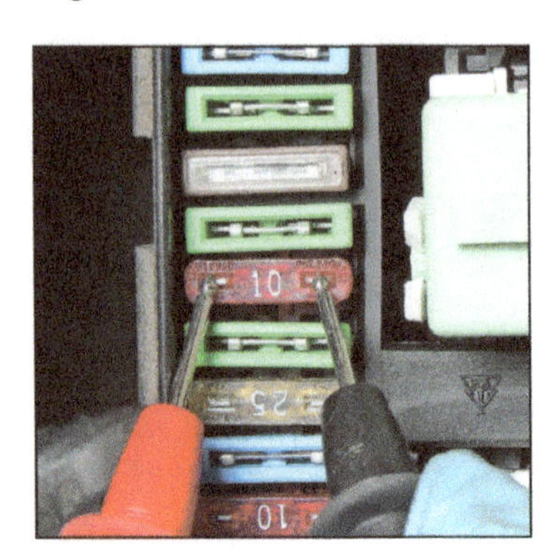

Testing fuse continuity
Page 10-6

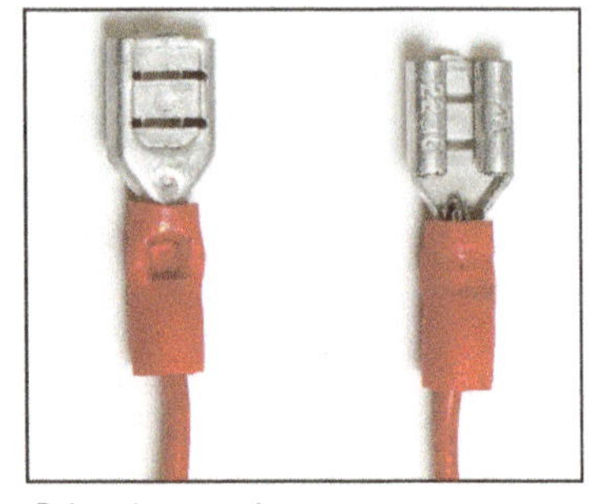

Crimping a wire connector
Page 11-7

by **ROB SIEGEL**

and the Bentley Publishers Technical Team

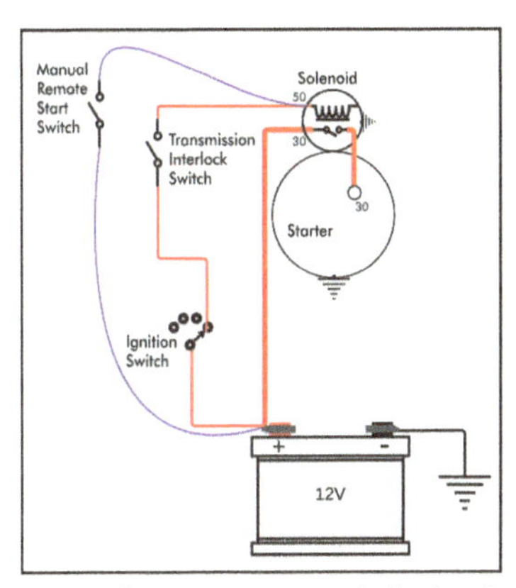

Manual remote start switch circuit
Page 15-7

VRT electronic ignition
Page 18-3

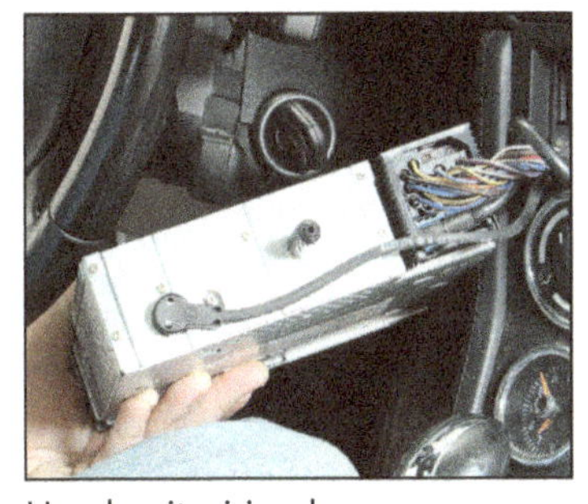

Head unit wiring harness
Page 22-4

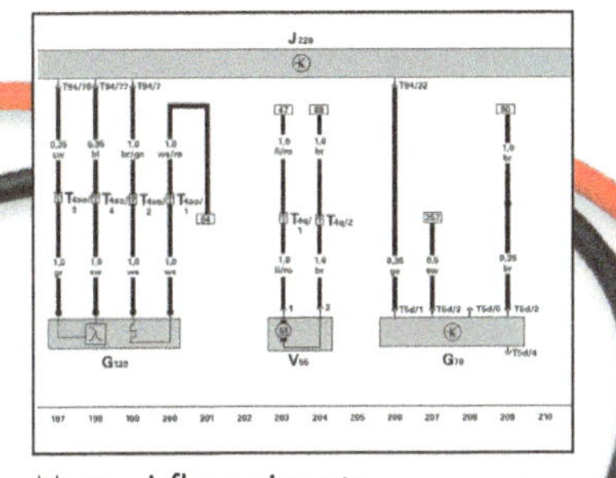

Mass airflow pinouts
Page 34-6

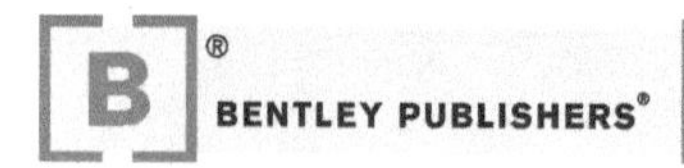

Bentley Publishers, a division of Robert Bentley, Inc.
1734 Massachusetts Avenue
Cambridge, MA 02138 USA
800-423-4595 / 617-547-4170

Information that makes the difference®

BentleyPublishers®.com

Technical contact information

We welcome your feedback. Please submit corrections and additions to our technical discussion forums at:

`http://www.BentleyPublishers.com`

Updates and corrections

We will evaluate submissions and post appropriate editorial changes online as updates or tech discussion. Appropriate updates and corrections will be added to the book in future printings. Check for updates and corrections for this book before beginning work on your vehicle. See the following web address for additional information:

`http://www.BentleyPublishers.com/updates/`

WARNING—Important Safety Notice

Do not use this manual for repairs unless you are familiar with automotive repair procedures and safe workshop practices. This manual illustrates the workshop procedures for some maintenance and service work. It is not a substitute for full and up-to-date information from the vehicle manufacturer or for proper training as an automotive technician. Note that it is not possible for us to anticipate all of the ways or conditions under which vehicles may be serviced or to provide cautions as to all of the possible hazards that may result.

We have endeavored to ensure the accuracy of the information in this manual. Please note, however, that considering the vast quantity and the complexity of the service information involved, we cannot warrant the accuracy or completeness of the information contained in this manual.

FOR THESE REASONS, NEITHER THE PUBLISHER NOR THE AUTHOR MAKES ANY WARRANTIES, EXPRESS OR IMPLIED, THAT THE INFORMATION IN THIS MANUAL IS FREE OF ERRORS OR OMISSIONS, AND WE EXPRESSLY DISCLAIM THE IMPLIED WARRANTIES OF MERCHANTABILITY AND OF FITNESS FOR A PARTICULAR PURPOSE, EVEN IF THE PUBLISHER OR AUTHOR HAVE BEEN ADVISED OF A PARTICULAR PURPOSE, AND EVEN IF A PARTICULAR PURPOSE IS INDICATED IN THE MANUAL. THE PUBLISHER AND AUTHOR ALSO DISCLAIM ALL LIABILITY FOR DIRECT, INDIRECT, INCIDENTAL OR CONSEQUENTIAL DAMAGES THAT RESULT FROM ANY USE OF THE EXAMPLES, INSTRUCTIONS OR OTHER INFORMATION IN THIS MANUAL. IN NO EVENT SHALL OUR LIABILITY, WHETHER IN TORT, CONTRACT OR OTHERWISE, EXCEED THE COST OF THIS MANUAL.

Before attempting any work on your vehicle, read **01 Safety** and any **WARNING** or **CAUTION** that accompanies a procedure in the manual. Review the **WARNINGS** and **CAUTIONS** each time you prepare to work on your vehicle.

Your common sense and good judgment are crucial to safe and successful service work. Read procedures through before starting them. Think about whether the condition of your car, your level of mechanical skill or your level of reading comprehension might result in or contribute in some way to an occurrence which might cause you injury, damage your car or result in an unsafe repair. If you have doubts for these or other reasons about your ability to perform safe repair work on your car, have the work done at an authorized dealer or other qualified shop.

Part numbers listed in this manual are for identification purposes only, not for ordering. Always check with your authorized dealer to verify part numbers and availability before beginning service work that may require new parts.

Special tools required to perform certain service operations are identified in the manual and are recommended for use. Use of improper tools may be detrimental to the car's safe operation as well as the safety of the person servicing the car.

The vehicle manufacturer will continue to issue service information updates and parts retrofits after the editorial closing of this manual. Some of these updates and retrofits will apply to procedures and specifications in this manual. We regret that we cannot supply updates to purchasers of this manual.

This book is prepared, published and distributed by Bentley Publishers, 1734 Massachusetts Avenue, Cambridge, Massachusetts 02138 USA.

ISBN 978-0-8376-1751-0 Editorial closing 03/2016 Job code: BHME-01-b001

Library of Congress Cataloging-in-Publication Data

Title: The Hack Mechanic guide to European automotive electrical systems / by Rob Siegel and the Bentley Publishers Technical Team.
Other titles: Guide to European automotive electrical systems
Description: Cambridge, Massachusetts : Bentley Publishers Technical Team, [2017] | Includes bibliographical references and index.
Identifiers: LCCN 2016007822 | ISBN 9780837617510 (pbk. : alk. paper)
Subjects: LCSH: Automobiles, Foreign--Electric equipment--Maintenance and repair--Handbooks, manuals, etc. | Automobiles--Europe--Electric equipment--Maintenance and repair--Handbooks, manuals, etc.
Classification: LCC TL272 .S5155 2017 | DDC 629.25/40287--dc23
LC record available at http://lccn.loc.gov/2016007822

Manufactured in the United States of America

Preface

So your European car has electrical problems. At least you think your European car has electrical problems. Maybe it won't start. Maybe the battery keeps running down and you don't know why. Maybe it starts but dies. Maybe one tail light is out, or worse, intermittent. Maybe you're not having problems now but you own a model that has a reputation for having electrical problems and you want to inoculate yourself with a shot of knowledge. Maybe your car has thrown an oxygen sensor code and you want to know how to test the sensor to determine if it's really bad.

Or maybe it's not about problems at all. Maybe you want to install a cooling fan big enough to suck a schnauzer off the sidewalk or headlights bright enough to blind a moose at half a mile and you're wondering if the 30-year-old speaker wire you found in the basement will work.

But what do we mean by *electrical problems*? And how can this book help you? In truth, since so many components in your car run off electricity, the term "electrical problems" can mean almost anything. But from a practical standpoint, this book will help you classify a problem into one of the following rough set of possibilities:

- Battery-related problems
    - The battery has run down and won't start the car.
    - The battery was run down, was recharged or jumped, and starts the car, but it's still weak, and that causes problems in a post-1990 electronics-laden vehicle.
- Charging-related problems
    - The alternator/regulator has malfunctioned and caused the battery to run down.
    - The battery runs down overnight for some unknown reason.
    - The combination of the battery and the alternator/regulator are functioning below specification, and that causes problems in a post-1990 electronics-laden vehicle.
- Component failure problems (an isolated component such as a switch, motor, relay, sensor, or actuator has died)
- Circuit failure problems
    - A wire has broken off a connector, or a connector isn't making contact.
    - A wire or connector on either the power side or the ground side is too corroded to carry the required amount of current (when it's on the ground side, people call it a "bad ground").
    - A power wire is shorting to ground and blowing fuses.
    - A power wire is shorting to another power wire, causing things to be on that shouldn't be on.
    - A wire is broken somewhere where you can't easily see it.
- Module failure problems (an electronic module that controls several components has died)

This book's goal is to help you understand how electricity works in your car well enough to be able to troubleshoot these problems. You'll develop an appreciation for the fact that alternators aren't meant to recharge dead batteries, and thus many mysterious electrical problems, particularly on a car with computers in it, vanish once the battery and alternator are up to snuff. You'll learn that, while wires can break or rub off their insulation and short to ground, it's far more likely for failures to be at connection points than in the middle of wires. You'll be able to safely add new electrical circuits for things such as stereos, gauges, and, yes, that schnauzer-sucking cooling fan and moose-blinding lighting.

You'll also come to understand the difference between static signals (those that present the same value all the time) and dynamic signals (those that change, either slowly or quickly, either regularly or irregularly). You'll learn how to apply this to troubleshoot common sensors and actuators such as oxygen sensors, crankshaft position sensors, ABS sensors, fuel pumps, and fuel injectors by understanding what kind of signals they either output or expect. You'll learn which of these devices require a simple inexpensive multimeter to troubleshoot, which require a more sophisticated automotive multimeter capable of detecting dynamically changing signals (square waves and sine waves), and why nothing beats an oscilloscope.

You'll learn about the digital end of the automotive world, the proliferation of computers and modules in cars, and where scan tools fit into all this. And you'll learn that if multiple systems are doing weird things, there may be a module problem, in which case the odds are you're not the first person to experience it, and a web search of a user forum is likely to point you in the direction of the offending module.

And, seriously, about using that 30-year-old speaker wire . . . you'll learn why using the thin wire to connect a heavy-duty electrical device is incredibly dangerous and could literally burn your ride to the ground.

While the book's content is particularly germane to European cars, which share a common underlying electrical DNA in the form of Bosch components and adherence to a terminal numbering scheme referred to as the DIN, the topics are, in fact, common to all cars.

So hang on. It's going to get electric.

Abbreviations and Acronyms

Below is a list of abbreviations and acronyms used in the book, along with their meanings.

Abbreviation or Acronym	Meaning
A	Amperes
A/C	Air Conditioning
AFR	Air/Fuel Ratio
A/F	Air/Fuel (Sensor)
ABS	Antilock Braking System
AC	Alternating Current
AGM	Absorbed Glass Mat
Ah	Amp-Hour
amp	Amperes
AMR	Anisotropic Magnetoresistive
ASC	Automatic Stability Control
AWD	All Wheel Drive
AWG	American Wire Gauge
BCI	Battery Council International
BSD	Bit Serial Data
BST	Battery Safety Terminal
BTDC	Before Top Dead Center
CA	Cranking Amps
CAN	Controller Area Network
CARB	California Air Resources Board
cat	Catalytic Converter
CCA	Cold Cranking Amps
CDI	Capacitive Discharge Ignition
CEL	Check Engine Light
CKP	Crankshaft Position (Sensor)
CMP	Camshaft Position (Sensor)
CNP	Coil Near Plug
COP	Coil On Plug
CRT	Cathode Ray Tube
CTS	Coolant Temperature Sensor
DC	Direct Current
DI	Distributor Ignition
DIN	Deutsches Institut für Normung
DIY	Do It Yourself
DLC	Data Link Connector
DME	Digital Motor Electronics
DMM	Digital Multimeter
DOHC	Double Overhead Camshaft
DPDT	Double Pole Double Throw
DPST	Double Pole Single Throw
DTC	Diagnostic Trouble Code
DVM	Digital Voltmeter

Abbreviation or Acronym	Meaning
DVOM	Digital Volt Ohm Meter
ECM	Engine Control Module
ECT	Engine Coolant Temperature
ECU	Electronic Control Unit
EEPROM	Electronically Erasable Programmable Read-Only Memory
EFI	Electronic Fuel Injection
EGR	Exhaust Gas Recirculation
EI	Electronic Ignition
EIA	Electronic Industries Association
EMI	Electromagnetic Interference
EN	European Norm
EPA	Environmental Protection Agency
ETA	Electronic Throttle Actuator
ETM	Electrical Troubleshooting Manual
EVAP	Evaporative System Monitor
FTP	Federal Test Procedure
FWD	Front-Wheel Drive
GND	Ground
HID	High Intensity Discharge
hp	Horsepower
HPDI	High-Pressure Direct Injected
HVAC	Heater Ventilation Air Conditioning
Hz	Hertz
I	Current
I/M	Inspection and Maintenance
IAC	Idle Air Control
IBS	Intelligent Battery Sensor
IDI	Inductive Discharge Ignition
IM	Inspection And Maintenance
ISO	International Standards Organization
JIS	Japan Industrial Standard
KAM	Keep Alive Memory
kg	Kilogram
kHz	Kilohertz
km	Kilometer
kW	Kilowatt
LED	Light Emitting Diode

Abbreviation or Acronym	Meaning
LIN	Local Interconnect Network
mA	Milliamp
MAF	Mass Airflow
MCA	Marine Cranking Amps
MFR	Multifunction Regulator
MIL	Malfunction Indicator Light
MIL-S	Military Specification
MMI	Multimedia Interface
MOST	Media Oriented Systems Transport
MS	Millisecond
mV	Millivolt
MY	Model Year
NC	Normally Closed
NMHC	Non-Methane Hydrocarbon
NO	Normally Open
NOx	Nitric Oxide
NTC	Negative Temperature Coefficient
NVLD	Natural Vacuum Leak Detection
O2	Oxygen Sensor
OBD	On-Board Diagnostics
OE	Original Equipment
OEM	Original Equipment Manufacturer
OL	Over Limit
OSHA	Occupational Health And Safety Administration
P/N	Part Number
PCV	Positive Crankcase Ventilation
PROM	Programmable Read Only Memory
psi	Pounds Per Square Inch
PSM	Porsche Stability Management
PVC	Polyvinyl Chloride
PWM	Pulse Width Modulation
R	Resistance
RAM	Random Access Memory
RCM	Reserve Capacity Minutes
RFI	Radio Frequency Interference
RMS	Root Mean Squared
ROM	Read-Only Memory

Abbreviation or Acronym	Meaning
rpm	Revolutions Per Minute
RV	Recreational Vehicle
RWD	Rear-Wheel Drive
SAE	Society of Automotive Engineers
SCR	Selective Catalytic Reduction
SOC	State of Charge
SOF	State of Function
SOH	State of Health
SPDT	Single Pole Double Throw
SPST	Single Pole Single Throw
SRS	Supplemental Restraint System
SUV	Sports Utility Vehicle
TCM	Transmission Control Module
TDC	Top Dead Center
TIS	Technical Information System (BMW)
TDR	Time Domain Refectometer

Abbreviation or Acronym	Meaning
TPS	Throttle Position Sensor
TVS	Throttle Valve Sensor
ULEV	Ultra Low-Emission Vehicle
USB	Universal Standard Bus
VAC	Volts AC
VAG	Volkswagen Audi Group
VDC	Volts DC
VDI	Voltage Drop Index
Vloss	Voltage Loss
VOM	Volt Ohm Meter
VRT	Variable Reluctance Transducer
VVT	Variable Valve Timing
W	Watt
WIS	Workshop Information System (Mercedes)
WSS	Wheel Speed Sensor

1

Safety

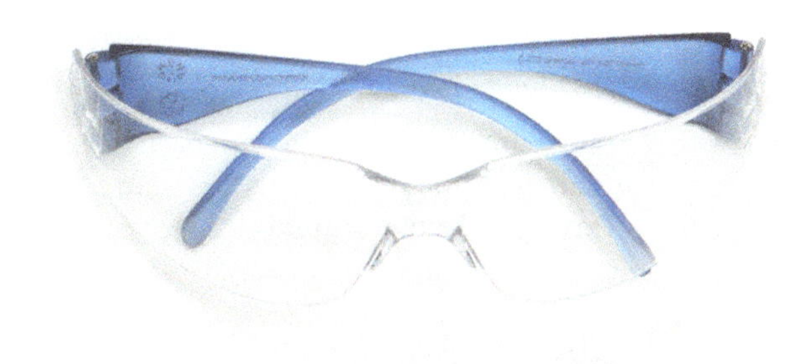

General

This chapter discusses practices and procedures that are intended to prevent electrical shock and injury posed by the hazards of automotive batteries and automotive electrical circuits. Your common sense, sound reasoning, and good judgment are crucial to safe service work.

Vehicle electrical circuits are designed with occupant and technician safety in mind. However, there are inherent dangers when working around electricity. Good workshop practices can help ensure safety.

High-Voltage Hybrid and Electric Vehicles

This book applies specifically to gasoline and diesel (combustion) vehicles. **This book is not applicable to high-voltage hybrid or electric vehicles.** High-voltage automotive circuits can be dangerous and even lethal. Specialized training and high-voltage test equipment is required to service the high-voltage side of a hybrid or electric car.

 **WARNING —**

Hybrid and electric vehicles use high-voltage electric motor/generators, high-voltage electrical circuits and wiring, and other potentially dangerous high-voltage components. Voltages in hybrid and electric vehicle circuits can be as high as 650 volts (AC or DC), depending upon the vehicle manufacturer. This is a lethal voltage level.

Safety as a State of Mind

Always make personal safety your top priority. Watch what you are doing and always stay alert. If something seems too dangerous to do, it likely is.

Figure 1 Think about the work you will be undertaking. Are you mentally and physically up for the task? Are you in the right state of mind?

Do not work on your car if you are tired, under the influence of drugs or alcohol, when you are not feeling well, or if you are in a bad mood. Your common sense and good judgment are crucial to safe and successful electrical work.

Think about whether the condition of your car, your level of electrical and mechanical skill, or your level of reading comprehension might result in personal injury, damage to your car, or an unsafe repair. If you have doubts for these or other reasons about your ability to perform safe electrical work on your car, have the work done at an authorized dealer or other qualified shop.

A good practice is to check that the key is off and preferably out of the ignition and on your person. This not only helps to be make sure live circuits are turned off and are able go to sleep, it also prevents anyone from starting the engine or turning the key while you are working.

12-Volt Battery Hazards

The car's 12-volt battery and 14.2-volt charging system voltage will not normally produce enough current flow to cause a severe electric shock, but since there are dangers present, including high-voltage circuit dangers (ignition and High Intensity Discharge [HID] headlights, for example), it is a good idea to be mindful around car batteries and all live circuits. Prior to undertaking any electrical repairs or testing, read the Warnings and Cautions in this chapter and included in other chapters.

Figure 2 Safety step one: Turn off and remove the ignition key.

Although shock hazard from a 12V car battery is minimal, there are potential battery-related dangers to be aware of:

- Battery acid: Battery acid is highly corrosive and can burn your skin on contact.
- Explosive hydrogen gas: A spark can cause a battery to explode violently, sending battery fragments and acid all over your skin.
- Sparks (arcing) between a car battery terminal and other metal parts: This can cause the metal to get hot enough to cause burns.
- Dead shorts: If a cable, a wrench, or other metal parts get shorted across the terminals of a good battery, it can start a fire.
- A frozen or swollen battery is a dangerous battery. Always allow a battery to thaw before servicing or charging. If the battery is swollen due to overcharging or venting issues, use extreme care to prevent any sparks. Mishandling a swollen battery can lead to an explosion (see **Figure 3**). Use common sense when working with a swollen or frozen battery.

Figure 3 A frozen or swollen battery is an explosion hazard. *Photo by Bertho Boman*

Voltage, Current and the Human Body

When your body is exposed to a live electrical circuit, the circuit voltage can cause an electric current to pass through your body (a shock). Major factors that affect the severity of the shock received when your body is part of an electrical circuit are:

- The magnitude or amount of current flowing through the body. Higher voltages "push" with greater force and make more electric current flow.
- The path of the current through the body.
- The length of time that the body is part of the circuit and is exposed to the current.

Other factors that may effect the severity of a shock are:

- The voltage applied to the circuit. Voltages above 60 volts DC (direct current) can push enough current through the body to be dangerous. The effects of AC (alternating current) voltages can be three to five times as great as DC voltages. 25 volts (rms) AC can be a hazard.
- The level of moisture or humidity in the surrounding environment.
- The phase of the human heart cycle when the shock occurs.
- The general health and condition of the person experiencing the shock.

Effects on the body from an electrical shock can range from a barely perceptible tingle to severe burns and immediate cardiac arrest. Also, involuntary reactions to a shock may cause additional injuries from striking nearby objects.

How Much Current Is Too Much?

In general, all wires should be considered live and potentially dangerous unless you absolutely know otherwise. When working on vehicle electrics, exercise care to avoid contact with both positive and negative live uninsulated wiring terminals at the same time.

Although it is not possible to list exact injuries that would result from exposure to a given amperage, general conclusions can be drawn. The table below shows the effects of increasing electric currents on the body based on 60-cycle AC house current.

Probable Effects of Electric Currents on the Human Body
(assumes a shock of 1 second duration from an AC 60-cycle voltage)

Current level	Probable effects
1 milliamp	Perception level. Slight tingling sensation. Still dangerous under certain conditions, especially wet conditions.*
5 milliamps	Slight shock felt, not painful but disturbing. Average person can let go. However, strong involuntary reactions to shocks in this range can lead to injury.
6 to 30 milliamps	Painful shock, muscular control is lost. Commonly referred to as the “freezing current” or the “let-go” range.
50 to 150 milliamps	Extreme pain, respiratory arrest, severe muscular contractions. Individual cannot let go. Can be fatal.**
1 to 4.3 amps	Ventricular fibrillation: the rhythmic pumping action of the heart ceases. Muscular contraction and nerve damage begins to occur. Fatal injuries likely.***
10 amps	Cardiac arrest, internal organ damage, severe burns. Death probable.

*Wet conditions contribute to low-voltage electrocutions.
Dry skin has high electrical resistance, but wet skin has much lower electrical resistance.
Dry skin: current = volts/ohms = 120/100,000 = 1 mA (this is a barely perceptible current level).
Wet skin: current = volts/ohms = 120/1,000 = 120 mA (this current level is sufficient to cause ventricular fibrillation).

**When muscular contractions caused by electrical stimulation (shock) do not allow the victim to let go, even relatively low voltages can be dangerous. The degree of injury increases with the length of time that the body is exposed to the current. 100 mA for 3 seconds is equivalent to 900 mA for 0.03 seconds. Both levels can cause ventricular fibrillation.

***High-voltage electrical energy greatly reduces the body’s resistance by quickly breaking down and puncturing human skin. Once the skin is punctured, electrical resistance is greatly reduced, resulting in massive current flow.

(Courtesy OSHA, www.osha.gov)

Warnings and Cautions

Before attempting any work on your vehicle, read **WARNINGS** and **CAUTIONS** in this chapter, as well as any **WARNINGS** and **CAUTIONS** that accompany a procedure in this book.

WARNINGS —
See also **CAUTIONS** *on next page*

- The ignition system produces high voltages that can be fatal. Avoid contact with exposed terminals. Use extreme caution when working on a car with the ignition switched ON or the engine running.
- Do not touch or disconnect any cables from the ignition coil(s) while the engine is running or being cranked by the starter.
- Connect and disconnect ignition system wiring and test equipment leads when the ignition is OFF.
- Before operating the starter without starting the engine (for example when testing compression), disable the ignition system. One good way is to remove the engine management main relay.
- Most cars from the late 1980s are equipped with a supplemental restraint system (SRS) that automatically deploys airbags and pyrotechnic seat belt tensioners in case of a frontal or side impact. These are explosive devices. Handled improperly or without adequate safeguards, they can be accidentally activated and cause serious injury.
- Dress properly. Never wear baggy clothing. If you have long hair you should tie it back so that it doesn't make contact with rotating parts. The same is true of jewelry and anything else that dangles loose from the body.
- Remove rings, bracelets and other jewelry so that they cannot cause electrical shorts, get caught in running machinery, or be crushed by heavy parts.
- To ensure safety, clothes should cover the whole body. When appropriate, safety glasses, hand protection, dust masks, and other protective gear should be used. Work boots or steel-toe boots should be worn as appropriate as well.
- Always wear personal protective equipment according to the type of job and tool you are going to be using. Always wearing safety glasses when working on your vehicle is a good safety practice. Minimize the damage to your hearing by investing in a pair of earplugs or earmuffs. Tools and power equipment can be loud, especially in a garage or workshop. Also wear a dust mask and non-skid safety shoes when needed.
- Battery acid (electrolyte) can cause severe burns. Flush contact area with water, then seek medical attention.
- Lead acid batteries give off explosive hydrogen gas during charging. Keep sparks, lighted matches, and open flames away from the top of the battery. If escaping hydrogen gas is ignited, the flame may travel into the cells and cause the battery to explode.
- Disconnect the battery negative (–) terminal whenever you work on the fuel system or the electrical system. Do not smoke or work near heaters or other fire hazards. Keep an approved fire extinguisher handy.
- Aerosol cleaners and solvents may contain hazardous or deadly vapors and are highly flammable. Use only in a well-ventilated area. Do not use on hot surfaces (engines, brakes, etc.).
- Due to risk of personal injury, be sure the engine is cold before beginning work.
- Do not reuse fasteners that are worn or deformed. Many fasteners are designed to be used only once and become unreliable and may fail when used a second time. This includes, but is not limited to, nuts, bolts, washers, self-locking nuts or bolts, circlips, and cotter pins. Replace these fasteners with new parts.
- Catch draining fuel, oil, or brake fluid in suitable containers. Do not use food or beverage containers that might mislead someone into drinking from them. Store flammable fluids away from fire hazards. Wipe up spills at once, but do not store the oily rags, which can ignite and burn spontaneously.
- Friction materials (such as brake pads and shoes or clutch discs) contain asbestos fibers or other friction materials. Do not create dust by grinding, sanding, or by cleaning with compressed air. Avoid breathing dust. Breathing any friction material dust can lead to serious diseases and may result in death.
- Greases, lubricants, and other automotive chemicals contain toxic substances, many of which are absorbed directly through the skin. Read the manufacturer's instructions and warnings carefully. Use hand and eye protection. Avoid direct skin contact.
- The air-conditioning system is filled with chemical refrigerant, which is hazardous. Make sure the system is serviced only by a trained technician using approved refrigerant recovery/recycling equipment, trained in related safety precautions, and familiar with regulations governing the discharge and disposal of automotive chemical refrigerants.
- Read each procedure and the **WARNINGS** and **CAUTIONS** that accompany the procedure. Also review posted corrections at ***www.BentleyPublishers.com/updates/*** before beginning work.

⚠ CAUTIONS —
See also **WARNINGS** *on previous page*

- If you lack the skills, tools and equipment, or a suitable workshop for any procedure described in this book, leave such repairs to a qualified shop or an authorized dealer.
- Manufacturers are constantly improving their cars and sometimes these changes, both in parts and specifications, are made applicable to earlier models. Any part numbers listed in this manual are for reference only. Check with your authorized dealer parts department for the latest information.
- Before starting a job, read instructions thoroughly, and do not attempt shortcuts. Use tools appropriate to the work and use only replacement parts meeting the manufacturer's specifications.
- Be mindful of the environment and ecology. Before you drain the crankcase, find out the proper way to dispose of the oil. Do not pour oil onto the ground, down a drain, or into a stream, pond, or lake. Dispose of waste in accordance with federal, state and local laws.
- Connect and disconnect a battery charger only with the battery charger switched OFF.
- Label battery cables before disconnecting. On some models, battery cables may not be color coded.
- Disconnecting the battery may erase fault code(s) stored in control module memory. It is recommended that you check for fault codes using special diagnostic equipment prior to disconnecting battery cables.
- If a normal or rapid charger is used to charge the battery, disconnect the battery or remove it from the vehicle in order to avoid damaging the vehicle.
- Do not quick-charge a lead acid battery (for boost starting) for longer than one minute. Wait at least one minute before boosting the battery a second time.
- Do not quick-charge an AGM battery and do not exceed 14.6 volts during charging.
- Sealed or "maintenance free" lead acid batteries should be slow-charged only, at an amperage rate that is approximately 10% of the battery's ampere-hour (Ah) rating.
- If the battery begins producing gas or boiling violently, reduce the charging rate. Boosting a sulfated battery at a high charging rate can cause an explosion.
- Before doing any electrical welding on a car, disconnect the battery negative (–) terminal (ground strap). Make sure ignition is switched OFF before disconnecting the battery.
- Use pneumatic and electric tools only to loosen threaded parts and fasteners. Do not use these tools to tighten fasteners, especially on light alloy parts. Use a torque wrench to tighten fasteners to the tightening torque specification listed.

Safety Equipment

The possibility of injuries caused by an electric shock, splashed chemicals, flying foreign objects, and fire can be reduced by using proper safety equipment.

Eye Protection

From chemical splashes to foreign object projectiles, serious eye injuries happen when you least expect them. Grease and oil splashes, burns from steaming coolant, and flying metal chips are all potential hazards when working on your car.

The best way to protect yourself from eye injuries is to wear protective eye-wear and be aware of the dangers around you. Always keep your safety eye-wear in good condition and replace if damaged.

Eye protection can be safety goggles or safety glasses. **Safety goggles with wraparound sides** should be worn if sprays or liquids are involved or when working near the battery. If there is a chance that something could be splashed or sprayed in the eye during your work, or if poisonous liquids are involved, a pair of safety goggles is the best choice. See **Figure 4**.

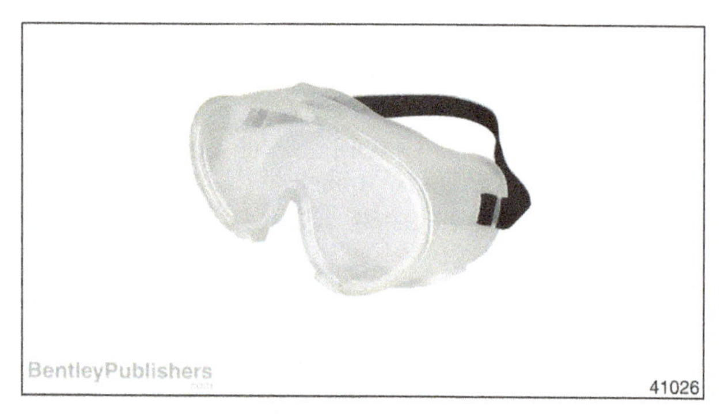

Figure 4 Wear safety goggles when working with sprays and liquids, or when working near the battery.

Safety glasses are the first line of defense from flying foreign objects. Both the frame and lenses are designed to be high impact resistant. Safety glasses won't normally alter your vision or scope of view and are light-weight. A good quality pair of safety glasses will have an antislip rubberized nose piece and temples. See **Figure 5**.

Figure 5 High-impact resistant safety glasses.

Hand Protection

While protecting the eyes is critically important, it's actually the hands that encounter the most abuse in the garage.

Not only do gloves protect the hands from sharp sheet metal and other skin-gashing hazards, they protect the skin from caustic chemicals. Skin is porous. The relationship of long-term exposure to chemicals and solvents and kidney, liver, and nerve damage are well documented. Dermatitis (inflammation of the skin) is a common aliment among technicians that are exposed to greases and solvents.

As with most other shop tools, there are specific gloves that work well for specific jobs. If you are simply trying to keep your hands clean and prevent chemical absorption, then the disposable "exam" glove is the glove of choice. They are tight fitting and provide great tactile sensitivity. And the best part is that clean-up is quick and easy.

Disposable gloves are available in vinyl, latex, and nitrile. Vinyl gloves can work well, but are not terribly durable. Latex gloves also work well, but they degrade quickly in the presence of gasoline. In addition, because they are a natural rubber product, some people have allergic reactions when wearing them.

Nitrile gloves are reported to be three times more puncture resistant than vinyl or latex gloves. Try to find the heavy-duty 9 mil gloves versus the common 6 mil. See **Figure 6**.

For an even higher level of protection, consider mechanic's gloves. They are made of a thick substantive material, so the trade-off is a minor loss of dexterity. They are resistant to most chemicals, have insulating properties, and offer lots of grip. However, grease and oil will eventually ruin them. See **Figure 7**.

If using mechanic's gloves while working around automotive ignition systems, be certain to select the ones with a nitrile coating for electrical insulation.

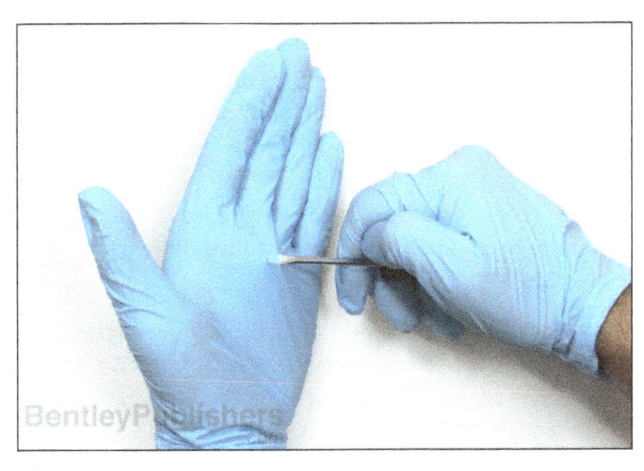

Figure 6 Nitrile disposable gloves are up to three times more puncture resistant than latex gloves.

Figure 7 Wearing gloves in the shop is a growing trend among professional technicians. Mechanic's gloves offer the best protection, but be prepared to lose dexterity when working with small objects. *Photo, Mechanix Wear, Inc.*

Safety Footwear

The importance of wearing protective footwear is obvious. The right pair can protect your foot from injury, slip hazards, cold temperatures, and even electrical charges. Safety footwear is available in many styles, including sneakers and fashionable shoes. See **Figure 8**.

Figure 8 Safety work boot.

Hearing Protection

Noise is one of the most common causes of hearing loss. A momentary loud noise, such as shot from a shotgun, can permanently damage your inner ear. Repeated exposures to loud noise over extended periods presents serious risks to your hearing. Hearing loss occurs gradually over time, so the importance of hearing protection is often overlooked.

Be aware of hazardous noise levels, such as engine noise, power tools, and powerful car stereos. Wear earplugs or earmuffs whenever you are working around loud equipment. Foam earplugs are available from a drug store. Earmuffs can be purchased at sporting goods or safety equipment stores. Permanent hearing loss due to loud noise is preventable. See **Figure 9**.

Figure 9 Basic hearing protection (left to right: earmuffs, tethered plugs, and foam plugs). All are appropriate hearing protection for automotive work.

Fire Extinguishers

Use Class C or Class D fire extinguishers for fires caused by either electricity or battery chemicals. The extinguishing agent in Class C extinguishers is non-conductive and is therefore the best choice for electrical fires. Class D fire extinguishers are designed for use on flammable metals. If a Class C or Class D extinguisher is unavailable, water is a suitable alternative. See **Figure 10**.

Figure 10 Have a fire extinguisher handy as a safety precaution.

Working Under the Car

Lifting the Car

For work that requires raising the car, the proper jacking points should be used to raise the car safely and avoid damage. There are normally four jacking points from which the car can be safely raised. The jack supplied with the car should only be used at the four side points specified by the manufacturer—normally just behind the front wheel or just in front of the rear wheel. Consult the vehicle's owner's manual for jacking points and jacking specifics. See **Figure 11**.

Figure 11 Jacking points (**arrows**) for a BMW.

 **WARNING —**

When raising the car using a floor jack or a hydraulic lift, carefully position the jack pad to prevent damaging the car body. A suitable liner (wood, rubber, etc.) should be placed between the jack and the car to prevent body damage. Watch the jack closely. Make sure it stays stable and does not shift or tilt. As the car is raised, the car may roll slightly and the jack may shift.

Working Under the Vehicle Safely

- Disconnect the negative (-) cable from the battery so that the vehicle cannot be started. Let others know what you are doing.

 **CAUTION —**

*Prior to disconnecting the battery, read the battery disconnection notes given in the **Battery** chapter.*

- Using an automotive floor jack, raise vehicle slowly.

 **WARNING —**

When raising the vehicle using a floor jack or hydraulic lift, carefully position the jack on the jack pad to prevent damaging the vehicle body.

- Use at least two jack stands to support the vehicle. Use jack stands designed for the purpose of supporting a vehicle. Place jack stands on a firm, solid surface. See **Figure 12**.

Figure 12 Do not work under a lifted vehicle unless it is solidly supported on automotive jack stands.

- Lower the vehicle slowly until its weight is fully supported by the jack stands. Watch to make sure that the jack stands do not tip or lean as the vehicle settles on them.
- Observe jacking precautions again when raising the vehicle to remove the jack stands.

 **WARNING —**

A jack is a temporary lifting device. Do not use a jack alone to support the vehicle while you are under it.

Do not work under a lifted vehicle unless it is solidly supported on jack stands that are intended for that purpose.

Do not use wood, concrete blocks, or bricks to support a vehicle. Wood may split. Blocks and bricks, while strong, are not designed for that kind of load and may break or collapse.

Use care when removing major (heavy) components from one end of the vehicle. The sudden change in weight and balance can cause the vehicle to tip off the lift or jack stands.

Do not support a vehicle at the engine oil pan, transmission, fuel tank, or on the front or rear axle. Serious damage may result.

2

Battery

⚠ WARNING —

This book applies specifically to gasoline and diesel (combustion) vehicles. **This book is not applicable to high-voltage hybrid or electric vehicles.** *High-voltage automotive circuits can be dangerous and even lethal. Hybrid and electric vehicles use high-voltage electric motor / generators, high-voltage electrical circuits and wiring, and other potentially dangerous high-voltage components. Specialized training and high-voltage test equipment is required to service the high-voltage side of a hybrid or electric car. Voltages in hybrid and electrical vehicle circuits can be as high as 650 volts (AC or DC), depending upon the vehicle manufacturer. This is a lethal voltage level. Do not use the instructions in this chapter to test the battery in a hybrid or electric vehicle!*

The Car Battery

Click.

It's the moment we all dread. Maybe you have a carload of antsy kids. Maybe you're late for a meeting. Maybe it's freezing out. But you get into the car, turn the key, and... click. Or, even worse, you turn the key and hear nothing. No click. Silence.

You've most likely got a dead battery. Of course, it could be bad battery cables or corroded terminals. We're going to tell you how to sort it all out.

After your tires, your battery is the most likely component of your car to strand you (what, you think it's a coincidence that AAA does mainly tires and batteries?).

On pre-1975 cars, the battery doesn't need to do much more than spin the starter motor and act as a buffer for the alternator. But on cars with lots of computer modules, all sorts of electrical glitches can be caused by a battery in less than perfect health.

The battery is the vehicle's main power source. It starts the car. But once the car is running, it is the alternator that is designed to handle all the electrical loads fitted to the car AND keep the battery fully charged under normal driving conditions. As we go through this chapter, we'll help you understand the difference between "car won't start" and "battery keeps running down."

How a Battery Works

A car battery is designed and constructed to set up an internal charge imbalance. With a lead acid battery, plates of two dissimilar metals, lead and lead peroxide, are submerged in water and sulfuric acid. The resulting chemical reaction triggers a migration of oxygen molecules from one plate to the other. This creates a surplus of charge on one plate and a deficiency of charge on the other. The difference in charge is voltage. When you get into the chapter on **Circuits (How Electricity Actually Works)**, we'll define voltage as charge imbalance, but don't worry about it right now.

But what you *do* need to understand right now is that your car doesn't actually have a 12-volt battery. We call it a "12-volt battery," but that's wrong. And I'm not talking about the 6-volt batteries on some oddball vintage cars. Seriously, "12-volt car battery" is a misnomer. *A car's battery consists of six individual cells, each of which should be outputting 2.1 volts* (see **Figure 1**). So a 12-volt car battery is actually 2.1 x 6 = 12.6 volts. As you'll see in a moment, the difference between 12 volts and 12.6 is highly significant.

Six 2.1V cells = 12.6V

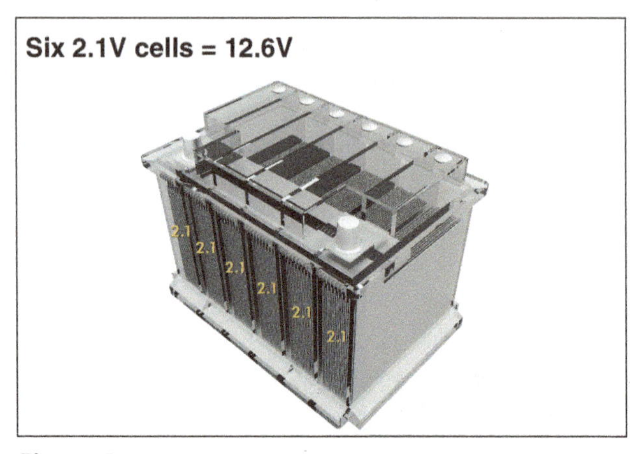

Figure 1 A six-cell "12-volt" automotive battery measures 12.6 volts when fully charged. *Copyright Yurily Merzlyakov*

Checking Battery Voltage with a Multimeter

Battery Voltage Test

See Test Set-Up on page 2-3

Battery troubleshooting starts with examining the health of the battery. The first thing we're going to do is check battery voltage with a multimeter. This is like visiting the doctor and having your temperature and blood pressure taken. It's the indispensable step you don't skip. Later in the chapter we'll tell you how to load-test a battery, which is a different, more thorough test. But checking battery voltage with a multimeter takes only seconds and puts the necessary information at your fingertips.

Take your multimeter out to your car. You do have a multimeter, right? If you don't, run to the store and, for the purpose of this test, buy the cheapest one they sell. For right now, any meter that measures voltage will do.

- Plug the red lead into the multimeter's positive socket, and the black lead into the negative one.
- Set the multimeter to read DC volts; this is typically the symbol with the "V" and two parallel black lines above it. If the multimeter does not have autoranging capability, set it to the 20-volt scale.
- Now, find your car's battery. If you don't know where it is, consult your owner's manual. In the good old days, the battery was under the hood, but on many cars, for purposes of weight distribution, the battery may be in the rear of the car behind an access panel. See **Figure 2**.
- Make sure the ignition key is OFF. Better yet, make sure it isn't even in the ignition.

Battery Voltage

See page 2-2 for step-by-step procedure

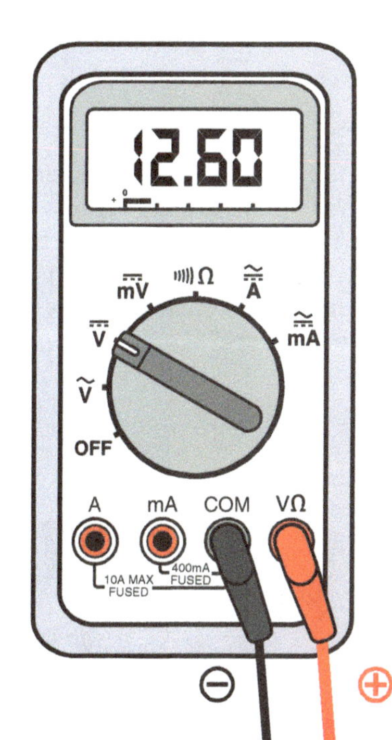

Measurement Across Battery Terminal

Test Conditions

Dial set to:	V DC Volts
Red input terminal:	VΩ
Black input terminal:	COM
Red lead:	battery (+)
Black lead:	battery (−)

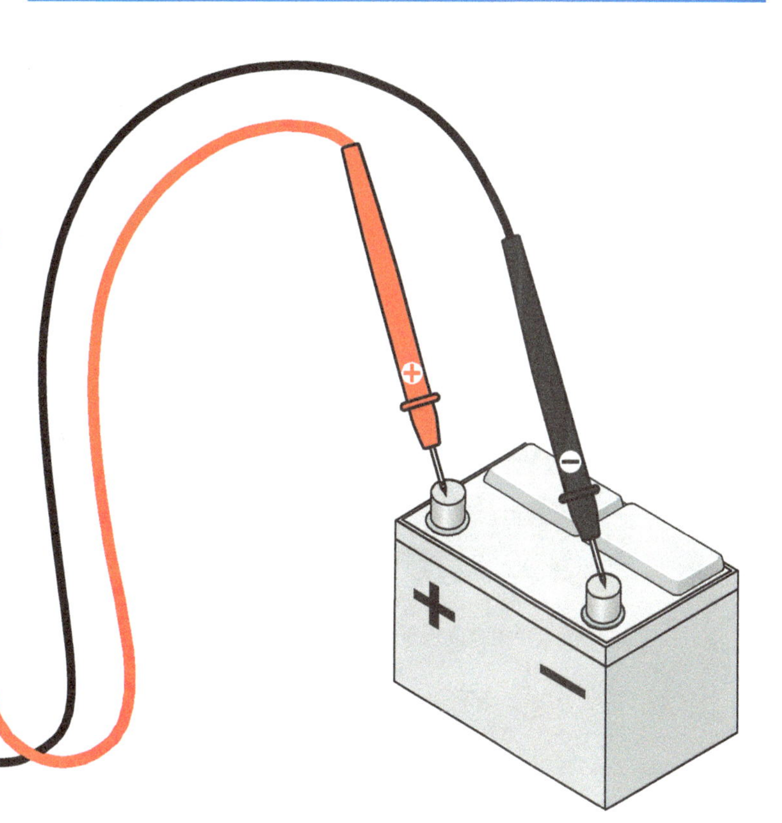

CAUTION —

Connecting a meter across a battery this way is safe ONLY for a voltage measurement on a gasoline or diesel vehicle! It is dangerous and should not be performed on a hybrid or electric vehicle!

Do not attempt to measure resistance or amperage across the battery. It will damage the meter.

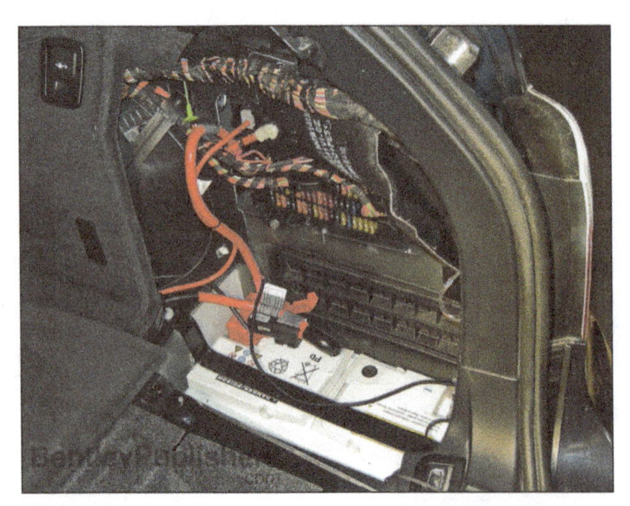

Figure 2 Battery location varies. On this 2007 BMW 530XiT, the battery is in the rear, behind an access panel.

- Touch the red multimeter probe to the positive battery terminal (marked with a big **+** and typically having a red cable), and the black probe to the negative terminal (marked with a big – and typically having a brown or black cable). See **Figure 3**.

NOTE —

Don't worry if you have positive and negative switched at this point. The multimeter will simply read a negative number instead of a positive one. You won't hurt anything.

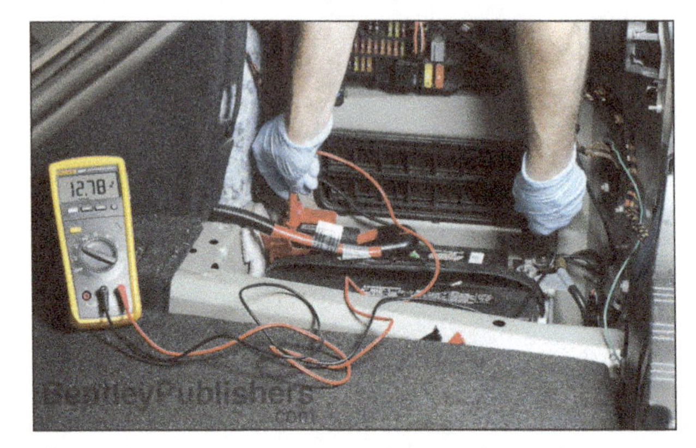

Figure 3 Measuring battery voltage with a multimeter. Press leads firmly against battery posts to get a good connection.

- What does the multimeter read? If the battery reads at least 12.6 volts (see **Figure 4**), in theory you're the owner of a fully charged battery.
- With each drop of 0.2 volts, your battery is discharged by 25%. See the **Battery Discharge Table**. At 11.8 V, the battery is considered fully discharged.

Battery Discharge (engine off, key off)	
Voltage	**% Charged**
12.6 or more	100%
12.4	75%
12.3	50%
11.8 or less	0%

Understanding that a fully-charged battery should read 12.6 volts is crucial. Web forums are full of posts from people whose car won't start and insist that their battery is fine because it's a 12-volt battery and it reads 12 volts and they won't accept friendly advice from others saying "dude, seriously, it's your battery."

Figure 4 Battery voltage should be about 12.6 VDC with engine off.

Now, assuming your battery is not dead, start the engine. With the engine running, measure the battery voltage again. It should read higher than 12.6 volts – somewhere in the 13.5 to 14.4 volt range. That's because the alternator is charging the battery by supplying it with current at a higher voltage. See **Figure 5**.

In just a few easy steps, we've dug down to and exposed the bedrock of the automotive electrical world. In the table below are the two numbers you need to know.

Healthy Battery Voltage Measurements	
Engine off	12.6 VDC
Engine running	13.5 - 14.4 VDC

And, from these two simple battery voltage facts, we can begin to tell you what to do when your car goes "click."

Figure 5 Battery voltage should be around **14 VDC with engine running**.

(Note that you can also get these readings quickly and easily from the cigarette lighter socket as described in the **Multimeters and Related Tools** chapter.)

Bad Battery

As a battery's resting voltage drops, particularly as it drops below 12.2 V (remember, that's 75% discharged), you can't expect it to start your car. So if your car won't start, and if the battery voltage is low, you've found the cause. Your battery's dead. Replace it, recharge it, or jump-start it (more on these options later in this chapter).

Bad Alternator

If the engine running voltage is not in the 13.5 V to 14.4 V range but is instead closer to 12.6 V, your charging system (alternator and voltage regulator) is not recharging your battery and you need to address the root cause. See the **Charging System** chapter for information.

 ***NOTE —***

The high electrical loads generated by many post-2005 luxury European cars require high charging rates. During high consumer load situations, the alternator needs high rpms to keep the battery charged. Low engine speeds and short trips will kill a battery in short order. If you own a luxo-barge, battery problems may, unfortunately, be more an issue of driving profile than alternator or battery malfunction.

How Dead Is Your Battery?

Knowing how to test whether your battery is adequately charged is important, but it's just the beginning of Battery Diagnostics 101. You still need to know if a dead battery can, or should, be resurrected. And even a battery that reads as charged on a multimeter may require further testing to assess its long-term reliability.

The difference between "click" and silence is, in fact, highly significant. "Click" means that the battery is at least trying to engage the starter motor, but the starter isn't turning the engine. Silence means that the starter isn't being engaged at all. Careful troubleshooting of both of these conditions requires stepping through a logic tree of possibilities, and we'll get to all of them in the next chapters.

But if the Dead Battery Event happens out on the road, here's how you can triage it into one of four categories:

Case 1
It's Swollen or Cracked or It Stinks Like Rotten Eggs – DON'T JUMP IT OR DRIVE THE CAR! This likely means the alternator is overcharging the battery and the battery is producing excess amounts of hydrogen sulfide gas. It's really obvious. It smells like rotten eggs. The excess gas often causes the battery case to physically swell. The case may crack. If you encounter either of these symptoms, do not drive or try to jump the car.

The battery may not have died yet, but it likely will, possibly in a catastrophic fashion. Hydrogen sulfide gas is explosive. If you smell it, jumping the car is exceedingly dangerous due to risk of spark. The safe way to deal with it is to put a new battery in it, and use the new battery to help test where the overcharging fault is. This is almost always due to a bad voltage regulator. See the **Charging System** chapter for more information.

Case 2
It's So Dead It'll Drag Down Your Car – Don't Even Think About Jumping It. If the battery is so weak that you can't get the headlights, fans, or dashboard lights to come on at all, or if the multimeter indicates the battery is completely discharged (less than 11.9 V), it's basically toast. You shouldn't jump it and drive on it unless it's an emergency, and you should only entertain recharging it with deep suspicion. Here's why.

- **Standard car batteries are NOT designed for deep drain and recharge cycles**. They are built to be continuously recharged by the alternator. They are not "deep-cycle" batteries. So if your battery was deeply drained, particularly to single-volt levels, you probably have damaged it. Yes, you can read how to de-sulfate the plates and resurrect a deeply discharged battery. Don't trust it. A good rule of thumb is, if a battery is less than two years old, it may recover from a deep-discharge event (you can recharge it and see how it behaves), but if it is older, it's done.
- **Alternators are NOT designed to recharge a flat battery.** On pre-1975 cars, you can often get away with jump-starting a flat battery and driving the car, but chances are high you that if you try this on a car

Deep-Cycle Batteries

The difference between a starting battery and a deep-cycle battery is that a starting battery is designed to be kept fully charged and to deliver brief bursts of high current, whereas a deep-cycle battery is designed to be regularly drained of most of its capacity. Deep-cycle batteries are used for recreational vehicles, fishing trolling motors, and other applications where they're the only power source available and they can be recharged at a later date. Everyone thinks those Optima red-topped batteries are deep-discharge batteries. They're not. The blue and yellow-topped ones are deep-cycle. The red one is a starting battery. A deep-cycle battery can usually be used as a starting battery (though it generally won't deliver quite as much current as a comparable starting battery), but a starting battery should not be used as a deep-cycle battery.

whose electrical system contains computers and modules, it will throw a hissy fit. The car may well start up and drive, but because of the electronics' sensitivity to voltage levels and spikes, weird things are likely to happen. The dashboard lights may flash, the car may have driveability issues, and it may die without warning. It's not safe. Don't do it. There's not a year that's a hard and fast dividing line for this. The newer the car, the more complex the electrical system. If a car has a flat battery, it may start up and drive when jumped, but may then act like it's possessed. Unless concerns over personal safety absolutely require jumping a flat lined battery in order to move a car from a highly dangerous location, you should always opt for installing a known good battery instead of jumping a drained one.

- **Batteries do not last forever:** Every battery should have a date code on it. Check it. If the battery is four or more years old, and it's dead, it's a good time to replace it. There's probably no simpler thing you can do to improve the reliability of your car than to install a new battery.

Case 3

It's Not Old and Shows Signs of Life – Jump it, and Recharge When You Get Home: If the battery voltage is low – say, 12.2 volts – but sufficient to make the starter click and run the car's headlights and fans, you can try jump-starting the car. The correct procedure for this is explained in detail in the Jump Starting section below.

But you need to find out the reason why the battery is low. Either the battery itself is going bad, isn't being properly charged by the alternator, or something caused the battery to drain while the car sat. Without recharging and load-testing the battery (both of which are covered later in this chapter), you don't know which. And unless you know that you did something specific to drain it, like leaving the headlights on, you really don't want to take the chance of it running down again.

At a minimum, after you jump it, you should test the battery voltage with the engine running. If the running voltage level is in that magic 13.5 to 14.4 V range, then you can probably drive the car home or to a service station and recharge and load-test the battery. (Be aware, though, that the computer and electronics in cars are very sensitive to voltage levels, and when the battery is low, weird electrical gremlins may appear.)

But if the voltage level with the engine running is still low, a malfunctioning alternator is likely the cause of the dead battery. Don't drive the car like this! It WILL strand you! Either have it towed to a repair shop, or tow it home, read the alternator chapter in this book, and troubleshoot the alternator yourself.

(That having been said, if you are driving a pre-1975 car with primitive electrics, you can in fact nurse a dead alternator home by swapping in a freshly-charged battery and driving with a minimum electrical load—i.e., no headlights or fans. But, on a computerized car, you can't get away with this except for very short distances. The low voltage levels from the bad alternator will play havoc with the car's electronic systems.)

Battery Cables

As part of checking battery health, it is important to inspect and maintain the battery cables. If the connector on the positive battery post has a visible mound of corrosion, that's an indication that the battery could be generating hydrogen sulfide gas because it's being overcharged by a faulty voltage regulator. See **Figure 6.**

Figure 6 Corrosion on the positive battery terminal may be an indication of overcharging.

Overcharging can be checked by measuring the voltage across the battery with the engine running. If the charging voltage is substantially greater than 14.4 V, the regulator is faulty. See the **Charging System** chapter for more detail.

The corrosion mound may also be an indication that the battery is nearing the end of its useful lifetime. See the **Load Testing a Battery** section later in this chapter. However, if the charging voltage is okay and the battery load-tests as good, then the terminal corrosion may be caused by a small leak in the seal between the post and the battery's case, which is little more than a minor annoyance. The corrosion mound itself can be cleaned off with a mixture of baking soda and water, or, actually, Coca-Cola.

WARNING —

When disconnecting a battery always always always disconnect the negative terminal first! Never ever ever disconnect the positive terminal before the negative is unhooked! *If, while you're unscrewing the positive battery clamp, your wrench accidentally touches the body of the car, the battery discharges through the wrench to ground. This could burn you badly, possibly melt the wiring harness, and even damage electronics. Once the negative terminal is unhooked, there's no longer a closed circuit for current to flow through, so if the wrench accidentally touches ground, it doesn't matter.*

When reconnecting the battery, reverse this order – connect the positive first, then the negative!

Cleaning the Connectors

When you have the battery removed for recharge or replacement, it's a good idea to inspect the inside surfaces of the cable clamps. The surfaces should be shiny and free of corrosion. Remember that when you try to start the car the entire electrical load is passing through that thimble-sized area. Buy one of those battery post and cable cleaning tools at an auto parts store. Be sure to buy the kind of tool that uses a brush and bristles, not the kind that employs angled blades that shave metal. The shaver type can remove so much metal that the clamps can no longer be tightened.

CAUTION —

Always disconnect the negative terminal before cleaning the positive terminal or post.

Undo the battery clamps (negative first, reconnect negative last), and clean the battery posts and the inner surface of the clamps. See **Figure 7** and **Figure 8**.

Corrosion or ill-fitting clamps will prevent a car from starting. A combination of cleaning, retightening, or sometimes even just tapping the clamp down on the post with a a block of wood and a hammer can get a stranded motorist going again.

Figure 7 Clean battery terminals using the brush of the battery post cleaner (**arrow**).

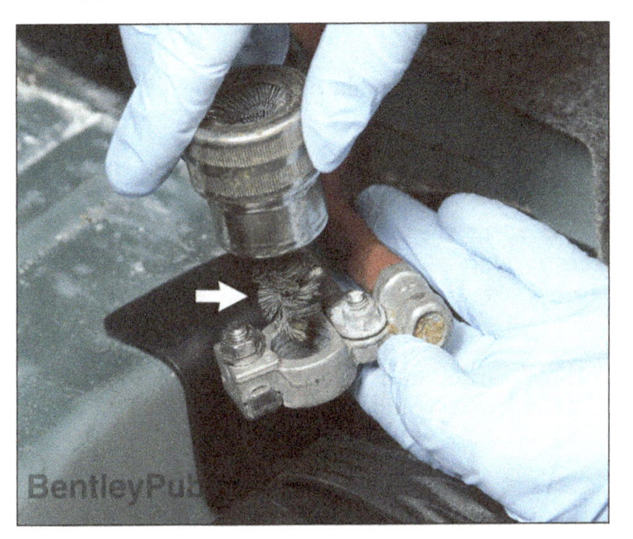

Figure 8 Clean the clamp using the top brush under the cap of the battery post cleaner (**arrow**).

Replacing Battery Cables

If, after cleaning, the clamps still don't make good contact, the battery cables should be replaced. This is trivial on pre-1975 cars, as the cables are relatively short and simple, and the only choice is whether to pay money for OEM cables versus using inexpensive generic ones. However, many newer cars, particularly those with trunk-mounted batteries, have battery cables which are integrated with the car's wiring harness, and replacement may include some expensive and invasive work to install a new rear-to-front cable.

Note that some European makes have a special battery "safety" cable that disconnects the positive cable from the electrical system in the event of an accident. BMW and Audi are two marques that have such battery terminals. See **Figure 9**. In the event the starter doesn't operate – especially after an impact – check the electrical integrity of the positive battery cable at the battery first.

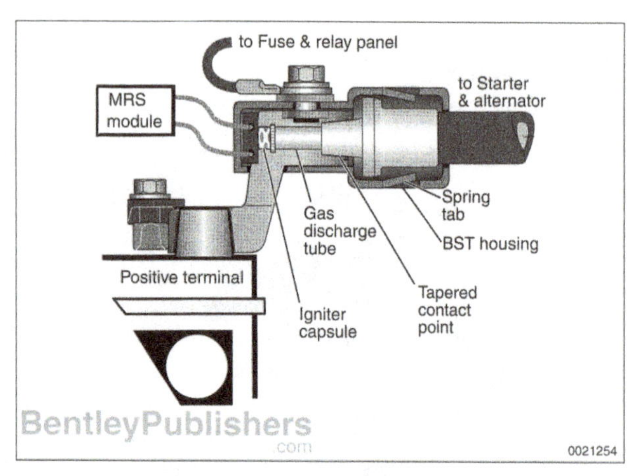

Figure 9 BMW Battery Safety Terminal (BST). In the event of an accident, an explosive charge disconnects the main power cable from the battery terminal. This was first used on the 1999 BMW 3 Series.

Safety positive cables are a great idea in the spirit of occupant safety, but if blown, replacement is expensive. All the more reason to go gently with the terminal cleaner.

Late model BMWs use an Intelligent Battery Sensor (IBS) integrated in the negative battery terminal. See **Figure 10**. Use care when removing the negative terminal on cars so equipped. The IBS is easily damaged with excessive force.

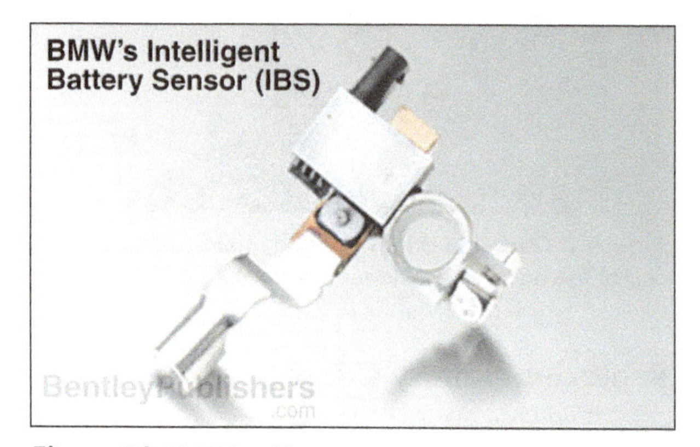

Figure 10 BMW Intelligent Battery Sensor (IBS). It communicates battery information to the ECM. This sensor is fragile; remove it gently. *Photo, HELLA KGaA Hueck & Co.*

Charging a Battery

You've been instructed several times in this chapter to charge the battery. How? With what?

Years ago, guys would talk about sitting around the garage and killing a six-pack while the battery charger fed a constant current into their battery and they hung around to measure the voltage and shut off the charger at the right time. But these days, human monitoring isn't necessary; it's all done with so-called smart chargers that measure the battery's voltage and resistance and adjust the charging accordingly. See **Figure 11**.

Figure 11 This three-stage "smart" charger measures battery voltage and resistance and adjusts the charging current automatically. The charger indicates when the battery is fully charged.

These state-of-the-art chargers cycle through distinct stages of battery charging. The first stage is referred to as the *bulk charge* or *constant current charge stage*. This is where most of the action is if a battery is badly discharged. The charger's current (the number of amps) is held constant, and the battery's voltage increases until the battery is charged to about 70% to 80% of its capacity.

The *absorption* or *topping charge* stage is next. Unlike the bulk charge which was performed using a constant current with increasing voltage, the absorption stage uses a constant voltage and lets the current decline until the battery is nearly fully charged.

Lastly is the *float charge* stage, which essentially compensates for the battery's tendency to self-discharge when unused. During float charge, the charge voltage is reduced to about 13.5 volts, and the current is moderated to fall off to just a trickle. For this reason, a single-stage charger that is only a float charger is often called a *trickle charger.*

 **CAUTION —**

Chargers get hot, particularly during bulk charging. It's good practice to leave them on the garage floor rather than in the vehicle trunk if charging for an extended period.

A battery charger will give you some indicator of its success. It may be a simple "green good red bad" indicator light, or a set of bar lights showing the charging amperage as it falls to near zero. If a battery has been deeply drained to single volt levels, sometimes a "smart" charger won't see enough of a load to begin the bulk charge and it'll immediately register as bad on the charger.

Choosing a Charger

So, what kind of charger should you buy? Like most things, it depends how much capability you want, how often you'll use it, and how much you're willing to pay. Remember that if something says it's a "trickle charger," it's usually *only* a trickle charger and not intended to recharge a dead battery. If a trickle charger is all you need, fine, but they're not terribly versatile.

The problem is that "smart charger" can mean anything. Really, what you want is a microprocessor-controlled charger that has the documentation to prove that it is a *three-stage charger* that accomplishes the bulk, topping, and float charging stages described above. Don't confuse some of the older chargers that say "60 amp boost / 10 amp charge / 2 amp trickle" with a "three-stage charger." These are typically three distinct settings with no microprocessor control automatically switching between them. And you really shouldn't be doing the 60-amp boost thing anyway, as that much current is likely to shorten battery life and can easily overcharge the battery.

Some chargers explicitly say "three-stage charger," but many just say "smart charger" or "microprocessor-controlled charger," and you need to read the documentation to see what they actually do. The problem is that if it's an inexpensive product, there's unlikely to be detailed documentation. The adage that "if it feels cheap, it's cheap" definitely applies. When a charger is cranking 10 amps into a battery during the bulk charge phase, it is going to get quite warm, so something in a metal case that has the ability to dissipate the heat is likely to be a more credible product than something that claims 10 amps but is in a plastic case.

Plus, if you read about this stuff on battery and charger web sites, you'll learn that there's a set point or threshold where the charger transitions from the bulk charge to the topping charge, and this set point varies with battery composition and temperature, so *really* what you want is a *temperature-controlled three-stage charger* that explicitly says it's good for both flooded cell and AGM batteries.

Because of the proliferation of computerized control modules which can cause a small battery drain in a car even when the car is parked, many owners have experienced a dead battery if the car has not been driven frequently. If this happens overnight, there is a problem that should be fixed (see the **Energy Diagnosis and Parasitic Drain** chapter), but if it happens after a week or two or three, it's often considered simply part of owning a modern electrically complex car. For this reason, many vehicle manufacturers sell their own trickle charger that plugs into the cigarette lighter, and recommends its use if the car is not driven frequently.

There's nothing wrong with leaving the battery connected to the car while it's being charged. In fact, on cars with battery management systems, the owner's manual may specifically instruct you to not disconnect the battery while charging, as leaving it connected may allow the car's computer to capture the charging history and use it as part of the battery management profile. However, if you're trying to revive a deeply discharged battery, it's advisable to remove it from the car, or at least disconnect the negative terminal, simply as a precaution.

Lastly, note that if a battery is visibly damaged (swollen or cracked), it shouldn't be charged at all; it should be carefully removed and replaced with new.

Maintaining Battery Voltage During Diagnosis

For all sorts of electrical diagnosis work, the key has to be turned to the ignition setting. The problem is, the battery can quickly go dead if the key is left on too long and the engine isn't running to spin the alternator to recharge the battery. The key-on current draw can be ridiculously high–20 amps and more–on a loaded-up European flagship.

If the engine isn't running and you need to do extended key-on diagnosis work, you may need a robust power supply unit clamped to the battery. See **Figure 12**. This is particularly important if you are re-programming (flashing a control module). The subject of module programming is outside the scope of this book, but many control units have been irreparably damaged because the battery voltage went too low during flashing. Clean (AC ripple-free) current at a steady voltage (~14 volts) is critical when flashing sensitive electronic control modules.

In simple terms, a good power supply unit can be thought of as a battery that won't go dead. The professional-quality power supply shown below is one such tool capable of delivering a steady 70 amps of current. Most battery chargers produce dirtier power and simply aren't designed to maintain high current for extended periods. If you don't have access to equipment like this, reprogramming is probably best left to a professional, and key-on electrical troubleshooting may need to be limited to short sessions that do not deeply discharge the battery.

Load Testing a Battery

With so much riding on your battery, how do you know it's good or bad? Earlier in the chapter you learned that if the battery reads at least 12.6 volts, in theory you're the owner of a fully charged battery. Is that it? Well, no. To use logic-speak, 12.6 volts is a necessary but not a sufficient condition. Unfortunately, it is possible for a battery to charge up to 12.6 volts but not be able to deliver the sustained amperage needed to start the car. What you need to do is load-test the battery with a high-amperage load and check voltage.

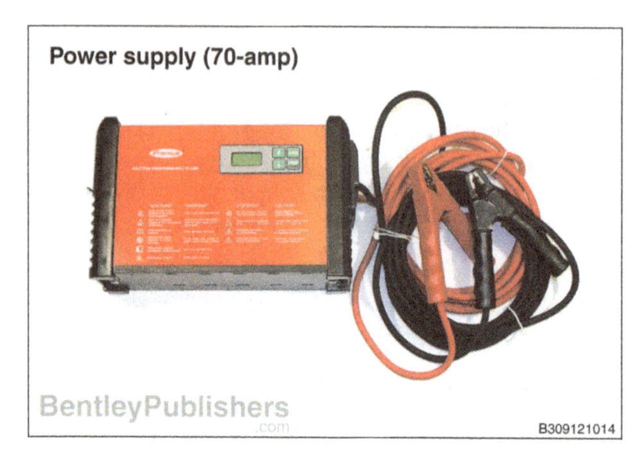

Figure 12 This professional 70-amp power supply has the power to manage high current demands while the key is on and the engine is off.

Let's back up a step. Generally, the first thing you do to test a battery is to fully charge it with a high-quality smart battery charger. The charger may not succeed, in which case the battery has failed even before the start of the test. Also, note that a fully charged battery, whether charged from a charger or from a drive in the car, may read above 12.6 volts. You should turn the car's high beams on for two minutes to use up any surface charge that may be on the plates, then turn them off, then wait two minutes before load-testing.

What is "load testing?"

We'll go into the concept of "load" further in the next chapter, but when we say "load," we mean "something that uses electric current." Some devices, like small electric lights, use only tiny amounts of current, whereas others, like the car's starter motor, use very large amounts of current. "Load testing a battery" means attaching a device that pulls a large known amount of current from the battery, and seeing what happens to the voltage of the battery while the large amount of current is being drawn.

Here's the reason you need to do it: If the car won't start, you can't always be sure if the problem is the battery, the cables, or the starter motor. Therefore, it's advisable to remove the battery and connect it directly to a "load tester," which is a test device that draws a large known amount of current from the battery. That way, you're isolating the battery to find out if it's capable of generating the kind of current that the starter motor needs.

Four Ways to Load Test a Battery

1. Get a Free Load Test

You can take the battery to an auto parts store that sells batteries. Many will load-test it for free. Of course, they are in the business to sell new parts, so there may be an inherent bias. To do it right, the battery should be fully charged, so results can vary based on state of charge. If you have a questionable older battery and you've bothered to remove it from the car and drive it to your parts store, go ahead and spring for the new battery.

2. Load Test by Using Your Car

You can use your car itself to perform a load test. It's actually quite easy.

- Begin with a fully charged battery.
- Burn the headlights for two minutes.
- Disable the car from starting by finding the car's fuel pump fuse or relay and unplugging it, or, on a car with distributor ignition, unplugging the wire to the center of the distributor and grounding it.
- Clip the leads of your multimeter (and an inexpensive multimeter is still fine) onto the battery terminals, or have a friend hold them there.
- Crank the starter for ten seconds while you or your friend watch the multimeter. The reading shouldn't fall below 9.6V at 70 degrees (this acceptable voltage level drops by 1/10 volt with every ten-degree drop in ambient temperature). See **Figure 13**.

Figure 13 Crank the engine for 10 seconds with the ignition disabled. On a healthy battery, voltage shouldn't fall below 9.6 volts.

Note that the ability to perform this test assumes that your starter will in fact crank your engine over, which may not be the case if you have a more complex no-start problem like a bad starter or cabling and want to rule out the battery.

However, if the starter *does* crank the engine over, one of the nice things about this test is that it helps to separate battery health from starter health. That is, if the starter cranks but quickly slows way down, the problem could be a weak battery or a bad starter. If the battery voltage drops below 9.6V, you know it's the former, not the latter. If the voltage stays up but the starter gets sluggish or stops, the starter should be investigated.

3. Load Test with a Carbon Pile Tester

Load testers generate a large electrical load and have an onboard voltmeter to measure and display voltage drop while that load is applied. The smaller handheld testers, often referred to as "toasters," contain coils of wire that draw current (and make heat). The bigger benchtop units referred to as "carbon pile testers" use stacks of carbon discs to draw current and dissipate heat. See **Figure 14**.

Ideally, you want to test a battery at half its cold cranking amp (CCA) rating for 15 seconds. For example, for an 800 CCA battery, you'd want a tester able to deliver 400 amps. Most of the "toaster"-style testers are only rated to 100 amps, making them incapable of reaching the half CCA threshold, but the carbon pile testers are typically rated to at least 500 amps, making them the better choice. Like the load test above, you want to verify that the voltage doesn't fall below 9.6 while the load is applied.

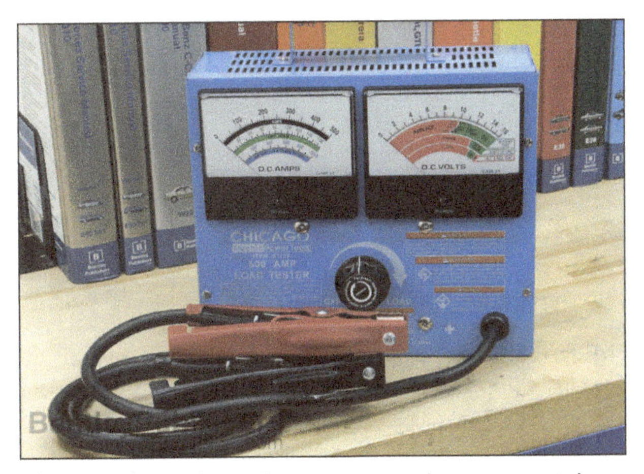

Figure 14 A carbon pile tester puts a large current draw on the battery, simulating starting conditions. Use extreme care when load testing. This includes proper eye and hand protection. Always follow the toolmaker's instructions.

4. Test with a Conductance Tester

While technically not load testing, conductance testers get at the battery health a different way. They deduce the degree of sulfation of the battery plates by sending a small AC signal through the battery, then measuring a portion of the AC response. This is a way to measure how much power the battery is capable of delivering.

Tales From the Hack Mechanic: Buying a Battery Tester

As with any consumer-quality tool, what might be wholly unsuited for professional duty may be perfectly adequate for twice-yearly usage. But is it worth buying a tester instead of just ponying up $150 for a new battery? I go through enough cars that testing the batteries and passing them to lightly-driven car duty is cost-effective, so I recently bought a $79 conductance tester. I tried it on new and known-bad batteries, and it correctly diagnosed them as such, so I've begun using it to triage non-fatally run-down batteries. We'll see how enamored I am as time goes on. If it tells me a run-down battery is good, and I recharge it, and it strands me the next day, my conductance tester may end up in the garbage.

Checking for the loss of conductance is a great way to quickly gauge the condition of the battery. As a battery ages, its internals corrode and break down and the electrolyte can dry out. This results in a loss of conductance. Conductance is an indication of the battery's state of health (SOH).

Consumer-quality conductance testers are now available in the magic sub-$100 range, and, in theory, they don't require the battery to be fully charged to do a test. See **Figure 15**. 5 milliohms is considered a reading for a healthy battery. No tool can guarantee that a battery that passes a test won't die tomorrow, but if you see battery resistance go from 5 milliohms one month to 100 milliohms the next, and if that reading is accompanied by slower cranking, and if the battery is three years old, it adds to the preponderance of evidence reinforcing a decision.

Figure 15 The conductance tester can be a reliable predictor of battery end-of-life. The 3.36 milliohm reading indicates the battery has a healthy low resistance.

Jump Starting

This section will walk you through the basics of jump starting a car, but bear in mind that no battery-jumping procedure can be truly universal. ***You should check your car's owner's manual first.*** Because of rear-mounted batteries, battery safety terminals, intelligent battery sensors, and other more sophisticated battery monitoring systems, some manufacturers have installed jump terminals under the hood to be used for jumping, and say in the owner's manual that direct connection to the battery is *verboten*.

Some owner's manuals specify that the car is not to be jumped at all because of potential damage to the electrical system, and instead mandate the use of a portable battery pack. Some say that the car should only be the jumper but not the jumpee. Now, you can read reports online that say either a) this is a load of hooey, or b) "I know someone who knows someone whose ECM got fried." So the best advice is this: **If your manual says don't do it, don't do it!**

As stated earlier in this chapter, if the battery is so drained that the dashboard lights don't even come on, jump-starting and then driving a computerized car is an iffy proposition, so this procedure assumes that the battery is slightly but not fully drained.

The Jumper Battery Pack

Before we get into traditional jump starting, there is one option you should be aware of. Portable jump packs have been extensively used by car dealerships for many years. However, a new generation of lithium-powered jump packs (see **Figure 16**) are small and light, and possess enough cranking amps for several attempted vehicle starts. In addition, they can be used to recharge batteries in cell phones and other portable devices. Very handy.

Figure 16 A small portable lithium jump pack can deliver up to 300 amps of peak power to jump-start your car. *Photo, Juno Power*

To Jump-Start a Dead Battery

To use another vehicle to jump-start a car, make sure both vehicle #1 (good battery vehicle) and vehicle #2 (dead battery vehicle) have the same electrical system voltage (that is, don't jump a 1960 VW Beetle having a 6-volt electrical system with a 12-volt car) and about the same battery rating in amp-hours (that is, don't jump a Smart Car that has a tiny little battery with a Ford F350 with dual batteries).

- Put on eye protection.
- Put vehicle #1 (good battery) close to but not touching vehicle #2 (dead battery).
- Make sure both vehicles are in park or neutral with the handbrakes set. Make sure all electronics in both vehicles are switched off.
- Take a high-quality set of jumper cables and lay them on the ground with none of the ends touching each other.
- Attach one of the red clamps to the positive (red) battery terminal of vehicle #1. See **Figure 17**.

NOTE —

You may have to wiggle the clamps and pull the clamps apart to get them to bite for a solid connection. The more corroded the battery terminals are and the colder the outside temperature is, the harder it is to get a good connection.

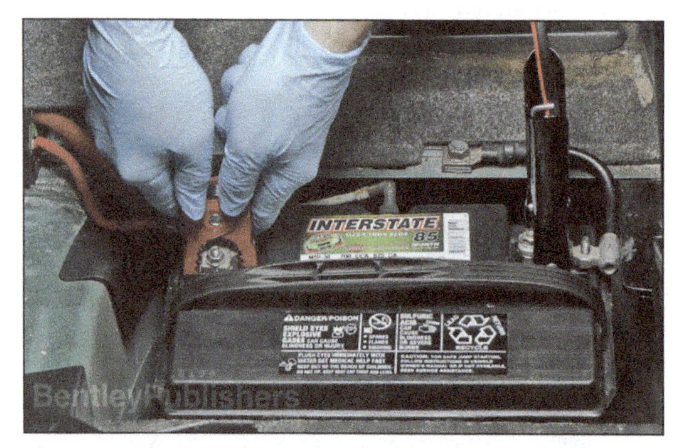

Figure 17 Once jumper cables are connected, make sure the clamps solidly bite by pulling the tops of the handles apart.

- Attach the red cable clamp at the other end to the positive (red) battery terminal of vehicle #2, repeating the "biting" process.
- Attach one black clamp end of the jumper cable to the negative (black) battery terminal of vehicle #1, repeating the "biting" process.
- Attach the other black clamp end to a bolt or bracket on the frame or engine block of vehicle #2. See **Figure 18**.
- Verify that all jumper cables are free and clear of the rotating fans in either engine compartment.
- Turn the key of vehicle #2 to the "ignition on" setting only. Do not start the engine. Check that the dashboard lights come on.

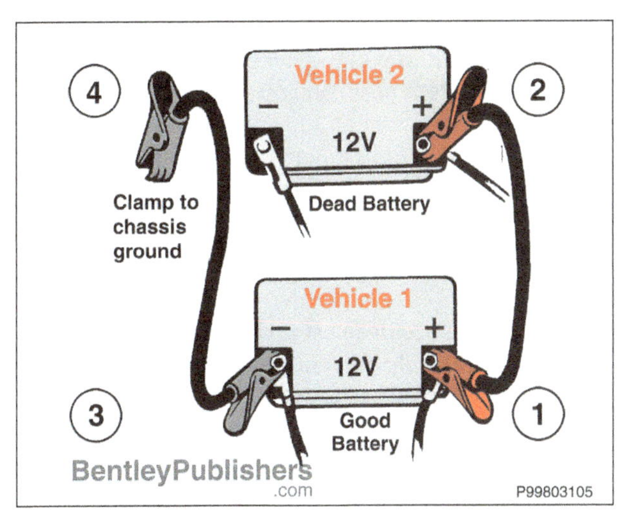

Figure 18 When jumping a dead battery, clamp the last cable connection (**4**) to chassis ground. Do not connect it to the battery. This is because you don't want to create a spark anywhere near the battery.

 ***WARNING —***

Do not attach the last negative terminal to the battery. Making the final connection will often result in a spark. This is normal, but you don't want the spark to occur near the battery, which could be venting explosive gases.

- If the dashboard lights do not come on at all, or are dim, then there is not a good electrical connection between the vehicles. Disconnect the fourth clamp, re-bite all others, re-bite the fourth clamp, and try again.

NOTE —

If you can't get the dashboard lights to come on no matter how hard you try, something else is wrong, such as a blown high-current fuse, or battery safety terminal. Jumping is not going to help start it.

- Try to start vehicle #2 (dead battery). This "passive jump" uses only #1's battery and not its charging system. This lessens any potential risk to vehicle #1's electronics and alternator. Most of the time, with small vehicles, this works fine.
- If vehicle #2 doesn't start, turn the key all the way back to the off position, re-bite everything as described above, and try again.
- If vehicle #2 still doesn't start, turn the key all the way back to the off position, then start vehicle #1 (which provides vehicle #2 with the output of vehicle #1's alternator). Let #1 idle for a bit, and try starting vehicle #2 again.
- If vehicle #2 still doesn't start, turn the key all the way back to the off position, leave vehicle #1 running, and raise the idle on vehicle #1 to 2,000 rpm to increase the output of its alternator for about a minute. Let the #1 return to idle and give it one more try.
- If this gets vehicle #2 running, let them both settle into an idle, and remove the battery cables from both vehicles in the reverse order in which you installed them.

Buying a New Battery

For the most part, any battery can be used to start and run any car. (No? You don't think so? What exactly do you think you're doing when you're jump-starting a car?) But it's not safe to drive unless the battery is the right size and securely screwed down to the battery tray. But if a battery fits securely, either by itself or with an adapter, and nothing hits the top when you close the hood or trunk, you should be okay.

On an older car with a battery under the hood, you don't really have to worry about too much else. An older car will tolerate a wide range of battery ratings. The battery is naturally vented to the outside. Flooded cell or AGM types are both fine.

If the car is newer and has an interior-mounted battery, be sure the replacement battery is sealed (AGM) or can be vented. This is usually just affixing a little rubber hose, but make sure to do it.

You're really best advised to buy the same type and rating as the original battery. If that's an 850 CCA AGM battery, that's what you should buy.

 ***NOTE —***

On late model BMWs and Audis, see the **Battery Registration / Coding** *section at the end of this chapter prior to replacing your battery.*

Battery Ratings

Batteries used to be rated by their *cranking amps* (CA) which is a measure of the amount of current the battery can provide at 32 degrees F. It's the same thing as *marine cranking amps* (MCA). CA was replaced by *cold cranking amps* (CCA), which lowers the temperature of the specification from 32 degrees F to zero degrees F. More specifically, CCA specifies the current – the amperage – that a lead acid battery can deliver, continuously, for 30 seconds, at zero Fahrenheit, until the battery is depleted to 7.2 volts.

Although this standard doesn't seem terribly useful (who drags a battery down to 7.2 volts?), it has emerged as the standard; every battery you buy will have the CCA rating stamped on it.

A far more useful number is the *reserve capacity* or *reserve capacity minutes* (RCM), which is the number of minutes a battery will allow you to pull a 25-amp load until it is at 10.5 volts. 10.5 volts is significant in that it is a number commonly used to denote a fully discharged 12-volt battery.

Whatever the pluses and minuses of the CCA standard are, it provides a method of comparison between batteries. The larger your car's engine, the older the car, and the colder the climate you live in, the more important it is that you don't shortchange the CCA rating.

Group Number, Dimensions, and Post Configuration

The Battery Council International (BCI) has a set of BCI group numbers that determine the physical dimensions and other aspects of a battery's configuration. The group number may not be printed on the car's original battery, but it can be looked up by any aftermarket battery manufacturer. See **Figure 19**. For example, a BMW Z3 takes a group 48 battery, as do many other cars. The group 48 battery is 11" long by 6 15/16" wide by 7 1/2" high.

Figure 19 An original group 48 battery with recessed posts installed in a 2012 Volkswagen GTI. A replacement battery should have the same group number and same (or at least very close) CCA rating.

However, it takes more than just dimensions for a battery to fit correctly. Also contained in the group number are:

- Top posts, raised or recessed posts, top threads, or side threads.
- Positive or negative terminals on left or right as viewed from the top down.
- If it has a hold-down ledge on the bottom and what the shape of that ledge is.

To continue using a group 48 battery as an example, in addition to its size, it has recessed top posts, and the positive one is on the right when the battery is oriented so they're both along the bottom, and has a hold-down ledge on both long sides of the battery.

If life were easy, the group number would unambiguously determine which posts are on the battery and guarantee a perfect fit. Unfortunately, this is not the case. Batteries of the same group number can, in fact, have different post configurations. Worse, if top posts are specified, not all top posts are the same size. According to the BCI, the familiar top battery posts in fact come in four flavors – Society of Automotive Engineers (SAE), Deutsches Institut für Normung (DIN), European Norm (EN), and Japan Industrial Standard (JIS) "pencil posts."

On all four, the positive post is slightly larger than the negative post to help prevent accidental incorrect cable hookup, and a 1:9 taper is used. The SAE post, also commonly called an "automotive post," is by far the most common. See **Figure 20**. If you want to know more about battery fitment, group ratings and post specs, search online for "BCI Battery Replacement Data Book." You'll find a thick download full of mind-numbing battery details.

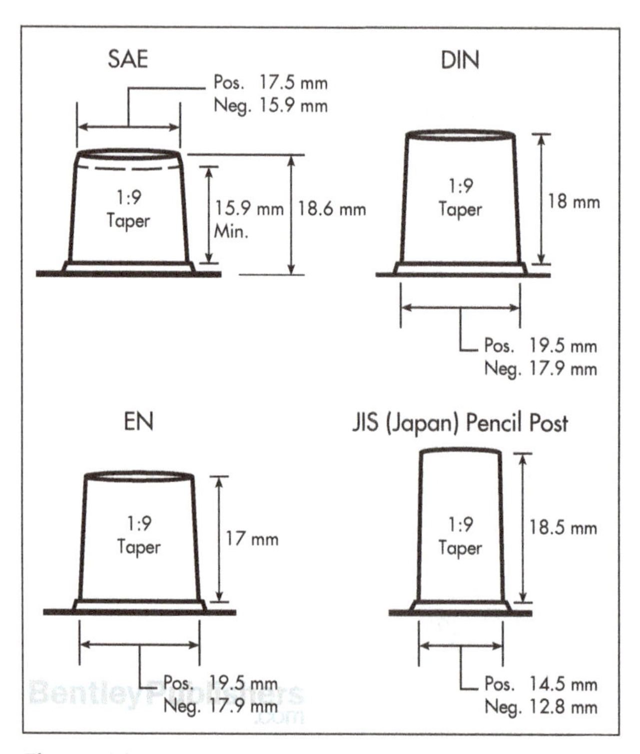

Figure 20 Automotive battery post dimensions. The most common is the SAE, aka automotive post.

Can a battery with a different group be used? It's possible and you might get lucky. But often you'll find the battery is slightly too wide, or the hold-down ledge is the wrong shape, so it sits on top of the curled ledge of the battery tray rather than under it. It's not safe to drive a car like this. If you hit a good-sized bump, the battery can bounce and short circuit against the underside of the hood. Plus, it's possible to muscle a battery into the tray or well only to find the positive and negative terminals are reversed, and if you rotate the battery around 180 degrees, the cables won't reach. For all these reasons, it's best – and often flat-out necessary – to use a battery with the correct group number.

So, while the group number should answer the dimensional question of "will it fit," it's not, however, the entire story. What's not covered under the group number are the battery technology and the battery rating, so let's talk about those.

Flooded Cell Versus Absorbed Glass Mat (AGM)

Traditional flooded cell batteries (**Figure 21**) utilize lead plates immersed in liquid electrolyte (sulfuric acid and water). Flooded cell batteries originally had individual twist-off caps on the six individual cells, allowing you to check the level of the electrolyte in the cells, and replenish if necessary with distilled water.

Because few customers want to deal with the possibility of splashing sulfuric acid on their car's paint or their skin, manufacturers began producing so-called maintenance-free batteries with more durable construction that minimizes electrolyte loss. These typically feature plugs that need to be pried up instead of caps that can be twisted off, so maintenance-challenged would perhaps be a more accurate term.

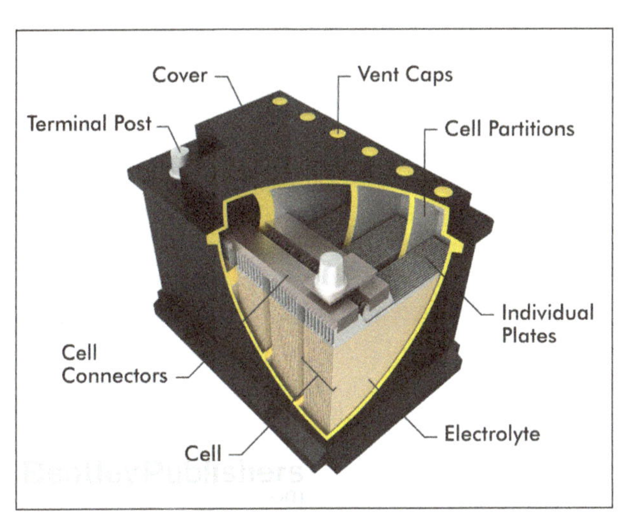

Figure 21 The traditional flooded cell car battery. *Copyright, Yuriy Merzlyakov*

The other main type of automotive battery other than flooded cell is the valve regulated battery, so called because an internal valve controls absorption of internally generated gasses. The two types of valve regulated batteries are gel cells and Absorbed Glass Mat (AGM). It is the latter – AGM – that has gained widespread use for automotive applications. See **Figure 22**.

AGM batteries use a fiberglass mat between the lead plates, with the electrolyte completely absorbed into the fiberglass. For this reason, there is no free liquid in an AGM battery, and they thus actually are sealed and maintenance-free, at least in the sense of never needing water.

AGM batteries have become popular due to their sealed nature, their resistance to damage from vibration, and their ability to retain charge longer while sitting on the

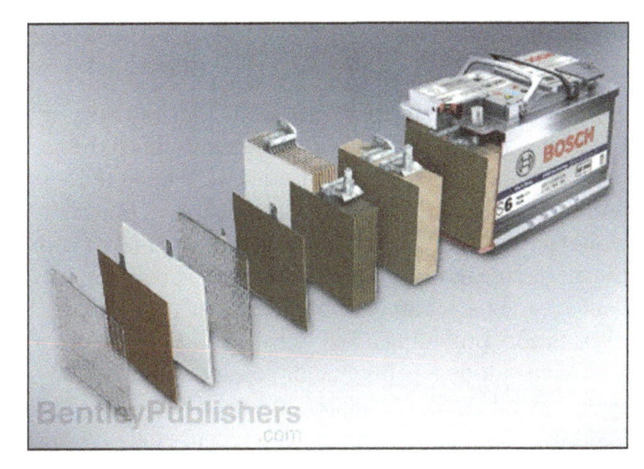

Figure 22 Absorbed Glass Mat (AGM) battery. This is the battery of choice for start/stop systems. *Photo, Robert Bosch GmbH*

shelf. If you have a newer car such as a BMW with Start/Stop, it came with an AGM battery. Replace it with an AGM battery of the same specifications. AGM batteries are more expensive than their flooded cell counterparts.

Lithium Ion

Lithium ion batteries hold the promise of weight savings and high power delivery, but cost has been high. However, as of this writing, they're finding their way into non-hybrid passenger cars through both original equipment (OE) and aftermarket channels. See **Figure 23**.

The lithium battery weighs about half as much as a comparable AGM battery and has the intelligent battery sensor built right into it. And in the automotive aftermarket, lithium ion batteries are available in standard BCI-code sizes. Right now, the cost is only justifiable in racing applications where weight is everything, but, as is the case with any emerging technology, cost is expected to come down.

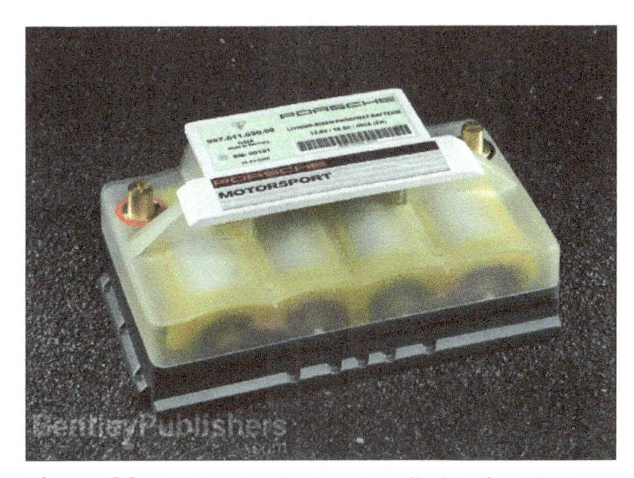

Figure 23 A lithium ion battery installed in the 2015 Porsche 911 GT3. In a high-performance race car, every ounce of weight matters. A lithium ion battery typically weighs half as much as its lead acid counterpart. This one is a whopping 22 pounds lighter than a lead acid 60 amp-hour battery. *Photo, Porsche Cars North America*

Battery Venting

Non-sealed (non-AGM) batteries may vent a small amount of hydrogen sulfide gas while they're charging (and a not so small amount of hydrogen sulfide if they're being overcharged). When batteries were mounted under the hood, this was rarely a concern, as the gas would pass harmlessly into the surrounding air, but with the increasing tendency to place batteries in the trunk or beneath the back seat to improve the car's weight distribution, providing a vent for the possible release of hydrogen sulfide has become important. As such, if a battery is replaced, it's important to verify that the venting is compatible with the car. Often, new replacement batteries include a "venting kit" which is just a length of flexible rubber hose to allow you to connect the battery's vent to the plumbing already in place in the car. See **Figure 24**.

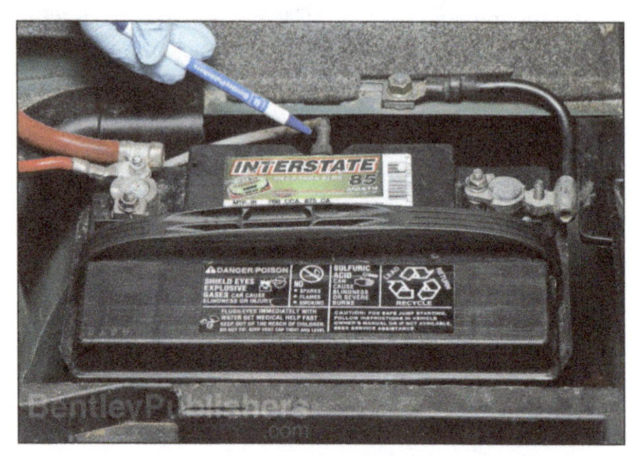

Figure 24 The battery vent is there to vent hydrogen sulfide gas into the atmosphere. Hydrogen gas is explosive and highly corrosive. Make sure the vent is correctly routed.

It's commonly stated that AGM batteries don't emit gas and thus don't need to be vented. It's a little more complicated than that. When they're properly charged by a correctly functioning alternator, any gas generated is internally recombined via the valve-regulated design. However, if the alternator or regulator malfunctions and overcharges the battery, excess gas can be generated which is vented through a safety valve that keeps the battery case from cracking and the battery from possibly exploding. The safety valve, however, is not a vent that a hose is attached to.

Battery Brands

As is the case with major appliances, there appears to be only a handful of independent domestic battery manufacturers. As of this writing, Johnson Controls makes batteries labeled as Diehard (Sears), Kirtland (Costco), DuraLast (Autozone), AC Delco, Interstate, Walmart, Advanced Auto Parts, and more, and also owns Optima. East Penn Manufacturing and Exide also produce batteries labeled with their own labels as well as private labels. In addition, Odyssey reportedly makes the AGM battery labeled as the Diehard Platinum.

Because of this label engineering, the recommendation is to shop for capacity and technology. That is, there may be little difference between flooded-cell-low-maintenance 850 CCA Group 48 batteries from vendors A and B; they may literally be the same battery with different labels. However, there is a difference between flooded cell and AGM, and if your late model car calls for an 850 CCA AGM battery, that's what you should buy.

Battery Registration/Coding

If your car has active battery management software, a new replacement battery needs to be "registered" or "coded" to the vehicle. These late model vehicles monitor battery condition and adjust charging voltage accordingly. The system is capable of knowing that battery's state of charge (SOC), state of health (SOH), and state of function (SOF). It has the ability to shut down non-essential consumers if the battery voltage drops too low, even in the middle of the night when you're sleeping.

BMW adopted this technology on most of their models from about 2004, but not all. One exception is the 2004-2010 BMW X3, which never received advanced battery management software. Audis equipped with the Multimedia Interface (MMI) beginning in the early to mid 2000s also appear to require battery coding AND an Original Equipment Manufacturer (OEM) Audi battery to carry out the required coding procedure. The coding process requires you input the battery's OEM serial number into the scan tool.

Why do you need to do this? In simple terms, registering or coding a new battery tells the engine management computer that the old weak battery is gone and a new charging strategy should be implemented on the fresh robust battery. Not registering a replacement battery can result in overcharging and shorter battery life. Note that it is important that the replacement battery has the same construction and CCA rating as the original battery. That is, if the original battery was an 850 CCA AGM battery, that's what you should replace it with. You can retrofit a different battery using a BMW scan tool, but you get very few retrofit options.

How do you do register a battery? You'll need a manufacturer-specific scan tool. This can be done by an authorized dealer service center or a qualified independent workshop with the proper equipment. Or you can do it yourself if you have a consumer-level scan tool capable of performing this function.

Now that you understand the basics of your car's main electrical power source, in the next chapter, we'll take a look at how electricity actually works.

3

Circuits (How Electricity Actually Works)

Voltage, Current, Resistance, and Load

Let's all admit something. When you encounter a long definition of Ohm's law, do your eyes glaze over? You're not alone. Really, no one cares about Ohm's law. No one wakes up excited to apply it. Kirchhoff's law? Even worse. And wiring diagrams? No one wants to read them. It's like reading spaghetti.

But here's the thing. You can't safely troubleshoot an electrical problem or modify an existing electrical circuit without at least some level of understanding of what's going on.

So think of this as your broccoli. (Please, no angry letters from vegetarians.) You have to eat one piece in order to have a shot at the pie. Maybe two. Trust me. Eat the damned broccoli. We'll have pie at the end. It'll be really good.

What's the Least I Need to Know?

- Most of the individual circuits in your car are series circuits – they have one load device such as a light or an electric motor, along with wires, switches, and connectors, strung together in a series, with one end connected to the car's positive battery terminal, and the other end to the body of the car, which is connected to the negative battery terminal.
- Ideally, only the load device has any substantial resistance. But as things age and corrode, the wires, switches, and connectors all can develop unwanted resistance. This unwanted resistance effectively steals energy from the circuit, making it so that the load device can't do its thing as well (the light isn't as bright, the motor doesn't spin as fast).
- In spite of what you've heard, resistance is usually a good thing. Resistance is what opposes the flow of current and lets the load device convert electrical energy to light, or a make a motor spin, or whatever it's supposed to do.
- In spite of what you've heard, it's misleading to say that resistance generates heat. It's really the flow of current that generates heat. And the important questions are whether or not there's supposed to be so much current that it generates the heat, and how that heat is supposed to be dissipated.
- It's when you have *unwanted resistance* in corroded wires and connectors that resistance is a bad thing. And there, yes, resistance *does* generate heat, because the corrosion is not a valid load device. It doesn't light up. It doesn't spin around. It just saps voltage and generates heat, with no well-engineered way for the heat created by the power consumed to be dissipated.
- If you care about the mystery of how a circuit uses all of its voltage and thus how the individual devices on the circuit are able to drop the voltage by certain amounts, it's really just a consequence of a) the current having to be the same everywhere in a series circuit, b) the current being determined by the sum of all of the resistances, and c) the individual voltage drops being proportional to the individual resistances. But you have to live with it for a while to really understand it.

What Is Electricity?

What is electricity, really? It doesn't matter. It's as relevant as asking what water is when you're simply thirsty. For our purposes, we're going to be pragmatic and say that electricity is charge. You can have charge stored up in a battery, but it is really *the flow of charge* that is useful. We are going to use what's known as *conventional flow notation* that assumes charge is flowing from the positive battery terminal, through the circuit, and returning to the negative battery terminal. You can read other books and learn that the actual exchange of electrons – valences and so forth – begins at the negative battery terminal and flows toward the positive (this is called *electron flow notation* and is used mainly by scientists), but from a practical standpoint, that doesn't matter either. We'll concentrate on what's practical and useful, and describe things in a way that is informal but still correct.

Water analogies are often employed to describe electricity, with voltage as the height of a waterfall, current as the amount of water flowing, and resistance as the diameter of a water pipe. Voltage is also often defined as electrical pressure, where "volts push amps through ohms." If these are helpful, use them. But understanding electricity really hinges on understanding the nature of resistance, so we'll talk a lot about that.

Simple descriptions of the key terms needed to understand electrical circuits are contained in the **Key Terms in Understanding Electrical Circuits Table**.

Key Terms in Understanding Electrical Circuits			
	Definition	**Units**	**Symbols and Abbreviations**
Voltage	The difference in charge measured at two points	Volts	V
Current	The rate at which charge is flowing	Amperes ("Amps")	I (in Ohm's law) A (on multimeter)
Resistance	A material's tendency to oppose the flow of charge	Ohms	R (in Ohm's law) W (on multimeter)
Load	An electrical device in a circuit that has resistance and draws current	none	none

Range and Precision of Units

Voltage

In a car, most of the time we're dealing with a constant voltage from the 12-volt battery (well, the battery is really more like 12.6 volts, and actually it's more like 14.2 volts with the alternator running, but, still, basically constant). The exception is the ignition system which may generate 10,000 to 40,000 volts at the spark plugs.

For the most part, the useful precision in voltage is about 1% of the nominal reading, meaning, on a 12-volt system, the useful precision is about 1/10 of a volt. So, when talking about battery voltage, there is a big difference between 11 volts and 12 volts, there's a minor difference between 12.4 and 12.6 volts and, for the most part, there's no useful difference between 12.6 and 12.61 volts.

Amperage (Current)

Unlike voltage which, other than the ignition, is nearly constant in most of a car's electrical circuits, amperage varies widely. To throw out some sample numbers, big electric motors like the starter motor draw a lot of current – hundreds of amps for short bursts. Fan motors may draw 20 amps. Small electrical loads like dome lights draw a fraction of an amp.

The prefix *milli* is frequently used to denote thousandths. Numbers less than 1 amp are often expressed in milliamps, or thousandths of an amp. For example, 0.130 amps = 130 milliamps, which is usually written as 130 mA. So-called parasitic drains (users of current that run the battery down when the car is at rest) should total less than 50 mA, but even that much current can run a weak battery down. Total resting current below 10 milliamps is considered low.

Because of the wide range of amperage values, one can't make blanket statements about precision in the same way we did with voltage. A reasonable expression of useful precision in amperage may be closer to 10%. That is, if a fan motor is spec'd at 20 amps, and instead it's drawing 22 amps, the fact that it's 10% high may not be significant.

Resistance

Because the resistance of electrical wire is low but the resistance of electrical devices is typically high, the resistances encountered in working on auto electrics are very wide. For small resistances, milliohms are often used to denote thousandths of an ohm. For example, ten feet of 10-gauge wire has a resistance of 0.0102 ohms, or 10.2 milliohms. Using the omega symbol, this also may be written as 10.2 mΩ. For large resistance values, the prefixes *kilo* for thousands and *mega* for millions are often used. For example, the resistance of spark plug wires is approximately 10,000 ohm, or 10 kΩ, per foot.

The useful precision of a resistance reading depends strongly on what the application is. That is, if you have one versus two feet of spark plug wire, that will double your resistance from 10 kΩ to 20Ω, but for a plug wire, that's not significant. In contrast, certain sensors like temperature sensors and throttle positioning sensors present a variable resistance, and a reading being off by 10% means that the car's computer will think the engine is 10% hotter, or the throttle is open 10% further, than it actually is.

Electrical Circuits Overview

Current Path – Why Many Things on Your Car Are Connected With Only One Wire

Above, when we said "it is really the flow of charge that is useful," we embedded the concept of a circuit in our definition of electricity. A circuit is a path for electricity to flow along, starting at the positive terminal of a battery, passing through something that uses the flowing charge to do something, and ending at the negative battery terminal. Yes, again, we know that the electrons actually flow the other way, and again, it doesn't matter. In a car, because the body and frame are almost always metal (the exceptions of fiberglass-bodied cars and Morgans with ash frames notwithstanding), the negative terminal of the battery is connected to the body, allowing the body of the car to be the ground path for the electricity to flow through (the exceptions of early

British cars with positive ground electrical systems notwithstanding).

Thus, many devices on the car appear to be connected using only a single positive wire. If a device is bolted directly to the body of the car (or to the engine, which in turn is connected to the body of the car), then the "wire" you're not seeing that completes the circuit is the body of the car acting as the ground path. In the case of a starter motor, there literally is no ground wire – it grounds to the engine block through the case of the starter – but sometimes a device has a very short wire connecting its negative terminal to the car's body. Where devices are attached to something fully insulated – for example, a windshield washer pump that attaches to a plastic bottle – there need to be two wires.

The Concept of a Load Device

The idea that a circuit includes "something that uses the flowing charge to do something" is absolutely crucial. That "something" is commonly called a *load device*. A load device can be an electric motor, a light bulb, or any number of other devices that are engineered to use the 12 volts supplied by your car's battery. What it *can't* be, though, is just a length of standard circuit wiring. Because that will burn. We'll come back to this point repeatedly as we step through the other sections in this chapter.

Tales From the Hack Mechanic: The Electromagnet, Part 1

When I was a kid, I read that I could build an electromagnet by wrapping a coil of wire around a steel bolt and connecting both ends of the wire to a battery. So I did. You know what? The wire got hot, burned my hand, and then melted. Because what I'd built was, basically, a short circuit. Despite the fact that the wire was wrapped around the bolt, there was nothing in the circuit to use the flowing charge, so *a lot* of charge flowed. In fact, I imagine that the electromagnet would've worked really well for the one second it was in existence before the wire melted.

As we just explained, my electromagnet ***needed a load device***. Though I didn't completely understand the concept, I figured out that I needed to put something in the circuit to actually use the flowing charge to do something other than just heat up the wire. I simply added a light bulb to the circuit, between the battery and the coil of wire. Once I did, the electromagnet actually worked without burning my hand.

Circuit Diagrams, Wiring Diagrams, and Conceptual Illustrations

One of the confusing aspects in all of this is that electrical circuits are often drawn in three different ways. A true *circuit diagram* is used to design, prototype, build, duplicate, and troubleshoot a piece of electrical hardware, and uses a set of standardized symbols for voltage sources, resistors, capacitors, and other electrical components. A circuit diagram generally shows you the entire electrical circuit, is intended to give you everything you need to calculate voltage, current, and resistance at any point in the circuit, and is rarely needed for a car unless you're trying to do something like troubleshoot a 1970s radio or a bad tachometer.

A *wiring diagram* also uses a set of standardized component symbols, but includes symbols that are more at the modular level. That is, you rarely care, on a wiring diagram, what an individual resistance is, unless there is a big component that is an actual resistor, such as a ballast resistor in an ignition system or a variable resistor that controls fan speed. A good wiring diagram typically gives you detailed information on actual wire sizes and colors and the connector types and locations, and tells you the location where the circuit is grounded to the body of the car.

Wiring diagrams for 60s and 70s cars used to show you the entire car on one page, giving you a pulled-back view of how components in the car are interconnected (see **Figure 1**). However, with the complexity of modern cars, a single page can no longer hold a wiring diagram for a entire car. In fact, a single page rarely draws out an entire circuit; it more often is a blow-up of a portion, with "comes from" and "goes to" notations at the ends.

In this book and others, we'll often mash circuit diagrams and wiring diagrams together and draw what's really more of a *conceptual illustration* where we show the circuit made by something like a car's horn, drawing the connections to the battery and ground without detailing specifics like the fuse number, wiring harness connection, and ground point location. Because it's hugely redundant to make three separate drawings for the same thing, this and any other book will usually only draw things once. It's usually clear enough from the context whether the drawing is a conceptual illustration (it usually is), a wiring diagram (sometimes), or a true circuit diagram (rarely).

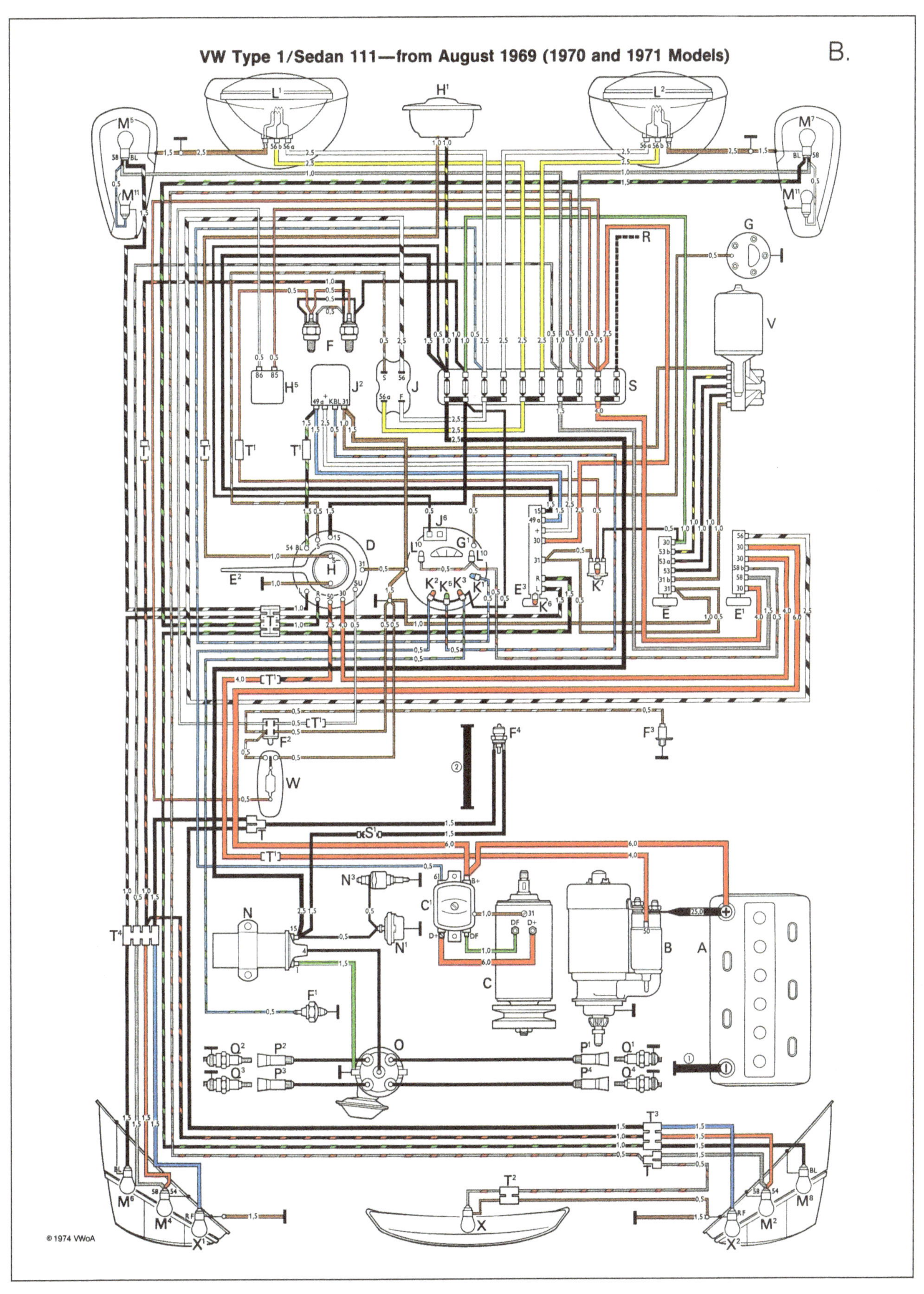

Figure 1 A wiring diagram for a 1971 VW Beetle, showing all of its components and wiring on a single page.

How Resistance Is Drawn in Diagrams

One bit of confusion is that, while the standard symbol of an electrical component that is an actual physical resistor is a jagged squiggly line as shown in **Figure 2**, we're not aware of a standard way to draw resistances that aren't actual resistors (for example, the resistance of a device, or resistance due to corrosion). So as not to cause additional confusion, we'll only use jagged squiggly lines to draw actual resistors. We'll use standard device symbols for things like motors, and list the resistances above or below them. For resistances of unwelcome loads or circuit wiring, we'll depict them using blank boxes with the resistance values inside.

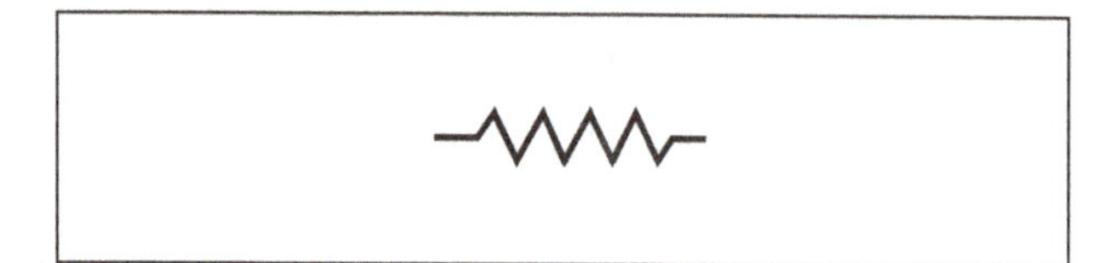

Figure 2 The standard symbol used for resistors.

Simple Circuits

Let's draw our first diagram, and just this once, we'll draw it two different ways. In **Figure 3** we show a circuit diagram of just about the simplest possible circuit of a 12V source powering a light bulb with a resistance of 12 ohms, using the standard electrical symbols. (In fact, light bulbs change their resistance as they warm up, but let's not worry about that right now.)

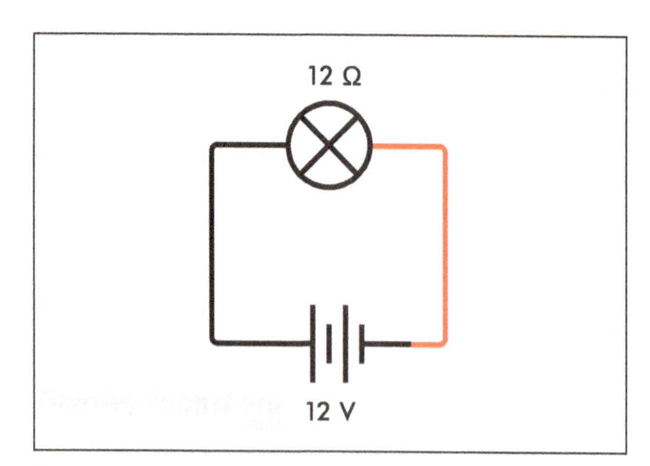

Figure 3 The simplest possible circuit, drawn as the kind of circuit diagram we usually won't show.

In **Figure 4** we draw the same thing as a conceptual illustration, using an automotive-style battery, and a light bulb instead of a symbol of a light bulb.

Now, just to get you used to what you'll be seeing throughout this chapter, instead of a 12-ohm light bulb, in **Figure 5** we show a generic 12-ohm "load device."

However, note that very few circuits on a car actually have wires going to the negative battery terminal. The

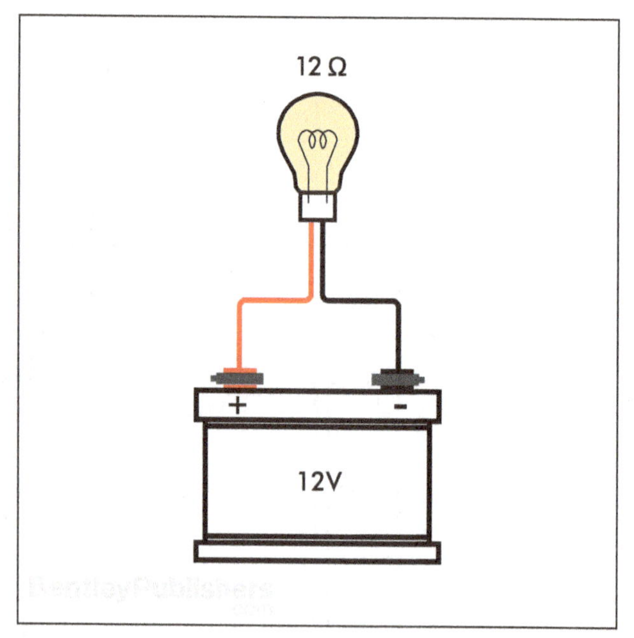

Figure 4 The simplest possible circuit, drawn as the kind of illustration we'll try to show.

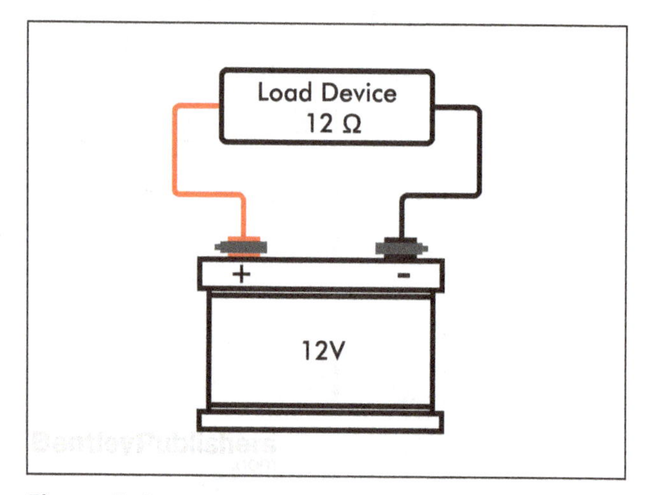

Figure 5 The same, drawn instead with a generic load device.

negative battery terminal is, as we said, attached to the body of the car, allowing the body to be used as the ground path. For this reason, you will often see circuit illustrations drawn using two ground symbols, one on the negative battery terminal, and the other on the negative side of the load device, as shown in **Figure 6**. Note that this not only simplifies the wiring of the car, it also simplifies the wiring diagram, as now only one line needs to be drawn for each circuit, plus a small ground symbol. To a certain extent, this is the dividing line between a circuit diagram (more **Figure 3**) and a wiring diagram (more like **Figure 6**). But make no mistake, **Figure 6** still depicts a circuit.

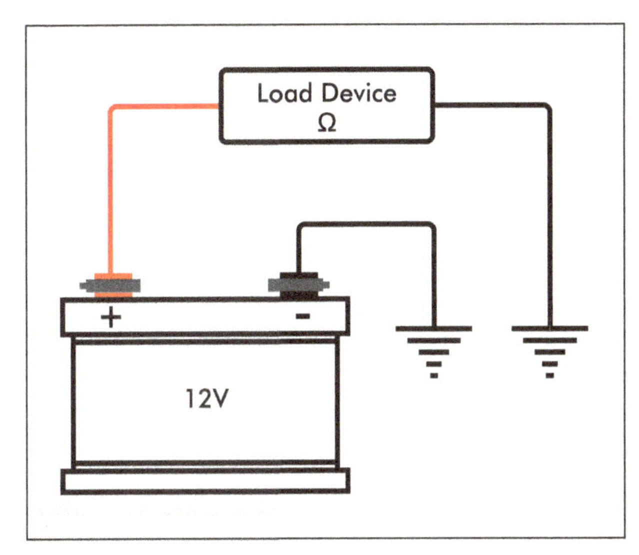

Figure 6 The simplest possible circuit depicting the ground path through the body of the car. Most wiring diagrams are drawn approximately like this.

Ohm's Law and Why Resistance Is Crucial

You might ask the question: If you just connect the two battery terminals together with a wire, is that a circuit? Well, that would look like **Figure 7**:

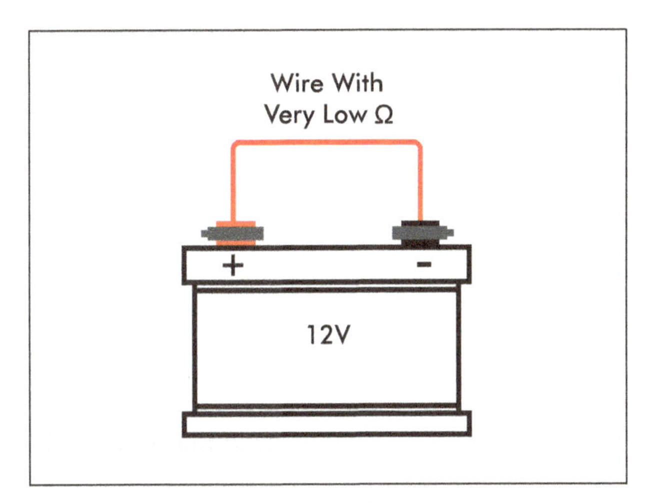

Figure 7 A circuit with no load device is a short circuit. This is dangerous! Don't EVER do this!

Is this a circuit? Yes, but it's a short circuit. It's worse than useless. It's incredibly dangerous. Here's why.

A useful circuit needs three things – a source of voltage (the battery), a path (the wiring and the body of the car), and a load device to offer resistance to the flow of current and perform useful work. The short circuit pictured in **Figure 7** is technically a circuit, but it's not a useful one because it doesn't include a load device. The circuit wiring itself has very low resistance, and by itself does not constitute a valid load device.

We're going to learn Ohm's law for the very reason that we need it to answer this question of why a short circuit is not a valid circuit. Answer it, and you'll develop a deep understanding of electricity and how to work with it safely in an automotive environment.

Ohm's law relates voltage, current, and resistance in a simple equation. It is usually written as V = I*R (voltage = current * resistance). What Ohm's law says is that, in most automotive circuits, since battery voltage is nearly a constant 12V, if the resistance is low, then the current must be high, and vice versa. Put another way, current and resistance are inversely proportional.

Some quick terminology.

- "I" is a silly letter to use for current, but it is the standard. Note that, on the multimeter, current isn't labeled "I" but is instead labeled "A" for "Amperage." It's the same thing.
- Similarly, note that, on the multimeter, resistance isn't labeled "R" but is instead labeled with an omega symbol (Ω).
- There are presentations of Ohm's law that try and align the symbols completely with multimeter symbols, writing it as "V = A*Ω," but we're going to stick with the widely used "V = I*R."

You'll see Ohm's law written in three different ways, depending on which two variables you know and which one you want to solve for. To be clear, these are not three different equations. They're the same equation, rearranged three times, to express an electrical value you don't know using two that you *do* know.

Three Different Ways of Writing Ohm's Law

V=I*R (voltage = current times resistance)

R=V/I (resistance = voltage divided by current)

I=V/R (current = voltage divided by resistance)

The third equation, I = V/R, is worth looking at. It is, in fact, where most of the action is. This says that, if voltage is constant, the lower the resistance is, the higher the current is. Think about that. To produce lots of current, all you need to do is lower resistance. Have you ever seen a car get consumed by fire from an electrical short circuit? It's frightening how quickly it goes from smoke to full-on conflagration. *This is a direct result of Ohm's law.* The kind of wire used to connect electrical devices in a car has very low resistance, so if both ends of a wire are simply connected to the battery, with no load device to make use of the flowing charge, that battery is going to produce an outrageously high amount of current. Enough to burn your ride to the ground.

Scared yet? Want to understand it maybe a little? Of course you do.

There's a related formula for power in watts (Watt's law), also expressed three different ways depending on which two of the three variables you know:

Watt's Power Law

P = I*V (power = current times voltage)

P = I2*R (power = current squared times resistance)

P=V2/R (power = voltage squared divided by resistance)

Our First Ohm's Law Example

So let's do our first Ohm's law calculation on the circuit with the light bulb in **Figure 5**. We know the voltage (12 volts). We're told the resistance (12 ohms). So, using Ohm's law in the form I=V/R (current = voltage divided by resistance):

- Voltage = 12 volts
- Resistance = 12 ohms
- Current = V/R = (12 volts) / (12 ohms) = 12/12 = 1 amp

Using Watt's law in the form P=I*V (power = current times voltage):

- Current = 1 amp
- Resistance = 12 ohms
- Power = I*V = (1 amp) * (12 ohms) = 12 watts

Pretty easy, huh? Nothing to be afraid of.

We now know enough to do the calculation of the current generated by a short circuit. You can look up the resistance per foot of standard gauge wiring. Let's take 10-gauge wire with resistance of 0.00102 ohms per foot. So 10 feet of this wire should have a resistance of 0.0102 ohms. Since we know voltage and resistance, we use Ohm's law in the form I=V/R:

- Voltage = 12 volts
- Resistance = 0.0102 ohms
- Current = V/R = (12 volts) / (0.0102 ohms) = 12/0.0102 = 1176 amps

That's a huge amount of current, probably five times what your starter motor draws when it's cranking.

Let's calculate the power. Since we know voltage and resistance, let's use Watt's law in the form $P=V^2/R$ (power equals voltage squared divided by resistance):

- Voltage = 12 volts
- Resistance = 0.0102 ohms
- Power = V^2/R = (12*12)/.0102 = 14417 watts

Over 14,000 watts of power. That's equivalent to the peak power of a generator big enough to power a commercial construction site. Your insulated battery cables certainly aren't designed to dissipate that kind of power. That's why you get smoke and flames.

The point is that, yes, connecting a short length of low-resistance wire across the battery terminals *does* make a circuit, and causes *a lot* of current to flow, but it's *a circuit that's not engineered to handle the current*, and thus it's one of the most dangerous things you can do with a car battery.

So any valid (i.e., non-short) circuit *must* have a valid load device. It's the thing that uses the flowing charge in order to do something. It's what was missing in the electromagnet referred to in the sidebar. A load device can be a light bulb which produces both light and heat. It can be an electric motor which produces mechanical motion and a little heat. It can't simply be a wire connected across the battery terminals because Ohm's law mandates that the wire's low resistance will result in spectacular amounts of current that will melt it in short order.

Why Resistance Is Often Misunderstood

Many people have heard that resistance generates heat, and are thus surprised when they read the above description about how low-resistance wire will get really hot. It is true that resistance generates heat, but it's a misleading statement. A better way to think of it is this:

- The flow of current generates heat.
- For a given current, if the resistance is higher, more heat will be generated.
- In an improperly designed circuit, if the current is higher than the load device can handle, more heat will be generated than the device and the circuit are able to get rid of.
- But because resistance opposes the flow of current, load devices are usually designed to have a moderate to high resistance precisely in order to limit current flow, which in fact reduces the heat generated.

But aren't electric heaters just made of wire? Isn't a car's cigarette lighter *exactly this*, just a piece of wire that gets wicked hot? Sort of, and sort of. Here's the thing. A heater needs to be something *designed to be in a circuit*, not something that's accidentally there due to a short and happens to get hot.

In the case of an electric heater or a cigarette lighter, yes it's "made of wire" but it's not "just made of wire." It's a small section of high-resistance wire that a) is designed to handle the current drawn at that resistance and to dissipate the heat generated, b) is surrounded by thermal barriers to make sure that heat doesn't burn up the rest of the car, c) has been thoroughly tested and is carefully fused (indeed, the cigarette lighter is designed to pop out when it gets too hot) and d) has regular low-resistance wire leading up to it in the circuit. A lot of engineering has to go into any device you connect to your car battery in order for it to be safe.

Tales From the Hack Mechanic: The Electromagnet, Part 2

Let's go back to my electromagnet example. Aren't electromagnets coils of wire wrapped around an iron core? Yes. Isn't that the same as the electromagnet I made? Yes, but with a lot of engineering and design. A real electromagnet is designed with a long length of wire – the longer the wire, the higher the resistance, so it draws a manageable amount of current. Don't real electromagnets still get hot? Yes. They have heat sinks to dissipate the heat, and may have a limit on the time they can continuously remain on, beyond which they'll overheat. But... *it's still just wire, right?*

Yes, but think about it. Light bulbs, electric motors, ignition coils... they're all basically just wire with a well-engineered mechanism for heat dissipation. The filaments in light bulbs run inside of a vacuum because if they didn't, the red-hot glow, combined with the oxygen in the air, would cause them to burn. Electric motors are designed to be cooled by the surrounding air, or in the case of electric fuel pumps, the surrounding gasoline. (This is true. The gas actually cools the fuel pump's windings. There's no oxygen, so it doesn't catch fire.) Ignition coils are surrounded by oil or air and are designed to be pulsed on and off.

At some level, they're all "just wire," but *they're designed and engineered to dissipate the heat they produce*. That's fundamentally different from simply connecting a wire across the two battery terminals.

The point is that any electrical device – any "load device" – you connect to your car's battery has hopefully been designed and tested so that, when 12 volts are applied, the current drawn by the device doesn't cause it to generate more heat than it can dissipate.

Static Versus Dynamic Resistance

When we say "resistance," we can be talking about *static resistance* (the resistance of an actual physical resistor, or a load device, or the wires themselves, or unwanted resistance caused by corrosion), or about the total *dynamic resistance* of a circuit. There is a subtle distinction between the two.

We defined resistance as "a material's tendency to oppose the flow of charge." That means that, technically, in order to have resistance, current has to be flowing. In the **Common Multimeter Tests** chapter, you'll learn that, to measure resistance, a) the circuit has to be unpowered, b) you conduct the resistance measurement on a portion of the circuit (usually one component), and c) the multimeter applies a small voltage to that component that causes current to flow through it. Thus, even though this is sometimes called static resistance, in order to measure it, current still needs to be flowing.

Now, you can power a circuit, or you can measure the resistance of a portion of the circuit, but you can't do both at the same time, So, when an entire circuit is powered by the car's electrical system, you *can't* connect a multimeter and measure the circuit's resistance. The best you can do is measure the voltage and the current and then *calculate* the dynamic resistance using Ohm's law.

Although, as per our first Ohm's law example above, it is resistance that determines the amount of current that flows, the fact is that, in a real circuit, current and resistance affect each other dynamically. As current flows, the circuit heats up. Increasing temperature usually increases resistance, causing less current to flow.

People rarely say "static resistance" and "dynamic resistance" because it's usually clear from the context which one they're talking about. Just remember that you only ever directly measure static resistance with a multimeter, and that dynamic resistance is a consequence of the amount of current that actually flows.

Conservation of Energy in a Circuit and "Dropping the Voltage"

We've seen that resistance is the necessary consequence of the presence of a load device as it is performing work. It's when the source of the resistance is not a load device (for example, when the terminals of your battery are all corroded) that it causes problems.

Here's another key to understanding resistance. Even when you think you have only one load device in a circuit, you're wrong. You don't. You never do. Your circuit, at a minimum, also contains a fuse, a switch to turn the load device on and off, plus wiring, plus the connections between the device, the switch, and the wires (**Figure 8**). That makes it a series circuit where all of these things are components in the circuit. And each has a resistance. Normally, the resistances of everything except the load device are, or should be, negligible. But if there are unexpected resistances from these other components, the circuit can malfunction.

So why doesn't every electrical circuit instantly drain the battery and burst into flame? Because all of the charge doesn't flow; only some of it flows. The total amount of charge that flows in the circuit (the current), and thus the total energy given up, are governed by the resistance of the circuit. If the circuit is a well-engineered one, it will have a valid load device with a reasonable resistance that will allow a reasonable amount of charge (current) to flow, and will convert the charge's energy to some other form of energy.

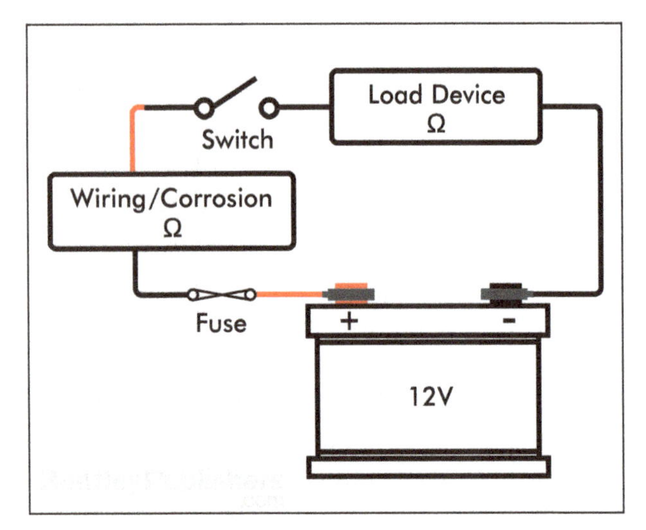

Figure 8 The minimum real-world circuit.

Energy in a Circuit

If electricity is the flow of this vague thing called "charge," then voltage is the amount of energy per charge. The battery is what gives the charge this energy. When the charge leaves one battery terminal, passes around the circuit, and returns to the other terminal, the charge's energy is used up. All of it. No energy is left over. That's what it means for a charge to complete a circuit between two battery posts that have a voltage difference. The energy in each charge must all be used up in the charge's trip around the circuit.

That other form of energy might be light to illuminate a road, electromagnet to turn a fan, acoustic to play music, or heat to defrost the rear window.

Regarding heat, all electrical circuits generate a certain amount of heat as they perform work and convert electrical energy into some other useful form, but it's only when the load device is, in fact, literally a resistor (or unintended resistance such as with a corroded connector) that *all* of the energy is going into heat.

You may read, perhaps misleadingly, that voltage is "used up" by a load device. It's not that voltage gets used up, strictly speaking. As we've explained, it's the *energy in the flowing charge* that gets used up. The charge is given energy by the battery, and when the charge completes the circuit, with the positive terminal at 12V and the negative terminal at 0 V, by the time the flowing charge is at the negative terminal, its voltage has to be at 0 V. For this reason, as we've said, you'll sometimes read that the load device "drops the voltage" across it.

As current flows through the circuit, each component that offers resistance lowers the voltage by an amount proportional to that resistance, in accordance with Ohm's law (V=I*R), in order to do the work the component is designed to do. Put another way, it is the act of using the flowing charge to do something that restricts the flow. No matter what the load device is, no matter what form the energy is converted into, the resistance of the load device causes the voltage to drop because charge has come into the device, has had its energy used, and has exited the device.

A somewhat weird counter-intuitive consequence of this is that, if the circuit does not complete its path to ground, the voltage is still there on the device because no charge has flowed to allow the device to drop the voltage. Try it. Go ahead. Connect one terminal of a light bulb or any other device to 12 volts, and measure the voltage between the other terminal and ground. It won't read zero like you probably expect from all the above conditioning about "dropping the voltage." It'll read 12 volts. The device can't drop the voltage if charge doesn't flow.

Kirchhoff's Voltage Law (below) mandates that, in a circuit, the sums of the voltages across all the components in the circuit have to add up to the battery voltage. So if there's not a load device in the circuit creating the necessary resistance to cause nearly all of voltage to drop as it moves through the circuit, it means that the wiring – the only other piece of the circuit – is going to have its own V=I*R component involved in dropping the voltage. And since wiring resistance is generally very low, the current flowing through the wires will be very high. At the end of this section, we'll step through a thought experiment that makes this more clear.

Kirchhoff's Laws

The good Dr. Kirchhoff actually handed down two laws to us – a *voltage* law and a *current* law. We're going to state them and then apply them to the different basic kinds of electrical circuits.

Kirchhoff's Laws

Kirchhoff's voltage law states that if you add up all of the voltages across the individual components in a circuit, they must equal the total voltage driving the circuit. As we just described, this is ultimately a conservation of energy principle.

Kirchhoff's current law states that the currents entering and leaving an electrical intersection must add up to the same amount. Like the Voltage Law, this is also ultimately rooted in a conservation of energy principal. This is applied slightly differently to series and parallel circuits.

In the literature, there is no such thing as *Kirchhoff's resistance law*. However, as a shorthand, we're going to call it that, because "the Ohm's law consequence of Kirchhoff's Voltage law as applied to series and parallel circuits" is too much of a mouthful.

Don't get hung up over the equations we're about to show. As we did with Ohm's law, we need to show you Kirchhoff's laws in order to help show why current flows the way it does.

Kirchhoff's laws are applied differently to different kinds of circuits, so we need to explain briefly what those kinds of circuits are.

Three Kinds of Electrical Circuits

There are three basic kinds of circuits: Series, parallel, and series-parallel.

1) Series Circuits

In a series circuit, there is only a single path for current to flow. If there are multiple devices along the path, they are daisy-chained end-to-end, one after the other, and the current flows through them "in series." Like old-school Christmas lights, if one of the devices fails, it creates an open circuit, which interrupts the flow of current and causes all of the devices to fail.

Figure 9 shows the classic series circuit diagram you'll see in every electronics textbook showing two resistors (here, 100 ohm and 50 ohm). In **Figure 10** we show an equivalent illustration showing load devices of the same resistances.

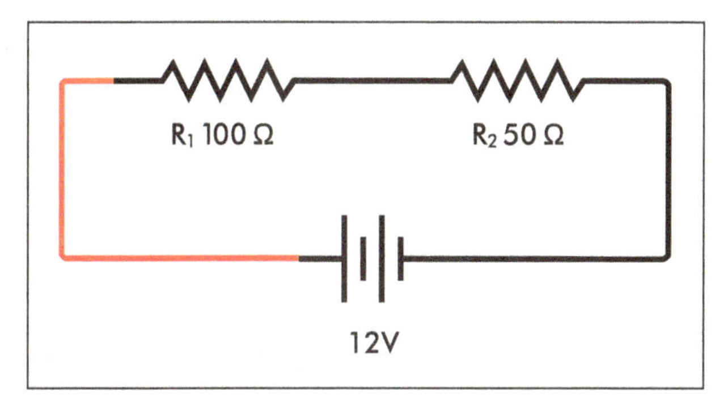

Figure 9 The classic series circuit diagram, completely unrealistic because no valid circuit has nothing but actual resistors in it.

Kirchhoff's Resistance Law for Series Circuits

In a series circuit, resistances simply add. The equation for this is:

$$R_T = R_1 + R_2$$

If you want to understand why, see the sidebar "Why Resistances Add in a Series Circuit."

Using our 100 and 50 ohm resistances above, the total resistance is thus:

$$R_T = R_1 + R_2$$
$$= 100 + 50 = 150 \text{ ohms}$$

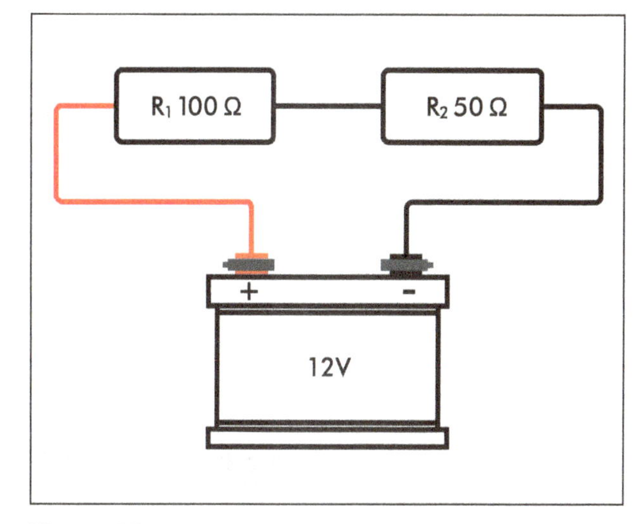

Figure 10 An illustration of the same series circuit using load devices instead of resistors.

Kirchhoff's Current Law for Series Circuits

The current law states that the currents entering and leaving an intersection in a circuit must add up to the same amount. However, in a series circuit, there are no intersections. So, instead, the current law takes the form that, **in a series circuit, the current is the same everywhere.** It has to be. It has nowhere else to go. This is enormously powerful, since it says that if you can calculate (or measure) the current at one place in a series, it has to be the same everywhere else in the circuit.

For example with two resistors, the current law says that the current flowing through both resistors is the same. Since we know the voltage and the total resistance, we can now calculate that current. Using Ohm's law in the form:

$$I = V/R$$
$$= (12 \text{ volts}) / (150 \text{ ohms}) = 12/150 = 0.08 \text{ amps of current}$$

Note that when you measure the voltage across a single component in a series circuit, the results may be unexpected unless you are aware that there are other components in series. For example, if you have a 12V circuit with a light bulb in it, you would expect the voltage across the light to be very close to 12V. If you measure it and find that it is instead 6V, you'd wonder where the rest of the voltage is going. If, however, you find that the light is part of a series circuit with two lights with the same resistance, then Kirchhoff's voltage law will tell you that this is correct – that each light should drop the voltage by 6V, adding up to the 12V total.

However, in truth, you very rarely see intended load devices wired in series, and the far more prevalent

Kirchhoff's Voltage Law for Series Circuits

Kirchhoff's voltage law states that the voltages across the individual components must add up to the total voltage. We can now calculate those individual voltages because:

- We know the individual resistances.
- In the current law, we said that the current flowing through both resistors is the same, and calculated that current.

So, using Ohm's law in form V=I*R for each of the two resistors:

$$V_1 = I*R_1 = (0.08 \text{ amps})*(100 \text{ ohms}) = 8 \text{ volts}$$

$$V_2 = I*R_2 = (0.08 \text{ amps})*(50 \text{ ohms}) = 4 \text{ volts}$$

Note that the individual voltages add up to the total 12V voltage, as it must.

application of Kirchhoff's voltage law to series circuits is that which we've been talking about – where, although there's only supposed to be a single load device in a circuit, resistance from corrosion at connections becomes an unintended load device. The voltage law tells you that, if the resistance at a corroded connection is eating up one volt, and if the circuit is powered by a 12V battery, the rest of the circuit only has 11 volts.

Why Resistances Add in a Series Circuit

To show that resistances add in a series circuit, we use Ohm's law in the form:

$$V = I*R$$

We can also express the total voltage in terms of the individual currents and resistances:

$$V_T = I_1*R_1 + I_2*R_2$$

Combining both equations:

$$I*R = I_1*R_1 + I_2*R_2$$

As you read in Kirchhoff's current law, the current has to be the same everywhere. Since all terms with I are equal, the I terms can be divided out of the equation. Therefore:

$$R_T = R_1 + R_2$$

2) Parallel Circuits

A parallel circuit has multiple paths for the current to flow. The fact that your car has a fuse box shows that the car is largely a collection of parallel circuits, where, if one fuse blows, the other circuits continue to work. To give a specific example, your left and right headlights are in parallel. If one dies, the other one still works.

At this point we're going to leave traditional circuit diagrams behind and go only with a representative illustration. **Figure 11** shows a parallel circuit with two load devices.

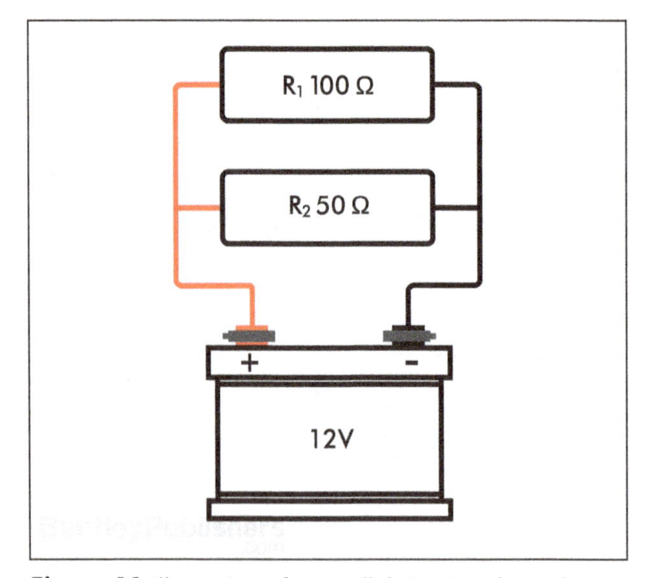

Figure 11 Illustration of a parallel circuit with two l oad devices.

Kirchhoff's Resistance Law for Parallel Circuits

In a parallel circuit, the resistances don't simply add; it's the reciprocals of the resistances that add. In other words, the relationship between the total resistance and the individual resistances is the scary-looking:

$$1/R_T = 1/R_1 + 1/R_2$$

If you want to understand why, see the sidebar "Why 1/Resistances Add in a Parallel Circuit."

Using our 100 and 50 ohm resistances above, the total resistance is thus calculated by using:

$$1/R_T = 1/R_1 + 1/R_2$$
$$= 1/100 + 1/50$$
$$= 3/100 \text{ ohms}$$

$$\text{Therefore, } R_T = 100/3, \text{ or } 33.33 \text{ ohms}$$

Note that, in the real hands-on world, you almost never have to do this calculation.

But note that this shows up an odd feature of resistances in parallel circuits – *the total resistance is always less than any of the individual resistances*. To many people, this is counterintuitive. However, if it helps to grease the intuitive skids, note that, with multiple parallel paths, electricity has more paths along which it can flow.

Note that continuity measurements in a parallel circuit can be misleading. If there's continuity, you can infer that the circuit is fine when, in fact, you may be seeing continuity across another leg of the circuit.

Kirchhoff's Voltage Law for Parallel Circuits

In a parallel circuit, the voltage across each component (or, I should say, across each parallel leg of the circuit) is the same. So, from a practical standpoint, with the voltage constant and the individual resistances known, it's the individual currents that can be calculated using Ohm's law in the form **I=V/R.**

Kirchhoff's Current Law for Parallel Circuits

In a parallel circuit, the currents in the different paths can draw different amounts of amperes. The law that the currents entering and leaving an intersection must add up to the same amounts is applied literally. We use that fact to calculate the total resistance.

Why 1/Resistances Add in a Parallel Circuit

If there are two currents, I_1 and I_2, in two branches of a parallel circuit, then the total current is their sum, or:

$$I_T = I_1 + I_2$$

According to Ohm's Law:

$$I = V/R$$

Therefore you can write this as:

$$V_T/R_T = V_1/R_1 + V_2/R_2$$

But because voltages are the same in each leg of a parallel circuit, the voltage terms can be divided out, leaving you with:

$$1/R_T = 1/R_1 + 1/R_2$$

Even though these equations look complicated, don't worry too much about parallel circuits. From a practical standpoint, you almost never need to do these calculations.

3) Series-Parallel Circuits

As the name implies, a series-parallel circuit (**Figure 12**) is a combination of the two. You can either say that series-parallel circuits are silly, or that your entire car is a collection of series-parallel circuits. (Actually, I think your car is a collection of series-parallel-series circuits, where the first "series" is a fusible link plus the undesirable resistances from the battery clamps and ground straps, the "parallel" part is all the separate circuits represented by the fuse box, and the second "series" is all the undesirable resistances from dirty contacts within the individual circuits.) But since the term is frequently used when discussing automotive electrical systems, we include it here. If you needed to solve for current or voltage at any point (and you almost never do), you'd break the circuit into its parallel and series components, and use the rules above individually.

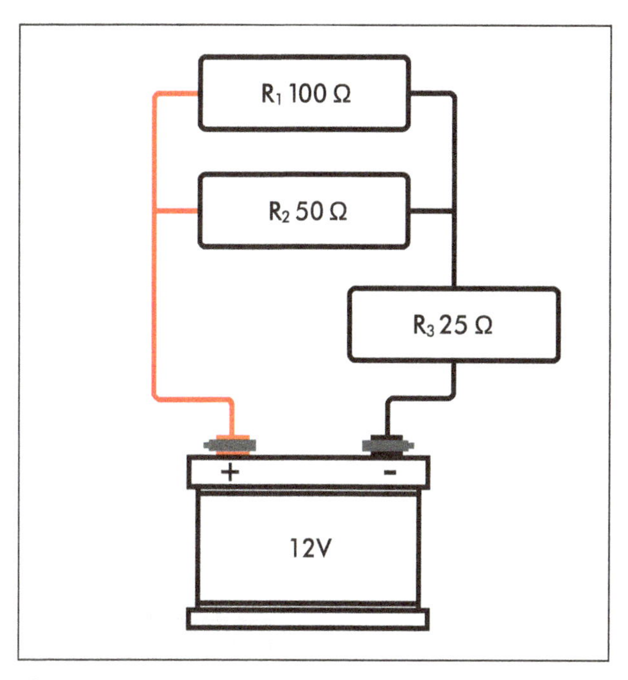

Figure 12 Illustration of a series-parallel circuit with three load devices.

The Five Kinds of Circuit Malfunction

Now that we've laid out the basics of a valid circuit, we can begin to describe what goes wrong. You may read that there are four kinds of circuit malfunctions: Open circuits, shorts to ground, shorts to power, and high-resistance failures. This is true, but for practical reasons, we add a fifth – component failure. We'll discuss them each below.

Open Circuit

An open circuit is technically not a circuit. There's a break somewhere preventing current from flowing all the way around the circuit. Thus, the device in the circuit doesn't turn on. A circuit with a switch that isn't closed (**Figure 13**) is an open circuit. However, when we speak of circuit malfunctions, it's usually because the circuit is *unintentionally* open. The break can be between the device and the positive terminal, between the device and ground, or somewhere within the device itself. Often the root cause is a wire separated from a connector or a solder terminal.

Short to Ground

A *short to ground* (**Figure 14**) is almost always what we mean when we refer to a "short circuit." Because the body of the car is used as the ground path, it is easy for a wire on the positive side of the circuit (between the battery's positive terminal and the load device) to fray its insulation, touch itself to the body of the car, and, without the resistance of the load device to limit the current and drop the voltage, draw a lot of amps. If the

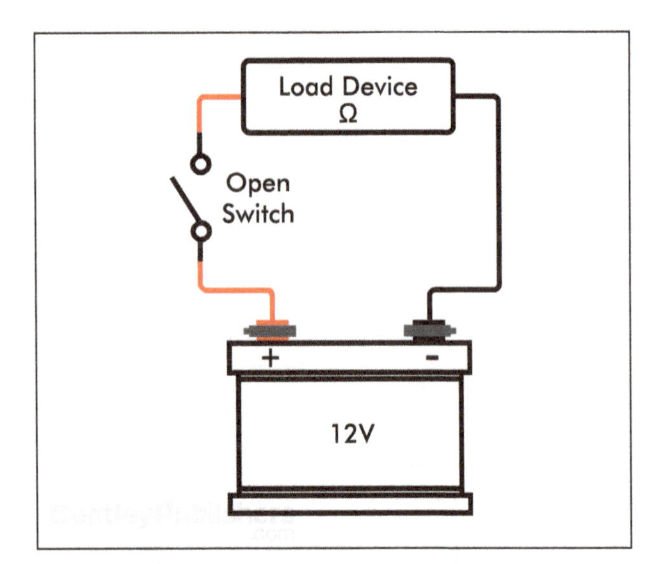

Figure 13 An open circuit is no different than a circuit with a switch that isn't closed.

circuit has a fuse, the fuse should pop before real damage is done. If not, the wiring will get red hot and burn off its insulation.

Often, the wire will melt and break the short circuit, creating an open circuit. If you're lucky, this will happen quickly, without affecting other wires and components. Unfortunately, that's rarely the case. If the wire runs inside of a larger wiring harness, it can wreak an incredible amount of damage. If the wire doesn't actually melt and break, current will continue to flow, causing an electrical fire.

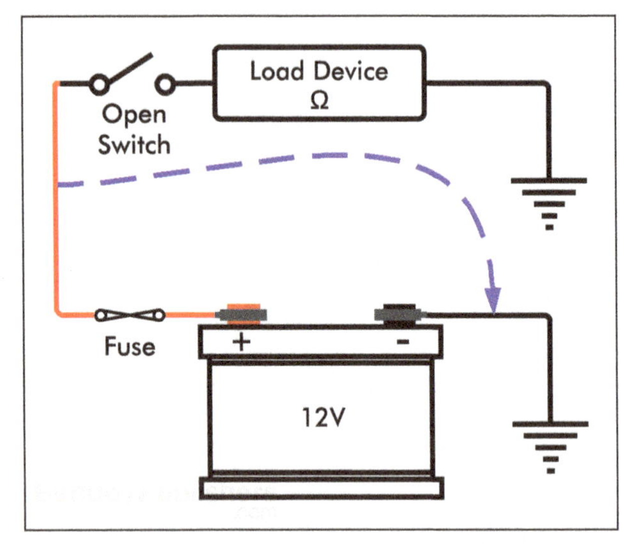

Figure 14 A short circuit (short to ground) depicted as a dashed purple line between the positive side of the circuit and ground.

Short to Power

A *short to power* (**Figure 15**), sometimes called *short to alternate feed*, occurs when something unintended is completing the circuit. It is typified by a device that keeps running. It can occur on either the positive or negative side of the circuit. For example, the power wire to the left rear directional might have rubbed off its insulation and be touching the power wire to the right rear directional, causing the left one to come on whenever the right one is switched on, even if its own switch is open.

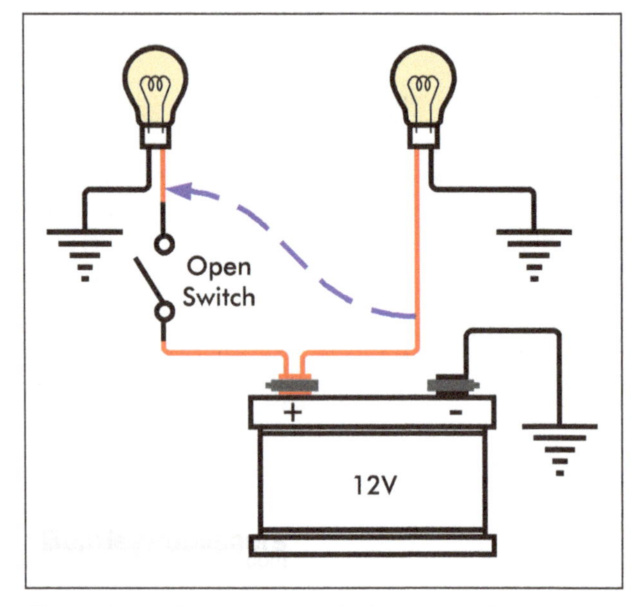

Figure 15 A short to power (dashed purple line) can power a circuit even if the switch to the circuit is open.

Component Failure

Part of the distinction between the electrical engineering view of circuit diagrams versus the practical automotive view is that most circuits in a car are physically spread out in the car. That is, the wiring for something like the tail lights literally runs the length of the car. Thus, in a car, there's the tendency to think of the "circuit" as the wiring. In truth, as we've learned, the circuit encompasses the components (the load devices) and the wiring as well as the source of power.

But thinking in terms of components and wiring is helpful in the following way. There really are only four fundamental kinds of circuit failures (open circuit, shorts to ground, shorts to power, and high-resistance failures, which we'll get to below). However, a component can also fail. From a troubleshooting standpoint, you often want to diagnose the problem as a wiring problem or a component problem. While it is true that a component failure is in and of itself ultimately one of these four (for example, a burned-out light bulb is literally an open circuit, with the burned filament preventing the circuit from being completed), the *action* that you take is different: You replace the bad component. We'll discuss this more in the **Troubleshooting** chapter.

High-Resistance Failure

Hopefully the book has given you a feel for the distinction between *intended resistance* in a load device that is used to convert the energy of the flowing charge into something desirable and useful, and *unintended resistance* in the form of corrosion, loose connectors, or broken strands in a length of wire that can sap the voltage and cause circuits to fail.

First, note there are a handful of devices that are basically just resistors and are made to be in a series circuit. For example, a ballast resistor is designed to be in series with an ignition coil. The knob that dims your instrument cluster is a variable resistor in series with all the lights, comprising what's actually a legitimate series-parallel circuit. And your fan speed switch is there to eat up some voltage and allow the blower to run at less than full speed. If you don't know these things are there you won't understand that they are *intentional* resistances and you'll be surprised when you troubleshoot the circuit.

A high-resistance failure occurs when circuit components such as switches, wires, and corrosion at connectors create enough resistance that they become unintended load devices. Any voltage lost across those resistances lessens the amount of voltage available to the intended load device – the starter motor, the lights, etc. – in the circuit. This is a very real phenomenon in the automotive world. Every connection point is an opportunity for corrosion to cause resistance and for that resistance to sap voltage (see **Figure 16**). The larger the amperage drawn by the device, the greater the potential for voltage loss.

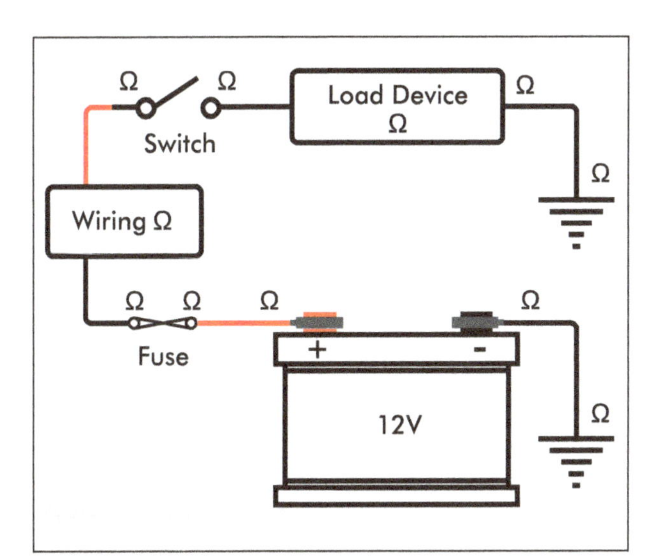

Figure 16 In a real automotive circuit, not only does the load device have resistance, but so does the wiring and every switch, fuse, and connection.

Let's walk through a specific example so you can see how even small amounts of unintended resistance can dramatically reduce the current available to the load device.

Let's say we have a battery supplying 12 volts to a starter motor. Let's assume we know that, during peak cranking, the starter motor needs to draw 240 amps. For the moment, we'll make this an ideal circuit where the only resistance is from the starter itself. See **Figure 17**.

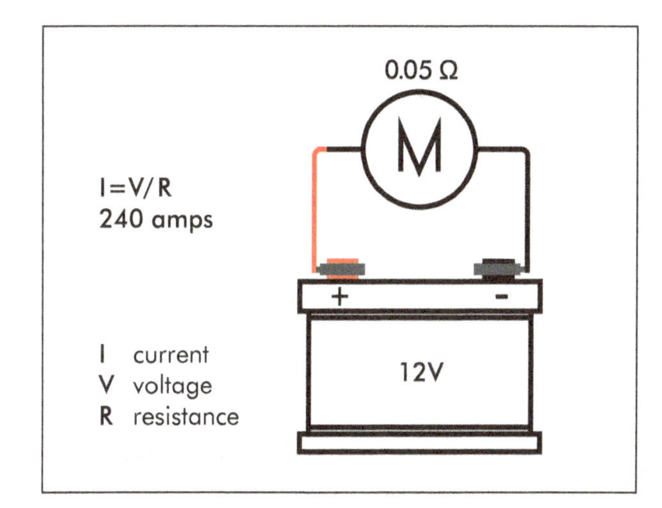

Figure 17 Ideal circuit with a starter motor drawing 240 amps.

Let's use Ohm's law to solve for that resistance. So:

- R=V/I (resistance equals voltage divided by current)
- R = (12 volts) / (240 amps) = 12/240 = 0.05 ohms

So, to be able to pull 240 amps from a 12V battery, the circuit must have a resistance of 0.05 ohms. Since in this ideal example, the starter is the only thing in the circuit, that must be the resistance of the starter. We've drawn that in **Figure 18**, using the circuit symbol "M" for a generic motor.

Now, let's look at the effect of even a small amount of unintended resistance from corrosion on a connector on the performance of this same circuit. Because this is not an actual resistor, we won't draw it as one; we'll draw it the same way we draw a load device, as a box with a labeled resistance value. See **Figure 18**.

Let's say we know that the resistance of this corrosion is 0.01 ohms. In reality, we don't know that, and we usually can't, as we'll get to below, but for illustrative purposes, just say that we do. This is now a series circuit with 0.05 + 0.01 = 0.06 ohms total resistance. But—and here's the part that's not well explained in most places—*because this is a series circuit, the current must be the same everywhere in the circuit*. So let's calculate the current. We know the voltage, and the total resistance, so we use Ohm's law in the form I=V/R:

- Current = (12 volts) / (0.06 ohms) = 12/0.06 = 200 amps

Knowing the current, we can now calculate the voltages across the individual resistances by using Ohm's law in the form V=I*R:

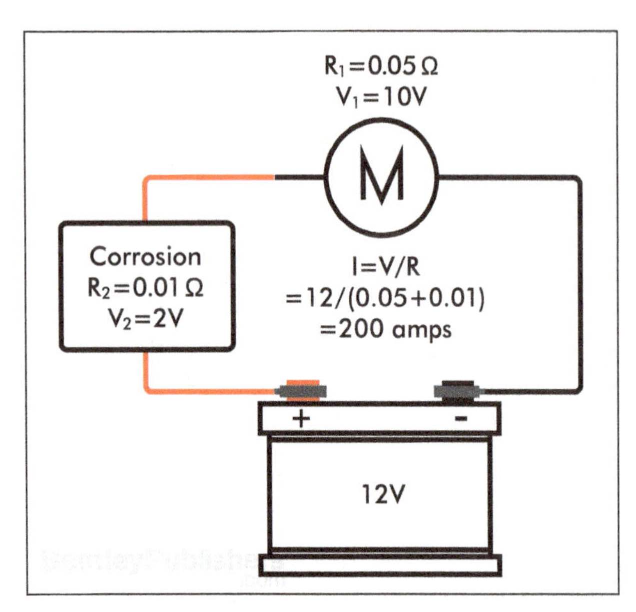

Figure 18 Adding 0.01 ohm resistance from corrosion reduces the current to the motor from 240 to 200 amps.

- V_1 = (200 amps) * (0.05 ohms) = 200*0.05 = 10 volts
- V_2 = (200 amps) * (0.01 ohms) = 200*0.01 = 2 volts

So, our tiny amount of resistance – just 0.01 ohms – has reduced the voltage available to the starter motor by two volts, and has cut the current available at the starter motor from 240 amps to 200 amps.

This is how unintended resistance in series with a load device, particularly a high-amperage low-resistance one, causes a circuit to malfunction.

Oh, and the power eaten up by this little bit of corrosion? Using Watt's law in the form P = I*V:

- Power = (2 volts) * (200 amps) = 2*200 = 400 watts

Four hundred watts. And where is that power going? The only place it can: Heat. That's enough power to heat a small room.

Voltage Drop Testing

In the previous section, we said that we can't know the actual resistance of a small amount of corrosion. This is because the multimeter isn't capable of accurately measuring resistances this small. But virtually any multimeter can measure the drop in voltage caused by that resistance. The gist of it is that you use the multimeter to verify that the load device is actually "dropping the voltage," and if it's not, to find out what is. We will revisit Voltage Drop Testing in more detail in the **Common Multimeter Tests** chapter, and will give a very specific example of a voltage drop test in the **Starter** chapter (which is where it really matters). We introduce it here because it fits with our discussion of how both intended and unintended load devices use voltage.

Three Quick Facts About Voltage Drop Testing

- A multimeter can't accurately measure very small resistances, so you can't use a meter to see if corrosion at a connection is creating resistance.
- But, since resistance creates a drop in voltage, you *can* use a meter to measure the voltage drop across a connection by placing the meter's probes on either side of the suspected connection.
- More detail is in the **Common Multimeter Tests** and **Starter** chapters.

In **Figure 19**, we show how to use a multimeter to measure the same voltage drop we calculated in the previous example. Putting the positive and negative multimeter leads on both sides of the starter will show the drop across the starter (10V as pictured). However —and this is the "how to actually do it" point of voltage drop testing—you can put the multimeter leads between the positive battery terminal and the load device and directly measure the voltage drop due to corrosion at the connections.

If the battery clamps and cables are perfect conductors, there'll be extremely low resistance, and the voltage dropped by that resistance will be near zero. But if there is substantial resistance, it will result in a measurable drop in voltage. You wouldn't know there's 0.01 ohms of resistance because the multimeter isn't accurate enough to measure that, but the meter certainly can measure a 2V drop in voltage. The same test can be performed on the path between the device and the negative battery terminal to look for a voltage drop from corrosion on the ground side of the circuit.

"Dropping the Voltage" Versus "Voltage Drop Testing"

Let's take this moment to clarify a confusing bit of terminology. We've said that a well-engineered load device is designed with a certain resistance to use up the energy in the circuit, and that, in the process, the load device has the effect of "dropping the voltage." In contrast, when discussing *unintended* or *undesirable* resistances in a corroded connector, we talk about conducting "voltage drop testing" to catch these voltage losses in the act. Fundamentally, they're the same thing, but the connotation is that one is desirable and the other isn't. That is, the resistance in the load device is *supposed* to be there – it's the designed-in part of the circuit – and is supposed to "drop the voltage." However, the resistance in a corroded connector is *not* supposed to be there. If there are voltage drops (noun) across switches or connectors, there's not as much voltage left for the real load device to drop (verb).

It's worth noting that the circuit analysis you see in textbooks (the use of Ohm's law to calculate values using

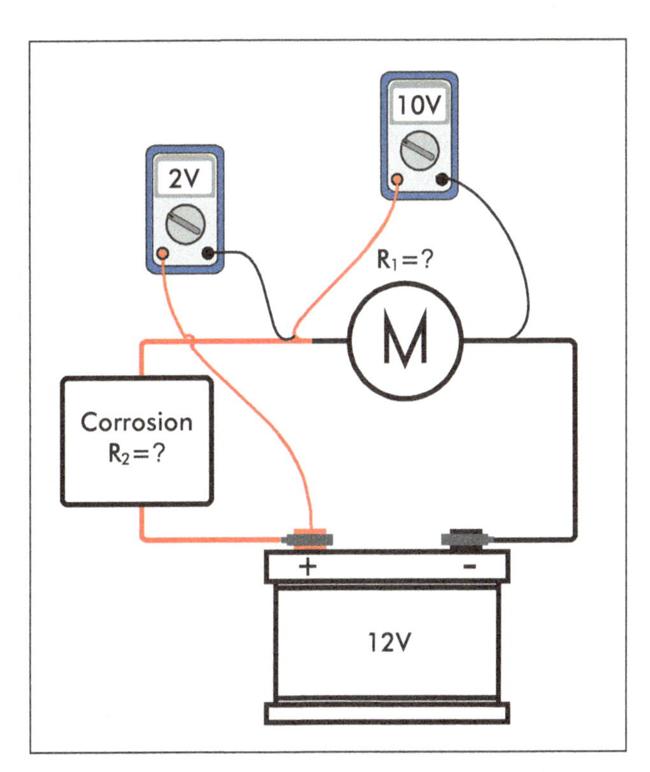

Figure 19 Using a multimeter to measure the voltage drop on either side of the load device, and between the battery and the load device.

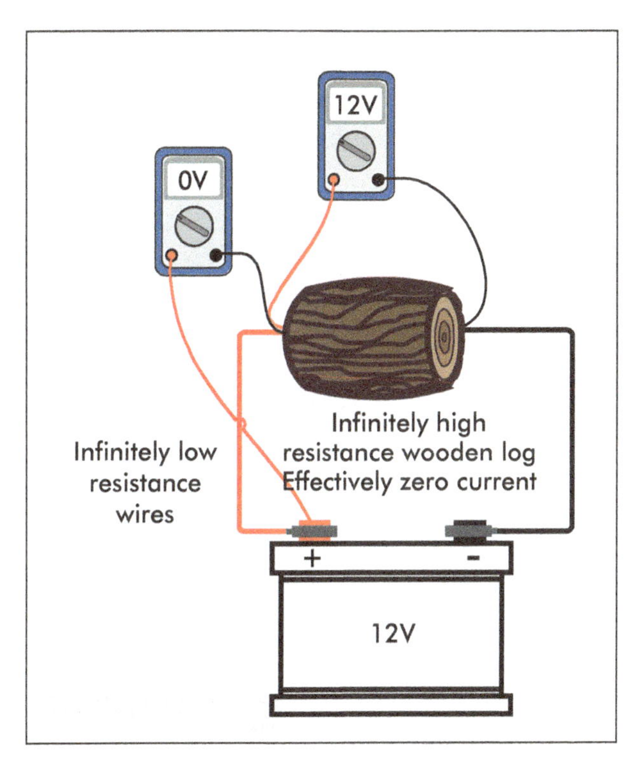

Figure 20 The thought experiment of how you get a valid voltage drop even over an object with ridiculously high resistance.

a drawing of a circuit) is often replaced in the real automotive troubleshooting world by voltage drop testing as you physically go around the *real* circuit with a multimeter.

How Does a Circuit "Know" to Drop the Voltage?

Earlier we discussed how the energy in the moving charge gets used up in making the trip around a circuit. But how does the load device "know" that it's the load device, and that it needs to drop the lion's share of the voltage, instead of leaving that job to the wires? Put another way, what percentage of the voltage are the load device and the wires each supposed to drop?

There's a thought experiment you can perform. Think of an open circuit as a load device with nearly infinite resistance, something like a wooden log. On your mental multimeter, slide the leads down from the battery terminals so they're right at the log and you're measuring the voltage drop across this item of ridiculously high resistance. It should still read 12V – that is, the log is dropping the voltage even though, because of the log's high resistance, essentially no current is flowing. Next, mentally perform a voltage drop measurement between the positive battery terminal and the "input" to the "device." This should show no loss, a drop of zero volts. These are both illustrated in **Figure 20**.

Now, let's jump to the other end of the spectrum of the thought experiment. If, in your mind, you simply connect a low-resistance wire between positive and negative terminals, we all agree that it'll burn up in short order. Earlier in the chapter, we gave an example using 10 feet of 10-gauge wiring having a resistance of 0.0102 ohms across a 12V battery, and showed, using Ohm's law, that the current generated would be 1176 amps, which would certainly melt the wire.

Because of the way that energy is used in a circuit, the wire would have to drop the voltage across its length; by the time you get to the negative battery terminal, the voltage *must* be at zero volts relative to the positive terminal. But now there's no longer a single discrete load to drop the voltage, so presumably, if you could measure the voltage along that wire at various points before it burned, you'd see a continuum of voltage values, as its resistance and the release of energy drops the voltage, presumably to 6V halfway down the wire, 3V three-quarters of the way, etc. This is depicted in **Figure 21**.

So:

- At one end of the spectrum, if the resistance of the load device is very high, the wires don't really matter because very little current is flowing, and the load device drops all the voltage across it.
- And at the other end of the spectrum, if the load device has very low resistance (or if there's no load device), the wires have to act like a load device, and the voltage is dropped continuously across the wires. If the wires are just standard circuit wires, they have

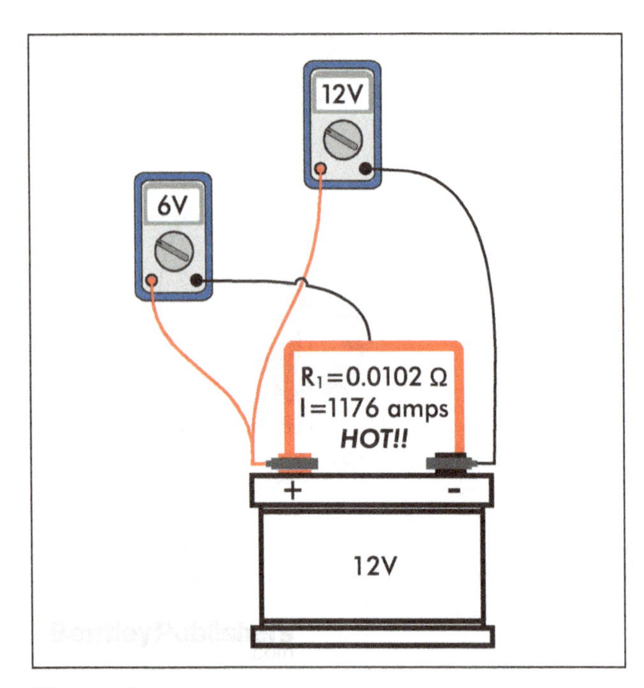

Figure 21 If there's no discrete load device, the wires themselves should drop the voltage continuously over their length. Until they melt.

very low resistance, and they allow so much current to flow that they burn up.

At what point is a load device so poor at doing its job of dropping the voltage that the wiring to and from the device has to shoulder the burden and basically becomes another load device? How is voltage really dropped by each component – including the wires – as current goes around a circuit? *What is really going on?*

Here's how you have to think about it.

- It's *all* about the resistances.
- The entire series circuit has to drop all the voltage.
- Each component drops the voltage by an amount proportional to what fraction its resistance is to the total series resistance (see sidebar).
- If the total resistance is high, only a small current will flow through both the load device and the wire, with the voltage dropped in proportion to their individual resistances.
- But if the total resistance is very low, a very high current will flow through both the load device and the wire.

Let me say it a different way. Consider the odd "look-ahead" quality of electricity, that it almost seems to "know" what is coming in the circuit ahead of it so only a portion of the voltage is dropped across a component if there's additional resistance yet to come. This occurs because, in a series circuit, the current has to be the same everywhere, and that current is dictated by the total resistance.

Why the Ratio of the Resistances Is the Same as the Ratio of the Voltage Drops

For a series circuit with two resistances, according to Kirchhoff's Voltage Law:

$$V_{Total} = V_1 + V_2$$

Expressing this using Ohm's law in terms of the individual resistances and currents for the load device (I_1 and R_1) and for the wires (I_2 and R_2) yields:

$$V_{Total} = I_1 * R_1 + I_1 * R_2$$

But, because, in a series circuit, the current has to be the same everywhere, the individual currents can be replaced by a single current I, yielding:

$$V = I * (R_1 + R_2)$$

If we use Ohm's law in the form I=V/R, this means that the current through the load device $I_1 = V_1/R_1$ must be the same as the current through the resistor $I_2 = V_2/R_2$. This yields:

$$V_1/R_1 = V_2/R_2$$

Or, put another way,

$$V_1/V_2 = R_1/R_2$$

In words, **the ratio of the voltage drops is the same as the ratio of the resistances.**

Think of it this way. When you turn on a circuit, the charges in the "front" of the current encounter the resistance in each component and slow down in response, which slows all the charges flowing behind them, and in an instant, a constant uniform current is created. From then on, what goes in one end comes out the other end. The current is behaving like an incompressible fluid. It propagates through the circuit incredibly quickly – at some large percentage of the speed of light – and then is evenly distributed. So it's not that electricity "looked ahead." It's that the pace of the traffic, if you will, was already established by the lead cars on the basis of how rough the road is.

If you could connect a load device directly to a battery without using wires, it would be guaranteed to drop 100% of the voltage across it. If instead you connect it to the battery with horrible corroded wires that lose 30% of the voltage (and they would have to be unimaginably bad to lose 30% of the voltage), the device will drop the remaining 70% of the voltage. But the voltage lost in the wires is just the symptom – it's the resistance that's the cause. It's the resistance that governs the current. Since the current has to be the same everywhere, that current is determined primarily by whatever component has the biggest resistance, and the voltage drops across the individual components are just the individual V=I*R consequences.

An Example Showing That a Load Doesn't "Know" How Much Voltage to Drop

Let's take a 12V circuit with a resistance of 1 ohm. By Ohm's law, this results in 12 amps of current. Let's assume the wiring is absolutely horrible and drops 30% of the voltage, leaving the device to drop the remaining 70%. You'll see that there's no magic here and nothing "knows" to drop anything by these amounts. Remember that, above, we showed that the ratio of the resistances is the same as the ratio of the voltages, so you can convince yourself that, for 1 ohm of total resistance and a 70/30 split, that means the resistance of the load device must be 0.7 ohms and the resistance of the wires must be 0.3 ohms, and in terms of voltage, the load device drops the voltage 8.4V and the wires drop it the remaining 3.6V. See **Figure 22**.

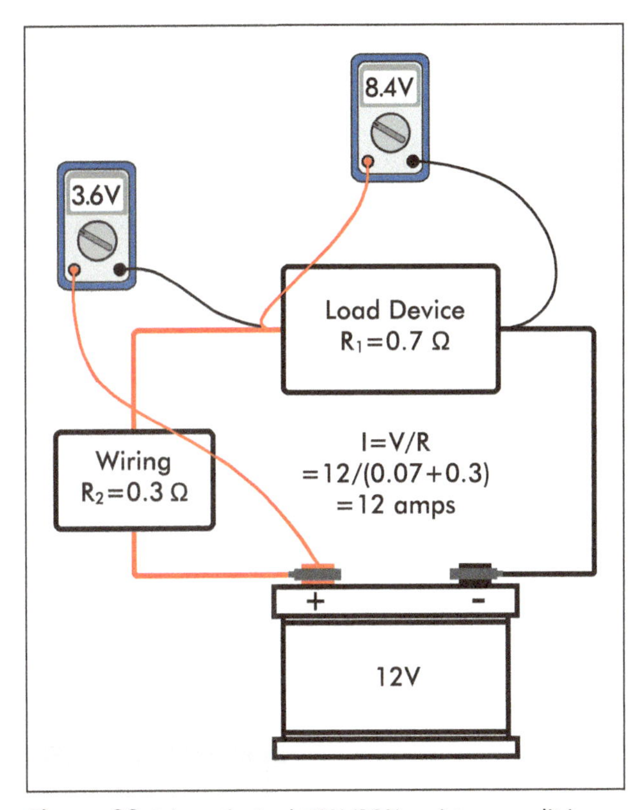

Figure 22 A hypothetical 70%/30% resistance split in a circuit with resistance 1 ohm. It's the total resistance that dictates the current. That current flows across the individual resistances, creating the individual voltage drops. If the wiring had no resistance, then the 0.7 ohm load device would've dropped the entire 12V, and according to Ohm's law, the current in the circuit would've been 17.14 amps instead of 12 amps.

Now, replace the horrible 0.3 ohm resistance wiring with good wiring (that is, assume near-zero resistance) and calculate the current through the load device. As must be familiar by now, using Ohm's law in the form I=V/R:

- Current = (12 volts) / (0.7 ohms) = 17.14 amps

Thus, as per **Figure 23**, by removing the high-resistance wiring, the current has increased from 12 amps to 17.14 amps. Moreover, because the 0.7 ohm load is now the only element in the circuit, it now drops all the voltage.

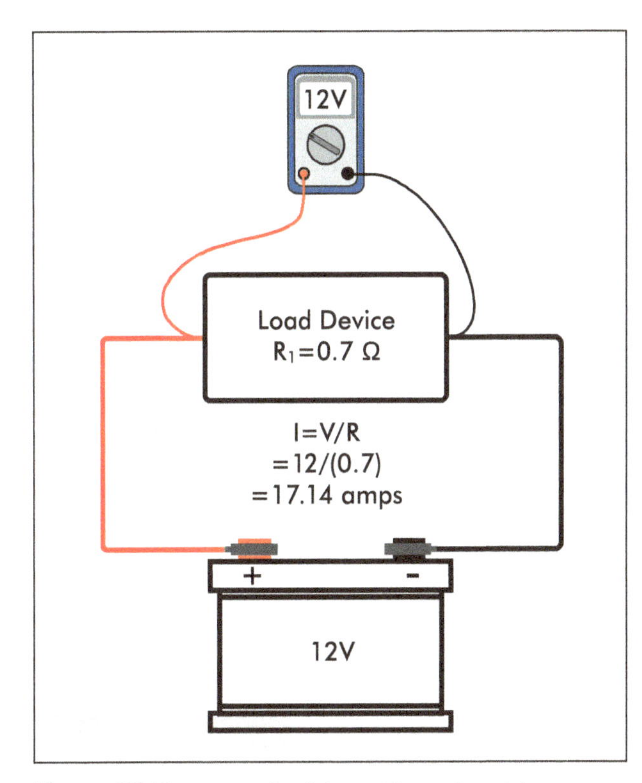

Figure 23 The same circuit but without the wiring resistance. Current through the load device is now 17.14 amps, and the load device drops all the voltage.

So how did the device go from dropping 70% of the voltage to dropping all of it? How did the current go from 12 amps to 17.14 amps? What magic is afoot here?

None. With the wiring's 0.3 ohm resistance as part of the circuit, it's the total resistance of 1 ohm that determines the current flow of 12 amps. Had the 0.3 ohm resistance in the wires not been there, by Ohm's law the current through the (now alone) 0.7 ohm device would not have been 12 amps. It would've been 17.14 amps, and it would've dropped ALL the voltage.

So how did the electricity "know" to drop the voltage 8.4 volts over the load device? It didn't. That's just what you get when you push 12 amps through 0.7 ohms of resistance. *All* of these individual voltage drops are nothing more than consequences of the current – which is set by the total resistance – not being high enough for the device to drop the entire voltage.

Now, in our hypothetical 70%/30% split, whether the wires are actually capable of carrying a given current and dissipating the heat created by the power is an entirely separate matter. Most of the time, when troubleshooting, you have the opposite – unexpectedly high resistance makes the current lower than it should be. In a proper circuit, they should be in balance.

A Calculation Using a Realistic Voltage Drop in a Starter's Wiring

Let's use a more reasonable resistance ratio and see what the consequences are on the wiring. Let's say the load device is the 0.05 ohm starter motor we played with above, and let's say we want to size the wiring so we have a 97%/3% voltage drop split, with 97% of the voltage dropped across the motor, and the remaining 3% dropped by the wiring, which is about the maximum voltage loss in the wiring you'd want to tolerate in a real high-current circuit. Using the $V_1/V_2 = R_1/R_2$ relationship we derived above, and knowing that V_1/V_2 must be in the ratio of 97/3:

$$R_1/R_2 = 97/3$$

Since R_1 is known to be 0.05 ohms:

$$0.05/R_2 = 97/3$$

Solving for R_2 :

$$R_2 = 0.001546 \text{ ohms}$$

So to achieve the 97%/3% voltage drop split, for a starter motor with a resistance of 0.05 ohms, the wiring needs to have a resistance of 0.001546 ohms. Later in the book, in the **Wire Gauge** chapter, we'll learn that wire of that resistance falls between 0 gauge and 1 gauge wire. In the real world, we'd choose the thicker 0 gauge wire.

With the resistance known, we now use Ohm's law in the form I = V/R. Using the sum of the two resistances, we calculate the current:

$$I = (12 \text{ volts}) / (0.05 \text{ ohm} + 0.001546 \text{ ohm}) = 12/0.06546 = 232.8 \text{ amps}$$

Thus, the current through the circuit would be about 233 amps. Calculating the individual voltages:

$$V_1 = (233 \text{ amps}) * (0.05 \text{ ohm}) = 11.65 \text{ volts}$$

$$V_2 = (233 \text{ amps}) * (0.001546 \text{ ohm}) = 0.35 \text{ volts}$$

From here we can calculate the power drawn by the starter and the wiring:

$$P_1 = I*V_1 = (233 \text{ amps})*(11.65 \text{ volts}) = 2714 \text{ watts}$$

$$P_2 = I*V_2 = (233 \text{ amps})*(0.35 \text{ volts}) = 81.55 \text{ watts}$$

Thus, 2714 watts are drawn by the starter, but, in addition to that, about 82 watts of heat has to be dissipated by the wiring. This is instructive. It shows that, during the normal operation of a starter, *even if there's no added resistive load from corrosion at a connection,* the wires between the starter and the battery have to dissipate about as much heat as what used to be a good-sized home stereo.

We could swing for the fence and combine this with the voltage drop example above and have three resistances (the starter, the wiring, and the corrosion), but by now you probably see that if a book shows you circuits with no load devices in them other than a resistor and says that wiring can be assumed to be a perfect conductor, it's doubtful the author has ever had to troubleshoot a slow starter motor.

Summing Up Resistance and Current

- The total series resistance of the load device plus the wiring plus any corrosion determines how much current flows.
- At one extreme, corrosion at a connection may increase resistance, limiting current to the point where a high-current device such as a starter motor won't work.
- At the other extreme, either through a short circuit or an improper device in the circuit, the total series resistance may become so low that the total amount of current becomes a problem, exceeding the capability of the wire to carry it and dissipate its heat safely.
- But it's the ratio of the individual resistances that determines how much voltage is dropped by the load device and how much is dropped by the wiring.
- For a high-current (low resistance) device, the wiring must be sized so that it is not the source of an unacceptable voltage drop, and also so that it can carry the current load and dissipate the heat.

Voltage Sag

We have one final topic in our electrical primer. Distinct from voltage drop is another phenomenon sometimes called *voltage sag*. Voltage sag can be demonstrated by measuring the resting voltage of 12.6V across a fully charged battery, and then, with the engine off, turning on a high-demand accessory like the headlights, measuring the voltage again, and seeing that the voltage is now substantially less, like 11.8V.

The voltage "sags" because the battery is not an infinite source of power. If the battery is large (has a high reserve capacity rating) and the electrical load is small (e.g., a 2W light bulb), then the voltage sag will be minimal. But the more current the load draws as a fraction of the rating of the battery, the bigger the sag will be. If you like thinking of voltage as electrical pressure, voltage sag can be thought of as the pressure drop that occurs when the circuit is "uncorked" and current begins to flow.

In the **Battery** chapter, we explained how, when the engine is running, the alternator puts out "charging voltage," but we didn't really explain why. The purpose is to compensate for this kind of voltage sag. In the **Charging System** chapter, you'll learn that the alternator and voltage regulator act together to keep the voltage high enough so that the sag doesn't preclude the electrical devices from doing their jobs.

Can Electricity in My Car Hurt Me?

The following **Probable Effects of Electric Current on the Human Body Table** was excerpted from the OSHA web site (and also appears at the end of the **Safety** chapter). The problem is that, if you look at these amperage numbers and compare them to the numbers we discussed above (such as 20 amps for a cooling fan), you'd conclude that an electric shock from the fan could kill you. And that's the wrong conclusion.

Here's what's going on. With an understanding of Ohm's law, we now know just enough to be, uh, dangerous regarding estimating the dangers of electricity.

Voltage, current, and resistance all hang together. We've seen that, for a fixed voltage, the amount of current that flows depends on the resistance. For you to experience electric shock, electricity needs to flow through your body. And there are fascinating articles on how perspiration and callouses on hands affect resistance and therefore affect the current that can flow through the body. ***But the numbers in the Current table below are not for an automotive 12 direct current (DC) system. They're for a 120 volt 60 Hz alternating current (AC) system.***

Probable Effects of Electric Current on the Human Body	
Current (milliamps)	**Probable Effect**
1 mA	Perception level. Slight tingling sensation. Still dangerous under certain conditions, especially wet conditions.
5 mA	Slight shock felt; not painful but disturbing. Average individual can let go. However, strong involuntary reactions to shocks in this range may lead to injuries.
6mA – 30mA	Painful shock, begin to lose muscular control. Commonly referred to as the "freezing current" or "let-go" range.
50mA – 150mA	Extreme pain, respiratory arrest, severe muscular contractions. Individual cannot let go. Can be fatal.
1A – 4.3A	Ventricular fibrillation (uneven, uncoordinated pumping of the heart). Muscular contraction and nerve damage begins to occur. Fatal injuries likely.
> above 10 amps	Cardiac arrest, internal organ damage, severe burns. Death probable.

Thermal Burns While Making a Connection: The Most Common Way to Be Hurt by Electricity in a Car

In addition to a short circuit or the "there's no valid load device so crazy amounts of current get drawn why on Earth were you doing this anyway" scenario typified by my electromagnet, there's a more mundane way to be burned by electricity in a car: Simply making a connection in a normal powered circuit.

A connection may be a ring terminal held in place by a bolt or screw, or a female quick-connect terminal (sometimes called a spade terminal) slid around its male counterpart, or some other mechanical coupling of two conductors. When the connection is secure and voltage is applied, the flow of current occurs through the entire surface area of both halves of the mated connector, an amount of metal that should be engineered to be able to dissipate the heat generated.

But if there is voltage already present on one of the connectors when the connection is first made, and if there is a path to ground, then when you touch the tips of both connectors together, *the current flows through just the tips*. The tips have much less surface area than the whole mated connector to dissipate the heat. Thus, if a circuit is already powered, and you do something like touch the edge of a ring terminal to a post, all of the current of the circuit flows through that small amount of metal on the edge. Because there's so little metal in edge-to-edge contact, it can't dissipate the heat as well as the whole connector, so it can get very hot very quickly. It may generate a spark. It may even micro-weld itself to the mating surface.

The worst case for this is if you touch a live bare multi-stranded wire to a battery terminal or to ground. Again, this is *not* a short circuit we're talking about; it's completing a *normal* circuit. All of the current will flow, or try to flow, through the small number of strands at the end of the wire. With so little metal to dissipate the heat, contact may even melt the thin strands that are trying to carry the entire current.

In other words, it can burn you.

For this reason, it is always much, *much* safer to cut power (to turn off the voltage in a circuit) before making any connection.

In general, low voltage DC does not present a shock hazard. This is not just our opinion. OSHA itself doesn't even call for equipment with less than 50 volts to be enclosed. (OSHA regulation 1910.303(g)(2)(i): "Except as elsewhere required or permitted by this standard, live parts of electric equipment operating at 50 volts or more shall be guarded against accidental contact by use of approved cabinets or other forms of approved enclosures.")

However, that doesn't mean it's "safe," because there are dangers other than shock hazard. There is burn hazard. A great way to think about it is presented in an OSHA training manual that describes three distinct burn hazards associated with electricity. We've summarized this in the **Burn Hazards Associated with Electricity Table** and added the third column. You'll see the first two are not likely to occur working on a 12V auto electrical system, but the last one (thermal burns) certainly can. It's *exactly* what happened when that foolish electromagnet without a valid resistance load burned my hand. It can happen anytime you touch a wire carrying a lot of current, particularly a bare wire that isn't designed to carry as much current as it is, with your bare hands.

There. The broccoli wasn't that bad, was it?

Burn Hazards Associated with Electricity

Type	Description	Danger at 12V DC?
Electrical Burns	Happen when electric current flows through tissues and organs	No
Arc Burns	Result from high temperatures when an arc flash event occurs	No
Thermal Burns	Happen when skin touches a hot surface	YES!!

By the way, we lied about the pie. We ate it all. This teaching stuff is hard work.

4

Multimeters and Related Tools

The Multimeter

The digital multimeter is to electrical work what the ratchet wrench and socket set are to mechanical work—the essential tool. You should be eating with it, sleeping with it, using it to prop open the pages as you read this book. It should be your friend, your lover, your constant companion.

In the **Battery** chapter, we jumped right in and threw you into the pit with a live multimeter. In the **Circuits** chapter, we went into the theoretical aspects of voltage, amperage, and resistance. In this chapter, we're going to talk about the multimeter, highlighting the various options available if you wish to purchase a multimeter. We'll also introduce you to some related electrical diagnostic tools. Then, in the next chapter, we'll walk you through common multimeter tests in a step-by-step fashion.

The term *voltmeter* was originally used to refer to an analog meter (meaning it had a swinging needle) that measured voltage and only voltage. These gave way to analog *multimeters* that were capable of measuring not only voltage but current and resistance as well. As calculators and watches turned digital, so did multimeters. DVM (digital voltmeter) and DMM (digital multimeter) refer to the same thing – a digital device that measures multiple electrical signals. Occasionally you see it written DVOM, which stands for Digital Volt Ohm Meter (though "Digital Volt-O-Meter" would be funnier). But the terms all mean the same thing – a handheld device capable of measuring voltage, resistance, and amperage. In this book, we'll steer clear of the acronyms and either use the term "multimeter" or just refer to "the meter." We may also occasionally slip and use the term "voltmeter" because the majority of the time, you are, in fact, using it to measure voltage.

What's the Least I Need to Know?

You need a digital autoranging multimeter with an audible beep for continuity. They're available for as little as $25. Buy one that fits your budget. And buy a test lead kit with alligator clips so you don't have to hold the leads in place. This gets you what you need for most of the work you'll need to do. If you also need to troubleshoot devices such as wheel speed sensors or camshaft positioning sensors that output a sine wave or square wave, you'll need an automotive multimeter that can detect periodic signals and measure frequency and duty cycle, or an oscilloscope. We cover these in the **Tools for Dynamic Analog Signals** chapter later in the book.

Anatomy of a Multimeter

While there are variations between brands and models, most multimeters share a common form factor and a similar look and feel. This general layout is shown in **Figure 2**. You interact with nearly every multimeter

Figure 1 A gaggle of multimeters (left to right: The Fluke 233/A with detachable display and automotive-specific test modes, introduced in 2009; a workhorse Fluke 85 III, a venerable Fluke 77, a Craftsman Pro, and the $5.99 special. The leftmost is pricey. The rightmost is so inexpensive it's disposable. The three in the middle are well-used. All of them are suited for working on your car's electrical system.

through four immediately visible areas. Working from the top down, there is:

- A numeric display at the top
- Pushbuttons to enable certain options
- A rotary dial to select the desired measurement
- Input terminals at the bottom into which to plug the red and black test leads

Note that, when we say that most multimeters have a similar look and feel, we are talking about autoranging meters. While we will touch on non-autoranging meters briefly, discussions in this chapter concentrate on autoranging multimeters.

The illustrations in this chapter and the next are based on a Fluke 88V multimeter, which for many years has been the industry-standard high-quality automotive meter. Your meter will look similar, but the options on the buttons and the settings on the rotary dial will be slightly different. If, after reading this chapter, you still have specific questions about what your meter will and will not do, you may need to consult the owner's manual for your meter.

That having been said, once you've used a multimeter to perform a voltage or a resistance measurement, you can usually pick up any other multimeter and easily do the same. Current (amperage) measurements are a little trickier, as the meter's configuration varies model to model. We'll discuss that in the **Measuring Current** section in the following chapter. Frequency measurements are the trickiest and most meter-specific of all. Those are covered in the chapter **Tools for Dynamic Analog Signals**.

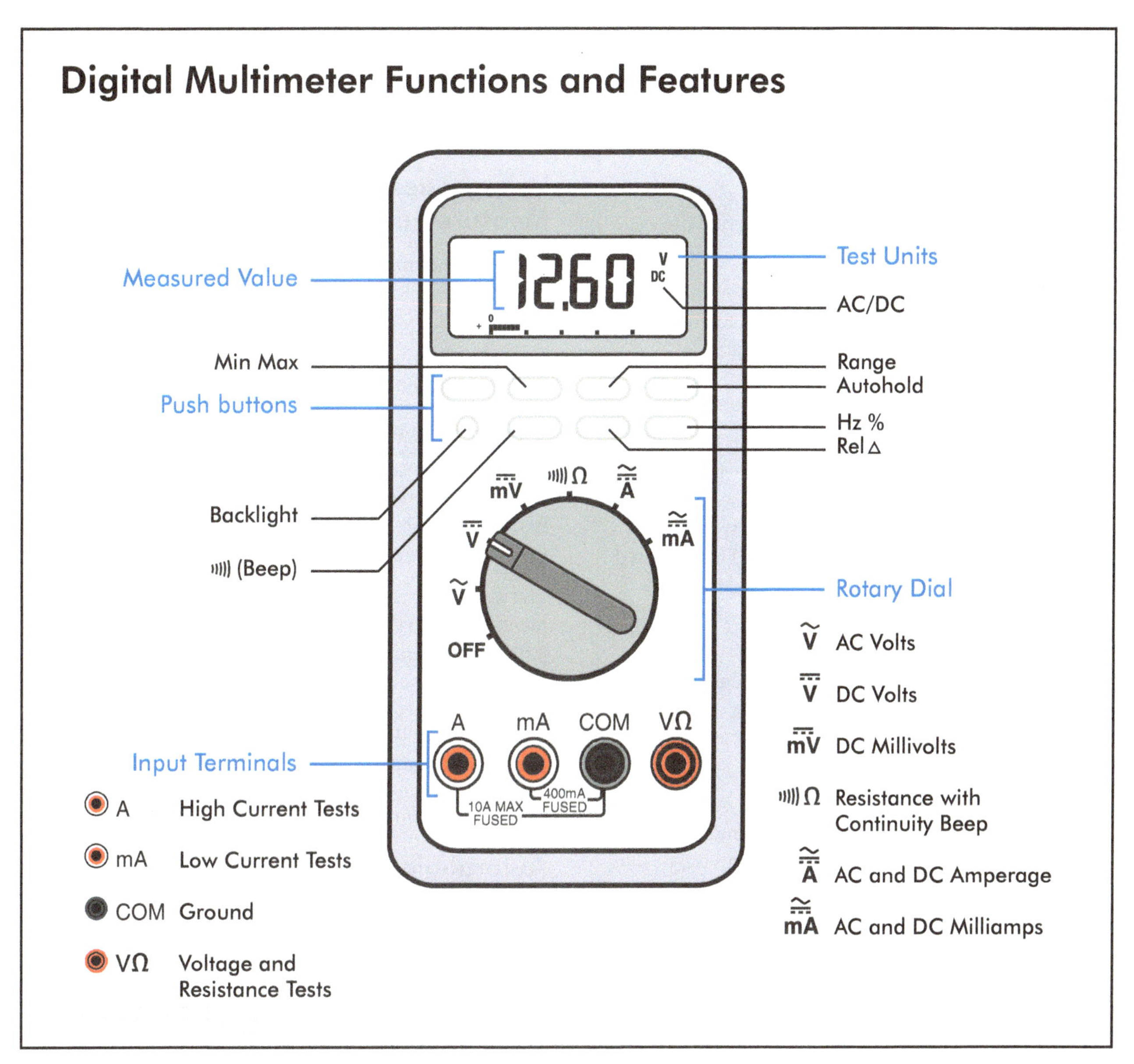

Figure 2 Functions and features of a typical digital multimeter.

Measured Value – The value being measured (here, 12.60) is shown in the display window at the top of the multimeter.

Test and Units – The multimeter's display will show the test (the kind of measurement) currently being performed, and the units for that test. In **Figure 2**, the test is the measurement of DC voltage, and the units are V for volts.

Pushbuttons – The location and functions of the pushbuttons vary substantially, but common pushbuttons include:

- A backlight button to turn on the light behind the display, making the measured value more readable in low light
- A min-max button for storage and review of minimum and maximum values of a measurement
- A "beep" button to emit an audible tone during a resistance measurement if the meter detects continuity (near zero resistance)
- A range button to allow manual selection of a range, even on an autoranging meter
- An autohold button to capture a reading when a stable reading becomes available
- A relative mode button to zero the display and store the current reading as a new zero point for subsequent measurements
- A Hz % button to measure the frequency of a voltage signal and report its duty cycle

Rotary Dial – The rotary dial is the primary location on the meter where the kind of measurement is selected. The exact functions and positions vary meter to meter, but as pictured, they are:

- AC Volts – measurement of the voltage of an alternating current (AC) signal.
- DC Volts – measurement of the voltage of a direct current (DC) signal.
- DC Millivolts – high-sensitivity measurement of the voltage of a direct current (DC) signal. Even on an autoranging meter, if a signal is less than approximately one volt, it is advisable to use the millivolt setting to obtain the most accurate reading.
- Resistance (continuity), sometimes with beep – measurement of resistance. Note that testing for continuity is testing the presence or absence of resistance. On some meters, the meter will always beep if continuity (near zero resistance) is measured. On others, the continuity test is a separate rotary dial setting, and/or requires the "beep" pushbutton to be pressed.
- AC and DC Amperage – high current measurement of both AC and DC amperage. This setting may use an approximately 10 amp fuse inside the meter, or may be unfused.
- AC and DC Milliamperage – low current measurement of both AC and DC amperage. This setting may use an approximately 400 mA fuse inside the meter, or may be unfused.

Input Sockets – There are typically four input sockets along the bottom edge of a multimeter. These are where the plugs ends of the test leads are inserted. One socket is black, and typically three sockets are red.

The plug end of the black lead should only ever be inserted into the black "COM" socket. It's labeled "COM" for "common," indicating that all settings use it.

The plug end of the red lead can typically go into one of three places:

- Red "VΩ" Socket – This is where to insert the plug end of the red lead for a voltage or resistance measurement, or, in fact, for any measurement except current measurements.
- Red "A" Socket – This is where to insert the plug of the red lead for a high current (e.g, 10A) measurement.
- Red "mA" Socket – This is where to insert the plug of the red lead for a low current (e.g., 400mA) measurement.

Some meters may have only two red input sockets, combing the high and low current measurement sockets (A and mA) into one.

Multimeter Features

Now that you've had a quick tour of the multimeter's basic anatomy, let's walk through the meter's functions in a more systematic fashion.

Voltage, Resistance, and Amperage Measurement

The overwhelming majority of the time you whip out a multimeter to do something on a car, you use it to do one of two things:

- Take a DC voltage measurement to verify the presence or absence of voltage.
- Take a resistance measurement to verify the presence or absence of continuity.

Not only will most every multimeter will do these two things, they are the two easiest things to do on a multimeter.

The next most common multimeter application is measuring current (amperage). For example, if your battery is running down overnight, you need to measure amperage to isolate what is draining the battery. So the multimeter needs to be capable of a DC amperage measurement. Be aware that there are some basic pocket-sized entry-level meters that do not include the ability to measure amperage, so if you're buying an inexpensive multimeter, be certain to check.

Autoranging

When meters were analog, they had multiple scales and ranges. You'd need to manually select the scale whose range allowed the needle to swing about 3/4 of the way across the gauge, as that provided the greatest accuracy for the measurement.

These days, even though nearly all new multimeters are digital, the least expensive meters still have multiple ranges, requiring you to manually select the range most closely matching what you're trying to measure by twisting the big rotary dial in the center of the multimeter. For a voltage measurement on a car, because you know you're nearly always measuring something around 12 volts, this isn't difficult; you simply select the scale closest to but not greater than 12 V, which is nearly always 20 V. For resistance, if you're simply performing a continuity test, the range doesn't matter, but if you're trying to perform an actual resistance measurement, it's a bit of a pain, as you might not know what the valid resistance reading is supposed to be. Plus, since each of the AC voltage, DC voltage, amperage, and resistance settings must then have the ability to select one of several ranges, that many settings on the rotary dial makes for a cluttered interface. This is shown in **Figure 3**.

For these reasons, all but the cheapest multimeters have *autoranging* capability, meaning they sense the signal being measured, and internally adjust their sensitivity to measure and display the signal as accurately as possible.

Autoranging is definitely something that you want. If the red $5.99 multimeter shown in **Figure 3** represents the low price point for a non-autoranging meter, we're only talking about perhaps another 15 or 20 bucks for an autoranging meter.

Note that amperage hasn't completely been tamed by autoranging. All but the nicest newest multimeters still have two separate ranges for amperage, with one typically in the 10-amp range, and the other typically in the 400mA range.

Figure 3 Because an autoranging meter (left) doesn't have a set of selectable ranges, the central rotary dial (arrow) has a much less cluttered interface than on a non-autoranging meter (right), and is thus much easier to use.

Audible Beep for Continuity

All but the very cheapest multimeters have the ability to, when you set them to measure resistance, generate an audible beep when the resistance is near zero. This lets you quickly verify the continuity of a wire (which is what you're most often interested in when you use the resistance setting anyway) without needing to look at the meter. It's something that you want.

Accuracy and Precision

Well, you're probably saying, what idiot would want to buy an inaccurate piece of equipment? It's more complicated than that. True accuracy and precision of any measuring equipment is a complex issue, depending on the analog electronics inside the equipment, the digital display, the quality of the test leads, how the measurement is actually taken, and other factors.

You'll see the precision of multimeters specified in terms of "display counts" or "counts." This is, simply put, the maximum value the meter can display, ignoring the decimal point. Regardless of whether the meter is autoranging or requires manual range setting, there's only a certain amount of resolution. Even the red $5.99 meter we showed is a 2,000-count meter, so in the "1,000" voltage range setting, it can display a range of from –1,000 to 1,000 volts, but the display will be rounded off at 1 volt. In the 200-volt range, it'll gain a decimal place, getting you from –200.0 to 200.0 volts. Going down to the 20-volt range, which is where you would be living for automotive work, it would give you a display of two decimal places (e.g., 20.00 volts). From a practical standpoint, this is sufficient precision; we rarely care about voltage precision past 1/10 of a volt.

But the *accuracy* of a meter is different question. Accuracy of voltage and amperage readings on multimeters are usually quoted as a percentage of the value being measured (you'll see the phrase "accuracy is 1% of value"). Sometimes the quoted accuracy changes depending on the range. For example, the 10A amperage setting is less accurate than the 300A setting. Resistance is sometimes quoted with the sensitivity in addition to the accuracy (that is, "sensitive down to 0.1 ohm").

When you buy a name-brand meter such as a Fluke, you're paying for a number of things. One of them *is* increased accuracy. You're also paying for the likelihood that the quoted accuracy is real, as opposed to a claim on an inexpensive product that is unlikely to be challenged in either a legal court or the court of public opinion.

The accuracy requirements of most automotive work, where you're simply checking for the presence and absence of battery voltage, are pretty forgiving. Basically, if a meter seems to give reasonable results (i.e., if you set it to read voltage, put the leads across the battery, and it reads about 12.6 volts), don't fret too much over questions of absolute accuracy. An exception is voltage drop testing, where tenths of a volt are, in fact, significant. And if, in addition to troubleshooting automotive electrical problems, you *also* fix radios, home electronics, automotive control modules – in other words, you do real electrical engineering, circuit analysis, and hardware repair – then absolute accuracy becomes a real issue.

Min/Max

There are times when you need to hook up a multimeter and do something while watching it, but you can't be two places at the same time. For example, in the **Battery** chapter, we discussed load-testing the battery by measuring the voltage across it while cranking the engine and verifying that the voltage doesn't fall below 9.6. This is difficult to do without a helper.

However, if your meter has Min/Max capability, it's trivial. The meter can capture a signal, then look through the captured data and tell you the minimum, maximum, and average values. On a Fluke 85, you simply connect the meter, set it to DC voltage, press the Min/Max button, do the test, then hit Min/Max once to show you the min, again to show you the max, and again to show you the average. Then you cancel recording by holding in the max/min button. See **Figure 4**.

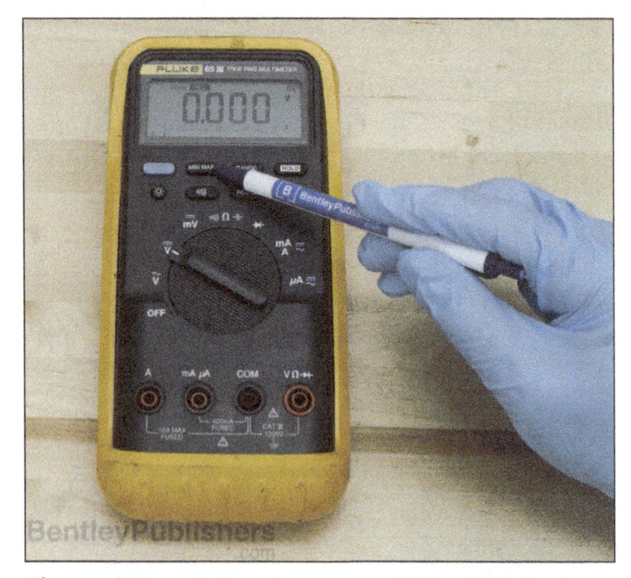

Figure 4 The min/max button on a Fluke 85 III. Very useful for load testing the battery using the starter.

Bar Graph

Some multimeters advertise a "real-time bar graph." What this means is that, at the bottom of the numerical display, it also shows you what's essentially a digital version of an analog needle in case that makes it easier to visualize maximum and minimum values. The bar graph can be useful for testing oxygen sensors. But do not confuse it with the real-time waveform graph you get on an oscilloscope.

Temperature Probe

It used to be that only expensive meters had a temperature probe (a digital thermometer) that could be plugged in to yield a direct temperature measurement. However, this feature has diffused downward so it is now available on many moderately priced meters. It's occasionally handy for measuring radiator or coolant neck temperature to know if a sensor is supposed to kick the electric fan on. These days, though, infrared thermometers are cheap; don't spend extra money on a meter just for this.

Why Shouldn't I Buy a $5.99 Multimeter?

As we pointed out, the el cheapo multimeter doesn't have autoranging, so it's better to pony up more like $20 and buy an autoranging meter.

But we cannot stress enough that *any* multimeter will do what you need it for most often, which is to test for the presence or absence of voltage, and the presence or absence of continuity. It's up to you how many additional features you want to pay for.

So, in fact, you *should* buy a $5.99 multimeter and leave it in the glove box of your car so you can use it to check the resting voltage of your battery when the car won't start, and the charging voltage once it's been jump-started. If your car is broken into and the meter is stolen, you can have a good laugh, then go out and buy another $5.99 meter.

Just also have an autoranging meter that you *don't* keep in the glove box.

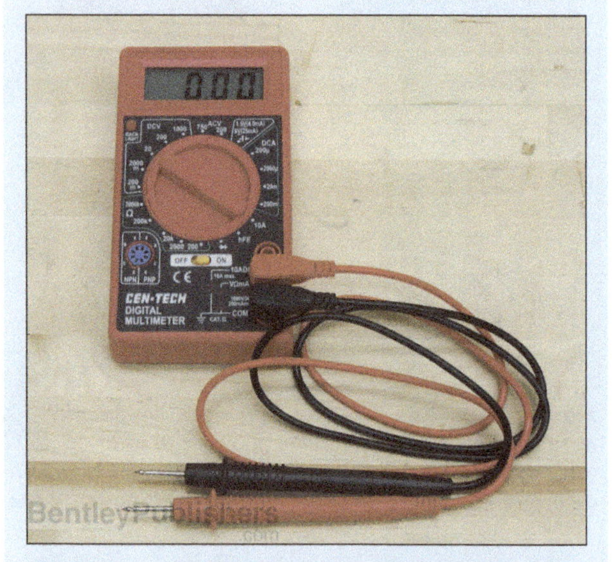

Figure 5 The $5.99 multimeter. If you drive an older enthusiast car with soul, don't leave home without it. Seriously.

True RMS

This is an add-on to the "accuracy" issue described earlier in this chapter. In order to reject noise and present the user with a stable value, multimeters take a number of readings and average them together. That's usually fine for DC signals, but because an AC signal varies with time, simple averages can produce erroneous and/or misleading readings if the AC signal is noisy. Because of this, better meters use a Root Mean Squared (RMS) technique to synthesize a more meaningful reading from noisy data. But it's not crucial to have RMS measurement capability in a 12V DC automotive environment.

Current Clamps

As we will describe in detail in the next chapter, in order to measure current (amperage) with a multimeter, you need to connect the meter in series with the circuit so that all of the circuit's current runs through the meter. There are, however, multimeters that get around this by making use of an inductive clamp, looking somewhat like a giant pinching thumb and index finger, to measure amperage indirectly without having to unplug anything.

These so-called *current clamps* (**Figure 6**) come in three flavors:

1) Clamps that are integral to the meter (called, not surprisingly, *clamp meters*) that *only* work inductively and do not have sockets for conventional test leads to be plugged in.

2) Clamp meters that are more like general multimeters and *do* allow the use of test leads, but with an integrated current clamp.

3) Clamps that are a plug-in accessory to an existing meter. These plug-in clamps typically measure the current and convert it to a voltage reading (for example, 1 millivolt per amp), so you plug them in as if they are test leads measuring voltage. If the meter then reads, for example, 14.7 millivolts, it means the clamp is measuring 14.7 amps.

Note, however, that the most common application of clamp meters is to check the current draw of household wiring and electronics. These are AC, not DC, devices. Thus, many if not most of the clamp meters *measure AC amperage only, and do NOT measure DC amperage at all.* Thus, if you buy a current clamp for automotive work, you need to check the specifications carefully.

Further, the main application for directly measuring amperage on a car is detecting parasitic drains in the tens of milliamps, and most clamping meters do not appear to have the sensitivity needed to do it, though Fluke's new model 365 comes close, listing its sensitivity as 0.1 amp. If you have a clamp meter that works with DC, you can use it to easily read higher-amperage values such as alternator output and the current draw from starters, fuel pumps, and other electric motors, but don't go spending grocery money on one.

Remote Display

The Fluke 233/A shown in **Figure 7** is available with a remote display that communicates to the body via wireless. This lets you, for example, troubleshoot a brake light by connecting the body to the brake bulb assembly, detach the head, take it with you inside the car, step on the brakes, and be able to see the voltage transmitted wirelessly by the body. Very handy.

Figure 6 A clamp meter. This one, the Fluke 365, has a detachable clamp. If you buy a meter with a clamp for automotive use, be absolutely certain it is capable of measuring DC current. Many don't. *Reproduced with permission, Fluke Corp.*

Figure 7 The Fluke 233/A with remote display.

Frequency Measurement

Everything we've discussed about multimeters so far pertains mostly to static (non-changing) measurements. But there's a whole class of sensors (ABS sensors, crankshaft positioning sensors, etc.) that generate data that changes over time in a periodic fashion, generally forming either a sine wave or a square wave. For these sensors, you need a tool capable of measuring frequency. We'll describe this type of data in detail in the **Dynamic Analog Signals** chapter, and describe the tools needed to measure it in the **Tools for Dynamic Analog Signals** chapter.

How Much Should I Spend on a Multimeter?

A car, for the most part, is a 12V direct current (DC) environment. Many of the added capabilities on expensive meters are to deal with things in the much more challenging AC environment. You don't need most of that added capability for the meat and potatoes of automotive electrical troubleshooting.

This book shows me poking around with Fluke multimeters. Fluke makes a great product. They're the BMW (or Porsche or whatever your European favorite car is) of multimeters. In addition to the meters themselves, Fluke's test leads and other accessories are of very high quality.

But I would be doing a disservice if I led you to believe that you must spend hundreds of dollars in order to begin doing any electrical work on your car. You don't. Really, it's no different from the situation with any other tool. Professional mechanics who use tools day in and day out have different requirements than do-it-yourselfers who use them occasionally.

Expensive, high-quality tools are wonderful. They feel great in the hand, they're unlikely to break when you use them, and documentation and customer service is typically superior to their inexpensive counterparts. But, for occasional use, pro-quality tools are not an absolute requirement. Don't hold off on buying a multimeter because you think it's Fluke or nothing and you don't have the money for the Fluke. You'll feel awfully silly when your car won't start, you jump it, and you have no way to tell whether or not the alternator is charging the battery.

And don't discount buying a used meter, if that's what your budget dictates. While the Flukes hold their value really well, you can find other good autoranging meters at yard sales for next to nothing. The Craftsman meter shown in **Figure 1**, for example, though 20 years old, would be a perfectly fine addition to anyone's toolbox.

Additional Leads and Clips

A multimeter typically comes with one set of basic three-foot probe leads with pointy tips. Two other attachments are almost immediately needed:

- Leads with alligator clips on the end to clamp onto things like battery and ground posts
- Leads with "plunger mini-hooks" that allow them to safely grip the ends of wires

These are typically available as part of an aftermarket test-lead kit (**Figure 8**). It'll typically come with a set of leads whose ends have sockets on them. Into the sockets you can plug standard probes, alligator clips, and mini-hooks. If you can accept a notch in quality below Fluke, test lead kids are inexpensive. It's something you need. Buy the kit.

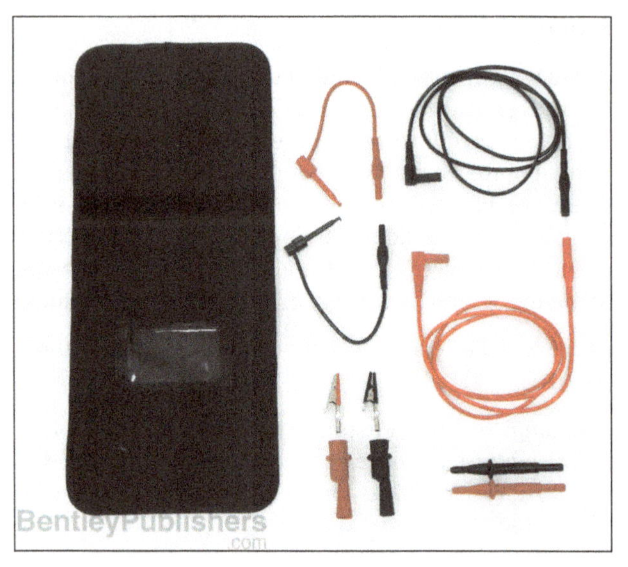

Figure 8 An inexpensive test lead kit.

In addition, at some point, you may find you need an extra-long set of leads to reach from the front to the back of the car. Commercially available multimeter leads seem to max out at six feet. To go farther, you'll need to cobble together some extension wires with clips on the far ends, and then clip the ends of the multimeter leads to the extension wires, taking care to use the little rubber insulating boots on the clips so they don't accidentally touch while you're stretching them 20 feet.

Back-Probe Pins

If you're using a meter to measure something whose electrical contacts are easily accessible, you simply employ the pointy probes or the alligator clips. If there's a connector in the way, you usually unplug it and measure voltage at the connector or resistance at the device. But often, as you'll see in the sensor-specific chapters at the end of this book, when troubleshooting a sensor, you need to leave the connector attached because it supplies power to the sensor, and without power, the sensor doesn't output its signal.

To get around this, you need to "back-probe" a connector, which means peeling back the weather-resistant rubber boot (if there is one) and inserting metal probes into the back of the connector so the probes touch the backs of the pins in the connector.

There's a long history to backyard mechanics using unbent paper clips or hat pins to back-probe connectors. It's a bad idea. It's too easy for the paper clips to swing around and touch each other, which can damage the sensor or, worse, the car's ECM. If you're in a bind and have nothing else, be sure to at least wrap the paper clip in electrical tape so it doesn't short against anything.

But what you should do instead is buy a set of back-probe pins. They're basically long needles that attach to the multimeter's probes. The better ones have insulated shafts so even if they touch, they won't short out (see **Figure 9**). To take a measurement while a device is connected, you gently work them into the back of a connector (see **Figure 10**). A test lead kit may even come with a set of back-probe pins. Make sure you have them. Don't rely on paper clips.

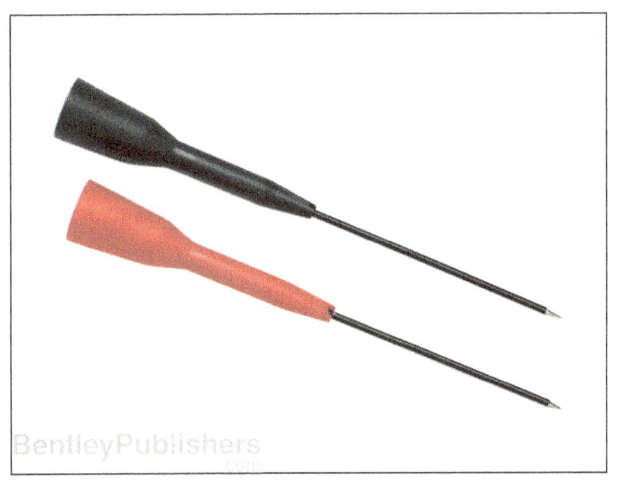

Figure 9 A set of insulated Fluke back-probe pins. *Reproduced with permission, Fluke Corp.*

Figure 10 Back-probing a connector.

On a closely related topic, some folks use wire-piercing probes to make a small puncture in the wire's insulation, then seal it up with clear nail polish. In general, we recommend you avoid this wherever possible and back-probe instead wherever you can.

Cigarette Lighter Voltmeter

As we've discussed in other chapters, the most important thing you can to do to insure the electrical health of your car is evaluate the resting (12.6V) and charging (about 13.5V to 14.4V) voltage of your car. While you can and should test both the resting voltage and the charging voltage right at the battery, sometimes charging problems only present themselves under actual driving conditions.

Fortunately, it is easy to build an adapter to connect a multimeter meter to the cigarette lighter socket and monitor voltage while you're driving. The connectors that go into the front of every multimeter are commonly known as "banana plugs" and are available at any electronic parts store or online. They are also available as a two-connector pair with a standard spacing that fits the main positive and negative socket on nearly every multimeter. Standard cigarette lighter plugs are also available. **Figure 11** shows an adapter cable built using these parts. Using this adapter, you can easily drive the car around, turn on lights, wipers, blower motor, and other accessories, and verify that the alternator continues to keep up with demand.

Figure 11 A homemade cigarette lighter attachment allows you to monitor charging voltage from inside the car.

Or, instead of building an adapter cable for the big multimeter, you can buy, for about ten bucks, a small digital voltmeter that plugs into the lighter socket. It looks like one of those little USB power adapters, except that instead of the USB slots it has a small LED voltage display. These devices are typically not terribly accurate, but they don't need to be, since what you're most often looking for is whether the voltage changes when the engine is running. If you see the voltage go up by about a volt and a half, then you know the alternator is charging the battery. If you don't, then it's not. See **Figure 12**. They're not a substitute for a general-purpose multimeter. You'll still want to keep a multimeter in the glove box.

Go ahead and play with your multimeter. Use it to do the battery check we described in the **Battery** chapter, verifying both the resting and charging voltages. Then put the multimeter (or its black sheep $5.99 cousin) in the recesses of your glove box. You'll be glad its there when, one dark and stormy night, the car won't start.

Figure 12 A voltmeter that plugs into the cigarette lighter costs about ten bucks and tells you the resting voltage (top) and charging voltage (bottom).

when, one dark and stormy night, the car won't start. When you check the battery, see 12.6 volts, go "a-HA! The battery's fine. It's probably just a loose battery clamp," bang on it like Fonzie and the car starts right up, you will be a god.

Other Electrical Diagnostic Tools

Test Lights

Test lights are inexpensive devices that look like a combination of an ice pick and a pencil flashlight with a wire and an alligator clip hanging off the back. You attach the clip to chassis ground and poke around with the tip of the test light. When there's voltage, the light illuminates. They're cheap, but with multimeters costing the price of a grande half caf skinny doubleshot mochachino, just buy the damned multimeter. It's nearly as easy to use, and it does much more.

Having said that, there are times when a test light comes in handy. For example, if you're trying to determine which of your 50 blade fuses has popped, and the fuse box is up underneath the glove compartment, a test light is a bit easier to use than a meter because, when you stick the pointy probe into the test tab in each fuse, it's easier to see the light go on, directly in your field of view, instead of having to look down at the screen of the multimeter. But don't buy one and skip buying a multimeter.

Circuit Testers

There is a class of automotive-specific electrical tools sometimes referred to as "circuit testers" or "circuit probes." These look like a cross between a test light and a multimeter. The best known brand is Power Probe. They have positive and negative cables that connect directly to the car's battery, and a pointy probe that can be stuck in places just like a test light or the leads of a multimeter. Voltage and other readings are displayed on the screen on the device's body.

Part of the utility is that you can hit a button and "inject power" or "inject ground" (e.g., the tool stops using the probe tip as a measuring device and instead connects it directly to battery positive or negative). This can be extremely useful during troubleshooting. For example, you can probe the positive terminal of a cooling fan that's not coming on. If you find there's no voltage there, you can "inject power" and find out if the fan comes on when supplied power. As with other nice-to-have tools, you shouldn't postpone buying a multimeter because you're saving up for a device like a Power Probe. If you have both, great.

Open and Short Circuit Finders

When troubleshooting a problem, you'll sometimes diagnose the cause as an open circuit or a short circuit, but will then need to find the location where the wire is broken or chafed and shorting out against the body of the car. You first try to locate these by visual inspection, but that may not be productive, as they may be located behind an access panel, or beneath the carpet, or where a wire goes through the firewall. Fortunately there are some cost-effective electronic aids that can help you if you get stuck.

The thing you need to be aware of is that when the ad copy for a product tells you it will find, or help find, shorts and open circuits, you need to understand exactly *how* it will help you do that. A multimeter will do those things if you punch holes in the wire's insulation and check the wire every inch. The key is finding a tool that helps you to do it quickly and non-invasively.

"Short Finder" (circuit breaker and Gauss gauge)

The first device is a somewhat old-school low-tech pair of tools sold as a set. A "short finder" or "short circuit tester" is a circuit breaker and a small handheld device (see **Figure 13**) that detects the flow of current without having to be wired into a circuit (a "Gauss gauge"). The circuit breaker is a small tube about the size of a roll of quarters, attached to several feet of wire with alligator clips at the ends. The Gauss gauge is a small analog gauge, slightly larger than a book of matches. The purpose of the Gauss gauge is not to tell you the amount of current, but rather to show you the direction of the current flow.

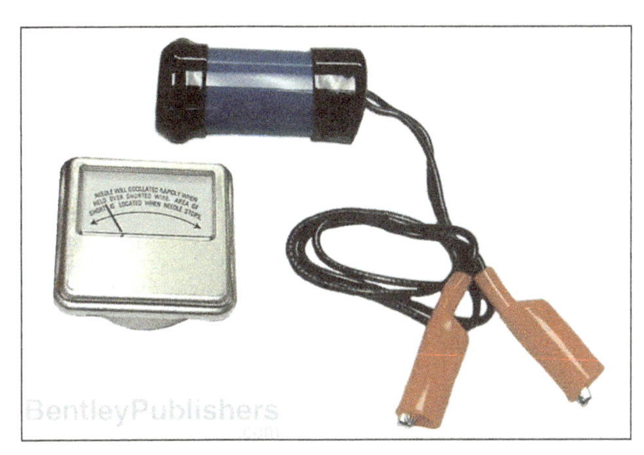

Figure 13 A circuit breaker and Gauss gauge-based short finder. *Photo from S&G Tool Aid Corp.*

To use the short finder:

- You connect the circuit breaker in place of the fuse that's blowing.
    - In a car with torpedo fuses, you simply clip the alligators onto the fuse box tabs.
    - In a car with blade fuses, you can either buy an inexpensive adapter that plugs in instead of the fuse and sticks up proud of the fuse box, or crimp a pair of short leads with male tabs at the ends to plug in where the fuse was.
- When the circuit breaker is in place and the circuit is energized, the breaker will briefly allow a small amount of current to flow before the short causes it to automatically trip.
- The breaker will take about 15 seconds to cool down. Then the process will repeat.
- During the short periods in which the circuit is engaged, you hold the Gauss gauge against the wire that's part of the circuit whose short to ground you're trying to chase.
- If current is still flowing, you're still "upstream" of the location of the short. You continue to move the Gauss gauge along the wire path that's causing the fuse to blow.
- The needle will change direction when you pass the short.
- You can also connect a test light or buzzer in series with the circuit breaker, so the light will flash or the buzzer will sound for the short periods that the circuit is energized so you know to pay attention to the Gauss gauge now.

An even older school variant of this involves using a compass instead of a Gauss gauge (the rapidly changing electric field from the breaker tripping would set up a magnetic field which would deflect the compass), but the whole breaker/Gauss gauge kit is only about 30 bucks.

Note that the short finder shown in **Figure 13** has two, uh, shortcomings. First, it requires the circuit to be energized and de-energized while it is shorted. Second, it won't help you find an open circuit. Addressing both of these issues leads you to look at the open and short tracers listed below.

Open and Short Circuit Tracers

Ideally, what you want is something that lets you "sniff" a fault for both open circuits as well as short circuits. There are several products called open and short circuit tracers that purport to do this.

Tracers employ a transmitter and a receiver. The transmitter typically powers off the car's battery, and has a probe that inserts into the connection on the load side of the blown fuse. In this way, since it is directly connected to both power and ground, it can sense the presence of a short between them. Some have both a light and an audible tone to alert you if a short to ground exists, letting you wiggle wires to force an intermittent situation into the open. In itself, this isn't anything you couldn't do with a multimeter, but the transmitter injects a signal into the wiring, and the handheld receiver then lets you hone in on the source of the fault, following the wire until the tone stops. See **Figure 14**.

To find these products online, the magic web search terms appear to be "open short circuit tracer or tracker or finder or detector," thereby distinguishing them from a "tester" that doesn't actually help you find anything. As of this writing, there appear to be two reputable manufacturers of these devices, with well-documented well-reviewed products. Each has a different operating mode for searching for an open circuit as opposed to looking for a short. Online reviews indicate both devices are quite helpful, but not magic wands. One manufacturer warns to disconnect any part of the system that might be sensitive to voltage and current pulses, including air bags and electronic control modules. Another admits that it may be difficult to follow a signal inside of a wiring harness.

In addition to these two > $100 products, there are a variety of inexpensive products, with no online user's manual and little user feedback, that appear to be little more than phone line tone generators with the RJ11 and RJ45 plugs removed and the word "automotive" printed on them.

TDRs

A time domain reflectometer (TDR) is a cable fault locating device that tells you the distance from the device to the break. They're designed for use with long lengths of cable, and are not really the right tool for automotive applications unless you're looking for a break in a cable in a 53' trailer. We mention them only so you don't buy the wrong tool.

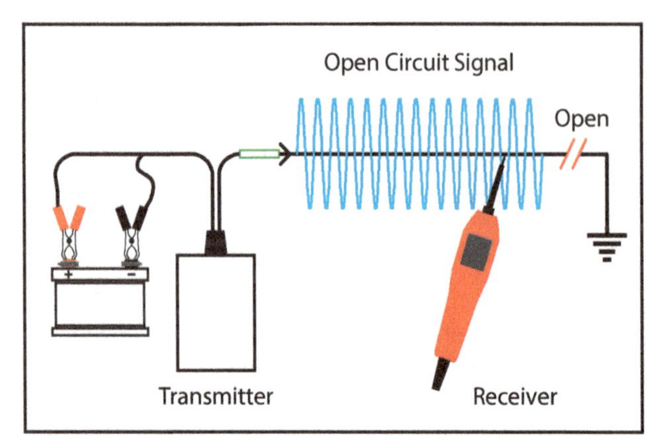

Figure 14 Open and short circuit tracers use a transmitter to inject a signal into a wire. A handheld wand helps you to "sniff" out the location of a break or short that causes the signal to stop.

Tone Generators

A tone generator is a type of cable tracker or tracer primarily used in the telephone and computer service world. Tone generators are designed to help identify which pair of wires in a complex bundle is connected to a certain wall jack. If you purchase a tone generator with sufficient sensitivity, you can also use it to locate cables behind walls and determine where a signal stops due to broken wires.

Could you use a tone generator to help find a break in a wire in a car? Yes. You could connect one of the transmitter's alligator clips to the fusebox connection to the broken wire (on the load side of the fuse), and use the receiver to follow the wire until you no longer hear the tone. Could you use one to find a short to ground? In theory, you can connect the positive lead as shown in **Figure 14**, and connect negative to ground, and probe until you lose the tone, which is presumably just past the short. But remember – the tone is much quieter when the circuit is shorted. So this isn't really what it's designed for.

So, if you already have a tone generator, the tone-following capability can be useful for locating a break in a wire. But if you're going to buy a new tool, the "open and short circuit tracers" mentioned previously are better options.

5

Common Multimeter Tests

Multimeter Tests

In this chapter, we're going to step through how to use a multimeter to perform the most commonly used tests. These include measurement of DC voltage, AC voltage, resistance, continuity, high current, low current, and voltage drop.

The illustrations shown are based on a Fluke 88V multimeter, which for many years has been the industry-standard meter. Your meter will look similar, but the options on the buttons and the settings on the rotary dial will be slightly different. If, after reading this chapter, you still have specific questions about what your meter will and will not do, you may need to consult the owner's manual for your meter.

Key Multimeter Concepts

- Internally, the multimeter measures voltage, amperage, and resistance in different ways. This is why the meter must be connected and configured in different ways depending on what you're measuring.
- The meter and the thing you're trying to measure must form an electrical circuit.
- Any time you use a meter to make an electrical measurement, it can change the very thing you're trying to measure (the classic problem of observation of the system interfering with the system). Modern multimeters are designed to minimize this effect.
- Voltage and current measurements are similar, in terms of what the meter and the circuit are doing, in that during these measurements, the circuit is powered. A resistance measurement is fundamentally different because the meter itself is sending out a small current to power the circuit.
- However, in terms of how you need to configure the meter, voltage and resistance measurements are similar because these both use the red VΩ socket, whereas performing a current measurement is fundamentally different because the red A or mA socket is used instead.

Related Chapters

- **Circuits: (How Electricity Really Works)**
- **Multimeters and Related Tools**
- **Tools for Dynamic Analog Signals**

Multimeter Configuration Table

Although we will go through all of the measurements in detail, we summarize the voltage, voltage drop, resistance, and current measurements in the **Multimeter Test Configuration Table**. The table also lists whether the circuit needs to be powered or unpowered, whether the measurement occurs in parallel or in series with the circuit, whether the test is occurring on the whole circuit or with the probes placed on one specific section of the circuit, and whether the plug of the red lead is inserted into the VΩ socket or the A or mA socket.

CAUTION —

It is both incorrect and dangerous to connect a multimeter in any fashion other than what is specified by the meter manufacturer. In particular, current is the only measurement conducted where the meter is connected in series and the circuit is powered. Performing any other measurement with the meter connected in series and the circuit powered may damage the meter, or at least cause the meter's internal fuse (if it has one) to blow.

Multimeter Test Configuration

Test	Circuit State	Parallel or Series	Whole or Section	Red Lead Position
Voltage	Circuit must be powered	Parallel	Whole	VΩ
Voltage Drop	Circuit must be powered	Parallel	Section	VΩ
Resistance	Circuit must **NOT** be powered	Series	Section	VΩ
Current	Circuit must be powered	Series	Whole	A or mA

5

Measuring DC Voltage

If you want to measure DC voltage, the multimeter must be connected in parallel with the circuit, and the circuit must be powered. This means that, most of the time, you don't need to disturb or unplug anything on the car. You merely connect the meter's black probe to ground, and touch the red probe to where you want to test for the presence of voltage. This allows a very small amount of the circuit's current to flow through the meter. In this way, the voltage is measured while disturbing the circuit as little as possible.

CAUTION —

Only use a digital multimeter to measure voltage across solid state (digital) electronics. Do not use an analog multimeter. For additional information, see **How a Voltage Measurement Works.**

How to Measure DC Voltage

See Test Set-Up on page 5-5

- Insert the plug of the red multimeter lead into the meter's socket that's labeled "V." It is almost always the same socket used to measure resistance, and thus is often labeled "VΩ."
- Insert the plug of the black multimeter lead into the meter's "COM" socket.
- Set the dial on the meter to measure DC voltage. This setting either says "DCV" or is labeled with a capital V with two straight lines over it. It does *not* have a wavy line over it. That's AC voltage.
- Energize the circuit (that is, turn on the thing that's supposed to be on).
- Touch or clip the tip of the black probe to a convenient ground (for example, the negative battery terminal, or bare unpainted metal).
- Touch the tip of the red probe to the place at which you're trying to determine the presence of voltage. In **Set-Up 5-A**, we show the red probe testing for voltage at the positive terminal of an ignition coil with the ignition key on.
- Read the voltage on the display of the meter. It should about 12.6V, and with the engine running it should read charging voltage (about 14.2V).

How a Voltage Measurement Works

Voltage is always measured in parallel with a circuit. Here's why. As you know from the **Circuits** chapter, if you connect a load device in parallel with a circuit, it gives the current a second path to flow. This increases the amount of current in the circuit. This in turn may affect the voltage in the circuit, which is the very thing we're trying to measure. This is sometimes referred to as the "loading effect of the meter." Therefore, to keep the measurement accurate, the current flow through the meter needs to be as small as possible. The way to limit current is to increase resistance.

For this reason, every voltmeter (or, we should say, every multimeter set to measure voltage) has a high internal resistance to reduce the amount of current flowing through the meter. The meter's reading is electronically calibrated for that small current. Vintage analog meters have this resistance expressed in terms of "ohms per volt" (for example, 20,000 ohms per volt) but on digital meters, it is expressed as a single number, usually 10 megaohms (check the owner's manual to be certain).

There are still a handful of places where an analog meter is useful. For example, when testing an oxygen sensor, some die-hard users prefer to see an actual physical needle swing between an upper and lower range rather than trying to read and remember high and low digital values. However, many people have read that analog meters aren't "safe" to use on digital electronics.

The problem is that digital circuits can be very sensitive to voltage levels and current draw. If you're, for example, measuring the 5V reference line supplied by a control module, you don't want the meter to draw additional current and blow a chip in the module. An analog meter might have an internal resistance as high as 20,000 ohms per volt, in which case it's probably safe to use, or as low as 1,000 ohms per volt, in which case it's probably *not* safe to use. Because the ohms per volt documentation is often missing, unless you know what the internal resistance is, it's best not to take a voltage measurement on digital electronics with an analog meter, and instead to use a digital multimeter with a 10 megaohm internal resistance.

So those old-school folks using a swing needle meter on an oxygen sensor should make sure that their meter has an internal resistance of least 20,000 ohms per volt.

In **Set-Up 5-A**, we show voltage being measured in parallel with the circuit. What if there's no "circuit?" What if we just want to measure the voltage across the battery? Is it safe to do this? Yes. It was, in fact, the very first thing we asked you to do in the **Battery** chapter.

Because of the high internal resistance of the meter, it's perfectly safe to use the meter to measure voltage directly across the battery's positive and negative terminals.

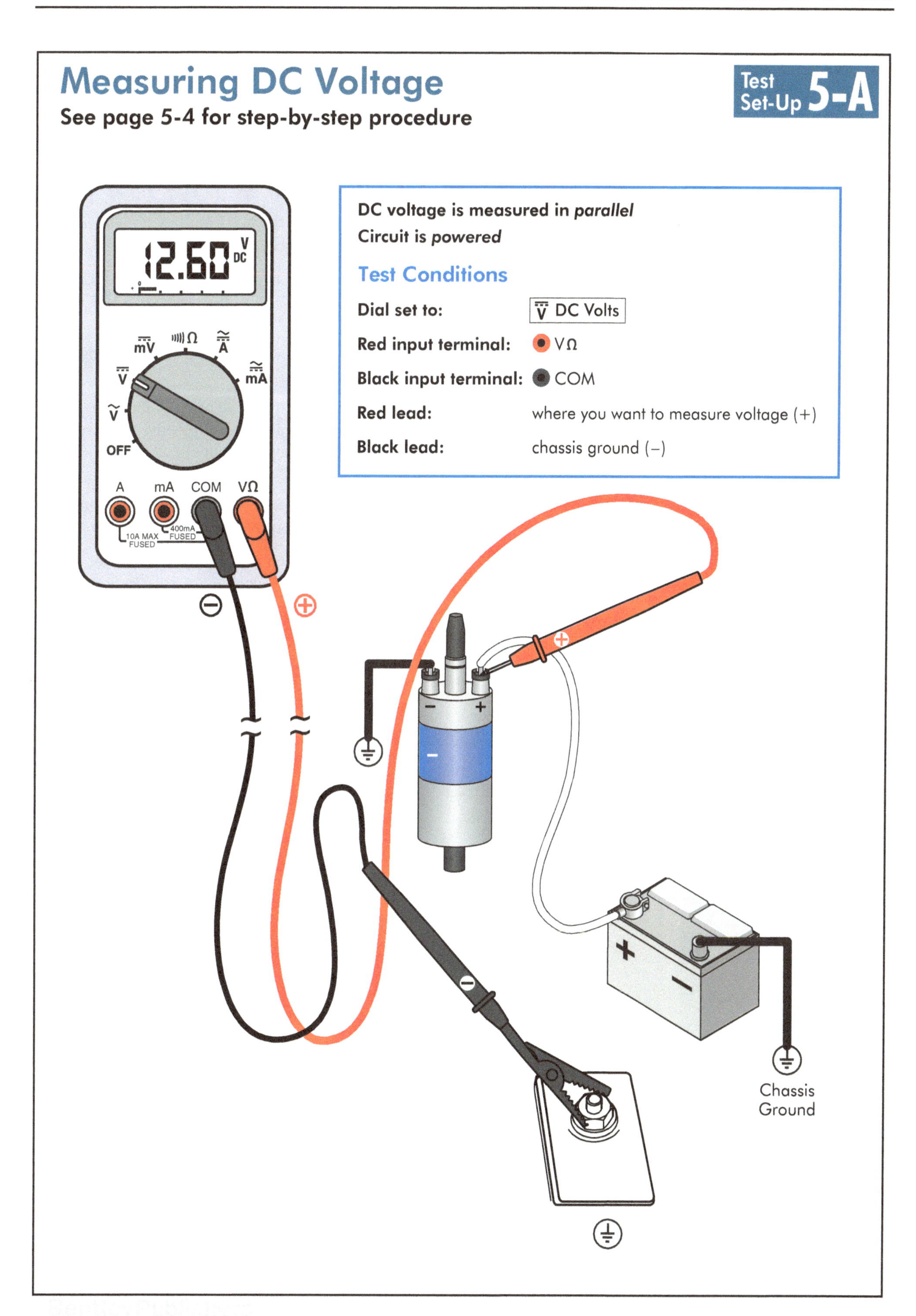
Measuring DC Voltage
See page 5-4 for step-by-step procedure
Test Set-Up 5-A
DC voltage is measured in parallel
Circuit is powered
Test Conditions
Dial set to: V DC Volts
Red input terminal: VΩ
Black input terminal: COM
Red lead: where you want to measure voltage (+)
Black lead: chassis ground (–)
12.60 V DC
mV
Ω
A
mA
V
OFF
A
mA
COM
VΩ
10A MAX FUSED
400mA FUSED
Chassis Ground

5

Measuring AC Voltage

There isn't really "AC current" in your car. The AC measurement is used mainly for one of two reasons – to check for unwanted AC ripple in the alternator, or to confirm the presence of the sine wave signals from variable reluctance transducer (VRT) sensors used on ABS sensors and camshaft position sensors in 1980s and 1990s cars.

To measure AC voltage, the multimeter is connected in the same way as the DC voltage measurement, in parallel with the circuit, and the circuit must be powered. This means that, most of the time, you don't need to disturb or unplug anything. You merely connect the meter's black probe to ground, and touch the red probe to where you want to test for the presence of voltage. This allows a very small amount of the circuit's current to flow through the meter, allowing the voltage to be measured while disturbing the circuit as little as possible.

How to Measure AC Voltage

See Test Set-Up on page 5-7

- Insert the plug of the red multimeter lead into the meter's socket that's labeled "V." It is almost always the same socket used to measure resistance, and thus is often labeled "VΩ."
- Insert the plug of the black multimeter lead into the meter's "COM" socket.
- Set the dial on the meter to measure AC voltage. This setting either says "ACV" or is labeled with a capital V with a wavy line over it.
- Energize the circuit (that is, turn on the thing that's supposed to be on).
- Touch or clip the tip of the black probe to a convenient ground (for example, the negative battery terminal, or bare unpainted metal).
- Touch the tip of the red probe to the place at which you're trying to determine the presence of voltage (as pictured, the output terminal of the alternator).
- Read the voltage on the display of the meter. It should show the AC voltage. In **Set-Up 5-B** we are showing how to check for alternator AC ripple.

Note that this is not 120VAC wall current. Any AC signal on a car is small. AC ripple on the alternator output should be a fraction of a volt. The AC sine wave from a VRT sensor may be between 1 and 5 volts.

Measuring AC Voltage

See page 5-6 for step-by-step procedure

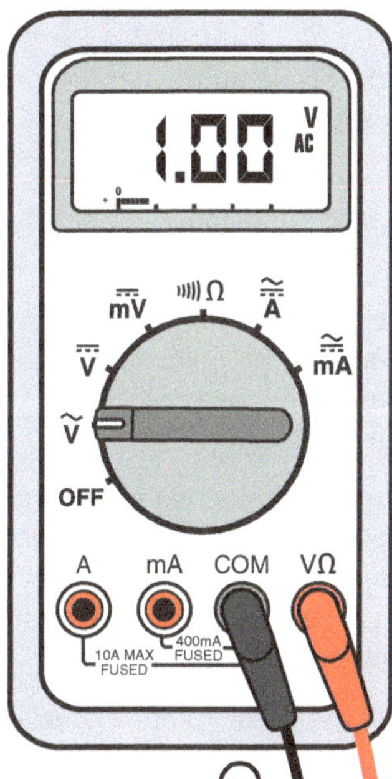

AC voltage is measured in *parallel*
Circuit is *powered*

Test Conditions

Dial set to:	Ṽ AC Volts
Red input terminal:	VΩ
Black input terminal:	COM
Red lead:	where you want to measure voltage (+)
Black lead:	chassis ground (–)

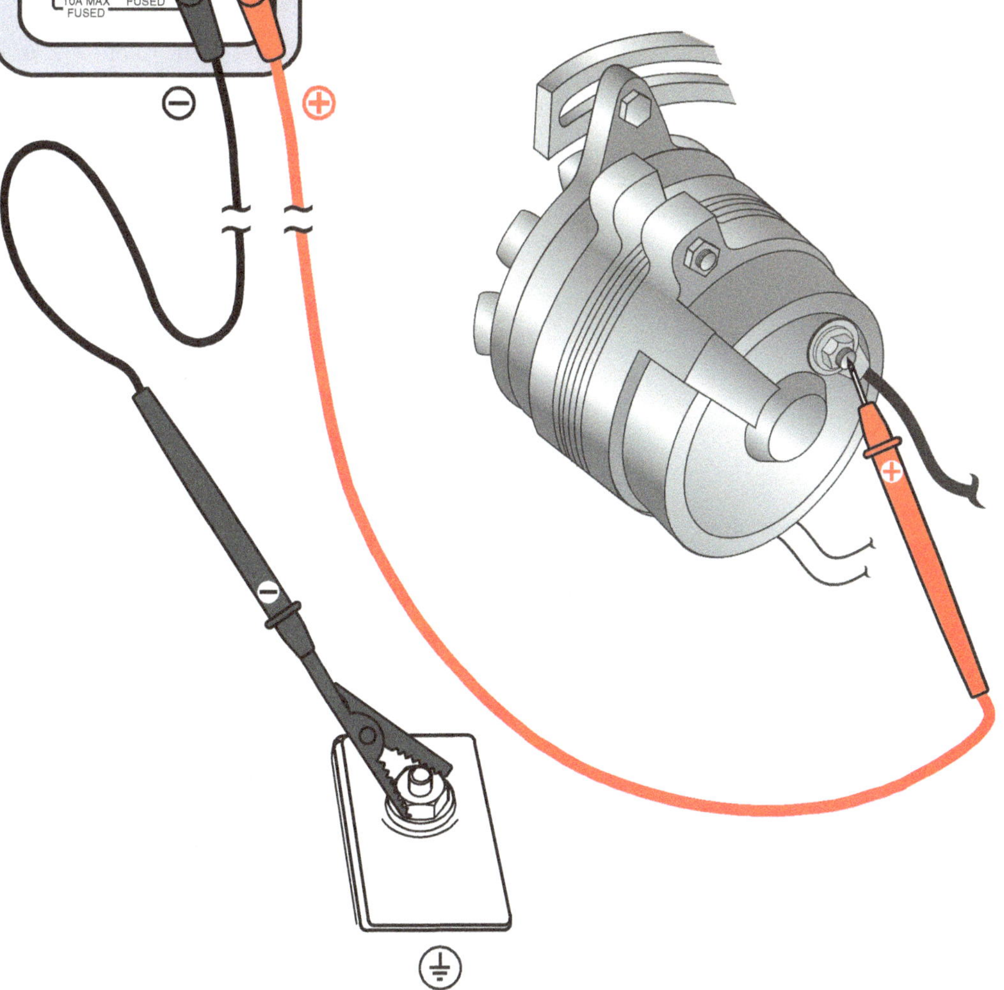

Measuring Resistance

After voltage measurement, resistance measurement is the next most common automotive application for the multimeter. Temperature sensors, for example, are typically thermistors (resistors whose values change with temperature). Throttle positioning sensors often use resistance track devices with a rotating arm. Relays contain wound coils of wire. The typical resistance of devices like this may be in the single ohms to thousands of ohms – easily measured by the meter.

The resistance measurement has two main differences from the voltage measurement. First, the circuit *must be unpowered*. If it isn't, the reading won't be correct, and you could damage the meter. Second, whenever possible, the section of the circuit you're measuring should be isolated from the rest of the circuit. That is, if there are wires leading to the device whose resistance you want to measure, it's best to unplug those wires. This is because you want to be sure you're measuring the resistance of the device and not of the wiring of the car.

CAUTION —

Because a resistance measurement is an active measurement that applies a small current between the probes, a resistance measurement should not be performed on solid state (digital) electronics, as there is a possibility that the small current applied may damage solid state electronics.

How to Measure Resistance

See Test Set-Up on page 5-9

- Insert the plug of the red multimeter lead into the meter's socket that's labeled "Ω." It is almost always the same socket used for voltage, and thus is often labeled "VΩ."
- Insert the plug of the black multimeter lead into the meter's "COM" socket.
- Set the dial on the multimeter to measure resistance. This setting either says "Resistance" or has the Ω symbol on it.
- With the two probes not touching, look at the meter's display. It should say "OL" for open circuit – no continuity. If it doesn't, the meter isn't set up correctly to measure resistance.
- Test the meter by touching the two meter tips together. The meter's display should read very close to zero. If it still says "OL," one of the probes isn't seated in its socket, or one of the probes is faulty, or a probe wire is broken.
- Touch the tip of the red probe to one end of what you're trying to measure resistance of. Touch the tip of the black probe to the other end of what you're trying to measure resistance of.
- Read the resistance on the display of the meter. If it says "OL," it is registering an open circuit – no continuity. Otherwise, it is measuring the resistance of the circuit path. In **Set-Up 5-C**, we show how to measure the primary resistance of an ignition coil.

How a Resistance Measurement Works

A resistance measurement is fundamentally different from a voltage or a current measurement because it is an active measurement. Multimeters measure resistance by using their internal battery to apply a small current to produce a known voltage, between 0.3 and 1.0 volt, at one probe. They then measure the resulting voltage drop at the other probe, and calculate the resistance using Ohm's law in the form that resistance is voltage divided by current.

Measuring Resistance

See page 5-8 for step-by-step procedure

Test Set-Up 5-C

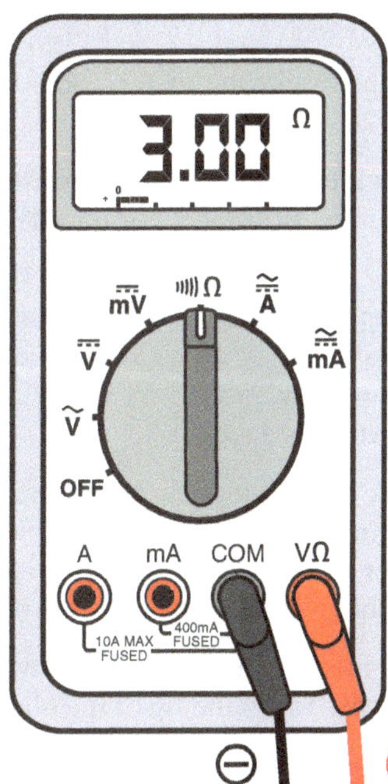

Resistance is measured in *series* on a section of the circuit
Circuit is *unpowered*
Device is isolated from circuit

Test Conditions

Dial set to:	Ω Resistance
Red input terminal:	VΩ
Black input terminal:	COM
Red lead:	One side of device
Black lead:	Other side of device

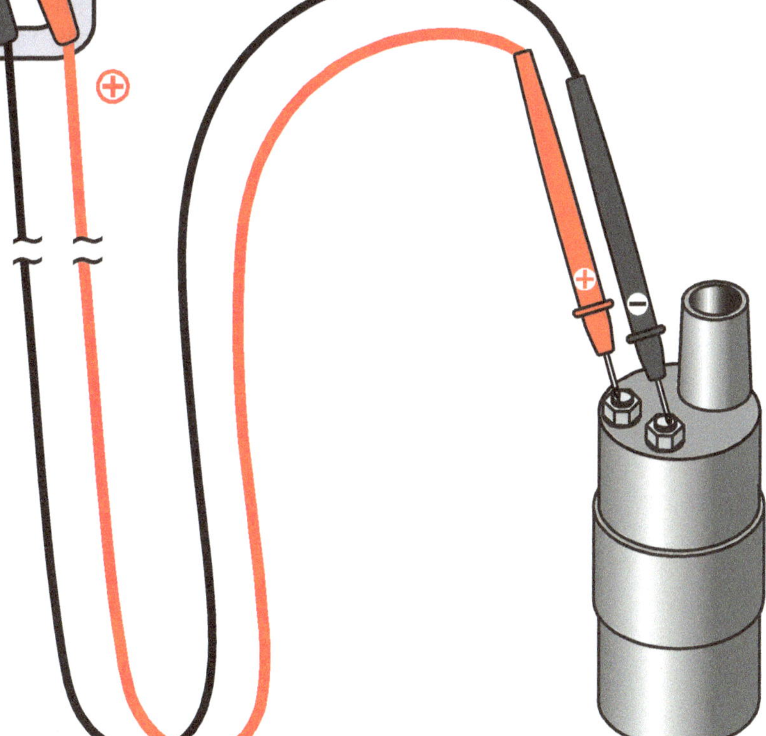

Measuring Continuity

Most of the time, when we set the multimeter to measure resistance, what we're really interested in doing is taking a binary continuity measurement – that is, determining whether or not a section of wire electrically connects to what we think it does, or a switch actually turns on when we throw it. Much of the time, this is all you can do anyway because the actual resistance of most electrical wiring is so low that it falls below the ability of most meters to measure it accurately. The meter can, however, easily determine if a length of wire has continuity or not.

The set-up for a continuity test is identical to that for the resistance measurement. However, enabling an option on the multimeter makes the test easier. Most multimeters have a "beep" function that emits an audible tone when continuity is detected (when the measured resistance is near zero). It's handy to have the beep function enabled when checking continuity. It is usually obvious how to turn it on by simply looking for a beep symbol on the front of the meter.

How to Measure Continuity

See Test Set-Up on page 5-11

- Insert the plug of the red multimeter lead into the meter's socket that's labeled "Ω." It is almost always the same socket used for voltage, and thus is often labeled "VΩ."
- Insert the plug of the black multimeter lead into the meter's "COM" socket.
- Set the dial on the multimeter to measure resistance. This setting either says "Resistance" or has the Ω symbol on it.
- With the two probes not touching, look at the meter's display. It should say "OL" for open circuit – no continuity. If it doesn't, the meter isn't set up correctly to measure resistance.
- If your meter has a "beep" function, turn it on, and test the beep by touching the ends of the probes together. The meter should beep and the display should read very close to zero. If not, check the leads.
- Touch the tip of the red probe to one end of what you're trying to check. Touch the tip of the black probe to the other end of what you're trying to check.
- In **Set-Up 5-D** we show the meter on the left connected to a closed switch, reporting continuity with a low resistance reading (0.1) and an audible "beep." We then show the meter on the right connected to an open switch, reporting no continuity with an "Over Limit" resistance reading (OL) and no beep.

Note that testing a wire for continuity is identical – the left meter would read a low resistance value and would beep if connected to a continuous piece of wire, and would read "OL" and would not beep if connected to a broken wire.

If you want to see if a brake light is grounded (if the negative terminal of the light is connected to the body of the car), unplug the wire from the brake light's negative terminal, then connect one probe of the multimeter (it doesn't matter which one goes where for a resistance measurement) to that wire, and touch the other probe to the body of the car. If the meter reads about an ohm or less, then it's connected to ground. If it reads "OL," that's an open circuit, and the negative wire is *not* in fact grounded.

Note that a wire's continuity says nothing about its ability to actually carry the current required by a circuit for successful operation. See the immediately following section on **Measuring Voltage Drop**.

Measuring Continuity

See page 5-10 for step-by-step procedure

Continuity is measured in *series* on a section of the circuit
Circuit is *unpowered*

Test Conditions

Dial set to:	Ω Resistance
Press:	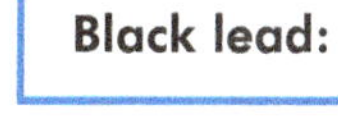 button
Red input terminal:	VΩ
Black input terminal:	COM
Red lead:	One side of switch or wire
Black lead:	Other side of switch or wire

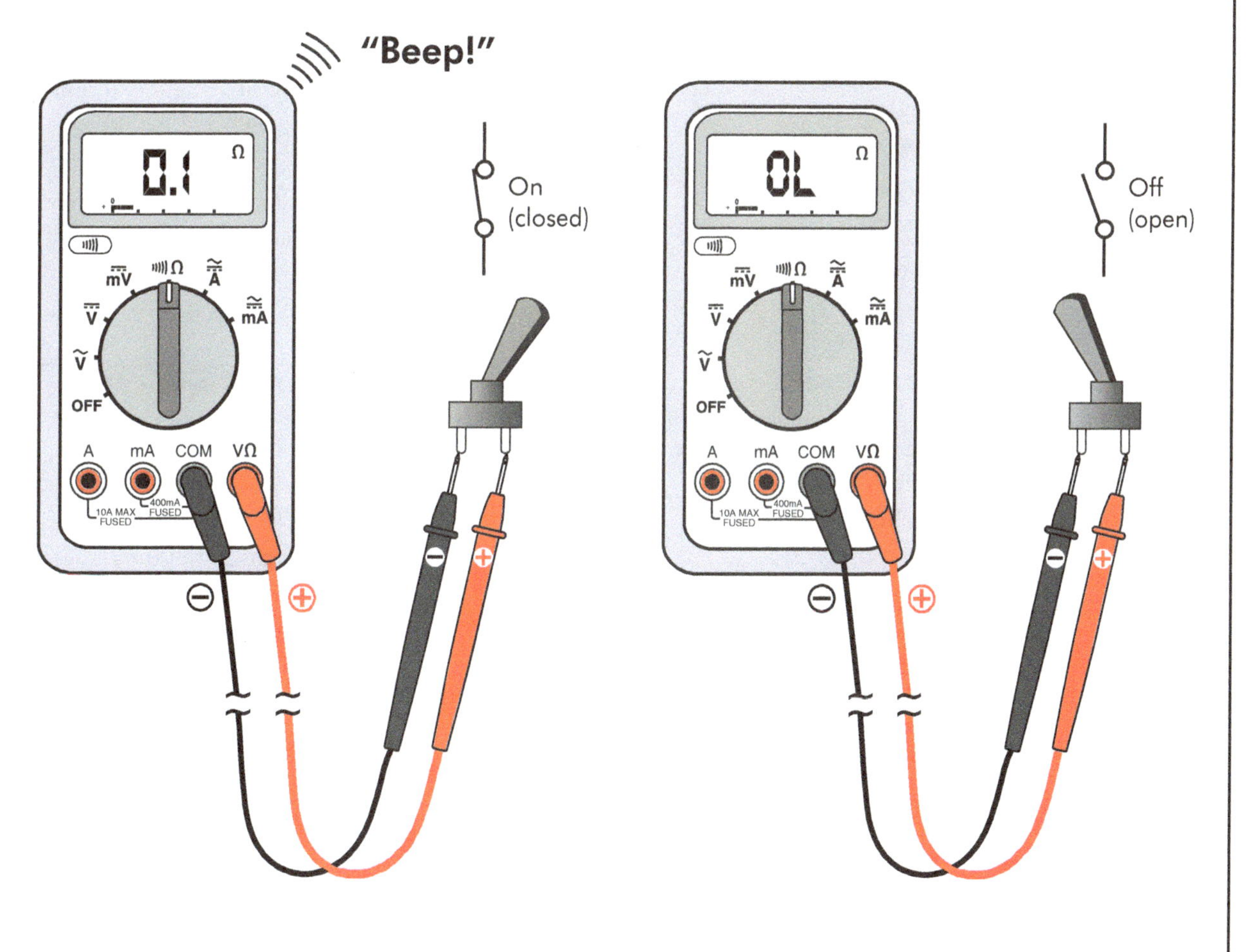

Measuring Voltage Drop

In the previous section, we said that the resistance of wire is so low that you can't accurately measure it with a multimeter. This makes it so you can't tell, with a resistance measurement, the difference between wire with all the strands present and wire where just a single strand is unbroken.

Figure 1 illustrates an extreme example. Sure enough, with only one strand of wire the diameter of a thread, the meter not only shows continuity, but also low resistance (0.3 ohms as pictured). But obviously, not a lot of current can be pulled through this single strand of wire. A high amperage load such as an electric motor would cause the thin strand to get very hot, increasing its resistance even further, resulting in a voltage drop across the damaged section. This in turn reduces the voltage and thus the current available to the device. The situation is the same if, instead of a single strand of wire, there is a badly corroded connection.

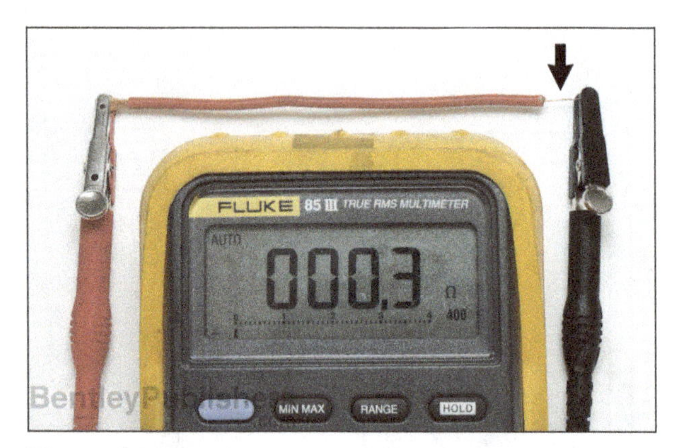

Figure 1 With only one strand of wire remaining on this wire (**arrow**), a resistance measurement still shows continuity, but the wire's ability to carry current is compromised.

The *voltage drop measurement* gets around this problem. Instead of directly measuring resistance in a section of a circuit, you measure the drop in voltage across that section of the circuit. In fact, you position the multimeter's probes similarly to the way you did for the resistance measurement, with the two test leads placed along a section of wire. However, unlike a resistance measurement, a) the circuit remains powered, and b) the section being tested remains in the circuit.

Voltage drop tests are most important with devices that draw a lot of current, such as the starter motor. The test can be conducted either along the positive side of the circuit (between the positive battery terminal and the device) or the negative side of the circuit (between the device and chassis ground). If everything is perfect and there is no resistance along the path between the probes, the voltage drop reading should be zero.

How to Measure Voltage Drop

See Test Set-Up on page 5-13

- Insert the plug of the red multimeter lead into the meter's socket that's labeled "V." It is almost always the same socket used to measure resistance, and thus is often labeled "VΩ."
- Insert the plug of the black multimeter lead into the meter's "COM" socket.
- Set the dial on the meter to the most sensitive DC voltage setting. This setting either says "DCV" or "mV." Because most voltage drops are small, you will get a more accurate reading using the mV setting even with an autoranging meter.
- Connect the red probe to the positive battery terminal.
- Connect the black probe to the device at which you're trying to measure the voltage drop (e.g., the positive terminal of the starter motor).
- Energize the circuit (e.g., crank the starter motor) and read the meter. **Set-Up 5-E** shows the meter connected between the positive battery terminal and the starter, with a voltage drop of 300mV.
- Repeat the test as necessary on the ground side of the circuit, looking for a voltage drop between the negative battery terminal and the negative terminal on the device.
- See the table of typical voltage drop values below.

Typical Normal Voltage Drop Values While Cranking	
Connection	**Maximum Drop**
Battery + post and starter post	800mV
Battery – post and starter case	800mV
Battery – post and chassis ground	500mV
Battery + or – post and battery terminal	100mV

NOTE —

There is a detailed example of a voltage drop measurement conducted on a starter motor in the **Starter** *chapter.*

For testing a device that draws lower amperage than a starter, a rule of thumb is that there should be virtually no voltage drop between a wire and its crimped connector, no more than 100mV at a ground connection, no more than 200mV across the length of a wire, and no more than 300mV across a switch.

Measuring Voltage Drop

See page 5-12 for step-by-step procedure

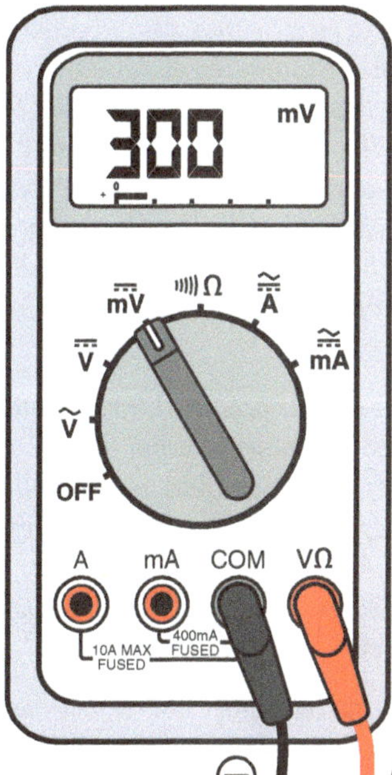

Voltage drop is measured in *parallel* on a section of the circuit

Circuit is *powered*

Probes located at points along circuit path

Test Conditions

Dial set to:	V DC Volts or mV
Red input terminal:	VΩ
Black input terminal:	COM
Red lead:	Along circuit path
Black lead:	Along circuit path

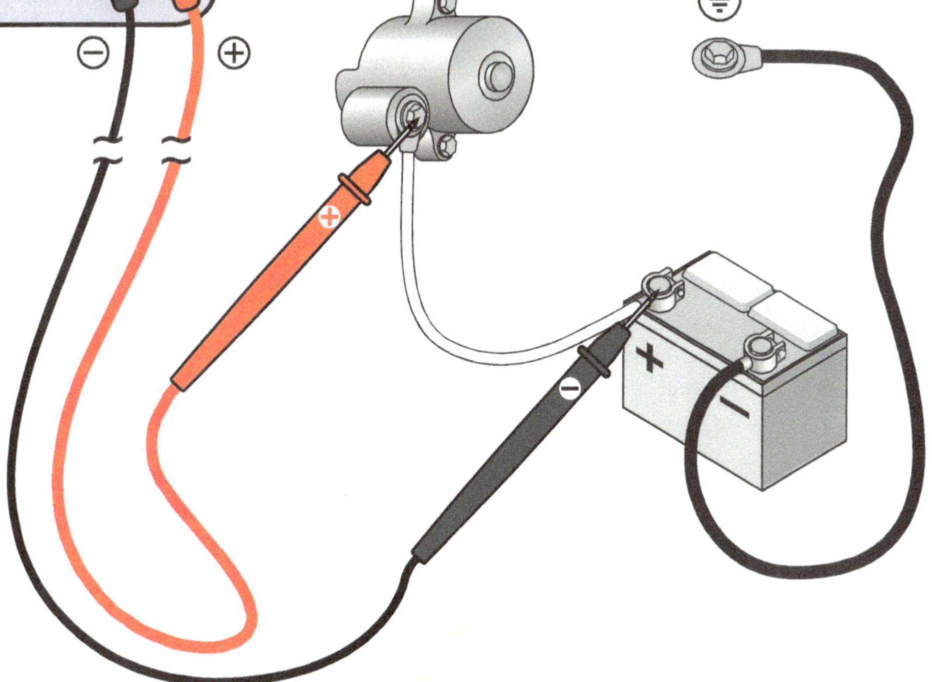

5

Measuring Current

Current measurements are less common than voltage and resistance measurements. They can be performed on individual devices, but they're mostly associated with parasitic draw tests (drains that cause the battery to go dead). See the **Energy Diagnosis and Parasitic Drain** chapter for additional detail on parasitic drains.

The words "current" and "amperage" are used interchangeably. The measurement is most commonly called a "current measurement," but the word "current" does not appear on most multimeters. The units of current are amperes, also called amps. The symbol for amps is a capital A. It is the capital A that appears on most multimeters.

Unless you have a current clamp to conduct a current measurement, the multimeter must be connected in series with the circuit, and the circuit must be powered. This means that all the current in the circuit is flowing through the meter, just as if it were a wire in the circuit.

 **CAUTION —**

Configuring a Multimeter to Measure Current (Amperage)

Meter settings for current measurement vary greatly from meter to meter. The input socket(s) for current measurements is almost always different than the voltage/resistance socket. Set-up variations also exist in the rotary dial, including separate positions for DC amperage and AC amperage and even high and low current. Be sure to set the meter up correctly before running current through it. It is easy to blow the internal fuses if you get things wrong. If in doubt, consult the manufacturer's user manual.

 **CAUTION —**

During a current measurement, all the current in the circuit is flowing through the meter. Many meters have a "high current" setting of 10 amps. 10 amps is not really all that high compared to the current that can be drawn by large electric motors and other devices in cars. If the current in the circuit exceeds 10 amps and the meter has an internal fuse, it will blow the internal fuse. If the meter does NOT have an internal fuse, it will damage the meter. Take care not to exceed the meter's amperage rating. See the meter's user's manual and the **Energy Diagnosis and Parasitic Drain** *chapter of this book for additional information.*

The way meter input sockets are fused varies meter to meter. Recent meters may have internal fuses protecting both settings, but on older meters, often only the low amperage setting has a fuse, leaving the high amp setting unfused. The two settings are usually clearly labeled as "FUSED" or "UNFUSED" on the front of the meter.

Because the circuit's current must not exceed the rating of the meter, you need to have some idea of how much current the circuit will draw. You can get some sense of this from the size of the fuse on the circuit. If, for example, the circuit has low-draw devices like interior lights and is protected by a 5A fuse, it should be safe to hook up a meter with a 10A rating. But if the circuit powers a large cooling fan that draws 25A and is protected by a 40A fuse, don't do it.

Obviously, don't ever try to directly measure starter motor current with a multimeter! It will blow the meter, or its fuse, instantly. The only way to measure the draw of a very high-current device like a starter motor is using a current clamp. If necessary, consult a repair manual or an enthusiast web forum for information on the expected current draw to make sure the measurement is safe for your meter.

How to Measure Current Through a Device

See Test Set-Up on page 5-16

- Verify that the expected current is within the current measurement range of the multimeter.
- Decide whether, on the basis of the expected current, to use the high current (e.g., 10A) setting, or the low current (e.g., 400mA) setting.
  - If you think the circuit you're measuring draws more than 400 milliamps, or if you don't know, use the high amperage setting. But if you think the draw might exceed 10 amps, and the high amperage setting of the meter is unfused, you could damage the meter.
- Turn the rotary dial to the desired setting for DC amperage. In **Set-Up 5-F**, we show this as the "A" with both the solid and wavy line over it. On this meter, it is the same switch position for both DC and AC amperage.
  - Amperage dial settings vary between meters. Some may say "DCA" or use a capital A with two straight lines over it. The DC amp setting will *not* have a wavy line over it. That's AC amperage.
- Insert the plug of the red multimeter lead into the meter's input socket for the desired amperage setting. It should have the "A" symbol (e.g., 10A or 400mA).
- Insert the plug of the black multimeter lead into the meter's "COM" socket.
- Disconnect the ground wire from the negative terminal of the device whose current draw you're trying to measure.
- Connect the red probe to the device's negative terminal. Connect the black probe to the ground wire that was connected to the negative terminal. Alligator clips are strongly recommended to hold the leads in place.
- Power the circuit and read the amperage on the display of the meter.
  - Depending on the meter, if the meter is set to the 400mA scale, the result may be displayed in milliamps (e.g., "200" means 200mA, not 200 amps).

How a Current Measurement Works

We said that, for a voltage measurement, a meter uses a high resistance to knock the current way down, and then uses the tiny remaining amount of current to measure voltage, calibrated for that high resistance. Well, what if current is actually the thing you want to measure? In this case, you *don't* want a high internal resistance. So, on a meter, when you turn the dial to select the amperage measurement, the 10 megaohm resistance is taken out of the internal circuit, and the reading has been calibrated for the less resisted current.

NOTE —

In the **Energy Diagnosis and Parasitic Drain** *chapter, we step through this process of detecting a parasitic drain in detail.*

Measuring Current (Amperage) Through a Device

Test Set-Up **5-F**

See page 5-14 for step-by-step procedure

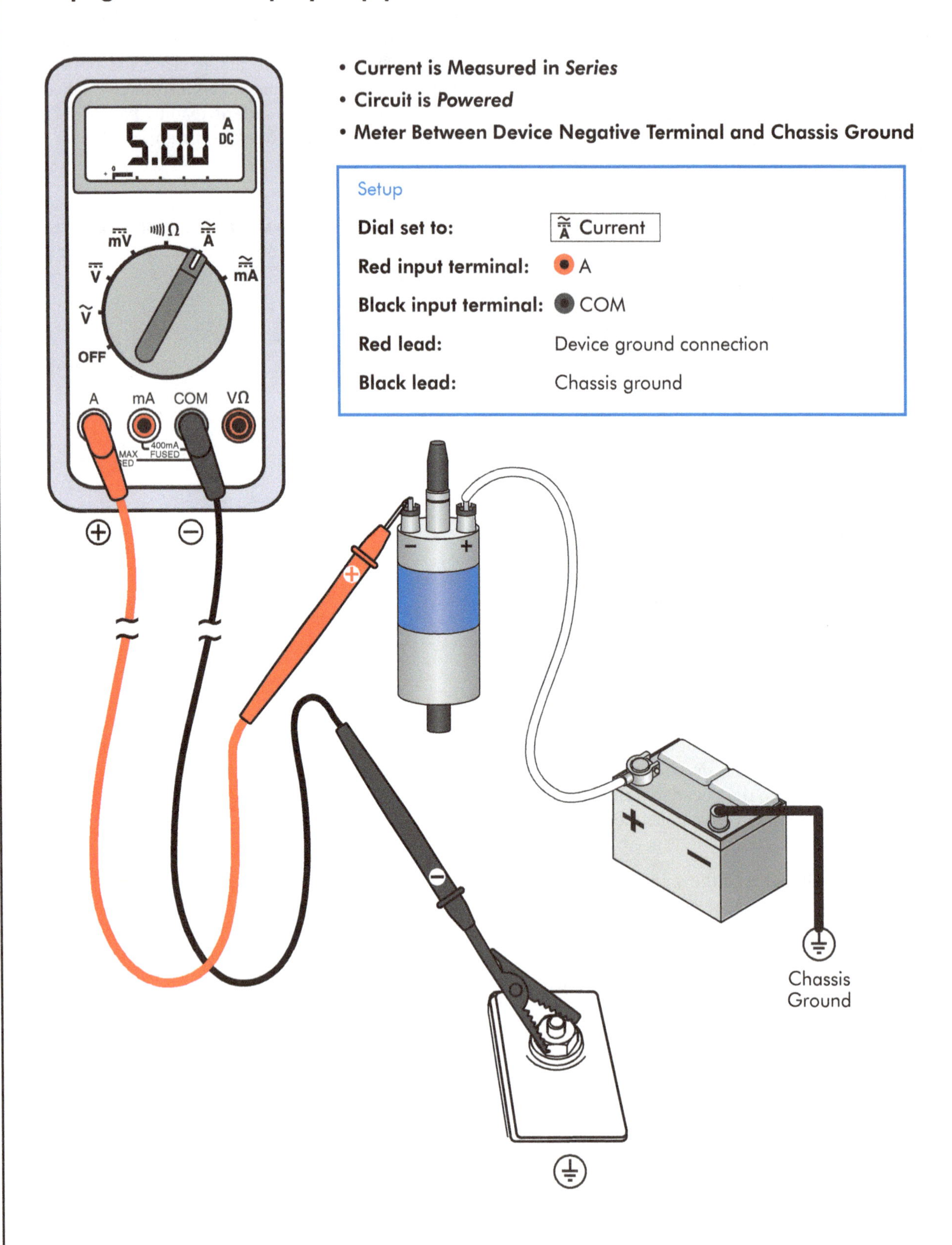

The DIN Numbering Standard

The DIN (Deutsches Institut für Normung)

The electrical systems in European cars have a similar look and feel, even across different brands, whereas the electrics in American or Japanese cars look very different. There are a couple of important reasons for this.

First, the electrical components in European cars are generally sourced from Bosch. The German company Robert Bosch GmbH has been manufacturing automotive electrical components since the late 1800s. Bosch is credited with developing and producing the first practical automotive magneto-based ignition system and, along with it, the first commercially viable high-voltage spark plug. By and large, European cars have Bosch electrical components, though the full uniformity of that claim has been diluted by the sourcing of components from other companies such as Siemens and Valeo. But it used to be that VW, Porsche, BMW, Mercedes and Audi all sourced their major electrical components—alternator, starter motor, ignition coil, distributor, etc.—from Bosch. This is one major reason why, if you move from under the hood of a Porsche 356 to a BMW 2002, you still feel surprisingly at home, even if you've gone from an air-cooled rear-engine to a water-cooled front-engine vehicle.

But the other part of the European electrical DNA, the commonality that underpins the comings and goings of electrons in European cars, is DIN 72552. The acronym "DIN" comes from *Deutsches Institut für Normung* (German Institute for Standardization). There are many DIN documents covering many standards. For example, having nothing to do with cars, we owe the letter-named paper sizes (A, B, C, and D) to a DIN standard. In the automotive stereo world, many people reference DIN 75490 without realizing it when they refer to the size of a radio as being "DIN" or "double-DIN." The round plugs used for old-style keyboards on PCs were commonly called "DIN" and "mini-DIN," described under DIN 41524.

However, it is the standards document named "DIN 72552" that is of primary interest to us. Officially published in 1971, "the DIN," as we simply refer to it, assigns a number to many commonly used automotive electrical terminals. The DIN standard is used by all German cars, all European cars we are aware of, and cars from other nationalities as well. It is incredibly powerful. For example, simply looking at the DIN, we see that "30" is "from battery + direct," and "15" is "battery + from ignition switch." Now, no matter what European car you look at, and what component you examine, if you see a terminal labeled "30," you know it should always be hot with voltage, and "15" should receive voltage when the ignition is on.

The DIN can be a lifesaver if you don't have a wiring diagram. For example, if the dimmer on the instrument cluster lights isn't working and you don't know which wire it is, you can consult the DIN and see the "panel light dimmer" is terminal 58d. You can then look on the back of the headlight switch or the instrument cluster and see if one of the terminals is labeled 58d.

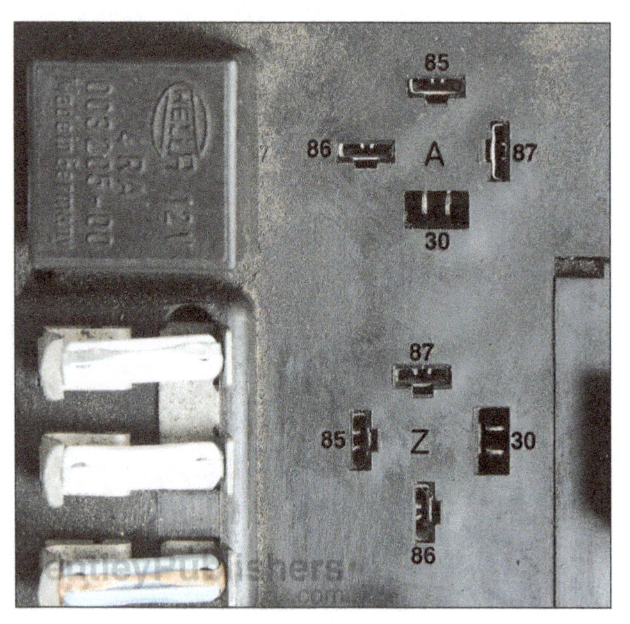

Figure 1 DIN standard numbering used on relays and other components is a major help in electrical troubleshooting.

As you'll learn in the **Switches and Relays** chapter, the fact that relays are numbered according to the DIN standard is so useful that it can elevate you to god-like status if you need to troubleshoot or bypass a relay. All of a sudden, unknown seemingly unrelated wires present a pattern and structure that makes them understandable, like driving past a grove of trees and viewing them from the angle that enables you to see that they're planted in rows.

Obviously, with the proliferation of electronics in cars, there are dozens of connectors that are new enough that they don't have DIN standard numbering. But when you do turn over a component and see the familiar DIN numbering, take advantage of it.

DIN Terminal Designations	
Contact	**Meaning**
Ignition System	
1	ignition coil, distributor, low voltage
1a, 1b	distributor with two separate circuits
2	breaker points magneto ignition
4	coil, distributor, high voltage
4a, 4b	distributor with two separate circuits, high voltage
7	terminal on ballast resistor, to distributor
15	battery+ from ignition switch
15a	from ballast resistor to coil and starter motor
Preheat (Diesel engines)	
15	preheat in
17	start
19	preheat (glow)
Starter	
50	starter control
Battery	
15	battery+ from ignition switch
30	from battery+ direct
30a	from 2nd battery and 12/24V relay
31	return to battery– or direct to ground
31a	return to battery– 12/24V relay
31b	return to battery– or ground through switch
31c	return to battery– 12/24V relay
Electric Motors	
32	return
33	main terminal (swap of 32 and 33 is possible)
33a	limit
33b	field
33f	2. slow rpm
33g	3. slow rpm
33h	4. slow rpm
33L	rotation left
33R	rotation right

DIN Terminal Designations	
Contact	**Meaning**
Turn indicators	
49	flasher unit in
49a	flasher unit out, indicator switch in
49b	out 2. flasher circuit
49c	out 3. flasher circuit
C	1st flasher indicator light
C2	2nd flasher indicator light
C3	3rd flasher indicator light
L	indicator lights left
R	indicator lights right
L54	lights out, left
R54	lights out, right
AC generator	
51	DC at rectifiers
51e	as 51, with choke coil
59	AC out, rectifier in, light switch
59a	charge, rotor out
64	generator control light
Generator, voltage regulator	
61	charge indicator (charge control light)
B+	battery+
B–	battery–
D+	dynamo/alternator diode+
D–	dynamo/alternator diode–
DF	dynamo field
DF1	dynamo field 1
DF2	dynamo field 2
U, V, W	AC three-phase terminals
Lights	
54	brake lights
55	fog light
56	spot light
56a	headlamp high beam and indicator light
56b	low beam
56d	signal flash
57	parking lights

DIN Terminal Designations	
Contact	**Meaning**
Lights (continued)	
57a	parking lights
57L	parking lights left
57R	parking lights right
58	license plate lights, instrument panel
58d	panel light dimmer
Window wiper/washer	
53	wiper motor + in
53a	limit stop +
53b	limit stop field
53c	washer pump
53e	stop field
53i	wiper motor with permanent magnet, third brush for high speed
Acoustic warning	
71	beeper in
71a	beeper out, low
71b	beeper out, high
72	hazard lights switch
85c	hazard sound on
Switches	
81	opener
81a	1 out
81b	2 out
82	lock in
82a	1st out
82b	2nd out
82z	1st in
82y	2nd in
83	multi position switch, in
83a	out position 1
83b	out position 2

DIN Terminal Designations	
Contact	**Meaning**
Relay	
85	relay coil–
86	relay coil+
Relay contacts	
87	common contact
87a	normally closed contact
87b	normally open contact
88	common contact 2
88a	normally closed contact 2
88b	normally open contact 2
Additional	
52	signal from trailer
54g	magnetic valves for trailer brakes
75	radio, cigarette lighter
77	door valves control

7

Switches and Relays

Switches

We first need to cover switches in order to understand relays. It'll seem like a weird detour, but trust me. You'll go *a-HA* and thank us. Really.

Most people are pretty familiar with the look, feel, and action of the different kinds of switches in cars. Traditional toggle switches, with the long thin baseball-bat-shaped stalks, are largely a thing of the past (though they're still used on MINI Coopers), and have largely been replaced by the push-on push-off variety exemplified by rear window defrosters. Rocker switches have become ubiquitous for controlling power windows. Rotary switches may be used for sound system and fan speed control.

Figure 1 Toggle switches on the dashboard of a 2008 MINI.

Momentary, Normally Open, and Normally Closed

Separate from the physical form factor is the question of what state the switch is "normally" in. Your horn is the classic "normally open" ("NO") switch. Press the switch and it momentarily connects two contacts to complete a circuit *(honk!)*. "Normally closed" ("NC") switches are less common, but you can find them for emergency stop functions in things like elevators and heavy machinery. Their normal state enables the running of the device, but you want to be able to hit the switch and kill power quickly. You can also envision an NC switch being useful on a high-security electronic door lock, where you want the lock to normally be in the "on" position, and where pushing the switch would momentarily cut the power and retract the lock.

Now, we just backed into use of the word "momentary," and it's important enough to formalize. Whenever you see "normally open" or "normally closed," this means that the switches are *momentary*, meaning they are in their normal state (open or closed) until physical action (pressing the switch) puts them into the other possible state (closed or open), and when the action ceases (releasing the switch), the switch returns to its normal state. But note that not all switches are momentary. You don't have to stand there and hold your wall light switch in the up position for the light to remain on. You'll occasionally see the words "maintained" and "latching" used to connote that a switch is not momentary, but this is not universal. For the most part, if a switch is not described as "momentary," it maintains its position without being held.

Poles, Throws, and All That

Next is the confusing sounding issue of poles and throws. It's actually pretty straightforward once you break it down. "Poles" refer to the number of separate circuits that pass through the switch. "Throws" refers to the number of paths (other than an open circuit) the switch can enable. There are four main variations you'll find in a car. See **Figure 2**.

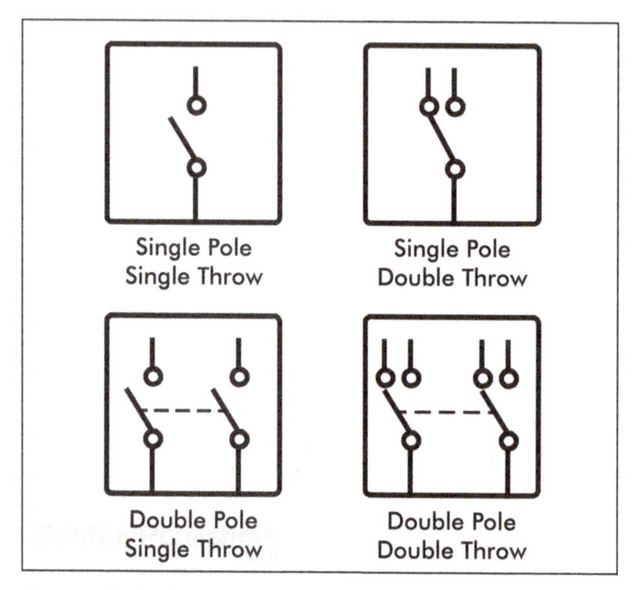

Figure 2 The four basic types of switches commonly found on a car.

SPST: Our horn switch is an example of a single pole single throw (SPST) switch, meaning it has one input (one pole) that, when closed, can take only one possible path (one throw), and flipping the switch merely connects and disconnects the one input from the one output. SPST switches are sometimes also referred to as "make or break" switches, since those are the only two possible conditions. A SPST switch may be a push-button, or a toggle, or even a rotary switch, but in all cases it will have two electrical terminals for connection. Its standard electrical symbol is depicted in the upper left corner of **Figure 2**. Note that if you were an electrical engineer drawing a diagram of a circuit that used a momentary SPST switch like a horn switch, you would use a different electrical symbol than this one. However, in the context of relays, this is the switch symbol used. If a switch is normally open, it is shown open as pictured. If it is normally closed, it is typically shown closed, often with the label "NC" on the diagram.

SPDT: The next level of complexity is the single pole double throw (SPDT) or "changeover" switch (upper right in **Figure 2**), where one input can follow one of two paths, but not both. Some SPDT switches such as used in a simple low beam / high beam circuit never shut the circuit off, and the switch merely selects between two circuit paths. But there are also SPDT switches with an off position, allowing neither path to receive power. The form factor of SPDT switches is typically toggle or rocker switches. They all have three terminals, one for the input and two for the possible output paths.

DPST: Double pole single throw (DPST) switches allow two independent circuits to be switched on and off simultaneously. They have four electrical contacts, providing input and output connections for both circuits. There's really no difference between a DPST switch and a pair of SPST switches. This is why the electrical symbol (**Figure 2** lower left) appears like a pair of SPST switches inside the same housing. DPST switches aren't used commonly in cars, but in household wiring they're employed because of the ability to simultaneously connect and disconnect the voltage and the neutral wires while keeping them isolated from each other.

DPDT: Lastly are double pole double throw (DPDT) switches, where each of two independent circuits are allowed to select between two different paths. Most power window switches are DPDT. Actually, to be more specific, most power window switches are DPDT momentary rocker switches.

Relays

Relays are the poster child for the usefulness of the DIN standard. You can take advantage of the fact that the terminals on most of the relays on your car are numbered according to the DIN, and thus you don't need to re-learn the numbering for each one. It's incredibly useful.

Until you understand them, relays seem like weird electrical entities that serve no useful purpose but to confuse amateur troubleshooters, but in reality they're very simple.

Open the hood of your car for a moment. You see the big fat cable going from your positive battery terminal directly to your starter motor? It's that size – usually two-gauge wire – because it needs to carry the hundreds of amps the starter may consume when you turn the key. So, think about it. Why isn't there a cable that size going all the way up to the ignition switch and back to the starter? (If you haven't actually checked, there isn't one.)

The answer is: *because of relays*. That starter draws a *lot* of current. You don't *want* all that current running through your ignition switch. You only want a cable going directly to the battery. So how do you turn the starter on and off? Imagine if, every time you wanted to start your car, you had to open your hood and close one of those big knife switches like out of a Frankenstein movie between your battery and starter, and then open the switch once the engine was running. It would be enough of a pain that, at some point, you might entertain using a pair of ropes to pull the switch closed and open. And then you would have the flash of insight and invent version 2.0 where, instead of ropes, you pull the switch closed using a small electromagnet, and install a spring that opens the switch if the electromagnet is shut off.

Enter the relay – an electrically actuated switch where a small amount of current is used to energize a little electromagnet which, in turn, closes and opens a set of electrical contacts through which a much larger amount of current can flow. See **Figure 3**.

Figure 3 The ubiquitous ISO mini relay with DIN-numbered terminals.

You can immediately appreciate why relays are useful for all sorts of high-current applications like, oh, say, running a starter motor: The high-current wiring can be kept short and close to the battery, and lower-current wiring can be used for the switch controlling the relay. When we get to the **Starter** chapter, you'll learn that its relay is a special kind called a solenoid, but don't worry about that right now.

A relay is just a switch where an electromagnet throws the switch remotely. Now that you understand switches, particularly the concepts of momentary, normally open, normally closed, SDST, and SPDT, you know almost everything you need to know in order to understand every relay used on your car.

The key things to remember are:

- The switch in a relay is a momentary switch with the same normally open and normally closed variants that a standalone switch can have.
- The switch in a relay has the same SPST and SPDT variants that a standalone switch can have. (There are DPST and DPDT relays, but they're far less common in cars than SPST and SPDT relays.)

There are two basic relay types – SPST 4-terminal and SPDT 5-terminal – that will comprise most of the relays in your car. There are a few variants of each of these, and a couple of oddballs that should make you exercise a bit of care should you need to go rummaging through your spare parts bucket, or accept a spare relay from a generous soul on the road. But understand these two basic types, and you will be Master or Mistress of Relays.

Physical Appearance of Relays

Nearly all of the relays on European cars built since the late 1970s have a very characteristic size and look. They are mechanical relays that use a small electromagnet to pull together a set of contacts. They're correctly referred to as ISO mini relays (that's the physical standard from the International Standards Organization [ISO], roughly a one-inch cube, actually 28mm x 28mm x 25.5mm). They're incorrectly referred to as Bosch relays (Bosch may have made the ones in your car, but they're also produced by any number of manufacturers), and informally referred to as "ice cube relays."

A relay's external case can be metal or plastic. They have four or five male quick-connect terminals on the bottom and, when used in an OEM application, usually mount in a molded plastic socket that has female counterpart for the four or five terminals. See **Figure 4**.

Figure 4 Older non-ISO relay (left) and ISO mini relay (right).

However, when relays are installed for a non-OEM application, the terminals on the bottom are often connected by wires that have individual female terminals crimped on them.

European cars built before the late 1970s often use relays with a larger form factor, more like 2" x 1", that do not sit in a socket and are instead fixed to a mounting face with a bracket and a screw. But the DIN labeling and the function of both the ISO mini relays and the older larger relays are the same.

As we keep saying, a relay is a device that uses a small low-current electromagnet to open and close a switch that's used to connect a high-current device. If you pop the plastic cover off an ISO mini relay, you can easily identify the electromagnet's coil and the small set of contact points. See **Figure 5**.

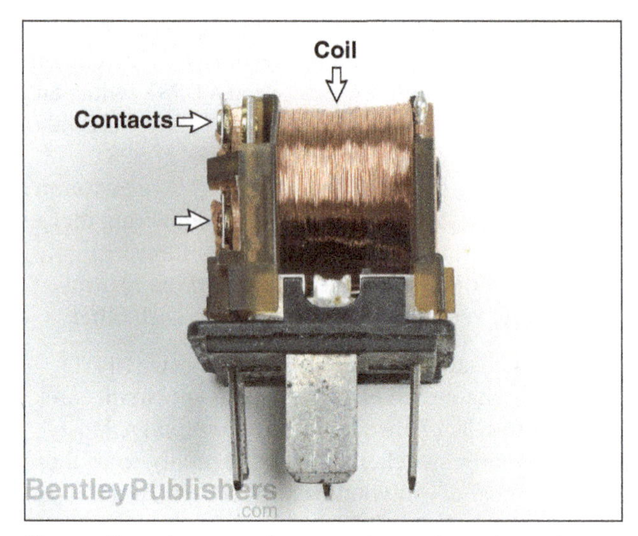

Figure 5 A relay is an electromechanical switching device. When power and ground are applied to the relay coil, the mechanical contacts complete (or open) a circuit.

Relay DIN Terminal Numbers

Here's what the DIN says. It's been annotated and the numbering has been re-ordered to make things clearer.

Table of DIN Relay Terminal Numbers		
Terminal	**Which Circuit?**	**Definition**
86	Low current (control)	Relay coil+
85	Low current (control)	Relay coil–
30	High current (load)	Line from battery positive terminal
87	High current (load)	Output, normally open
87a	High current (load)	Output, normally closed

Electrical Structure of Relays

Think of a relay as two separate circuits – the low-current control circuit (the electromagnet) and the high-current load circuit (the wiring to the device you actually want to turn on and off). Another way to think of them is as the "switcher" and the "switchee."

Terminals 86 and 85 are the + and – lines for the electromagnet (the "switcher" control circuit). When you apply a voltage to one of these terminals and ground the other, it energizes the electromagnet, which pulls the arm in the relay, which connects 30 to 87 (the "switchee" load circuit). See **Figure 6**.

That's really all it is. Well, almost all, and we'll get to that. If there's no diode (we'll get to that too), connections to terminals 86 and 85 can be reversed and it won't matter. Likewise, if it's a four terminal relay (yes, more on that later too), 87 and 30 can be reversed without consequence. But if the connections are confused between the low and high current sides, well, that's not good.

Figure 6 The DIN numbering of the terminals is present on the bottom of the relay.

Switching the Relay

What's missing in the DIN description is, to sound like Dr. Seuss for a moment, what switches the switcher. That is, something must complete the circuit through 85 and 86 to energize the electromagnet. Often that something is an old-fashioned mechanical switch, though it may be a control signal or a temperature sensor.

Let's start with a switch. Let's take the case of your car's horn. On the high current (load) side, relay terminal 30 goes to a connection to the battery, and relay terminal 87 is connected to the positive terminal of the horn itself. The negative terminal on the horn is grounded to the body of the car. On the low current (control) side, relay 86 goes to battery voltage, and relay terminal 85 goes to the horn button in your steering wheel, which is actually a switch connecting to ground. When you hit the horn button, that connects 85 to ground, completing the control circuit, causing current to flow from 86 through 85 to ground, energizing the electromagnet, pulling the contacts in the relay closed, and connecting 30 to 87, thereby supplying battery voltage to the horn. See **Figure 7**.

Note that, as pictured, the horn switch is on the ground side of the control circuit. This is the case with most cars.

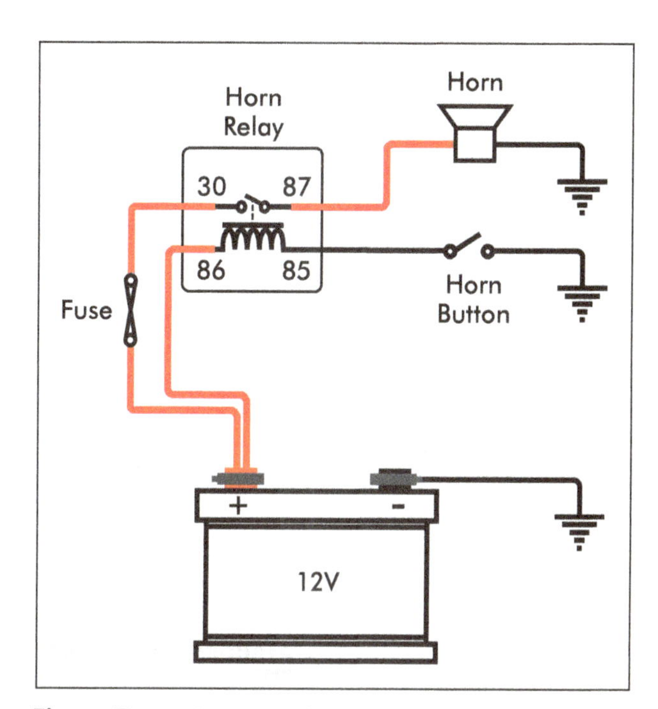

Figure 7 Sample wiring of a horn using a switch on the negative side of the relay.

With devices other than horns, though, it's more common for the switch to be on the hot (positive) side of the control circuit. Say you want to install a switch inside the car to manually turn on an auxiliary cooling fan, and you wisely want to use a relay to avoid having all the current from the fan going through the switch.

On the load (high current) side of the relay, 30 would go from the relay to a fused connection to the battery, and 87 would be connected to the fan motor's positive terminal. The negative side of the fan would be grounded to the body of the car. On the control (low current) side, 86 would receive battery voltage coming from the switch itself, and 85 would go to ground. Throw the switch, it connects 86 to 85 and energizes the electromagnet, connecting 30 to 87, supplying battery voltage to the fan, and making the fan turn. See **Figure 8**.

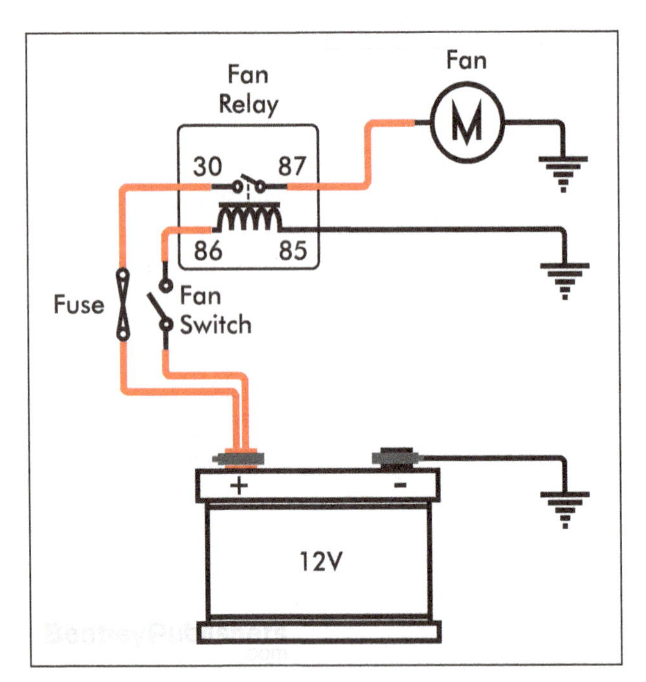

Figure 8 Sample wiring of a cooling fan on the positive side of the relay.

Note that in both of the above examples, we show a fuse between the relay and the battery on the high current side, and no fuse on the low current side. However, on existing circuits in a pre-1975 car, depending on the age and model, both sides, one side, or neither side of the circuit may be fused (it's actually pretty horrifying how much is unfused on a vintage car). But if you're wiring in a new relay-based circuit such as for our hypothetical fan, you should *always* fuse both sides of the circuit, because the high current side is, well, high current. A good practice is to wire it directly to the battery with a clearly visible fuse. But certain liberties can be taken with the low current side.

For example, because the current is low, it's not unreasonable to connect it to an existing fused circuit through either of the methods discussed in the **Adding a New Circuit** chapter.

Relay Current Consumption

As we know from the **Circuits** chapter, the higher the resistance of a device, the less current the device consumes. So the higher the resistance of the little coil of wire in the relay's electromagnet, the better. Though the resistance varies by manufacturer, mini ISO relays generally have a resistance of between 50 and 150 ohms.

Using Ohm's law in the form I = V/R, we can easily calculate the current. Assuming there's a nominal 14.2 volts of charging voltage present, the low estimate would be 14.2/150 or 0.095 amps, and the high estimate would be 14.2/50 or 0.28 amps. On the high current side, the load circuits of mini ISO relays are generally spec'd for up to 30 amps, but are available with higher amperage ratings. So, right there, you can see the utility of using a relay – it takes only a few tens of milliamps of current to switch tens of amps of current. It's clear why running a short thick wire with a fuse from a relay directly to the battery, and running a long thin wire from a relay to a switch inside the car, perhaps tied into an existing fused circuit, is so practical.

Type A and Type B Relays

ISO standards for mini relays appear to include the physical terminal arrangement labeled as Type B, where the pairs of control side and load side terminals are arranged opposite each other on the bottom of the relay (that is, the control side terminals 86 and 85 are opposite each other, and the load side terminals 30 and 87 are opposite each other). If you do a web search for "ISO mini relay," you will most often see the B terminal layout, which is appropriate if the layout is an actual standard.

However, there is another layout – Type A, in which terminals 86 and 30 are transposed. See **Figure 9**.

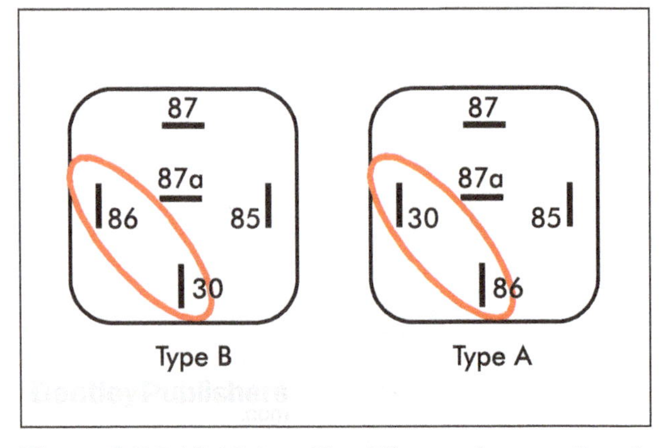

Figure 9 ISO Mini Relays. The difference between Type B and Type A relays is that 30 and 86 are reversed.

Upon looking into this Type A/B thing more, it's confusing and horribly documented. Even the terms Type A and Type B are not universal. The electronics manufacturer Hella lists relays in their catalog showing pictures of the layouts of each item (some A, some B) but with no descriptive nomenclature whatsoever. Apparently, unless you know the exact part number, you simply need to check the layout carefully.
In fact, knowing this, even if you read a part number off a bad relay to order its replacement, we'd *still* recommend you check the layout carefully.

What this means is this: The numbering of the pins is universal (it's prescribed by the DIN), but the arrangement of the pins is, apparently, not. So, in the section **How to Test a Relay** later in the chapter, when we describe *testing* the relays, the numbering is correct, but the actual terminal locations will need to be read directly off the bottom of your relay.

Reading the Relay Schematic

The relay schematic representation is usually printed or embossed on the side of the relay itself. There is a fair amount of variation in the way this schematic representation of the relay is drawn, but it should be possible to make out the electromagnet (it usually includes a coil of wire, sometimes also drawn with a bar, as we have done, depicting the electromagnet's iron core, but sometimes the electromagnet is just drawn as a little cylinder). The electromagnet is always the low current (control) side of the relay.

It also should be possible to make out the switch on the high current (load) side of the relay, and to tell whether it is a SPST or SPDT switch. As with stand-alone drawings of switches, the switches in relays should be drawn in their normal state. Since normally open switches are more prevalent, that is what you'll see most. Normally closed switches are sometimes explicitly labeled as NC and will have a terminal 87a.

Note that in the terminal layout illustrations:

- The numbering of the terminals on the relay is usually embossed directly into the relay, on the bottom, next to each terminal.
- The bottom of the relay is shown—not the socket it plugs into. The terminals in the socket will be the mirror image.

SPST Make-and-Break Relay

Most relays in your car are the SPST normally open variety. They have four terminals. (Some SPST relays may have five terminals with an additional terminal 87, which is internally connected to the adjacent terminal 87.) Putting a voltage across 86 and 85 energizes the electromagnet, which pulls the switch closed, which connects 30 to 87. See **Figure 10**.

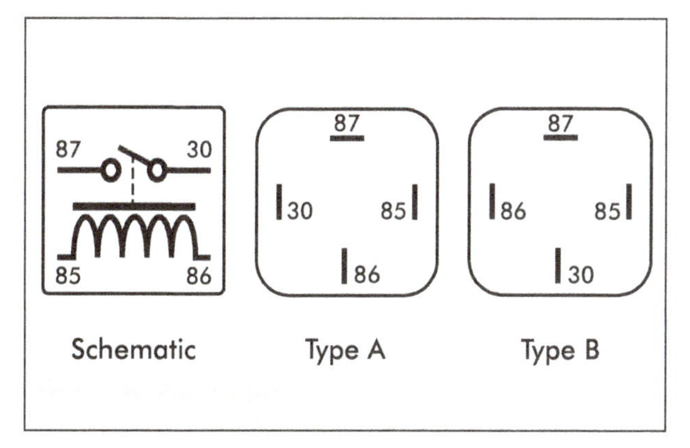

Figure 10 Four-terminal Single Pole Single Throw (SPST) normally open relay.

SPDT Changeover Relay

The second most commonly employed relay is the SPDT, changeover relay. Using a SPDT switch, terminal 87 functions exactly like it does in the SPST relay above. However, there's an additional terminal. Terminal 87a is the other terminal of the SPDT switch, and is normally closed, so when the electromagnet is not energized, 30 is connected to 87a. When the relay is energized, 87 connects to 30, and the 87a connection to 30 is broken. The lettering NC may or may not appear next to the 87a terminal on the wiring diagram and/or the diagram on the side of the relay. See **Figure 11**.

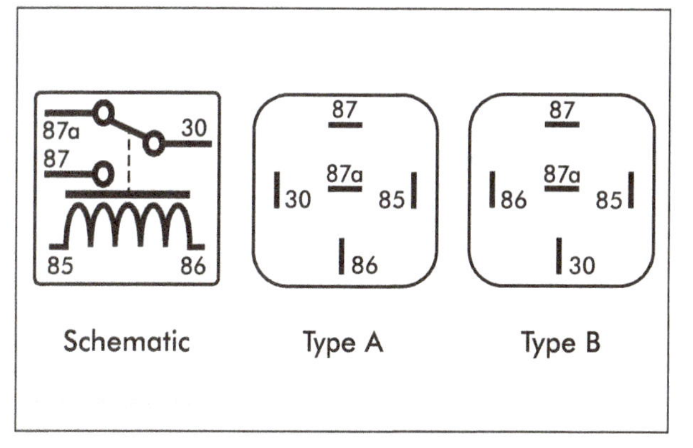

Figure 11 Five-terminal Single Pole Double Throw (SPDT) changeover relay (87 normally open, 87a normally closed).

Tales From the Hack Mechanic: Appreciating the Changeover Relay

I recently found an SPDT relay in a place that made me appreciate its use. This kind of relay is used when air-conditioning is installed in a BMW 2002 in a way that makes it so the standard blower fan and the A/C fan don't fight against each other. The wire that originally powered the blower fan is instead connected to relay terminal 87a. Since 87a is normally open, the blower fan has power, until the A/C is turned on and the relay is energized. When it is, power is switched from 87a to 87, which powers the A/C fan.

Often you will see 87 labeled in a DIN table as "input" (sometimes listed in other sources as "common"). This is highly misleading. Here's why. On an SPST relay, the wires to 87 and 30 can be transposed with no effect, since all the SPST switch does is make and break the connection between them. But on the SPDT relay in our example, there's one input (30) and two outputs (87 and 87a), so there is a clear input and output side. 30 is clearly the input side, with voltage then passed to either 87a (normally closed) or 87 (normally open). Seeing it like this, 30 is also clearly the "common" side, as it is common to both switch positions. Why the DIN lists this incorrectly (and we have checked in the Bosch Automotive Handbook itself), we do not know.

Not-So-Common Relays

SPST Normally Closed Relay

Sometimes you'll see a relay that has four terminals that, instead of being normally open, is normally closed. The giveaway is a center terminal labeled 87a (no 87), and the switch on the schematic being drawn as normally closed. See **Figure 12**.

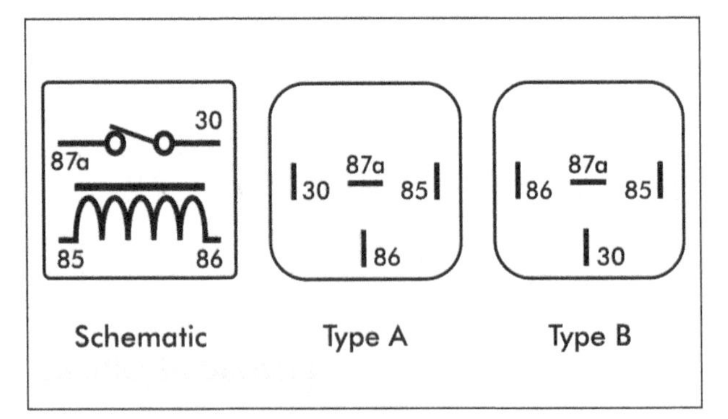

Figure 12 Single Pole Single Throw normally closed relay.

SPST Relay with Double Output

You will also see relays that have five terminals but are SPST relays with an extra connection on 87 (two terminals labeled 87). You can confirm this by checking for continuity between the two 87 terminals when the relay is unpowered.See **Figure 13**.

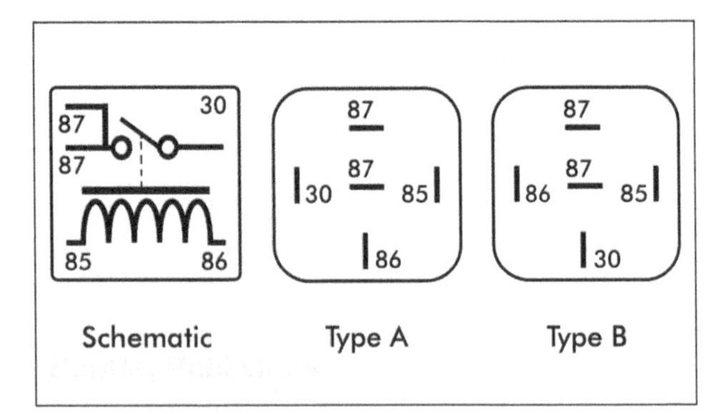

Figure 13 Five terminal Single Pole Single Throw normally open relay.

And Others

Simply going to the an on-line BMW catalog, selecting an E24 6 Series, and searching for "relay" reveals quite a variety of relay configurations in addition to those we've laid out above, including some with half-width terminals (see **Figure 14**). It also reveals a proliferation of resistor and diode symbols in the circuit diagrams. This requires scrutiny, below.

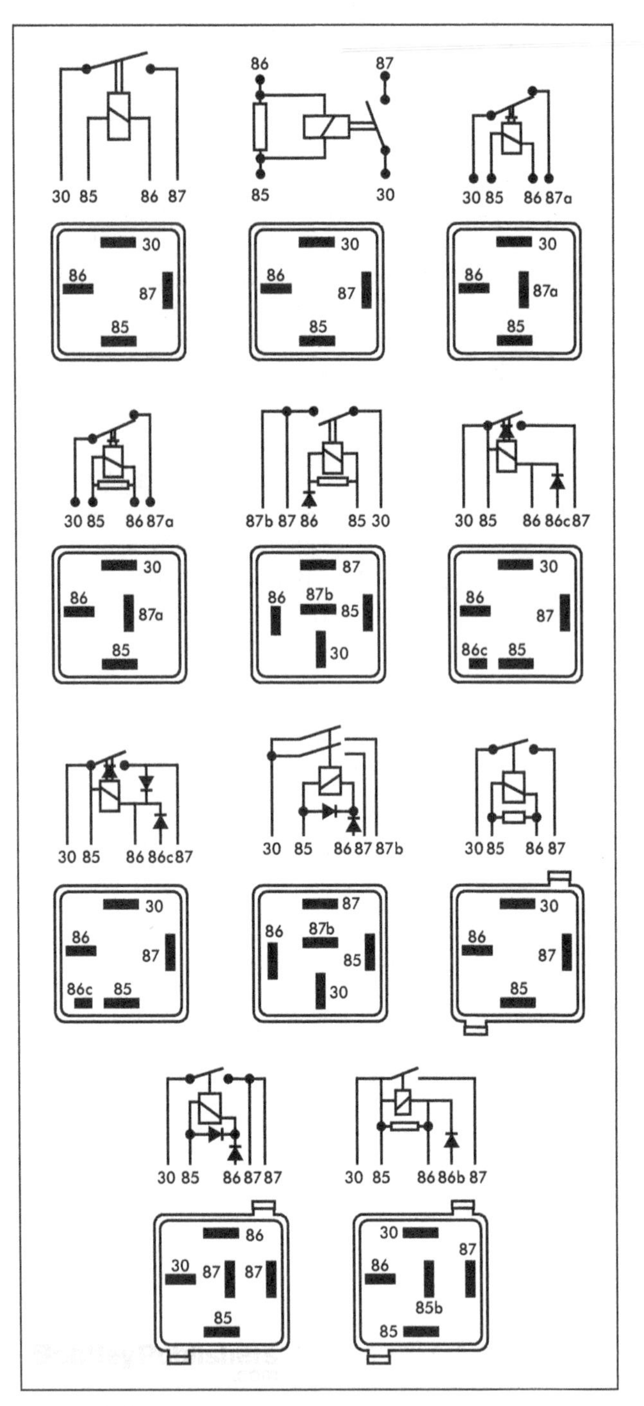

Figure 14 A sampling of relays used on a 1979 BMW 635CSi from the BMW parts catalog.

Resistors, Diodes, and Fuses in Relays

When you cut off the voltage to the coil of an electromagnet, it causes the magnetic field to collapse, which creates an electric field, which induces a short high-voltage surge in the coil windings. This applies not only to ignition coils, but to the coils in the relay's electromagnet as well. This momentary surge can cause a voltage spike, which can sometimes be heard as a pop if the radio is on.

While the pop is little more than a minor annoyance, the real concern is that the voltage spike can potentially harm sensitive electronics, particularly if they are located close to the relay. For this reason, many relays are fitted with either a diode or a resistor across the input and output of the coil. The diode is referred to by several names. It can be called a tamping, quenching, clamping, or flyback diode. The resistor is often called a "snubber" resistor. Both perform a similar function, which is to suppress the surge.

Diodes are electrical components that allow current to flow in only one direction. When used for surge suppression in a relay, a diode is often installed in parallel with the coil of the electromagnet, opposite to the coil's polarity, with the diode's anode (positive) end connected to the coil's negative connection (85), and the diode's cathode (negative) end connected to the coil's positive connection (86). In this way, while the coil is energized, no current will flow through the diode.

However, when the coil is de-energized, the collapsing magnetic field induces a voltage that's opposite in direction to the original voltage. Since it's opposite in direction, it can pass through the diode, which allows this voltage to go back into the winding of the coil and safely to ground. If the relay has a quenching diode or snubber resistor, it should be pictured on the mini schematic on the side of the relay. See **Figure 15**.

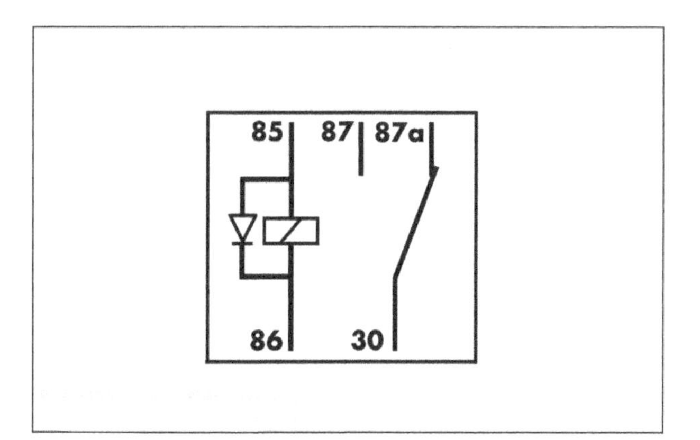

Figure 15 Relay with surge suppression diode. When voltage is applied to a relay coil, a magnetic field is created. When you remove the voltage the magnetic field collapses and creates a reverse polarity voltage surge. Adding a diode across the coil allows the surge to be dissipated in the resistance of the coil wiring.

Not all resistors and diodes are placed in parallel with the coil. Some are in series with one of the coil's lines. Some relays have two diodes, where one is in parallel and the other is in series with the coil. Some relays have a diode across the low and high current circuits. Some relays have both resistors and diodes. Ultimately, these are all noise reduction tricks, either to reduce the relay's noise on the other circuits, to reduce the effect of noise on other circuits on the relay, or both.

Lastly, there are relays with integrated blade fuses protecting the high current side. If a new circuit is being wired in, this allows one to get by without wiring in a separate fuse. In general, however, it's better practice to wire a new fuse directly to the battery, which makes it so the entire length of the new circuit wiring is protected.

Interchangeability of Relays

Because they all adhere (or should adhere) to the DIN standard, there is a high degree of interchangeability of ISO mini relays. If you're using a relay as part of a new circuit to turn on a power amp or a radiator fan in a pre-1975 car, there's very little risk in grabbing virtually any four-pin relay from your spare parts bin or the parts rack at an auto parts store. However, if you're replacing a bad relay in an electronics-laden car, care should be taken to be certain you're replacing it with one that's interchangeable.

- Be absolutely certain whether the original relay uses the Type A or B layout. Look at the numbering of the pins on the bottom of the original relay and compare them carefully to the new relay. Remember that the two layouts have 30 and 86 swapped relative to one another.
- The original relays under the hood of your car may be designed to be weathertight. Don't replace them with non-weathertight relays, even if the pinouts are equivalent.
- If you're replacing a five-terminal SPDT relay, be certain you're replacing it with another five-terminal SPDT relay and not a five-terminal SPST relay with a double output.
- If you need a four-terminal SPST relay and don't have one, but have a five-terminal SPDT relay, you can use it instead, as long as there's an open socket for the unused 87a NC terminal.
- If there's a quenching diode shown on the schematic of the relay you're replacing, you should replace it with another relay with a quenching diode.
- A relay with a quenching diode can replace one with a resistor. However, because the diode is sensitive to the polarity of the wiring of the coil whereas the resistor is not, terminals 86 and 85 must be wired as positive and negative, respectively. This should be the case on any original wiring on a European car, but with user-installed wiring, anything is possible.

- A relay with either a quenching diode or a resistor can replace one that originally had neither. In fact, it's generally preferable. And, given the choice, diodes provide better spike reduction than resistors. The same caveat as above on the polarity of 86 and 85 applies.
- If a relay has multiple quenching diodes, they're there for a reason. If a relay has gone bad and you've put in a non-quenched one to get home, replace it with the correct part as soon as possible.

How to Test a Relay

Now that you know everything about relays (and, really, you do), it's pretty obvious how to test one: Divide it into the low current (control) and high current (load) sides, and test both sides individually. Below we show the testing for a Type B relay where the terminal pairs for each side of the circuit are opposite one another. Remember that if your relay has the Type A layout instead, the terminal pairs for each side of the circuit are not opposite, but rather adjacent to one another.

Low Current (Control Side) Test

- Set your multimeter to measure resistance, and use it to check across the low current (control) side of the relay by measuring the resistance across terminals 86 and 85.
- It should measure between 50 and 150 ohms. See **Figure 16**.

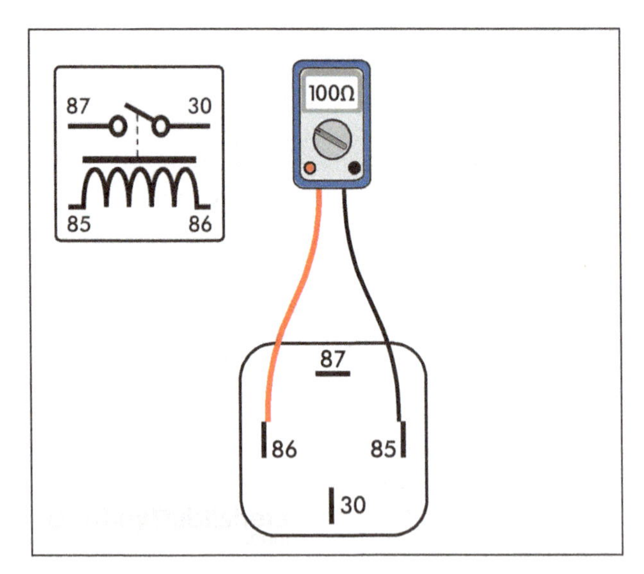

Figure 16 Resistance across terminals 86 and 85 should be between 50 and 150 ohms.

- If it shows no continuity, the windings in the electromagnet are likely broken.
- If it shows substantially less than 50 ohms, the electromagnet may be shorted.
- If the resistance is as specified, apply battery voltage to terminal 86 and connect terminal 85 to ground using a pair of test leads. (Note that, if the relay has a diode, powering it the wrong way will allow current to flow through the diode, which acts as a short circuit across the coil, which will pull too much current and blow the diode.) See **Figure 17**.

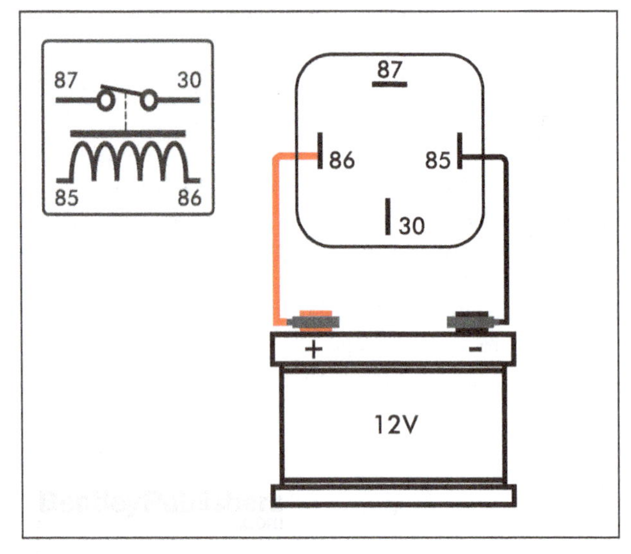

Figure 17 Applying power to 86 and ground to 85 should make the relay audibly click.

- When you energize the relay, you should hear and feel the relay click as the electromagnet pulls the relay contacts closed. If you hear and feel it click, the control side is working, and you can proceed to the next test. If you don't hear and feel it click, and you're certain you're testing it correctly, the relay is dead.

High Current (Load Side) Test

Without the relay energized:

- Use the multimeter to measure resistance across terminals 30 and 87. Since the relay is normally open, it should read as an open circuit – no continuity, "OL" on the multimeter. See **Figure 18**.
- If the relay is SPDT and has a pin 87a, measure the resistance across terminals 30 and 87a. Since 87a should be normally closed, you should detect continuity, meaning the multimeter should read well under an ohm. See **Figure 19**.
- Now, energize the relay (apply battery voltage to terminal 86, terminal 85 to ground) and measure the resistance across terminals 30 and 87. Since the switch should be closed, the multimeter should show continuity – resistance of well under an ohm. See **Figure 20**.

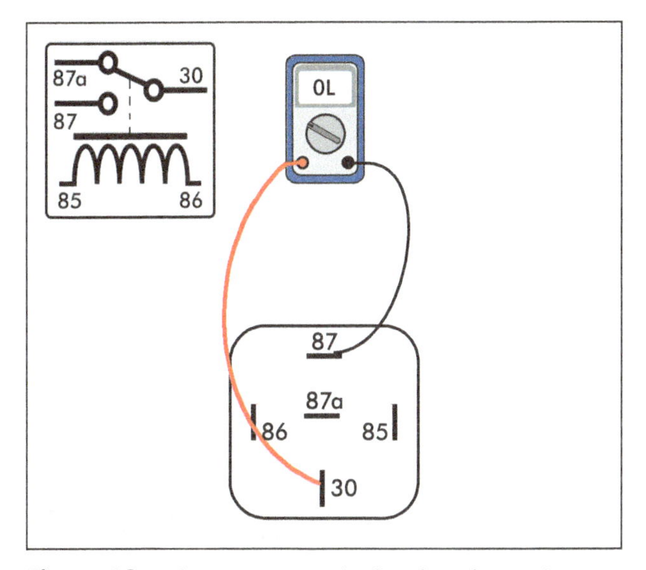

Figure 18 With no power applied to the relay, resistance across 30 and 87 should show an open circuit.

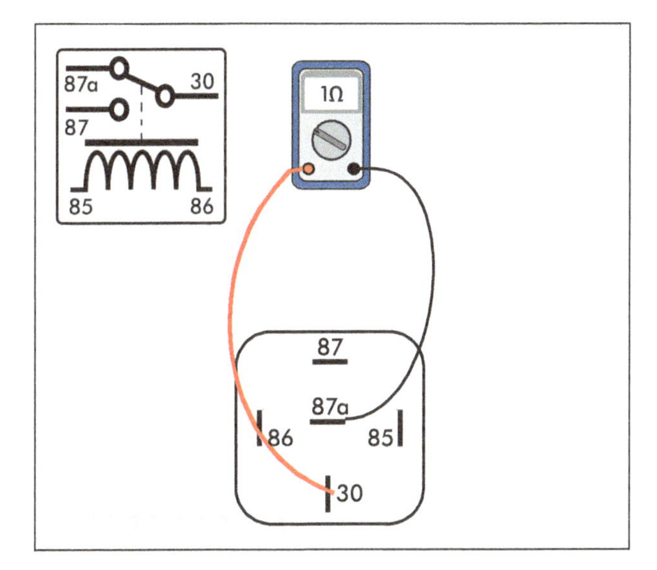

Figure 19 With no power applied to the relay, resistance across 30 and 87a should show continuity.

- If the relay is SPDT and has a pin 87a, measure the resistance across terminals 30 and 87a. Since 87a should now be open, it should read as an open circuit. See **Figure 21**.

Keep in mind that the relay contacts can become pitted with age, and that continuity across 87 and 30 isn't the same as the ability of the contacts to carry current (see the section on **Measuring Voltage Drop** in the **Common Multimeter Tests** chapter).

And that, my friends, is just about everything about testing relays. But you cannot go forth and walk on relay-related waters just yet.

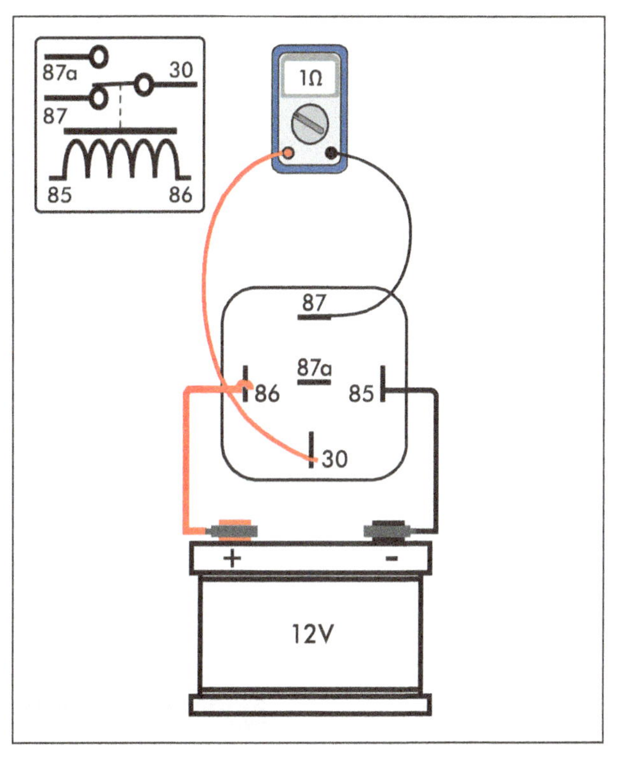

Figure 20 With the relay energized, measuring the resistance across 30 and 87 should show continuity.

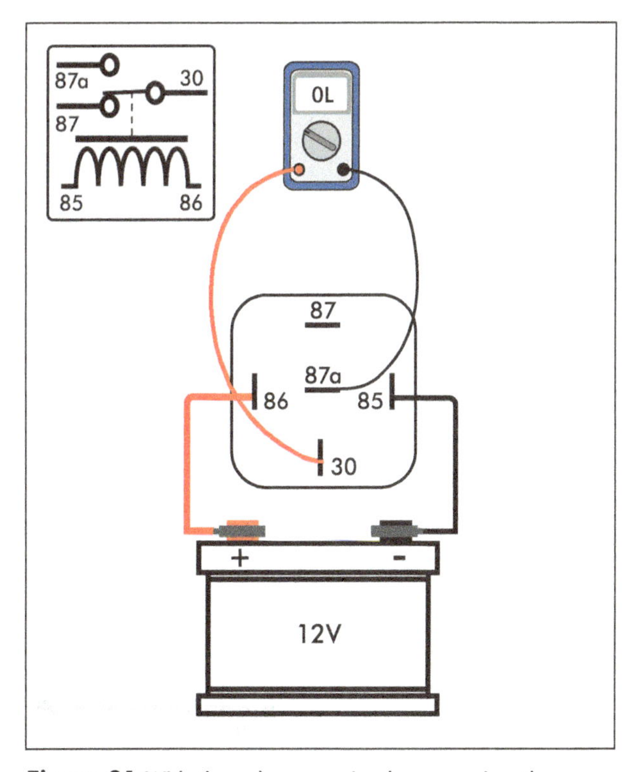

Figure 21 With the relay energized, measuring the resistance across 30 and 87a should show an open circuit.

How to Bypass a Relay

When you tested the relay above, you presumably determined if it was good or bad, and if it was bad, you presumably procured a good one. (If not, bad reader!)

But distinct from the relay itself is the wiring feeding the relay. The fact that all relays are similar means that all wiring feeding them is similar, and that the sockets that relays plug into can be tested in a similar way. Fortunately, the DIN makes it so that most relay wiring follows one of two configurations – it's usually either for the four-terminal SPST relay or the five-terminal SPDT relay.

Here, we're going to tell you how to bypass a relay, which requires an understanding of the socket the relay plugs into, which, in turn, puts a foot across the fuzzy line separating the relay from the circuit-specific wiring.

Let's reprint the DIN spec for relays so we can have the table at our fingertips. However, this time we're going to apply it not to the relay, but to the socket the relay plugs into.

DIN Specification for Relays		
Terminal	**Which Circuit?**	**Definition**
86	Low current (control)	Relay coil +
85	Low current (control)	Relay coil –
30	High current (load)	Line from battery positive terminal (fused) (actual "input" and "common")
87	High current (load)	Output, normally open
87a	High current (load)	Output, normally closed

Knowing the Socket Terminal Numbers

The terminal numbers are usually printed on the bottom of the relay, but you need to have them accessible *at the socket* to troubleshoot the circuit. If you're lucky, the numbers will be printed on the socket itself, but usually they're not. See **Figure 22**.

Fortunately, it's easy to figure out. Unplug the relay, hold it in your hand, turn it over, look at the bottom, find the terminal numbers, and orient the relay with terminal 87 at the top so it matches the pinouts we showed above. Note that this orientation should be invariant regardless of whether it's a Type A or a Type B relay. Now, take a piece of paper, draw the terminals that are on the bottom of the relay, with vertical and horizontal lines depicting the terminals, and write the terminal numbers, *mirrored left to right*, next to the lines, as shown in **Figure 23**. Note that the example shown is a Type A relay.

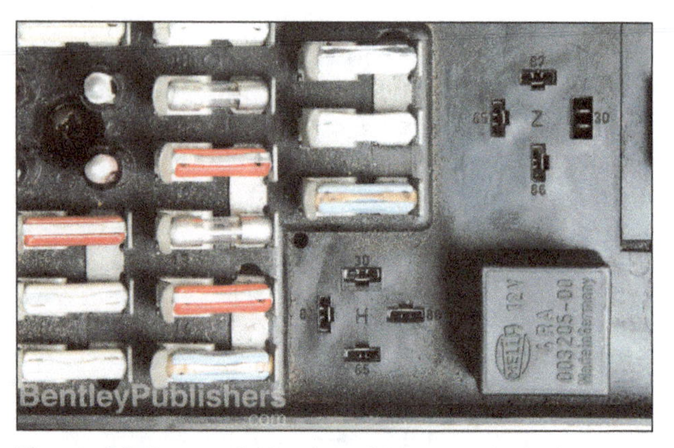

Figure 22 If you're lucky, the relay terminal numbers are printed on the relay socket, as they are for the horn and other relays of a 1979 BMW 635CSi.

Figure 23 Writing the mirror image of the relay terminal numbers on a piece of paper.

Next, check that you've gotten it right by placing the relay on top of the piece of paper and verifying that the numbers match. See **Figure 24**. If they do, you now have the terminal numbering for the socket the relay plugs into.

If the relay *doesn't* plug into a socket and is instead fed by individual wires and connectors, the steps are the same, except that you need to label the wires individually with tape, then pull them off the relay.

Just as we did when we troubleshot the relay, to troubleshoot the relay's wiring, we divide the circuit up into its low current (control) side and its high current (load) side. If the relay itself checks out as good but you don't hear the relay clicking, then there is very likely a problem on the control side.

Remember that there are "Type A" and "Type B" relays! In **Figure 23**, we show a Type A relay, but in all the circuit illustrations in this and other chapters, we show a Type B relay, where the terminal pairs for the low and high current sides are opposite each other. If the relay and its socket are Type A, the terminal pairs will be next to each other, not opposite each other. Check the numbering of the terminals on the bottom of the relay to be certain.

Figure 24 Laying the piece of paper with the relay socket terminal numbering next to the socket.

Using the Jumper Wire

When a relay is energized, the electromagnet pulls the switch closed, which, according to the DIN, sends power from terminal 30 to 87. The beauty is that, if there's voltage at terminal 30, *we can temporarily bypass the relay and connect 30 and 87 directly together with a jumper wire*. This is worth doing if for no other reason than it will permanently cement in your brain exactly how a relay in its socket works. See **Figure 25**.

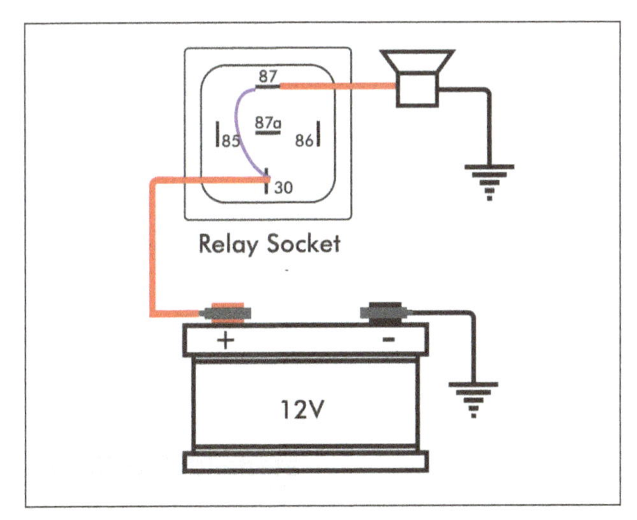

Figure 25 If terminal 30 has power, jumpering relay socket terminal 30 to 87 should sound the horn, run the fuel pump, or fire up whatever the relay socket is connected to.

So:

- First, verify that you have voltage at socket terminal 30. Depending on how the circuit is powered, you may need to turn the ignition key to the accessories or the ignition setting. If you never get voltage at terminal 30, that problem must be troubleshot before proceeding.
- Take a piece of 12 gauge or fatter wire that's about 6" long. Strip 1/4" off both ends. If you can take a pair of male quick disconnect terminals and crimp them to the ends of the wire and keep this in your electrical toolkit forevermore as your relay bypass wire, so much the better. If you crimp a fuse holder in the middle of it, you get a gold star.
- If you only need to test the circuit for a moment, stripped wire ends are fine, but don't try and jam them into the female connectors and leave them there. For anything other than a momentary on/off test, take a few minutes and crimp on the male connectors. You'll be glad you did.
- Connect one end of the jumper wire to relay socket terminal 30, the other to 87. The circuit should come alive. If it doesn't, the problem isn't the relay – there's some other problem in the wiring of the load side of the circuit. See **Figure 26**.

NOTE —

Be aware that as soon as you connect the jumper wire from 30 to 87, the circuit will come alive (if it's a horn relay, the horn will honk; if it's a fuel pump relay, the fuel pump will run).

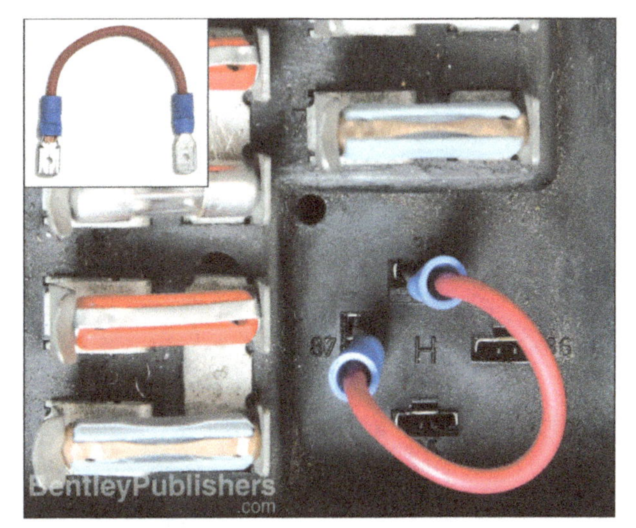

Figure 26 Jumper wire installed across terminal 30 and terminal 87 of a relay socket. Inset shows homemade relay jumper.

Now that you know how to do this, go to your car, look up where the fuel pump and auxiliary cooling fan relays are, pull them out, and jumper across each of them in this manner. Re-install the relay and put the jumper wire in the glove box. You now know how to do this when your car suddenly dies because the fuel pump has shut off or is running hot in traffic because the damned auxiliary cooling fan isn't turning on when it should.

Testing Non-DIN Relays

Most European relays follow the DIN standard, *most*, but not *all*. If they're ISO mini-relays, they nearly always follow the DIN standard, perhaps with some occasional special sauce. And there are relays that are not ISO minis that also follow the numbering standard (certain BMW fuel pump relays come to mind). There are, however, some outliers. For example, the terminal numbering of windshield wiper/washer relays and hazard relays in late 70s/early 80s BMWs bear no resemblance to the DIN numbers above. Technically, though, these are more than just relays, as they include the timers that control the intermittent wipers and flash the hazard lights, and they have their own DIN enumeration. However, the relay schematic is still usually printed on the side. Now that you understand the control side and the load side, you can look at the schematic and troubleshoot the relay in the same way.

Relay Testing Flowchart

Below is a flowchart that shows the connection between the basic steps we've discussed above. See **Figure 27**.

In words, what it says is:

- If the relay doesn't click, test it as per the procedure above. If it's bad, replace it.
- If the circuit still doesn't work, make sure there's voltage at terminal 30. If there's not, fix it.
- If the circuit still doesn't work, take the relay's control side out of the picture by jumpering across the relay.
- If the circuit still doesn't work, troubleshoot the load side.
- If the circuit still doesn't work, troubleshoot the control side.

In the next chapter, we'll step through an entire example with a horn.

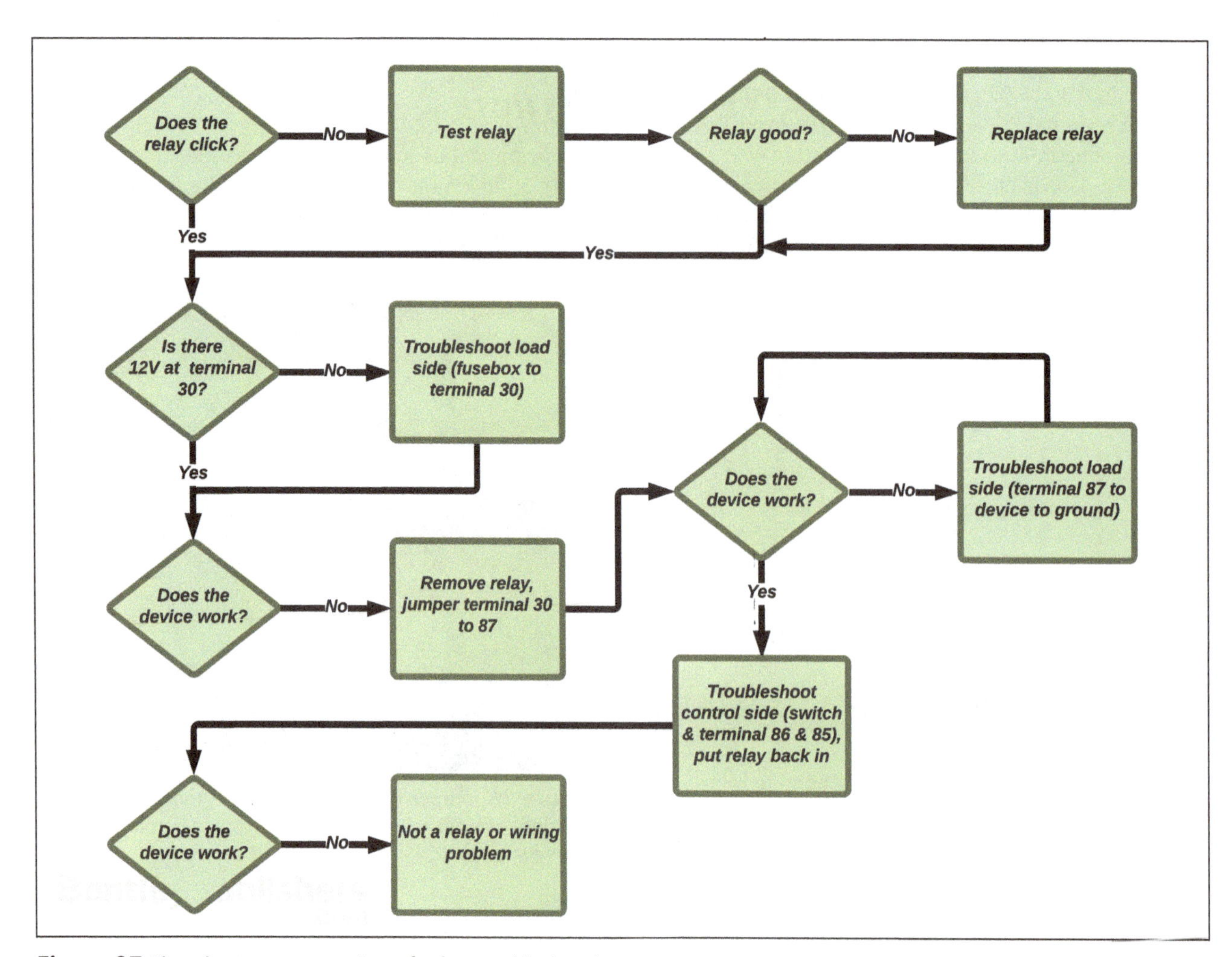

Figure 27 Flowchart representation of relay troubleshooting.

8

Troubleshooting

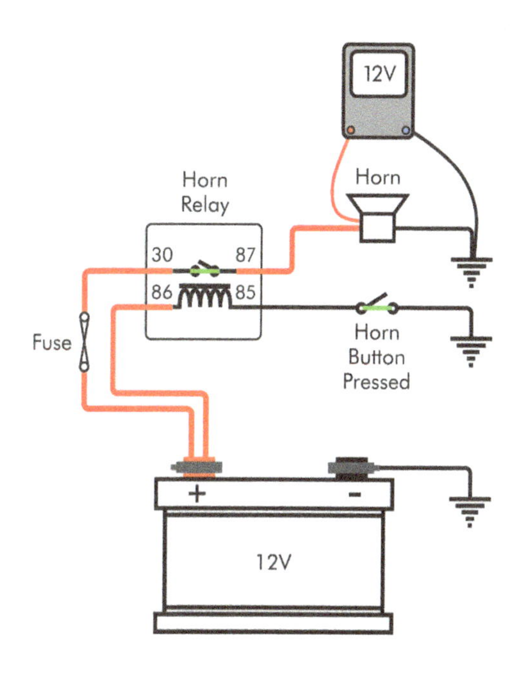

Circuit Troubleshooting

The image most people have of someone tracking down an electrical problem is a person hunched over an engine compartment with a wiring diagram spread out nearby and a multimeter in hand. It may well come to that, but that's not necessarily the best place to start.

In this chapter, we're going to take a simple circuit and methodically step through every inch of it as a troubleshooting example. The horn circuit has a load device (the horn), a switch (the horn button), a fuse, and a relay. It's a great set of ingredients for a first troubleshooting meal.

But before we sit down at that table, let's go through some general tips that apply to troubleshooting any device and circuit.

General Troubleshooting Tips

- **Remember the Five Kinds of Circuit Failures:** Recall from the **Circuits** chapter that there are five main kinds of circuit failures: Open circuits, shorts to ground, shorts to power, component failure, and malfunctions caused by voltage drops. At its most fundamental level, electrical troubleshooting is figuring out which one of the five you're chasing. Sometimes they cascade into each other. For example, a component may fail because it is internally shorted. This in turn will cause a fuse to blow, creating an open circuit.
- **Use Your Eyes:** If an open circuit isn't caused by a popped fuse, it's most likely caused by a detached connector or broken wire. Most short circuits are caused by wires that have had their insulation rub off and are touching ground (referred to as a short to ground). A thorough, inch-by-inch inspection along the wiring path will often reveal the cause.
- **Use a Multimeter:** Looking at the horn example that follows, you'll see that you can make it through step 2 (checking the fuse) and step 3 (wiring the device directly to the battery) without using a multimeter, but after that, the multimeter is the go-to tool. It's is so important that we gave it two chapters of its own in this book. If you don't own a digital multimeter, read those chapters, then go and buy one and become practiced in verifying the presence or absence of voltage, and the presence or absence of continuity.
- **Wiring Diagrams:** Learning how to read wiring diagrams takes some practice, but once you have the secret decoder ring, they can be very useful for troubleshooting circuit problems. Wiring diagrams are available in high-quality repair manuals such as the Bentley manuals. They are also usually available through the manufacturer's web site via a subscription-based fee service intended to dispense technical repair information to professionals.
- **Rule Out the Fuse:** You'll feel silly if you spend hours chasing an electrical problem and find out the root cause is a blown fuse. The newer the car, the more places fuses can be stashed (the BMW E39 5 Series, for example, has six fuse locations). Consulting the owner's manual, a repair manual, or an enthusiast web site will usually reveal the correct fuse location. Of course, if a fuse is blown, you'll need to find out why it blew, and we'll get to that, but one thing at a time. So, check the horn fuse. If it's an old-style torpedo fuse, twist it around in the fuse holder just to make sure the problem isn't corrosion around the edges of the fuse holder.
- **Find Everything on the Fuse or Circuit That Does Not Work:** If one device isn't working, check carefully to determine if other devices on the same fuse or circuit are not working as well. Is one tail light out or are both of them out? If the instrument cluster lights aren't working, are the other dashboard lights that are tied to the dimmer switch also not working? Such clues can help lead you to a common fault, such as the connector on the back of the fuse box itself being loose.
- **Can You Swap Sides?** Components that come in pairs – headlights, for example – offer a unique troubleshooting advantage in that if both quit working, odds are it's the wiring, but if one quits working, you can swap the components and see if the problem stays with the component or with the wiring. For example, most cars have two horns. Do both of them not work? If only one does, you can swap them left to right. If the problem stays with the horn (e.g., horn #1 doesn't work whether it's on the left or right side), it's likely a bad horn, but if the problem stays with the wiring (e.g., horn #1 doesn't work on the left side, but does work on the right), it's likely a wiring problem on the left side.
- **Check for Voltage on the Positive and Negative Side:** If the problem is an open circuit, voltage should be present right up to the point where the circuit is open, even on what is normally the negative side of the circuit. For example, if the ground wire to your horn has broken, there is no path to ground, so there is no current, so the voltage never is "dropped" by the horn like it normally would be, so measuring the voltage between the ground side of the horn and ground would show battery voltage.
- **Don't Forget that Voltage Sag is Normal:** If the engine is off and the alternator isn't charging the battery, it is perfectly normal to see less than 12.6V on a circuit when you power it. You may well see less than 12V. As we described in the **Circuits** chapter, the "sagging" of the voltage is due to the fact that the battery is a finite source of power. In fact, if you see exactly battery voltage on a circuit, it almost always means that there isn't any current flowing in the circuit.

- **Make Sure the Circuit Is Supposed to Be Powered:** Whatever thing you're troubleshooting that should be powered and isn't, make sure it's actually *supposed* to be powered. For example, if the horn doesn't honk when the key is turned to "accessories," maybe that's how it's supposed to work. Maybe you need to turn the key to "ignition" for it to work. Been there, burned daylight on that.

Lastly, note that, when we say, for example, "use the multimeter to measure voltage," the procedure for doing so can be found in the chapter **Common Multimeter Tests.**

The Sample Horn Circuit

Let's say your horn won't work (as opposed to not shutting off, which would be a different problem). **Figure 1** shows a simple diagram of a basic horn circuit showing power, the switch, the relay, the horn, and the wiring.

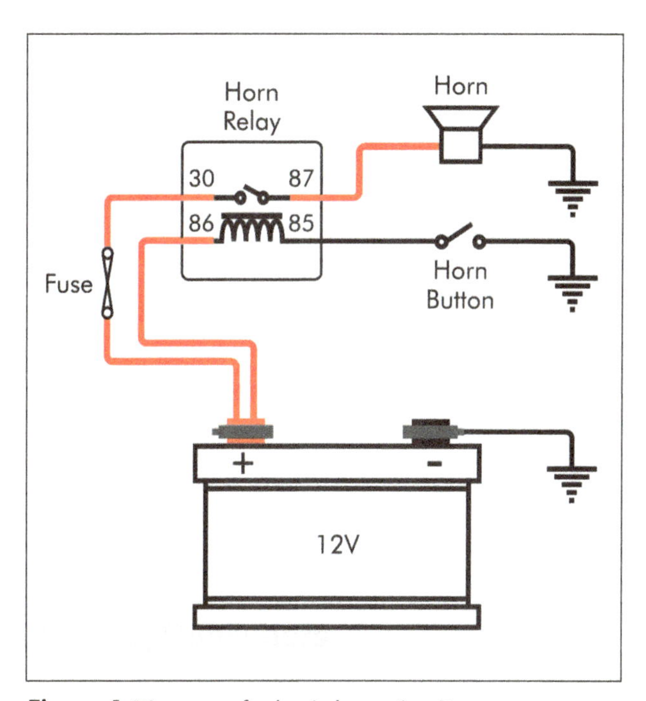

Figure 1 Diagram of a basic horn circuit.

As we discussed in the **Switches and Relays** chapter, most circuits with relays have the switch on the positive side of the control section of the circuit, but because the horn button grounds to the body of the car, the switch—the horn button—is on the negative side of the control circuit. Here's what happens when you press the horn button:

- The relay's control circuit is completed.
- Current flows from the battery, to 86 to 85, to ground, energizing the electromagnet.
- The electromagnet pulls the relay's internal contacts closed, completing the load circuit.
- Current flows from the battery, to 30 to 87, to the horn and then to ground. BEEP!

As we step through troubleshooting the horn circuit, we'll highlight the things that are specific to it because the switch is on the negative side, and comment on how the troubleshooting would be different on other positive-switched relay circuits.

Checking for Component Failure

Whether you're troubleshooting a horn that has gone silent or a cooling fan that stopped turning, you'll want to first know: Is the actual piece of hardware – the load device itself – dead? Because if it is, there's little reason to spend time testing the circuit wiring. The best, fastest way to do this is to temporarily wire the damned thing directly to the battery and see if it comes to life.

It is incredibly useful to have a twenty-foot-long set of 12-gauge red and black wires with alligator clips on both ends and a fuse holder spliced inline on the red lead. You can buy these or make them yourself. Then you can use them to connect things directly to the battery. This completely eliminates any of the car's fuses, relays, wiring, and ground connections, and tests the functionality of the device all by its lonesome self.

CAUTION —

One caveat to directly powering a component: Be certain the component you are checking is indeed powered by 12.6 volts. Most are, but not all are. Some components may use less, or even more than battery voltage. For example late model direct injection fuel injector voltage can range from 30 volts up to 120 volts, depending on the system.

So, to bypass everything else and directly test the horn in this fashion, as shown in **Figure 2**:

- Undo the existing connections to the horn (usually two spade terminals).
- Connect the red and black clip leads to the horn's positive and negative terminals, respectively.
- Verify the clip leads aren't touching each other or ground.
- Put an appropriately sized fuse in the fuse holder (8 amps should be plenty for the horn).
- Clip the positive lead to the positive battery terminal.
- Touch the negative lead to the negative battery terminal.

If the horn doesn't blare, or if the fuse immediately pops, you need to buy yourself a horn. But if, when wired directly to power, the horn works, then there's a wiring issue, meaning either the device isn't being supplied with power, or the ground is bad, or it's the relay, or the switch, or it's a voltage drop problem.

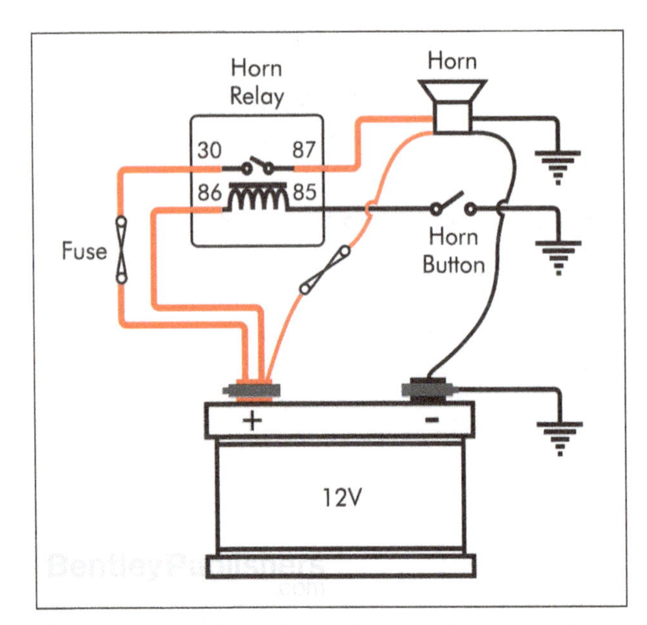

Figure 2 Test a device (here, the horn) by temporarily connecting it directly to the battery using a fused set of test wires.

Now, you can't always easily do the "wire it directly to the battery" trick. And not everything announces its life and functionality as clearly and unambiguously as a blaring horn or a spinning fan. But if you can take advantage of this, it can be a huge time saver.

Checking for Ground at the Connector

It makes sense to check the ground connection next, because a power problem can be due to several things, but a ground problem typically can only be a bad ground connection. See **Figure 3**. With the circuit off (that is, don't try and honk the horn):

- Set the multimeter to measure resistance.
- Touch one lead of the multimeter to the negative terminal of the horn, and the other lead to a convenient ground on the body of the car.
- There should be continuity, meaning the multimeter should read well under an ohm. If the multimeter shows no continuity (if it reads "OL" indicating an open circuit), the horn wiring has a bad ground.

As we said in the section above called **General Trouble-shooting Tips**, if there's a bad ground connection, the device will not have dropped the voltage across it, so there will be voltage on the device's ground terminal.

Thus, the other way to test this is:

- Set the multimeter to measure voltage.
- Touch one lead of the multimeter to the negative terminal of the horn, and the other lead to a convenient ground on the body of the car.

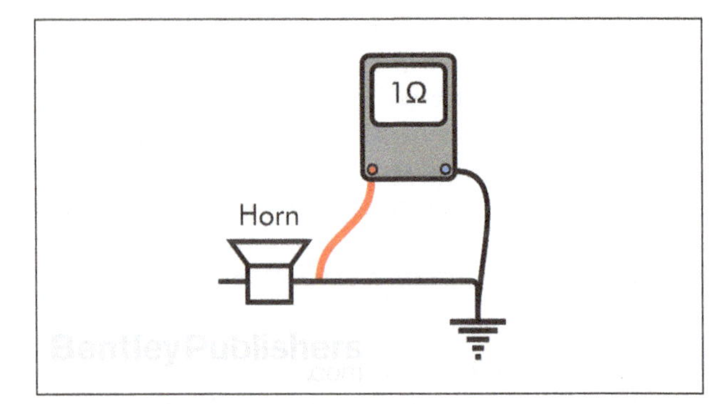

Figure 3 There should be continuity between the negative side of the connector and ground.

- Turn on the ignition and hit the horn button.
- If the multimeter shows battery voltage at the horn's negative terminal, the horn has a bad ground.

Ground connections are usually small ring terminals. Often, unscrewing the connection, cleaning corrosion off the ring terminal with a file, and retightening it fixes the problem. Or you may find, particularly on a vintage car, that with age and heat, the ground wire has become brittle and broken off from the connector. In extreme cases with brutally rusty cars, the metal around the ground attachment point may be completely rotted and broken away from the body of the car, making it so there is no ground return path. In any case, trace the negative wire's path to ground, fix it if it's bad, and retest.

Checking for Power at the Connector

If the ground is good, next check if there is power at the positive connector. See **Figure 4**.

- Set the multimeter to measure voltage.
- Connect the red lead of the multimeter to the positive wire going into the horn.
- Connect the black lead to any convenient ground on the body of the car. With the horn unpowered, the multimeter should read zero.
- Then put the multimeter where you can see it from inside the car and turn the circuit on, meaning, in this case, turn the key to ignition and hit the horn button. This will energize the relay.
- The multimeter should read battery voltage. But if the multimeter instead reads zero, then the horn isn't getting power, and you'll need to trace the wiring to see where the failure is.

If there is power at the connector but the horn doesn't sound, and you've verified the horn does sound when wired directly to the battery, either there could be a voltage drop issue, which we'll get to last, or there could simply be a mating problem at the connector.

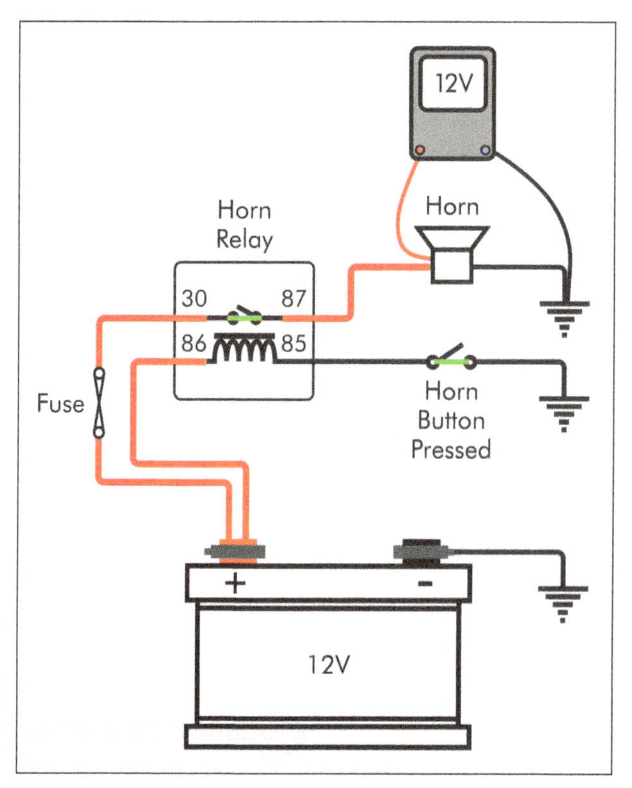

Figure 4 Check for power at the positive connector.

On older cars, as the plastic in connectors deteriorates, the pins or terminals in the connectors can sometimes get loose, making it so the male and female halves aren't coupling. It's common to find that one of the terminals isn't being held in place and is pushing out the back of the connector. I recently had an E36 3 Series BMW where this was the case (I'd installed a new cooling fan and it only turned on intermittently because of a loose terminal in the plastic connector housing).

Troubleshooting the Relay

If there's no power at the horn connector and you're not intimately familiar with the car, to continue troubleshooting you probably will need to consult a wiring diagram, or at least develop an understanding of the components and how they're connected. With our horn example, we saw in **Figure 1** that the horn is controlled by a relay that is switched on the ground side. So troubleshooting "no power at horn positive connector" leads inexorably to troubleshooting the relay and its wiring.

In the **Switches and Relays** chapter, we presented the following basic approach to troubleshooting a relay and its related wiring. Consult that chapter for more detail, including a flowchart, but the basic approach is:

- If the relay doesn't click, test it (see testing procedure in **Switches and Relay** chapter). If testing shows the relay is bad, replace it.
- If the circuit still doesn't work, make sure there's voltage at terminal 30. If there's not, fix it.
- If the circuit still doesn't work, take the relay's control side out of the picture by jumpering across the relay socket (30 to 87).
- If the circuit still doesn't work, troubleshoot the load side (87 to device).
- Put the relay back in. If the circuit still doesn't work, troubleshoot the control side.

Because testing the relay itself and jumpering across the relay are the same for every relay, these topics were covered in the **Switches and Relays** chapter, but because the other steps are more circuit-specific, we will cover them in detail here.

First, as always, let's remind ourselves of the DIN relay numbering in the table below.

DIN Specification for Relays

Terminal	Which Circuit?	Definition
86	Low current (control)	Relay coil +
85	Low current (control)	Relay coil –
30	High current (load)	Line from battery positive terminal (fused) (actual "input" and "common")
87	High current (load)	Output, normally open
87a	High current (load)	Output, normally closed

Power Coming Into the Relay

Remember that all a relay is doing is switching together the two terminals on the load side (30 to 87), which supplies voltage to 87 to turn on whatever it is you want to turn on. The DIN says that terminal 30 is the relay input terminal. It should have battery voltage, even without the horn button being pressed. If it doesn't, nothing is going to work. Referring to **Figure 5**:

- Set the multimeter to measure voltage.
- Do what is needed to energize the circuit (e.g., if the horn only works with the key turned to ignition, turn the key to ignition).
- Put the positive multimeter probe into terminal 30 of the relay socket.
- Touch the negative probe to any convenient ground.

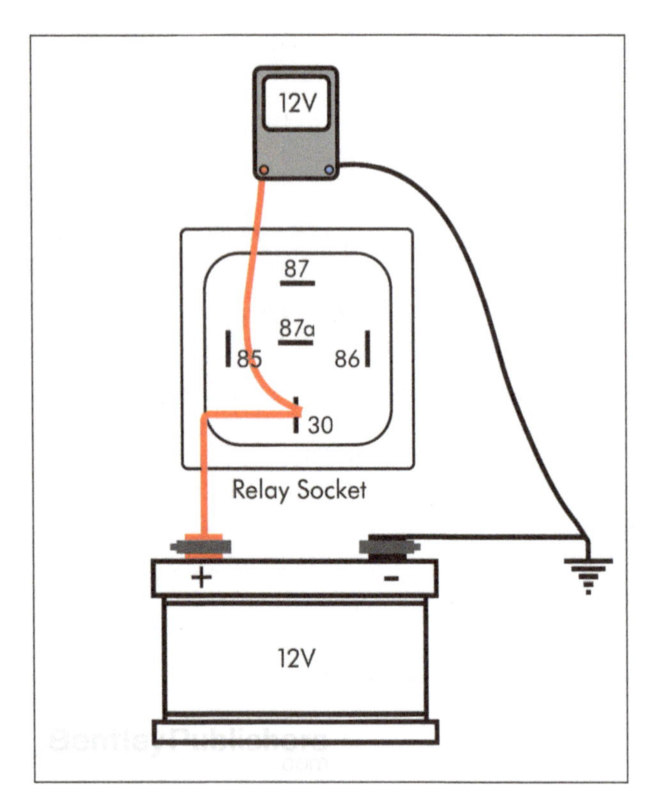

Figure 5 There should be battery voltage at relay socket terminal 30.

- If the multimeter doesn't register battery voltage, you'll need to trace the wire feeding terminal 30 back to the fuse box and find out why it isn't live.

Absence of voltage at terminal 30 could be caused by:

- The connector having been pushed through the bottom of the relay socket.
- The wire being broken off the bottom of the connector.
- The wire being broken in the middle.
- The other end of the wire being disconnected from the fuse box.
- The failure of some other component between the fuse box and terminal 30.

You'll need to trace the wires, identify the problem, and fix it.

Power Leaving the Relay to the Horn

We verified there was power coming into relay socket terminal 30, and if there wasn't, fixed it so there was. Now, as per the instructions in the **Switches and Relays** chapter, we will jumper terminal 30 and terminal 87. Next, redo the Check for Power at the Connector to check for voltage at the horn's positive connector. See **Figure 4**. If there's still no voltage there, it means voltage is going into terminal 87 but not making it out the other end at the horn. Check the continuity of the wiring between these two points as follows. See **Figure 6**.

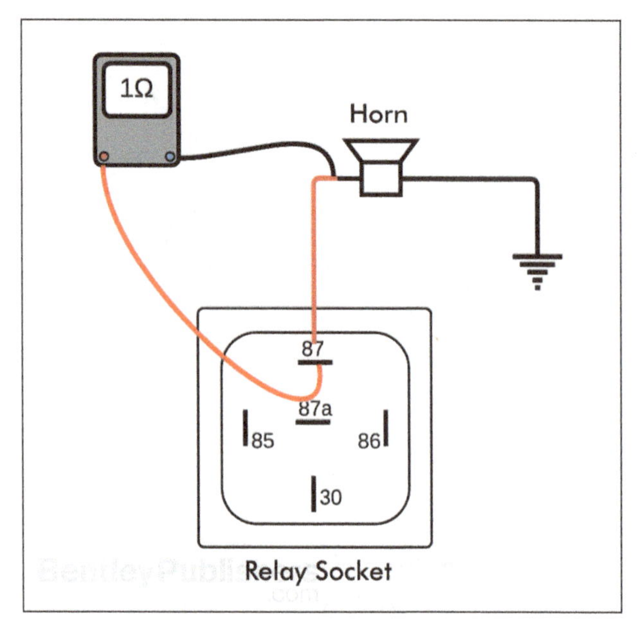

Figure 6 There should be continuity between 87 and the horn's positive terminal.

- Set the multimeter to measure resistance.
- Touch one lead of the multimeter to relay socket terminal 87, and the other lead to the positive wire coming into the horn.
- There should be continuity, meaning the multimeter should read well under an ohm. If the multimeter shows no continuity (if it reads "OL" indicating an open circuit), you'll need to trace the wire from 87 to the horn's positive terminal to figure out why there isn't continuity.

Lack of continuity between these two points could be caused by:

- The connector having been pushed through the bottom of the relay socket.
- The wire being broken off the bottom of the connector.
- The wire being broken in the middle.
- The other end of the wire being disconnected from the horn's positive terminal.
- The failure of some other component between the relay socket and the horn.

You'll need to trace the wires, find the problem, and fix it.

Troubleshooting the Relay Control Circuit

If we've gotten the horn to honk by bypassing the relay with a jumper, and if we've verified that the relay itself is good, then the problem must be in the horn control circuit. The DIN tells us that, to energize the relay's electromagnet, there needs to be voltage across pins 86 and 85. So, as shown in **Figure 7**:

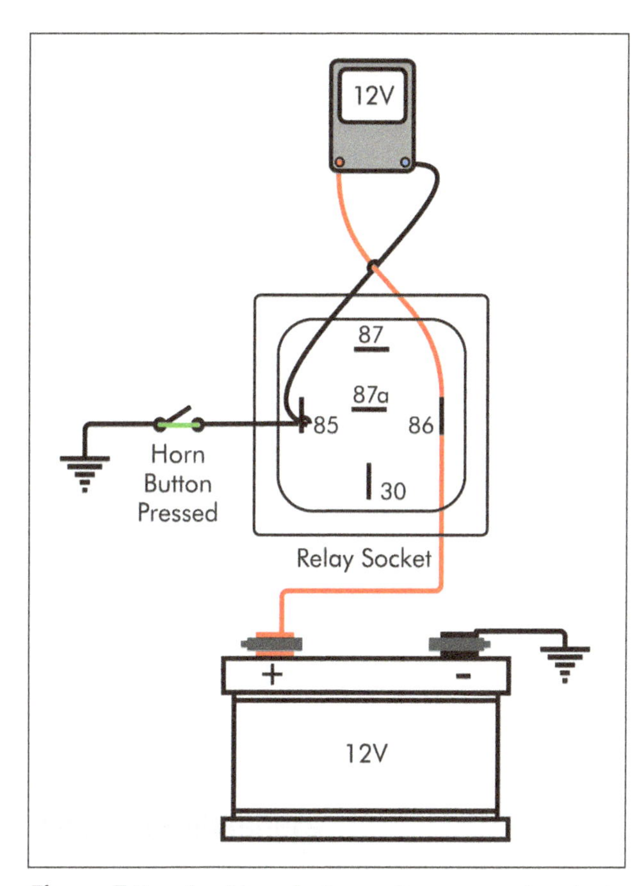

Figure 7 You should see battery voltage across terminals 86 and 85 in the relay's socket when the circuit is trying to energize the relay (in the case of a horn circuit, when the horn button is pressed).

- Set the multimeter to measure voltage.
- Put the positive multimeter probe lead into relay socket terminal 86, and the negative probe lead into 85.
- Turn the ignition on and engage whatever switch you would to energize the relay (for example, if testing the horn relay, hit the horn button).
- If the multimeter doesn't register battery voltage, then there's not voltage across 86 and 85 to energize the relay, and you'll need to figure out why.

If there's no voltage, the problem could be either on the positive side or on the negative side. They both need to be investigated. On a horn circuit, the switch (the horn button) is on the negative side (is connected to relay socket terminal 85), but on most other circuits, whatever is "switching the switch" is usually on the positive side (connected to relay socket terminal 86) instead.

Note that it would be unusual for the relay to click and not have it connect 30 to 87, since the clicking is the sound of the electromagnet presumably pulling the contact between 30 and 87 together, but stranger things have happened. Note also that, if the device works with the jumper but doesn't work with the relay, and the relay tests out good, the problem may be that one of the female terminals in the relay socket may be loose and backing out when the relay is plugged in. I've had this happen with headlight relays on BMW 2002s.

Voltage Side of the Circuit

With our horn circuit as we've pictured it in **Figure 1**, the positive side, 86, is connected directly to the positive battery terminal. Referring to **Figure 8**:

- Set the multimeter to measure voltage.
- Connect the positive multimeter lead to relay socket terminal 86.
- Connect the negative lead to ground (the body of the car).
- Turn the ignition on. You should see battery voltage on the multimeter.
- (Note that, if the horn switch was on the positive side of the relay, we'd also have you hit the horn button, but since the horn button is on the negative side of the relay, for this test, it doesn't matter. To test a different device that's powered off the positive side of the relay, you'd need to throw whatever switch turns the device on.)

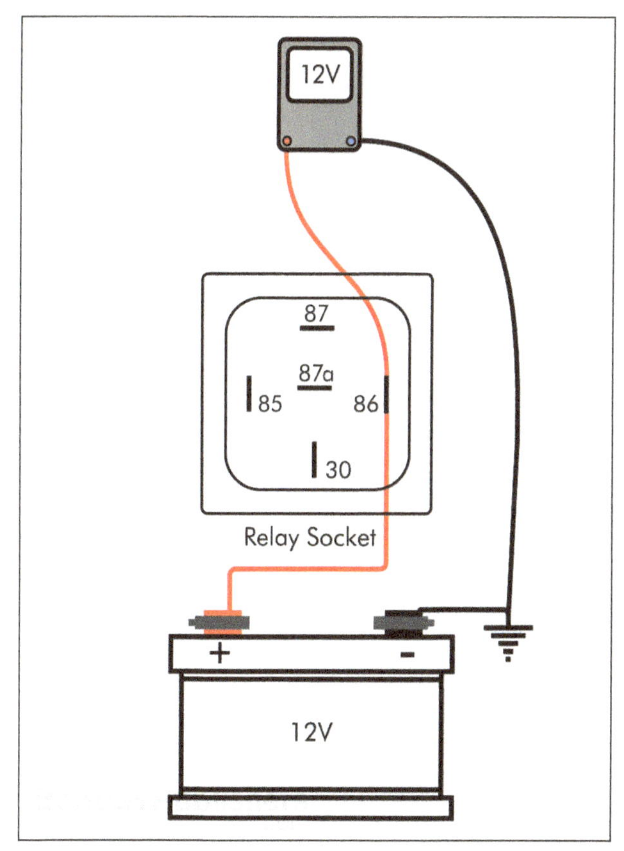

Figure 8 You should see battery voltage between terminal 86 and ground in the relay's socket.

8

- If the multimeter shows battery voltage, then the problem must be on the negative side. But if it doesn't, trace the positive wire from 86 back to the fuse box until you find the failure.

Ground Side of the Control Circuit

On the negative (ground) side, as pictured, relay terminal 85 goes to the horn button, which then connects to ground. We want to verify that pressing the horn button actually connects 85 to ground. Refer to **Figure 9**.

- Set the multimeter to measure resistance.
- Make sure the ignition is off.
- Connect the positive multimeter lead to relay socket terminal 85.
- Connect the negative lead to ground (the body of the car).
- Press the horn button.
- The multimeter should report continuity (in the neighborhood of an ohm).

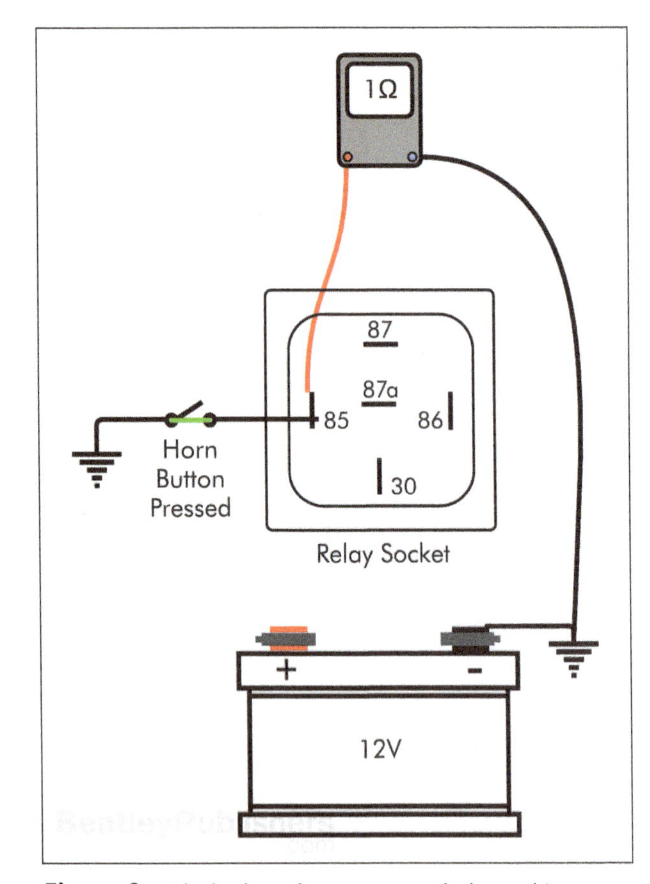

Figure 9 With the horn button pressed, the multimeter should show continuity between 85 and ground.

- If it doesn't, the problem could be in the switch (meaning the horn button) or the wiring between the switch and ground.

Testing the Switch

If we were talking about a circuit with a conventional switch, we'd simply use a multimeter set to measure resistance and place the leads across the terminals and see if it shows continuity when the switch is turned on, but the horn button is unusual in that there's an additional electromechanical component needed because the steering wheel has to be able to rotate around, so the horn switch can't be connected with a solid wire.

To accomplish this, on a car without an airbag, there is typically a circular metal contact ring a few inches in diameter attached either to the back of the steering wheel or to the steering column. Then there is typically a small spring-loaded electrical contact in the other location (i.e., if the ring contact is on the steering wheel, the spring-loaded plunger is on the steering column, and vice versa).

NOTE —

In 1989, all U.S.-spec cars were mandated to have airbags (many European cars began using them years earlier). As a vehicle safety system, the airbag needed a reliable connection. This is accomplished with what is known as the "clock springs" or "toroidal coils." The clock spring contains a ribbon cable that expands and contracts when you turn the steering wheel and maintains a continuous electrical connection to the air bag, horn, and various steering wheel controls.

For this example, we're going to use a very simple car – a 1972 BMW 2002tii with a single ring contact. For a car with a clock spring, a wiring diagram will be needed to determine the pin number in the multi-pin connector on the clock spring.

The contact ring (**Figure 10**) is mounted on the steering column and is connected, through the car's wiring harness, to relay terminal 85. The little plunger is mounted on the steering wheel and is connected to the horn button with a quick-connect terminal.

Horn buttons can have one or two wires coming out of them. One wire goes to the contact mounted on the back of the steering wheel, which should ultimately connect to relay terminal 85 via the ring contact and spring-loaded plunger (see **Figure 11**). If the steering wheel and its hub are made of conductive materials, then the horn button, when pressed, simply grounds through the steering wheel to the steering column and there is no second wire. But if there is not a conductive path from the steering wheel to the steering column, a second wire is employed to make contact with the steering column.

You can test a simple horn button as follows:

Figure 10 Relay terminal 85 is connected to the contact ring that sits behind the steering wheel...

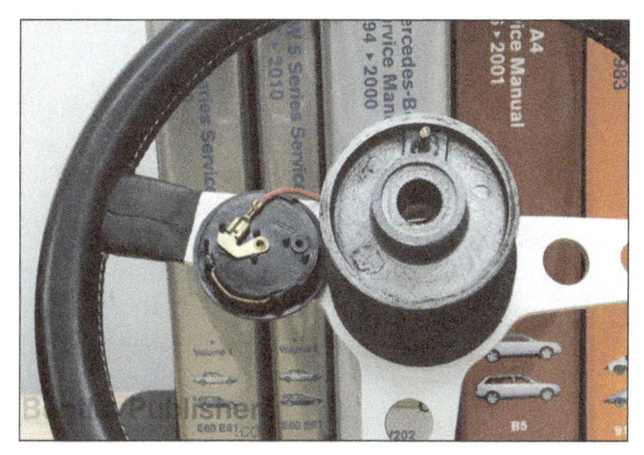

Figure 11 ...which is touched by the spring-loaded plunger, which connects to ground when the horn button is pressed (example from a 1972 BMW 2002tii).

- Pull it out from the center of the steering wheel (you can usually pop it out of the center of the wheel by getting your fingernails around it, or if necessary, by gently prying it out with a plastic or nylon pry tool).
- Use a multimeter to check for continuity between the wire and the side of the button (or the second wire) when the button is pressed.

On a car with an airbag and a clock spring, it's more complicated, because the switch is a much more integral part of the steering wheel. Check a repair manual for your make and model.

However, on a 40-year-old car, the problem is more likely to be the plunger and ring contact. It's fairly common in old cars for the little spring-loaded plunger to fail. It can stick, or the housing it slides in can crack and the plunger can fall out. Since the plunger is trapped between the steering wheel and the ring contact, the steering wheel must be removed to inspect it or address a problem. Even though pulling the steering wheel is trivial in a pre-airbag car, if you don't want to do it, you can test for continuity between the wire that the horn button connects to (the one connected to what's mounted on the back of the steering wheel) and relay terminal 85. But if there's not continuity there, to address a problem with the plunger, you will have to pull the wheel off.

Voltage Drop Test

Always remember that a continuity test of a wire, or a switch, or a relay, is only that – a continuity test. It does *not* test whether that component can carry current. For that, you need a voltage drop test. If the horn, relay, wiring, and button (switch) all appear to be functioning, but the horn still isn't sounding, it is possible that there is a voltage drop problem on the high current (load) side of the circuit. Voltage drop problems tend to affect devices that draw a lot of current. While a horn is not a high-current device like an electric motor, if the horn bleats slightly but doesn't erupt in full blare, voltage drop is a possibility. Conduct a voltage drop test on the high current (load) side of the circuit as shown in **Figure 12**.

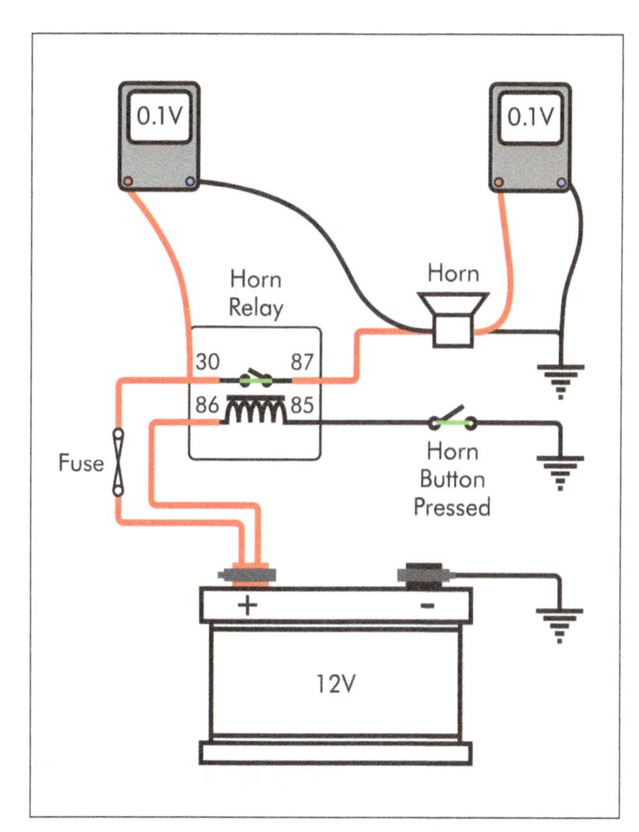

Figure 12 Conduct a voltage drop test if all wiring and components check out.

- Set the multimeter to measure voltage.
- Connect it between the feed to relay terminal 30 (or the fuse, or the battery itself) and the horn's positive terminal.
- Turn on the ignition, and press the horn button.

- If the multimeter is reading more than a few tenths of a volt, you may have found your problem.
- Hone in on smaller sections (for example, between 87 and the horn) to try and isolate exactly where the voltage drop is occurring. Often you'll find a corroded connector, or a crimp that has gone bad and the connector is hanging by a few strands of wire.
- The same voltage drop test can be done on the negative (horn to ground) side.
- Note that, with relays, it would be unusual to find a voltage drop problem on the low current (control) side, as the nature of voltage drop problems tends to affect high current circuits far more than low current circuits.

Finding a Short to Ground (The Fuse Keeps Blowing)

If, no matter what you do, the fuse that the circuit is on keeps blowing, you need to determine how and when it blows. Remember, we are using the horn as an example, but the device could be anything.

Does the fuse blow:

- Immediately (as soon as you put the fuse in)?
- As soon as you power the circuit (e.g., as soon as you hit the horn button)?
- Shortly after you power the circuit (e.g., shortly after you hit the horn button)?

What you do next depends on the answers to these questions.

Fuse Blows as Soon as You Insert It

If the fuse blows as soon as you insert it because a circuit on the fuse is shorted to ground, this is often called a "hard short" or a "dead short." The first issue is that there is often more than one circuit on a fuse. You'll need to determine which circuit is causing the fuse to blow.

- Look at a wiring diagram or the owner's manual of your car for the devices on the fuse that's blowing.
- Systematically disconnect those devices one by one from the fuse circuit until you find the device or circuit that is blowing the fuse.

Once you narrow it down to a particular device or circuit powered by that fuse, if the fuse blows as soon as you insert the fuse in its socket (or as soon as you turn the ignition switch and provide power to the fuse), it likely means that something in the always-on part of the circuit has a short circuit to ground. The cause is usually something simple, but that doesn't mean it is easy to find.

Finding a Short to Ground

The cause of a short to ground is usually insulation that has chafed off the outside of a wire, exposing the actual conductor inside and allowing it to touch ground, either by lying directly against chassis ground, or by contacting an exposed ground wire.

There is no magic bullet for finding a short to ground. You need to physically examine the wiring in the circuit to find where insulation has rubbed off.

If that doesn't work, or if the wire is difficult to follow inside of wiring harnesses, behind panels, and under carpets, there are short finding tools discussed in the **Multimeters and Related Tools** chapter, but these tools are not magic bullets.

To find a short to ground:

- Look at a wiring diagram for the circuit to find the portion of the circuit that is always on.
- For example, the wiring illustration for our horn shows that relay terminal 30 is connected to the fuse, so it would be necessary to inspect where terminal 30 connects to the relay socket and make sure the connector to terminal 30 isn't touching anything else, then trace that wire from there back to the fuse box. All along the way, visually inspect the wire for rubbed-off insulation where bare wire could be touching the body of the car.

The Ground Leg Trick

Actually, if the fuse blows as soon as you insert it, there *is* a trick to help you find the short to ground. If you have a dead short to ground, you will almost certainly find that the output side of the fuse holder of the blowing fuse is electrically grounded. You can use this to your advantage.

- Go to the chapter on **Circuit Protection (Fuses)** and look up the procedure to determine which side of the fuse is "hot" (is supplied voltage by the charging system). Let's say the left side is the "hot" or input side. That means the right side is the output side that goes to the devices.
- Pull the fuse out and insert a male spade connector into the output side of the fuse holder. This will give you something to electrically clip onto.
- Set the multimeter to measure continuity, connect one lead to the spade connector you just inserted, and the other to chassis ground. The meter should beep, indicating that the output side of the fuse holder is indeed shorted to ground.

- Now, with a wiring diagram to tell you what circuits and wires are on that fuse, begin laying your hands on and moving those wires. Wiggle sections of the wiring harness. Unplug connectors. At some point, the meter will stop beeping, indicating that you have lifted the short to ground. Carefully inspect the wiring path between whatever you're holding that made the difference and the fuse.
- There's a variation of this trick that involves, instead of using a multimeter to measure continuity, connecting a light bulb or a buzzer between the battery positive terminal and the spade terminal (ground), and wiggling wires until the light goes out or the buzzer stops. In this variation, the buzzer or light bulb is acting as a load device, providing resistance to reduce the flow of current, and changing the dead short into a viable circuit path. Both techniques are doing the same thing – using the unintended dead short to ground as the ground leg of a circuit.
- If this trick does not help you find the cause, you may need to try one of the short-circuit finding tools discussed in the **Multimeters and Related Tools** chapter.

Fuse Blows as Soon as a Switch Is Pressed

If the fuse blows when a switch is pressed (e.g., when you hit the horn button), it likely means that something in the switched-on part of the circuit is shorting to ground. To find it:

- Look at the wiring diagram for the circuit to find the portion of the circuit that is on when the switch is pressed.
- For example, the wiring illustration for our horn shows that the switched-on part is the line from relay terminal 87 to the positive connector on the horn. Check the connector at the relay socket, then trace the wire as it runs from 87 to the horn and look for where the insulation could be worn off, allowing bare wire to touch the body of the car.
- In addition, in our wiring illustration, for simplicity, we did not show a fuse on the hot side of the control part of the relay (between 86 and the battery), but it's possible this is connected to the same fuse. If the coil in the electromagnet in the relay is internally shorted, this could also cause the fuse to blow. The relay troubleshooting in the DNA section can be used to test the internal relay coil. Removing the relay, replacing the fuse, and reapplying power would test the theory.

Note that you can try "The Ground Leg Trick" in this scenario as well. You just need to remember to hit the switch to complete the ground leg.

Fuse Blows After the Circuit Has Been Powered for a Short While

If the circuit runs for a short while but then the fuse blows (this is unlikely with a horn, unless you really like to lean on your horn), it is possible that the device itself is drawing too much current. If the device is an electric motor, this could mean a temporary mechanical restriction (e.g., something is jammed in a fan), or something such as a worn bearing or a bad seal is causing the motor to use more current than normal. This is often an indication that failure of an electric motor is imminent. To test for this:

- If the device is a motor, and you can access it, with the power turned off, spin the motor with your fingers to feel if it is seized or if the bearing is loose. Replace it if it is defective.
- Set the multimeter to measure current (amperage).
- Unplug the negative wire coming from the device and going to ground.
- Use the multimeter as described in the **Common Multimeter Tests** chapter to measure the current drawn by the device, taking care not to exceed the rated current capacity of the meter.
- If the device draws more current than it should, replace it.

Finding an Open Circuit

If voltage isn't making it to the device or to the relay, then there is an open circuit. Something is interrupting the continuity. Like a short circuit, it is usually something simple, but that doesn't mean it is simple to find.

The cause of an open circuit is usually:

- A wire that has snapped off from a connector
- A connector that is not correctly seated
- A pin that has pushed out from the back of a connector
- A bad component somewhere in the circuit

The least likely possibility is that a wire is broken somewhere inside its insulation where you can't see it, but that is not impossible.

There are open circuit finding tools discussed in the **Multimeters and Related Tools** chapter, but these tools are not magic bullets.

While it is unlikely that a wire will break inside the middle of a straight section of wiring harness that's firmly anchored to the body of the car, broken wires *are* a known problem on certain cars, in certain locations where things move. For example, on a BMW E46 3 Series wagon, a section of wiring harness between the body and rear hatch is subjected to repeated bending strain, and broken wires are in fact common inside that one section. Broken or chafed wires are also not uncommon near the connectors that feed power seats. There can also be broken wires between the alternator and its external voltage regulator due to the rocking of the car's engine on its mounts.

It is also, unfortunately, possible that a wire has short-circuited to ground, burned inside the wiring harness, and separated, creating an open circuit. If this is the case, though, there is usually ample physical evidence of it. Even with a thick wiring harness with many wires bundled inside, amperage that's sufficient to completely part a wire will usually burn through or at least substantially discolor the outside of the wiring harness, not to mention engender a pervasive smell of fried insulation.

To find the failure point of an open circuit:

- Look at the wiring diagram for the circuit to find the portion of the circuit that is on when the switch is pressed.
- For example, the wiring illustration for our horn shows that voltage should be leaving relay terminal 87 and going to the positive connector on the horn. Check the connector at the relay socket, then trace the wire as it runs from 87 to the horn and look for where the wire could be broken, or a connector could be unplugged.
- If no broken wire is found, check any wiring harness that the circuit uses for the visual or olfactory presence of burning.
- If the cause is still not found, you may need to try one of the open-circuit finding tools discussed in the **Multimeters and Related Tools** chapter.

Wire Gauge and Current

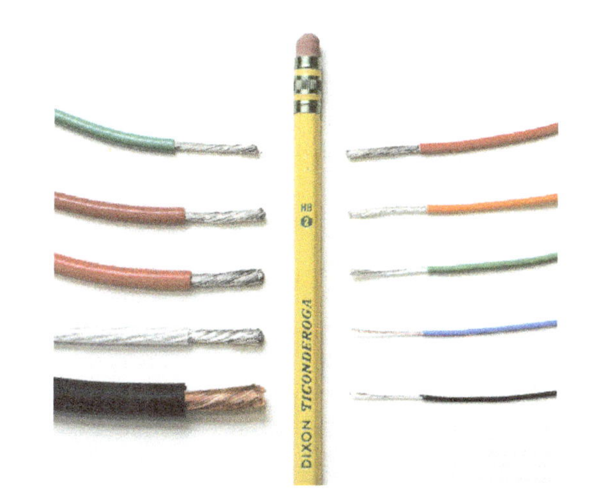

Wire

If you've been safely building cables and adding circuits for years, you're probably all set. But if you grab bits of wire at random and stuff one end beneath the tip of a fuse, you should probably find a fire extinguisher and read on.

In the **Circuits** chapter, you had some exposure to the concepts of voltage, current, and resistance and how they're related by Ohm's law without wading too deeply into physics. We're going to continue that basic approach here. For safety's sake, you do need to understand why you can't connect that 1,000-watt power amp with the hair-thin wire used for your home computer's speakers.

Heat

In understanding how to match the appropriate gauge wire with a specific current flow, it can be useful to make an analogy between the gauge (size) of a wire, and the size of a water pipe. Both need to be appropriately sized to carry their intended load. A 16" drainage pipe would be big, cumbersome, and wasteful to use as a drinking straw (and kind of gross, though it would deliver ample quantities of water to your face). Conversely, a soda straw is not going to carry the amount of water needed to supply a city. When you try to put too much water through it, at some point the straw is simply going to fail.

Without getting wrapped around the axle over the finer points of the analogy, it's the same thing with wire. Wire needs to be the right size to carry a given amount of electrical current. Bigger is better only up to a point – make it too big and you can't even fit the wire into the connectors at the end. Not to mention the added weight; imagine if every wire in your car were the size of the battery cables. But make the wire size too small for the electrical demand, and, like pushing a city's water supply through a soda straw, it will fail. And failure is serious business – high current generates high temperature until the wire gets red hot and burns off its insulation. It may burn an entire section of a wiring harness. It may touch directly to ground, turning your entire car into a space heater. So, to misquote Goldilocks, wire size needs to be just right.

Well, what size *is* right?

Multi-Stranded Copper

First, understand that virtually all automotive wire is copper, chosen because it is an excellent conductor of electricity (to improve on copper, one needs to go to silver, a bit more expensive). Second, understand that because solid copper wire (like the stuff your house is wired with) can crack when bent repeatedly, virtually all automotive wire is stranded – constructed from many small individual wires so it can easily bend into position and move and vibrate in the automotive environment.

Multi-stranded wire also, counter-intuitively, is capable of carrying more current than solid wire of the same diameter due to the *skin effect* where most of the current travels along or near the outside of a conductor. Lastly, unless we explicitly say otherwise, when we say "wire," we are talking about a single insulated piece of multi-stranded copper wire that carries a single electric signal. We are not talking about a cable consisting of several wires, or a coaxial cable that has a signal wire in the center with a ground shield wrapped around it.

Insulation

The wires inside your wiring harness would all instantly short together into a molten heap if the conductors were not insulated. The insulation on automotive wires is typically PVC, but the underhood wires may use insulation made of cross-linked polyethylene for improved heat performance. Certain wires you never have to deal with – wire that's wound around armatures in alternators and motors and part of coils in relays and ignition systems – appear to be bare, but are not. Commonly referred to as "magnet wire," this thin single-stranded wire has a transparent coating often referred to as enamel or varnish, but which is usually a thin polymer.

Shielding

In addition to insulation, certain wires have a foil shield to either prevent the transmission of noise into other circuits (as is the case with spark plug wires), or to prevent noise coming in, as is the case with cables connecting some crankshaft positioning sensors. With the advent of computers, modules, and communications buses in cars, wiring began using a twisted shielded pair configuration. These are what they sound like – pairs of wires twisted together like stripes on a barber pole, with either individual foil shields for each pair, or one shield for multiple pairs. Let's hope your automotive adventures don't lead you down the road of needing to splice into a communications bus.

The American Wire Gauge

The American Wire Gauge (AWG) system assigns a number to standard wire diameters. At first it's a little confusing because the smaller the AWG number, the bigger the wire, but you get used to it. See **Figure 1**.

There is also a metric wire gauge system that takes the metric cross-section size in square millimeters and rounds it down. See the **Common American Wire Gauge Sizes Table**. All dimensions are for the wire core only; they do not include insulation. Note that while European cars use metric wire sizes, wires are still commonly referred to by the equivalent AWG numbers. I don't know anyone who says their BMW has 32 square millimeter battery cables; most people just say "2 gauge."

Note also that, with the rising cost of copper, wire size has become a bit optimistic. That is, your 2-gauge

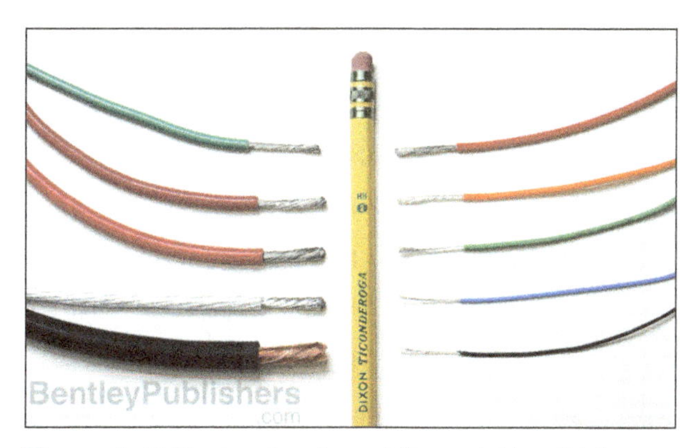

Figure 1 AWG wire sizes from 6 (lower left) to 24 (lower right) with a standard yellow #2 pencil shown for scale.

Common American Wire Gauge (AWG) Sizes			
American Wire Gauge (AWG)	**Wire Diameter (inches)**	**Equivalent Metric Wire Size (MM^2)**	**Rounded Metric Wire Size**
0	0.3249	53.5	52.0
2	0.2576	33.6	32.0
4	0.2043	21.2	19.0
6	0.1620	13.3	13.0
8	0.1285	8.37	8.0
10	0.1019	5.26	5.0
12	0.0808	3.31	3.0
14	0.0641	2.08	2.0
16	0.0508	1.31	1.0
18	0.0403	0.82	0.8
20	0.0320	0.52	0.5

battery cable may or may not actually have a copper diameter of 0.2576". It's not like there's a federal regulatory agency inspecting wire sizes and enforcing AWG labeling. However, if the wire is specifically advertised as "AWG 2 gauge," or uses the term "Full Spec," you stand a better chance of it being what it's supposed to be.

How Big a Wire Gauge Do I Need?

Example Using Ohm's Law and a Spreadsheet

Let's do an example. Let's say we want to wire an auxiliary cooling fan to the battery. The fan is spec'd as drawing 15 amps of current. The battery is in the back of the car, so we estimate we need 20 feet of wire. What wire size do we need to use?

Let's begin with a very quick Ohm's law reminder:

Ohm's Law Reminder

V = I*R ("Voltage equals current times resistance")

The first thing we need to do is decide how much loss we're willing to accept. A good rule of thumb is that you don't want more than a 3% loss of voltage (voltage drop) from the one end of the wire to the other. How does that translate into voltage, and how does that translate into wire size? If the alternator is functioning normally and supplying the car's electrical system with "charging voltage" of 14.2 volts, then:

- 3% = Maximum acceptable voltage loss (drop) over wire
- 97% = Percentage of voltage that should be transmitted
- 14.2 volts = Voltage at alternator
- 97% of 14.2 volts = 0.97 x 14.2 = 13.77 volts

If the alternator is supplying the car's electrical system with 14.2 volts, a 3% drop means that, at the end of the cable, we want 97% of 14.2 volts, or 13.77 volts.

So, how do we determine what size wire will satisfy a set of requirements such as this? It's not difficult, but there are a few steps.

The calculation follows the same principles as the method for calculating current using the resistances in the circuit that we showed in the **Circuits** chapter. The main difference is that here we know the current drawn by the load device, since it is specified in amps (or in terms of power) by the manufacturer. Rather than use the resistances to calculate the current, we're just accepting the current as a fixed value, and using it, along with the desired voltage drop ratio, to calculate what wire size we need.

This is the breakdown for the calculation:

- We know that the current draw of our example fan is 15 amps (presumably part of the spec for the fan).
- We know the length we need to run the wire is 20 feet (presumably we measured it).
- We can look up the resistance per foot of several standard AWG wire sizes in industry tables. (If it's listed as "resistance per thousand feet," divide the number by 1,000.)
- We can multiply the resistance per foot by 20 feet to calculate the total resistance of 20 feet of several different wire sizes.
- Using Ohm's law in the form "voltage equals current times resistance" (V = I*R), we can multiply the current (20 amps) times the resistance of each 20-foot length of wire, to calculate the voltage loss resulting from that current (15 amps) flowing through 20 feet of wire with that resistance.

Table of Voltage Drops for 20 Feet of Different Wire Gauges Carrying 15 Amps							
AWG	**Resistance**	**Length**	**Total Resistance = Resistance * Length**	**Current**	**Vloss = I * R**	**V in**	**% Loss = Vloss/14.2**
Size	(per foot)	(feet)	(ohms)	(amps)	(volts)	(volts)	(%)
0	0.0001	20	0.002	15	0.03	14.2	0.21%
1	0.000126	20	0.00252	15	0.0378	14.2	0.27%
2	0.000159	20	0.00318	15	0.0477	14.2	0.34%
4	0.000253	20	0.00506	15	0.0759	14.2	0.53%
6	0.000402	20	0.00804	15	0.1206	14.2	0.85%
8	0.00064	20	0.0128	15	0.192	14.2	1.35%
10	0.00102	20	0.0204	15	0.306	14.2	2.15%
12	0.00162	20	0.0324	15	0.486	14.2	3.42%
14	0.00258	20	0.0516	15	0.774	14.2	5.45%
16	0.00408	20	0.0816	15	1.224	14.2	8.62%
18	0.00652	20	0.1304	15	1.956	14.2	13.77%
20	0.01036	20	0.2072	15	3.108	14.2	21.89%

- We can express the loss as a percentage of the original 14.2 volts to calculate the % loss using the formula "% loss = (Vloss/14.2)."

Because of the repetitive nature of the calculation, this is easiest to do in a spreadsheet. We've calculated the **Table of Voltage Drops for 20 Feet of Different Wire Gauges Carrying 15 Amps** for our simple example.

If you look in the "Loss" column and read it up from the bottom, looking for the first number that is less than 3%, you come to 2.15%. If you then look left and across to the "AWG" column, you'll see that the 2.15% loss comes from using 10-gauge wire. So, if you want to run a device that draws a 15 amp load through 20 feet of wire, and have a voltage drop less than 3%, if you use 10-gauge wire it will not only do it, it'll beat your requirement, generating 2.15% loss instead of 3%.

Wire Size Worksheet

Although the wire size calculation is easiest using a computer and a spreadsheet, it can in fact be done using a worksheet and a calculator. This is a brute-force technique, but it is quite do-able. The following sample worksheet enables you to:

- Enter the length in feet of the wire in the "Length" column.
- Multiply the resistance per foot by the length in feet to calculate the total resistance for each wire size.
- Enter the current in the "Current" column.
- Multiply the current times the resistance for each wire size to calculate the voltage loss ("Vloss") resulting from that current flowing through that length of wire.
- Express the loss as a percentage of the original 14.2 volts to calculate the % loss using the formula "% loss = Vloss/14.2."
- Look at the % loss column for each wire size and determine the point at which the loss is less than our recommended 3%.

 **CAUTION —**

The VDI table is for COPPER WIRE ONLY. With the steadily rising price of copper, there has been interest in using aluminum wiring in cars. Just remember that, if this table tells you you need a certain gauge, it's for copper wire.

Wire Size Worksheet							
AWG	**Resistance**	**Length**	**Total Resistance = Resistance * Length**	**Current**	**Vloss = I * R**	**V in**	**% Loss = Vloss/14.2**
Size	(per foot)	(feet)	(ohms)	(amps)	(volts)	(volts)	(%)
0	0.0001					14.2	
1	0.000126					14.2	
2	0.000159					14.2	
4	0.000253					14.2	
6	0.000402					14.2	
8	0.00064					14.2	
10	0.00102					14.2	
12	0.00162					14.2	
14	0.00258					14.2	
16	0.00408					14.2	
18	0.00652					14.2	
20	0.01036					14.2	

The Voltage Drop Index Method

There is, in fact, an elegant way to determine the correct wire size without doing the math and filling out a table yourself. It's called the Voltage Drop Index (VDI) method, and was developed originally for alternative energy applications to allow rapid look-up of the wire size needed between a solar panel or wind turbine and a bank of storage batteries. You use the desired voltage, amperage, wire length, and percent permissible voltage drop to calculate a number called the Voltage Drop Index (VDI) with this very simple equation:

Voltage Drop Index (VDI) Method

VDI = (AMPS * FEET) / (2 * VOLTS * %DROP)

where:

- AMPS is the amperage of the load (or watts/volts if the load is expressed in watts).
- FEET is the one-way length of the wire in feet.
- VOLTS is the input voltage.
- %DROP is the percent maximum permissible voltage drop (e.g., for a maximum 3% loss, use 3).

You then look the VDI up in the table below, find the closest value, and read off the AWG size to the left of it. If your VDI falls between two rows in the table, round up to use the heaver sized wire, particularly if it's for an always-on application like a fuel pump. The "Ampacity" column lists, as a sanity check, the indicated fire hazard rating for the wire gauge as set by the National Electric Code. If the amperage of your load device is significantly higher than the number in the Ampacity column, keep moving up the column, stepping up to larger wire sizes, until you meet or exceed the amperage of the load.

Voltage Drop Index (VDI) Table		
Wire Size (AWG)	**VDI**	**Ampacity**
0000	99	260
000	78	225
00	62	195
0	49	170
2	31	130
4	20	95
6	12	75
8	8	55
10	5	30
12	3	20
14	2	15
16	1	not listed

For example, let's use the same numbers we used when we walked through the worksheet above – a 15 amp load, 20 feet of wire, voltage of 14.2 volts, and a maximum voltage drop of 2.15%. Plugging the values into the VDI equation, we get:

VDI = (AMPS * FEET) / (2 * VOLTS * %DROP)
(15 * 20) / (2 * 14.2 * 2.15) = 4.9

This is approximately 5, which when you look it up in the VDI table, maps to a 10 gauge wire, which matches our previous spreadsheet example very nicely.

Cool, huh?

Note that the wire length in feet is the *total* wire length for the circuit. If you're running 20 feet of wire to carry the voltage, and the chassis of the car is carrying the ground, then the length is 20 feet. However, if the ground is isolated from the chassis and you're running 20 feet of a *pair* of wires, one for the voltage and the other for ground, then the length is 40 feet.

You can always check the *actual* voltage drop with a multimeter. And if you have any doubt whatsoever, go up to the next largest wire size.

10

Circuit Protection (Fuses)

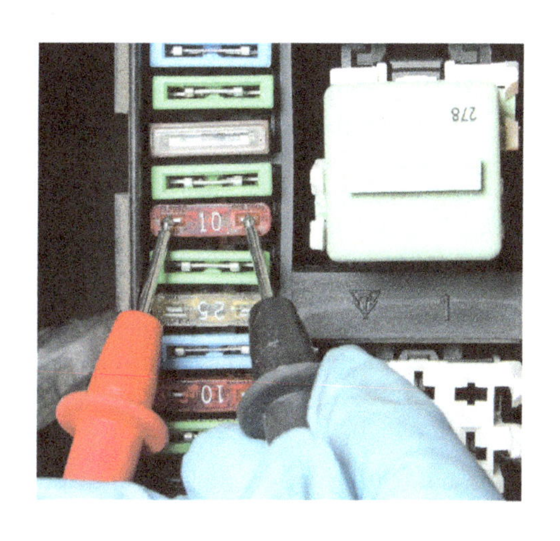

Circuit Protection

In the previous chapter on **Wire Gauge and Current**, we went through the calculation to right-size a wire to carry the current for a given electrical load. That's good – essential, even – but if the device drawing the current malfunctions and draws more than anticipated, or if the wire short-circuits to ground, you don't want the circuit to carry the current – you want to *stop* the current. That's why most of the car's wiring has some sort of circuit protection – either a fuse, a fusible link, or, in rare cases, a circuit breaker – to interrupt the current flow if it becomes too high.

In describing circuit protection, let's start with the pieces that we know are there – the fuses that protect the individual circuits, and the fuse box that holds them. Then we'll describe the overall layout of the electrical system, and fusible links that may or may not be interposed between the battery and the fuse box and the battery and the alternator.

Fuses

Fuses are by far the most common form of circuit protection. Though they vary in size and form factor, they all work basically the same way. They contain a thin piece of metal that narrows in the middle. The fuse is designed for the metal to melt at the narrow point if the current through it becomes too high. Fuses are usually sized so the normal current flow is about 80% of the fuse rating (e.g., a 20-amp fuse would be used to protect a circuit with 16 amps of current).

It's ironic that a car's wiring is sized to provide plenty of current capacity at low resistance whereas fuses are sized the opposite way – to ensure that, at a given current, enough resistance is encountered to generate heat sufficient to melt the fuse – but it's a necessary evil. Because of this, in high current situations, some voltage drop across a fuse is virtually unavoidable. Note that this is exactly why most battery-to-starter connections are unfused. Fusing is at odds with maximizing current flow while minimizing voltage drop.

Cylindrical Tapered European Ceramic "Bullet" or "Torpedo" Fuses

For the most part, the fuses that are in your fuse box come in two basic styles – cylindrical and blade. Cylindrical fuses can be glass or ceramic, and can have squared-off or tapered ends. If they're glass, the thin metal fused material is inside. If they're ceramic with tapered ends, the fuse material is wrapped around the ceramic body the long way, and the tapered ends sit in a fuse holder with pairs of tabs that have holes in them to squeeze each fuse at the ends. They're commonly called "European ceramic fuses," "bullet fuses," or "torpedo fuses" (see **Figure 1**) And, really, it's a poor design. It concentrates all electrical contact at the small area

Figure 1 The tapered European "bullet" or "torpedo" fuse.

where the tapers sit in the holes, providing an opportunity for a viscous cycle of corrosion generating high resistance generating heat. People experiencing intermittent electrical problems with European cars equipped with bullet fuses routinely give them a little twist to re-seat the tapers in the holes in the tabs. Periodic removal, inspection, replacement of the fuses, and cleaning of the holes with a small brush or a Dremel tool is a more thorough approach.

Typically one of the two tabs in the fuse holder is thick right-angle metal, whereas the other tab is a springy clip. If a fuse is loose, the top of the springy clip can be bent slightly inward to increase the pressure on the tip of the fuse and improve electrical contact. While bending, care needs to be taken, though, to keep the tip at a right angle to the fuse so electrical contact is spread around the entire fuse tip. To make matters worse, while the original fuses have a ceramic body that serves to dissipate heat, many of the cheaper aftermarket replacements have plastic bodies. Because plastic does not dissipate heat as well as ceramic, there are reports of the cheaper aftermarket fuses melting (that's the body of the fuse melting, not just the metal). Vintage BMWs, Porsches, and VWs all used 5A (yellow), 8A (white), 16A (red), and 25A (blue) fuses, but torpedoes are also available in 20 and 30 amp sizes for non-original applications.

Cylindrical Glass Fuses

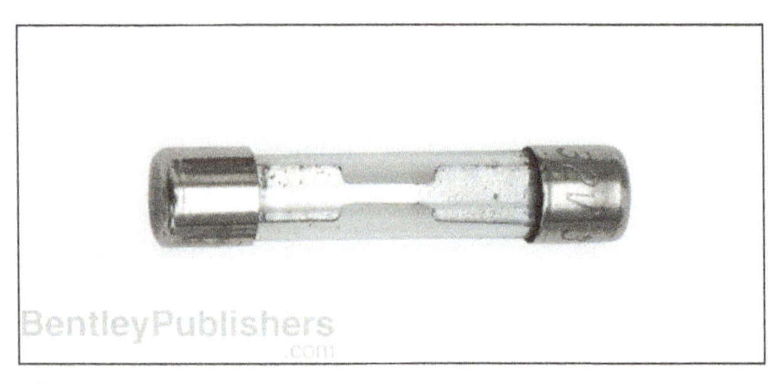

Figure 2 The traditional glass fuse.

Cylindrical glass fuses (**Figure 2**) with squared-off (non-tapered) ends were used on many pre-1975 American cars and all manner of electronics equipment. We are not aware of them being used on European cars, which is too bad, because it's an inherently better design than the bullet fuse – without a taper at the end, a cylindrical fuse is grabbed by the sides of the ends rather than end-to-end, resulting in larger surface area and more reliable electrical contact. Some folks prefer glass fuses because the fusing element is trapped inside the glass; it can't melt and wind up somewhere else. For this reason, there is demand and an aftermarket source for glass fuses with tapered ends that fit into holders designed for tapered ceramic fuses.

Blade Fuses

Figure 3 Left to right: low-profile mini, mini, regular, and maxi blade fuses. Micro fuses are pictured, but are too small to see. That's a joke.

So-called blade fuses (**Figure 3**) are superior to the cylindrical tapered fuse. The rectangular legs of the fuse make good firm contact with their female socket counterparts across their entire face, not just in a small corrosion-prone area. And the sockets are recessed into the fuse box, reducing exposure that can cause corrosion. The Germans changed over to blade fuses in waves during the 1980s. BMW, for example, introduced blade fuses in 1984 with the E30 3 Series, but the E28 5 Series and E24 6 Series continued using tapered ceramic fuses through the late 1980s.

The most commonly used blade fuse is the "APR" or "regular" sized blade fuse, which is 19.1 mm (about 3/4") long. The fuse tabs are essentially the same size as quick connect terminals, making it easy to, if necessary, splice a fuse into a circuit even without a fuse holder in an emergency. APR fuses are often referred to interchangeably as "ATO" and "ATC" fuses, but technically there's a difference – "ATO" fuses are open at the bottom between the fuse legs, but "ATC" fuses are closed, meaning the fused element is completely encased in plastic.

There are other blade fuses smaller and larger than the "regular" size. A "maxi" fuse is a large blade fuse that can carry a lot of amperage. BMW, Porsche, and VW all began using maxi fuses to protect high amperage circuits for normal current draws that would've been unthinkable during the era of torpedo fuses. At the smaller end, there are mini, low-profile mini, two-legged micro, and 3-legged micro fuses.

The Fuse Box

Most fuses are centralized in the aptly named fuse box (**Figure 4**), which may be under the hood, underneath the dash, in the glove box, or in the trunk. The electrical systems on cars up through the early 1970s were so basic they didn't need many fuses. Indeed, early VW Beetles and BMW 2002s used only six fuses to protect an electrical system so simple it could be sketched from memory on the back of a napkin.

Figure 4 Under-dash fuse box from a 1972 BMW Bavaria using torpedo fuses.

As electrical systems in cars became more complex, the number of fuses increased. As cars moved from having six bullet fuses to eight, then 12, then to blade fuses, the function of the fuse box itself changed from being merely a centralized place where the fuse holders were located, to being a mechanically and electrically engineered structure, designed to be waterproof (if it's under the hood), and to hold not only the fuses but also the rapidly proliferating number of electric relays controlling fuel pumps, fans, window motors, and more. See **Figure 5**.

Figure 5 The much more complex fuse box of a 2007 BMW X3, up inside the glove compartment.

This evolution is particularly striking on the Porsche 911 which, because it used essentially the same body from 1963 through 1989, crammed larger, more complex fuse boxes into the same physical space.

The Hot Side of a Fuse Box

You should think of it like this: Every fuse should sit between a source of current and a user of current. The fuse itself doesn't have a defined direction in which current flows, but the holder in which it sits certainly does. When a lone fuse and its holder are installed for something like a radio, the direction of current flow is perfectly clear – the wire from one end of the fuse holder goes to power (from, say, the DIN "15" line from the ignition), the wire at the other end goes to the radio, and the current then flows into the fuse holder, through the fuse, and out the other end of the holder to the radio.

But when a fuse is in a fuse box, it's not obvious which is the input side and which is the output side. Do you care? If you're simply changing a popped fuse, no, you don't. But if you want to understand the architecture so you can add a circuit to the fuse box, yes, you care a lot. And it's amazing how often people get this wrong.

If you look at a simple fuse box with a single row of fuses on a 60s or 70s European car like a VW Beetle, you'll see the pairs of tabs that squeeze the tapered fuse ends. But if you unscrew and lift up the fuse box (with the battery disconnected, of course) and look at the back, you'll see two rows of male connector tabs (see **Figure 6**). One row is the input side, the other row the output side. And, somewhat surprisingly, there's usually not an exact one-to-one correspondence between the spade connectors on the bottom and the tabs on the top. That is, there may be 12 fuses, but on the bottom of the fuse box, there may be 8 tabs on one side and 24 tabs on the other side.

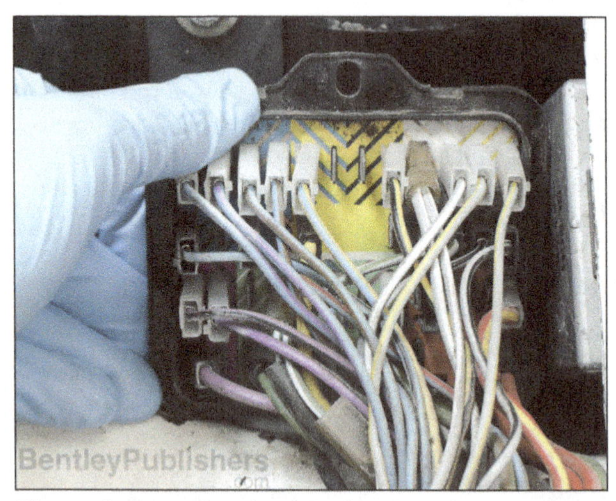

Figure 6 The underside of the fuse box of a 1972 BMW 2002tii. The cross-hatched color coding indicates which tabs are for which fuses.

If you want to add a new circuit (and we'll talk in more detail about this later), you need to know which side is which, because if you add the circuit to the input side, you've bypassed the fuse and wired the circuit directly to the battery, so the circuit has no protection, even though it's connected directly to a fuse. The current has to flow through the fuse in order to protect a circuit.

Figure 7 A side shot of the fuse box of a 1972 BMW 2002tii. On the fuse on the near right, you can see the thick red wire feeding the right side, indicating that that's the input side, or hot side, of the fuse.

So, which side is which? Well, the side to which a small number of thick red and yellow wires are attached should be the input side (see **Figure 7**). There are usually fewer inputs than fuses because the fuse box itself almost always bridges several of the inputs together. For example, a single wire from the battery, capable of carrying 32A, may power two 20A fuses (80% of fuse capacity, remember?).

The side with a larger number of tab connectors should be the output side. It's quite normal for multiple circuits to be connected to the output side of a fuse. There may be extra unused spade connectors to allow for additional circuits *provided they do not exceed the amperage rating of the input side of the fuse.*

Finding the Hot Side with a Multimeter

If there's any question which side is the input side, remove a fuse, then turn the ignition key to run and use a multimeter to check the voltage between both sides of the fuse holder and ground. The side that registers battery voltage is the input side. So be certain that any connection you make is on the other side so it is powered through the fuse. See **Figure 8**.

And, for the sake of all that is good and holy in this electrical world, don't EVER do what we show in **Figure 9** and power something by stripping the end off a wire and trapping it between the fuse taper and the fuse holder tab, even if you put it on the output side. Not only is it an inherently unreliable connection for the thing you're trying to hook up, it makes the fuse connection itself unreliable for the rest of the circuits on the fuse because the fuse is no longer sitting squarely in the holder. It's just wrong. Don't do it.

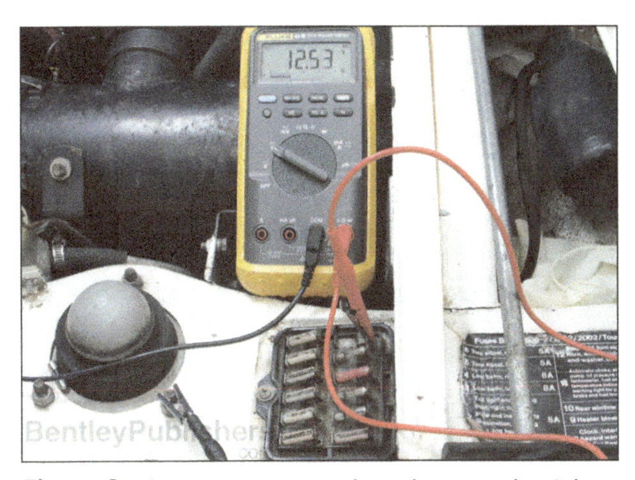

Figure 8 When we measure voltage between the right side of the fuse holder and ground, we see that, indeed, the right side is the input side of the fuse.

Figure 9 Don't EVER hook up a circuit by trapping a wire between the end of a fuse and its holder!

Bad Connections at the Fuse Box

As you saw in the pictures in the previous section of the underside of a fuse box, the connectors on the back are often simply push-on connectors similar to those used elsewhere in a car. It is possible for them become corroded or loose or disconnected like any other connector.

Some cars have fuse connections that are particularly fragile. The fuse box on a pre-1984 Porsche 911 has a single row of 20 bullet fuses that sits proud of the front left inner fender wall. Because it isn't recessed into a panel, the wires going into and out of the fuse box are plainly visible, which is at once convenient and terrifying. It allows you to see that, rather than affixing the input and output wires to the back of the fuse box with push-on connectors, the wires are simply passed through holes in the ends of the fuse holders, then are mechanically held and electrically connected with a screw (**Figure 10**). The tightening of the screw flattens the end of the wire. Over time, it's common for the screw-to-mashed-wire interface to corrode, increasing the resistance of the connection, and becoming a failure point.

Figure 10 Fuse box from a 1983 Porsche 911SC still has its wires connected with old-style screw-mash fittings, a potential cause of intermittent electrical problems. *Photo courtesy Weekend Rides LLC*

More Than One Fuse Box

All fuses are not neatly housed in one fuse box. This is especially true for post-1996 cars. The 1997–2003 BMW E39 5 Series, for example, has fuses in six separate locations:

- 5 fuses in the "E-box" (electronics box) in the engine compartment (**Figure 11**)
- 45 fuses in the glove compartment top panel
- 2 fuses behind the glove compartment
- 21 fuses above the battery in the luggage compartment
- 7 fuses above the positive battery terminal
- 8 fuses MEGA fuses under the passenger seat

Figure 11 Fuses outside "the fuse box" – the "E-box" (electronics box) in a 2007 BMW X3.

To identify the main fuse box locations in a specific model car, consult the owner's manual or the appropriate Bentley Repair Manual.

Inspection and Replacement of Fuses

With most of these types of fuses, you can tell, in theory, through visual inspection if they're blown – just look for a break in the thin metal. With tapered ceramic fuses, this is certainly true, since, without anything enclosing the fusing element, it's bloody obvious if the fuse has popped (it may be the *only* real advantage of tapered ceramic fuses). With blade fuses, though, it's more difficult. While they're designed so that the fusing element *should* be visible from above, in practice it's sometimes difficult to tell, particularly since the fuse box is often buried deep under the dash or in the glove compartment.

So you can do one of two things. You can pull the suspect fuse out (which, since they're packed in so closely, usually requires using the little plastic pulling tweezers most cars come with) and look at it from the side. Or you can test it in place for continuity with a multimeter – all blade fuses have small test points on the top. In contrast, a mega fuse may not have its conductive link visually exposed, and probably needs to be tested for continuity with a meter.

How to Test a Fuse with a Multimeter

Figure 12 A blown torpedo fuse (left) is completely obvious. It's a bit harder to tell with a blade fuse (right), since you need to see the melted section through the plastic.

When an old-school torpedo fuse blows, it's clearly visible because the metal fusing material is out in the open, but on a blade fuse, because the metal is enclosed, it's a bit harder to tell. See **Figure 12**. There are times when you may need to test the continuity of the fuse with a meter,

- Set the multimeter to test continuity.
- Touch the multimeter's leads to the ends of the fuse. See **Figure 13**. You can do this with the fuse in the holder. On the top of a blade fuse are two small test points. Position one multimeter probe on each. It doesn't matter which end goes where. What you don't want to do is hold the fuse and the tips of the multimeter's leads in your hands, because you can wind up measuring the resistance across your body instead of across your fuse, and think the fuse is good when in fact it's not.

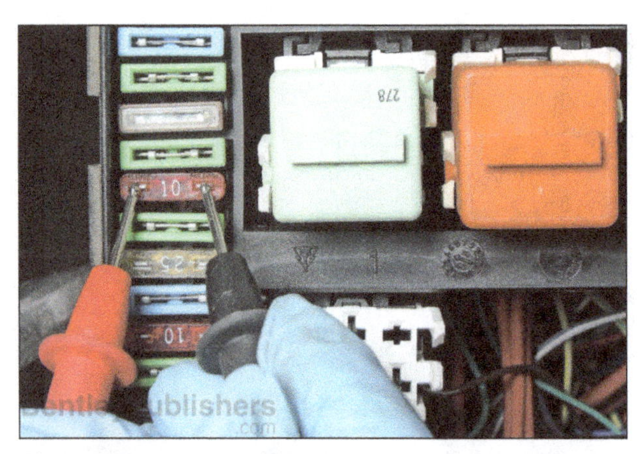

Figure 13 You can test the continuity of a fuse without removing it from its holder by using the two small test points on the top of the fuse.

- If the meter reads "OL," there is no continuity, which means the fuse is blown.
- Otherwise, the meter should beep and read a low resistance value – less than an ohm – indicating that there is continuity across the fuse, and thus that it is good.

If a fuse has blown, and blows again after replacement, something is wrong. Do not simply replace it with a higher amperage fuse. If it blows immediately, odds are something is shorted to ground. If it blows after a few seconds, it is likely powering something that is drawing too much current due to malfunction (a motor that is seizing, for example). If it blows after some larger amount of time passes, it's likely that corrosion is increasing resistance in a connection, or a high-use electric motor like a fuel pump is nearing the end of its lifetime. See the **Troubleshooting** chapter for more information.

Circuit Breakers

Distinct from melt-and-replace fuses or fusible links, circuit breakers, like the ones in the fuse panel in your house, trip and shut off but then can be reset. Use of these is not common on European cars, though the U.S.-spec BMW 3.0CS did have small circuit breakers on the power window motors, which always seemed like a curious admission that the window motors were troublesome even when new.

Why Not All Circuits Are Protected

At the beginning of this chapter, we said "that's why most of the car's wiring has some sort of circuit protection." Wait. *Most?* Yes. The older your car is, the greater the likelihood that not all the circuits are protected. We summarize the issues in the following box, then explain them in detail:

Key Concepts in Protection of High-Current Circuits

- Most cars have no circuit protection on the starter motor.
- Many cars have no circuit protection on the ignition switch, fuse box, or alternator.
- If there is circuit protection on the ignition switch, fuse box, or alternator, it usually employs fusible links or bolt-down fuses that are located near the battery.

If you're going to muck about with the car's electrical system, it is crucial to know enough about the basic wiring layout to understand this. A grossly simplified illustration of the electrical connections on a typical European car of the 1960s through the 1970s is depicted in **Figure 14**. The actual wiring of your car will vary substantially from this illustration, but it's there to show how it is possible—in fact, likely—that certain of the wiring has no circuit protection.

In this basic wiring figure, we see that there are two or three wires coming off the clamp on the positive battery cable.

- There's the big thick positive battery cable that is integral with the positive battery clamp. This goes directly to the starter solenoid. As we said earlier in the chapter, because of the starter's need for high current with as low a voltage drop as possible, ***there is typically no circuit protection on the thick cable between the battery and the starter.*** If this cable accidentally touches chassis ground, you'll have an arc welder on your hands. That's why you need to be *very careful* working on the starter and *always* disconnect the battery's negative cable before disconnecting the fat starter cable.
- There's a second, thinner cable that may be integral with the positive battery clamp or may be attached to the battery clamp with a ring terminal. It vanishes into the wiring harness and, if you look at a wiring diagram, you'll see that it typically gets split, with one branch going directly to fuse box to power always-on accessories such as the hazard flashers and dome light, and another branch supplying voltage to the ignition switch. The other fuses in the fuse box are then typically fed, directly or indirectly, from the ignition switch's positions (accessories, ignition, and start).
- We show the "accessories" and "ignition" switch positions each feeding several fuses that are jumpered

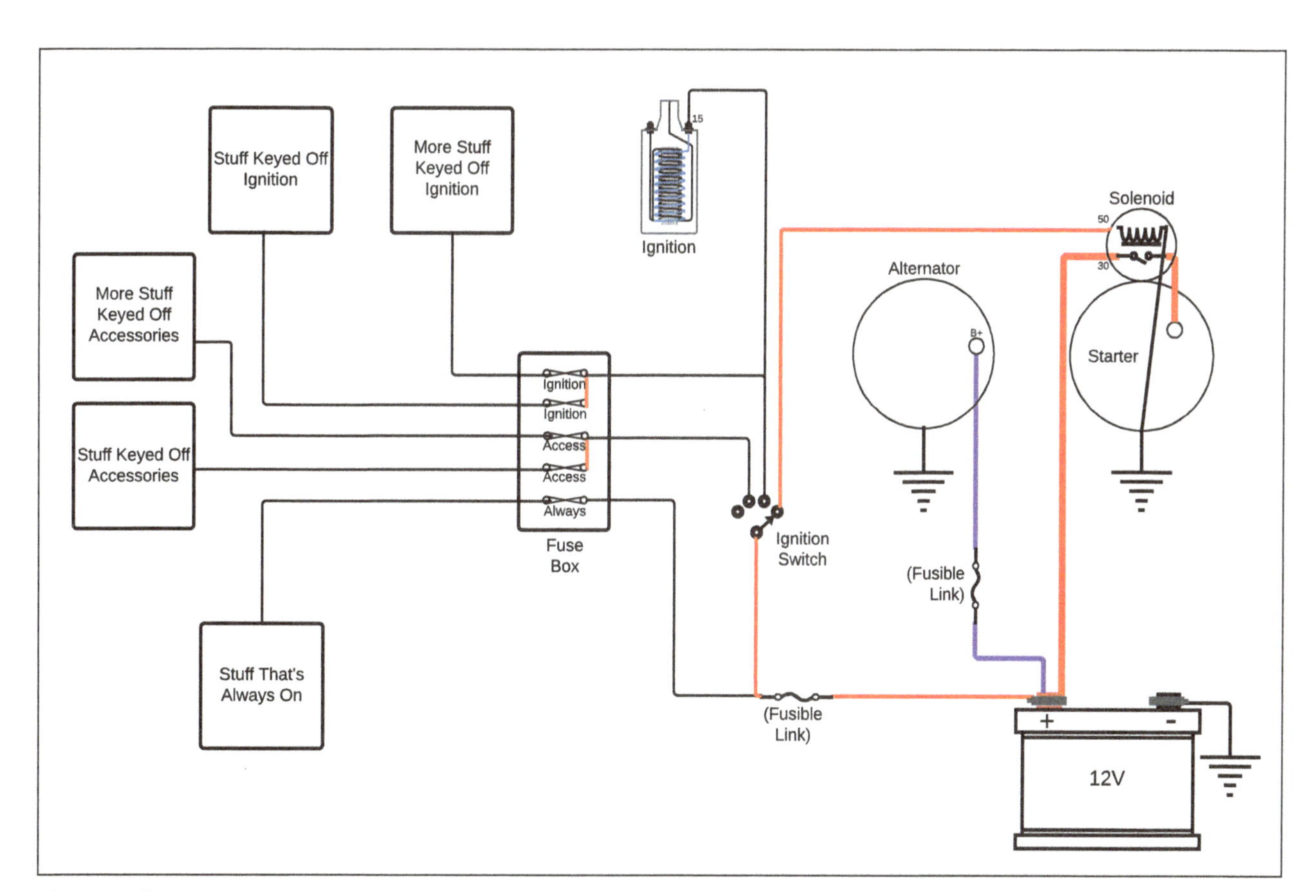

Figure 14 The layout of the basic wiring of a 1960s or 1970s European car. Two fusible links are shown. If neither fusible link is present, there is no protection on the fusebox or alternator wiring.

together on their input side. The ignition switch position may also directly feed the ignition coil in an unfused fashion as pictured. Similarly, the "start" position that feeds voltage to the starter solenoid is often unfused as pictured. But the main thing to realize is that ***there may or may not be any circuit protection between the battery and the fuse box and between the battery and the ignition switch***. This is the case on most pre-1975 cars. Touch these wires to the body of the car, and you'll get sparks until contact is removed, the wire melts, or the battery is discharged.

- The third wire that may or may not come off the positive battery terminal is the alternator "B+" wire. On some cars, the alternator's B+ terminal is connected directly to the positive battery terminal via its own wire that has a ring terminal on the end which gets bolted onto the battery cable clamp. On other cars, there may not be a separate wire to the battery; the alternator wire may be integral with the main harness and mate up with the always-on circuit at the fuse box. In either case, ***the alternator wire may not have any circuit protection***. This is the situation with many cars, and is why, if you're, for example, changing an oil filter and accidentally touch a wrench to the terminal on the back of the alternator, sparks fly but no fuse blows.
- In addition to a car's original wiring, if there are any other ring terminals connected directly to the positive battery clamp, odds are they power some non-original accessory. If they don't have their own fuse holder right there at the battery, they probably are not protected either.

Fusible Links

As per the basic wiring layout figure shown in the previous section, some cars do have circuit protection between the battery and the fuse box and alternator. What it is, what it's called, and how easy it is to replace all vary.

Certain 1980s and 1990s-era BMWs like the E30 3 Series and E34 5 Series have a fusible link that's integral with the cable from the battery to the fuse box as a last line of defense (**Figure 15**). Unlike a fuse, which is designed to be easily replaceable if popped, a *fusible link*, as described in this context, is riveted in place between ring connectors that are permanent parts of the battery cable. Thus, to replace the link, the insulation around it needs to be cut away, and either the rivets need to be drilled out, or the entire length of cable needs to be replaced – a non-trivial task as battery cables have become more integrated with a car's wiring harness.

If a car has a fusible link, it is typically located very close to the positive battery terminal, thus keeping the unfused section of wire as short as possible. For example, on a BMW E34 5 Series, the battery and a fusible link are located underneath the back seat. If a car has voltage across the battery terminals but no voltage at the fuses, and the battery cables are clean and tight, it's possible the car has a fusible link that has failed. The irony is that, instead of blowing to save the car from disaster, these links can fail when vibration causes hairline cracks, sometimes manifesting as vexing intermittent problems before failing completely.

Figure 15 Fusible link, buried inside the positive battery cable in the trunk on a 1987 BMW 325is.

MEGA and MIDI Bolt-Down Fuses

Another group of high-current fuse mechanisms are "bolt-down fuses," also known as "stud-mount fuses," that are held onto a pair of threaded metal posts with nuts. These are similar in their large physical size and high amperage rating to fusible links, but are designed to be more easily accessed and replaced than riveted fusible links hiding inside battery cable insulation.

Bolt-down fuses are commonly called "MEGA fuses" (**Figure 16**), but "MEGA" is in fact one of two commonly used sizes. The other is "MIDI." Their sizes are shown in the **Bolt-Down Fuse Sizes Table**:

To make matters confusing, both MEGA and MIDI bolt-down fuses are sometimes referred to as "bolt-down

Figure 16 A MEGA-sized bolt-down fuse, typically mounted on threaded posts.

Bolt-Down Fuse Sizes	
Type	**Size**
MEGA	50.8mm (2”)
MIDI	30mm (1.18”)

fusible links,” or simply “fusible links.” In addition, in certain sections of the automotive industry, “fusible link” sometimes means “fusible link wire,” a section of cable where the conductors in the wires themselves are designed to melt in a controlled fashion and the insulation is designed to contain any spark or flame and thereby prevent a spreading electrical fire. The take-away message is that if you need to replace a bolt-down fuse or fusible link in a car, you need to be very clear about exactly what the part is when procuring a replacement.

Comparing Fuses at the Battery on Two Cars

It will be instructive to compare the fusing that exists at the batteries of two BMWs– a 1999 Z3 and a 2007 X3.

Figure 17 shows the Z3. It has a single unfused thick red cable, approximately 1 gauge, directly feeding the starter solenoid. There is a single thinner unfused red cable, approximately 8 gauge, that connects to the fuse box and alternator, and through which all of the current of the car (except the starter motor) flows. There are no fuses or fusible links that protect the fuse box and alternator. Even though the Z3 is very much a modern computerized car, the basic wiring at the battery is just like a 40-year-old car.

Figure 17 On a 1999 BMW Z3, the positive battery terminal has a thick wire going to the starter, and one or more thinner wires powering the entire electrical system of the car. It's just like any 40-year-old car.

In contrast, **Figure 18** shows the battery and power distribution box of a 2007 E83 BMW X3. Because this is a newer, larger, more complex car with substantially more electronics generating higher electrical demand than the 1999 Z3, it has a variety of high-current fuses, resulting in a more complex set of battery connections. The thick red wire coming off the positive battery post is actually *not* the starter wire. The starter wire is the *very* thick red wire on the right side of the battery where it emerges from a protective plastic loom. It's *huge*, about the size of a thumb. That starter cable is connected to the Battery Safety Terminal (BST), which we'll revisit briefly at the end of this chapter.

Figure 18 A more complex set of battery connections on a BMW X3.

Let's concentrate on the red wire that comes off the positive battery post. Because this car's electrical systems pull a much larger load than those on the Z3, the thickness of this wire is nearly that of the starter wire in the Z3. It's connected to the green power distribution box. Inside the box shown in the next figure, there are three bolt-down fuses (**Figure 19**). We can now correctly describe these as one MEGA and two MIDI fuses.

Figure 19 The X3's power distribution box with one MEGA and two MIDI bolt-down fuses.

Thus, in the X3, there is circuit protection between the battery and the alternator and fusebox, and that protection is in the form of bolt-down MEGA and MIDI fuses. But note that MEGA and MIDI fuses can be used elsewhere in a car (for example, there's a whole row of them beneath the front seat of a BMW E39 5 Series).

Battery Safety Terminal

BMWs from the 1997 E39 5 Series and 1999 E46 3 Series onward, have a Battery Safety Terminal (BST). The BST has an explosive charge that, in the event of an accident, blows and severs the connection to the battery. Certain model Audis have an inertia sensing switch that performs the same function. It's worth nothing that, while this is not a form of current-based circuit protection, it *is* circuit protection and is there precisely because the starter and alternator wires often have no current-based circuit protection. In the event of an accident, crushed metal could cut one of these wires, lay it to ground, and cause a fire.

The BST severs this connection at the battery so that cannot happen. Note that, as pictured in the power distribution box of the X3, the other smaller red wire stays connected to the battery, allowing the car's ECU to unlock the doors and turn on the hazard light when it senses an accident has occurred.

A Final Word on Safety for Unfused Circuits

Hopefully we've given you a good overview of what fuses do, and perhaps most importantly, why it's likely that your car may have circuits that are, in fact, unfused. You now know why you need to be *very* careful poking around at the back of the starter motor, alternator, ignition, and main feeds to the fuse box. Lay these wires to ground, and there may well be no fuse to prevent thousands of amps from coursing through the wiring, ruining your precious ride in seconds.

11

How to Make Wire Repairs

Repairing Wires

Making wiring repairs consists primarily of splicing wires together. We discuss both of these in this chapter. In the following chapter, we discuss repairs specific to wiring harnesses.

Crimp-On Connectors

Whether you're splicing in a new electrical circuit or replacing a ring terminal that broke off the end of a 45-year-old alternator wire, at some point you're going to need to build something with wire and connectors. While computerized cars use a dizzying array of connectors, you'll find that you most often use only three or four simple connectors for repair and modifications.

Common Single Wire Crimp-On Connectors

Because a car uses the body as ground, when we repair a circuit or add a new one, most of the time we're dealing with single wires. For example, if we need to wire a new fuel pump or power amplifier, we rarely deal with positive and negative wires at the same time; instead we usually run a wire to connect the positive side to a voltage source, and run a separate wire to ground the negative side to the body of the car. For this reason, many repairs or modifications require single wire connectors. To ensure electrical integrity and robustness, crimp-on connectors are preferred. By far the most commonly used crimp-on connectors are quick disconnect terminals, butt splices, bullet terminals, and ring and spade post connectors. Ring terminals and butt splices span every available wire size down to 0000, but the wire size range for quick disconnect terminals is more limited. If the wire size is very small, you would not use a quick disconnect terminal; you'd use a crimp-on pin as part of a larger multi-pin connector.

Standard AWG Size Connector Colors

Even though European cars use metric wire sizes, it is common practice to repair them by tapping into the huge number of available AWG crimp-on connectors. There are three color-coded ranges of AWG sizes that account for nearly all low to medium amperage automotive wiring. See **Figure 1**. These are:

Color	AWG Size
Yellow	10–12
Blue	14–16
Red	18–22

This makes it extremely easy to select the right sized crimp-on connector. Let's describe the basic connectors first, then go into the crimping process.

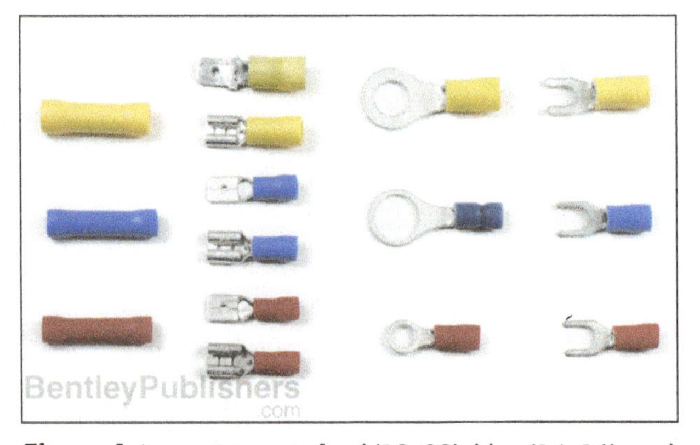

Figure 1 An assortment of red (18–22), blue (14–16), and yellow (10–12) AWG sized crimp-on connectors. Left to right: butt splice, quick disconnect, ring terminal, and spade terminal.

Quick Disconnect Terminals

The clumsily-titled "quick disconnect terminals" (also sometimes referred to as push-on connectors, flat blade connectors, and, erroneously, spade connectors) are used for a great variety of make-and-break connections. The male terminal has the form factor of the leg of a regular blade fuse with a 1/4" wide flat face. In addition to being available as crimp-on connectors, the male tab of the push-on is present on many parts of a European car, including the back of many ignition switches, starters, alternators, coils, and the undersides of nearly every relay. Both the male and female connectors normally have the metal connector exposed, but are available in insulated versions. Typically whichever side of the connector is supplying the power would have the insulation around it to prevent accidental grounding were it to be dropped against the body of the car while the power was on.

Butt Splices

Butt splices are cylindrical crimped connectors used to permanently join sections of wire. Their use is straightforward – strip the end off each piece of wire, insert both bare ends into the connector, and crimp. The generic inexpensive ones shown on the left in **Figure 2** are a bit on the bulky side. More importantly, they are *not watertight*, and thus are best suited for interior applications. Further, this kind of connector is blind – there is no window through which you can see where the wires actually are relative to the crimper.

Several higher quality versions of butt splice connectors are also available. In **Figure 2**, the connectors in the middle have a window through which the correct position of the wires can be seen through the integrated translucent insulation and verified prior to crimping. The highest quality butt splices are the small mil-spec (MIL-S-81824) aircraft-quality connectors shown on the right. These have a very compact barrel-crimp connector, barely larger than the copper wire itself, with a window like the connectors in the middle row, plus a separate heat shrink glue-lined tubular housing. When the housing is slipped over the crimped barrel and heated, the glue melts, creating a truly watertight connection. The downside is that, because these are mil-spec connectors whose shape is different from the generic ones, they require a professional-level crimping tool with a professional-level price.

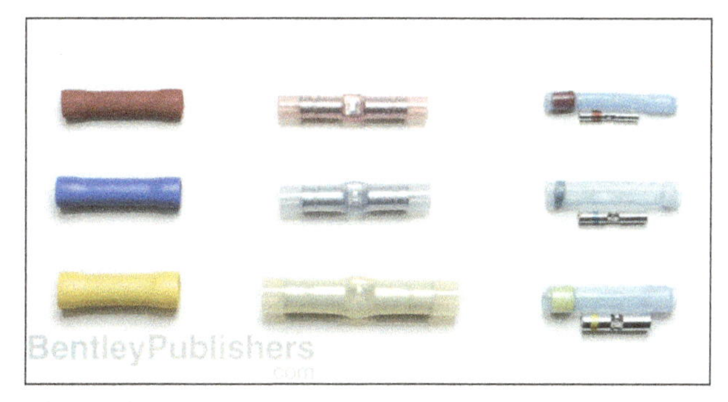

Figure 2 Three kinds of butt splice connectors – generic for interior applications (left), with integral heat-shrink tubing (middle), and small aircraft-quality connectors with self-sealing heat shrink tubing (right).

Tales From the Hack Mechanic: Knurled Nuts

What about using knurled nuts to join wires? The kind that are used in household and audio electrical connections? They're not really mechanically secure enough to use in a high-vibration environment like an automobile. In truth, I use knurled nuts all the time as a temporary measure when I'm figuring out how to wire something up. But once I figure it out, I replace them with butt splices.

Bullet Connectors

Bullet connectors (**Figure 3**) aren't as ubiquitous as quick disconnect terminals, but they serve a similar purpose of allowing make-and-break functionality. Unlike push-on connectors that are present on relays, starters, and other components, bullet connectors are rarely an integral mounted part of a piece of equipment. However, because their diameter is narrower than quick disconnects and because the female connector naturally surrounds and insulates the male, they are sometimes found in the middle of wires leading to components that may need to be removed. For example, the air-conditioning compressor is often attached to the wiring harness via a bullet connector. The female bullet is typically on the power side of a connecting pair, and is typically fully insulated to prevent accidental grounding. When mated, no part of the connector is exposed. Some folks routinely use a male/female pair of bullets instead of a butt splice, thereby allowing the joined connection to be dismantled easily if necessary.

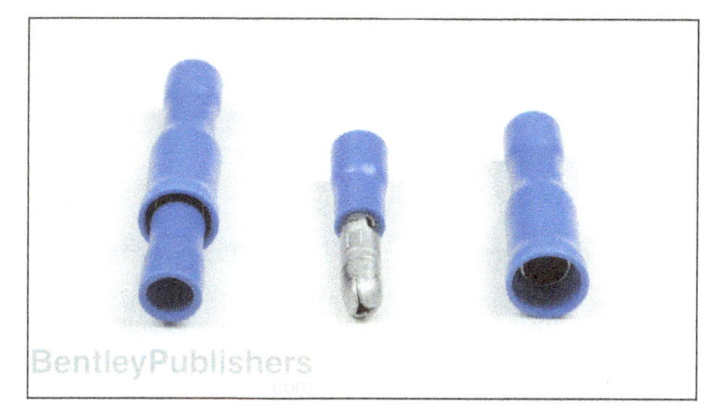

Figure 3 Male and female (middle and right) bullet connectors. When mated (left), no part of the connector is exposed.

Post Connectors (Ring and Spade Terminals)

Post connectors allow a wire to be attached to a threaded post or screw. On a car, they're used mostly for attaching wires to ground. There are three types: Ring terminals, spade terminals, and hook terminals.

Ring terminals have a hole in them and thus completely surround a post. This has the advantage of making it unlikely the terminal will accidentally come off, even if the nut or screw is loosened, but also makes it so that the nut *must* be completely removed in order to pull off the terminal. This can be maddening when you're lying on your back on the driver's seat with your head up under the dashboard trying to remove a ground connection from a threaded post deep behind the instrument cluster.

In contrast, spade terminals are like a fork with two tines, allowing you to loosen up the nut holding them and slide them out without taking the nut all the way off the post. It is easy to convert a ring terminal into a spade simply by chopping a notch in the ring with a wire cutter (see **Figure 4**), and many a do-it-yourselfer has done exactly this to facilitate an installation when there are multiple ring terminals already on a ground post in a tight spot to reach.

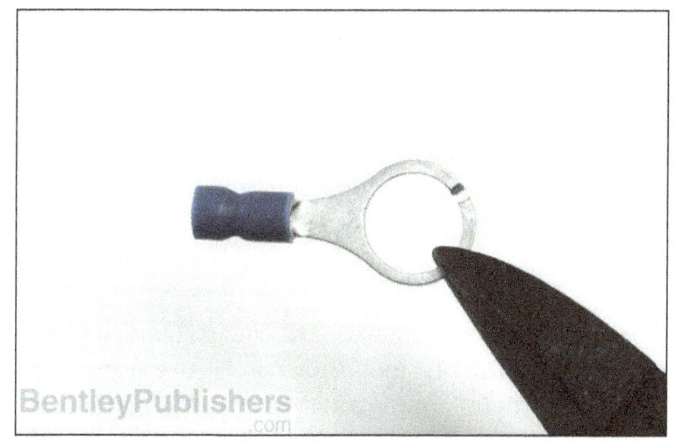

Figure 4 Making a spade post connector out of a ring terminal.

Hook terminals replace the "two fork tines" approach of the spade with a hook-shaped attachment. These are less commonly part of factory spec on European cars.

Fully Insulated, Insulated, and Non-Insulated

When a connector is described as "fully insulated," it means that the entire mating area is protected from accidental grounding. Fully insulated quick connect terminals, for example, are designed so neither half can easily ground out even when the two halves are pulled apart. See **Figure 5**. In contrast, "insulated" often simply means whether or not a connector has the color-coded plastic base. We'll return to this "insulated" issue when we discuss crimping below. Non-insulated connectors have no surrounding plastic at all and are simply bare exposed metal.

Figure 5 Fully insulated female (center) and male (right) quick disconnect connectors, and the mated pair (left).

Cutting / Stripping / Crimping Tools

To cut wire, strip wire, and crimp a connector onto a wire, you really need three different tools – one for each of those three functions.

But wait. What about those all-in-one cut/strip/crimp tools (**Figure 6**)? They probably threw one in for free with your set of AWG color-coded crimp connectors, right?

In a word, feh. They may or may not strip wire adequately. They may or may not crimp connectors adequately. And they're almost useless for cutting wire. It may be good enough in a pinch (bad pun), but you won't want to depend on it to make reliable electrical connections.

Cutting

Let's talk about cutting first. The cheap stamped steel jaws in the all-in-one tools shown in **Figure 6** are rarely sharp enough for cutting anything except the thinnest gauge wire even when new, and they dull really quickly. And they tend to mash the strands of wire, resulting in a flattened cut that doesn't fit cleanly into the barrel of the connector you want to crimp. To cut wire, you need wire cutters (also called diagonal cutters or wire cutting pliers). Having both a medium-sized and a small pair of wire cutters as shown in **Figure 7** is helpful. The bigger they are, the fatter the wire they'll cut, but the smaller they are, the more control you have cutting fine wires. The medium-sized set might not be heavy enough to chew through 2-gauge battery cable, but it's rare you need to do that. If you do, add a large pair to your shopping list.

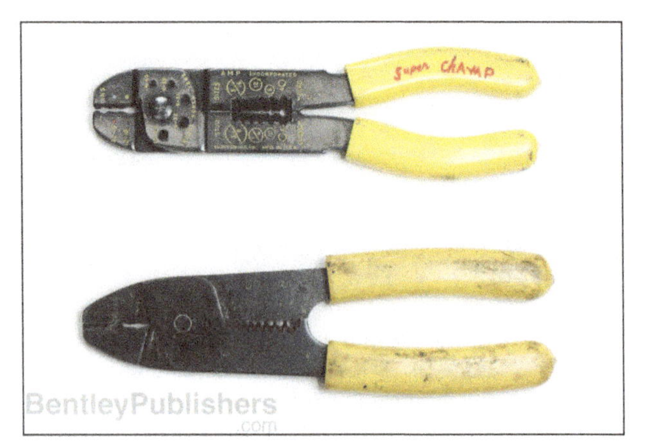

Figure 6 Two inexpensive multi-function cutting/stripping/crimping tools. The one on the bottom is the better tool due to the presence of the crimping stake and the stripping holes that completely surround the wire.

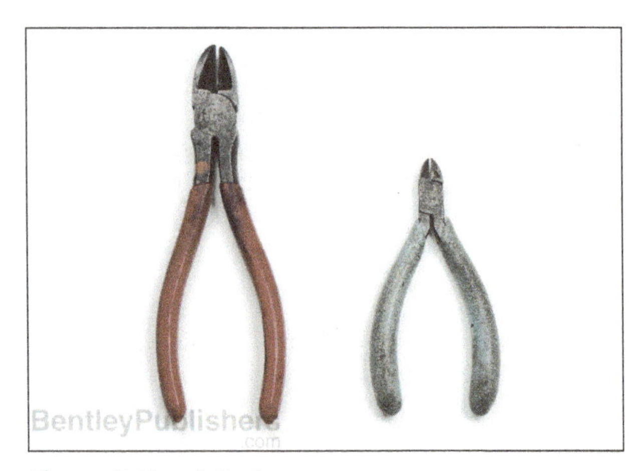

Figure 7 6" and 4" wire cutters.

Stripping

To strip wire, a dedicated pair of wire strippers is best. The two-headed Frankenstein-style ones on the left in **Figure 8** are particularly useful, as they grab and strip in one squeeze. The ones on the right are kind of nice in that they have both the metric and AWG wire sizes printed next to the holes.

However, the all-in-one tool may be fine for stripping depending on whether the stripping holes overlap to form circles without gaps. If you look closely at **Figure 6**,

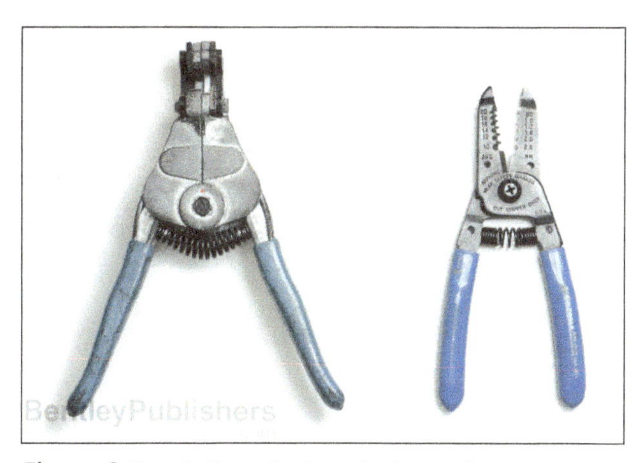

Figure 8 Two dedicated wire stripping tools.

you'll see that each has a set of stripping holes, but on the tool on the bottom, the stripping holes close all the way around the wire. On any stripping tool like this, you select the pair of holes that correspond to your AWG wire gauge, close them around the insulation about 1/4" from the end of the wire, squeeze the tool, then slide it off the end of the wire, pulling the short stripped piece of insulation with it (**Figure 9**). Then, with your thumb and forefinger, you gently twist the wire strands in a clockwise direction as you're facing them to remove any fraying and make them nice and tight so they fit into the barrel of the connector you're about to crimp. .

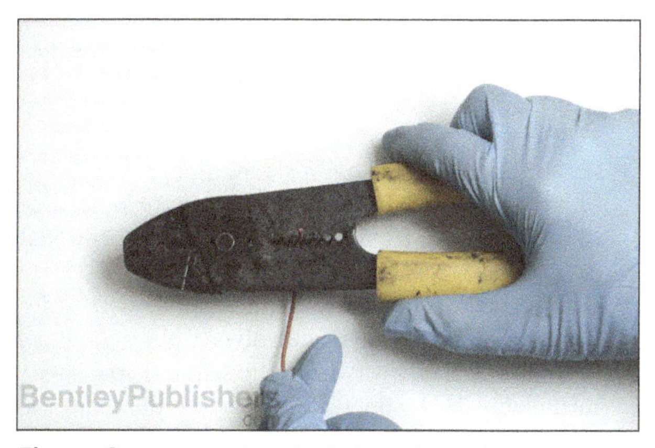

Figure 9 To strip, select the hole in the stripper that matches your wire gauge, squeeze to cut the insulation, then gently pull the wire out.

Note that, if you don't have a stripping tool, it is possible to strip wire using diagonal wire cutters by first gently using the cutters to partially cut into the insulation, manually turning the cutters around the wire, then trapping the wire between the partially open cutters, holding the wire, and moving the cutters forward. This takes practice. A stripping tool is much easier to use.

Whichever stripping tool you use, visually inspect the stripped end to make sure that you haven't torn or sliced off any wire strands (that the exposed stripped end isn't smaller than the portion inside the insulation). When you're stripping wire to crimp a connector on it, you really don't need to strip more than 1/4" of insulation because, when you put the stripped end into the barrel of the connector, the bare wire only needs to reach the other end of the barrel while the wire's insulation tucks inside the connector's insulation.

Crimping

To crimp a connector on the end of the wire, the all-in-one tool may or may not be adequate. If you look near the tip of the tool, you should see the three color-coded crimp openings, each corresponding to one of the three crimp-on connector sizes in the table above. Each opening really needs to have a stake, or prong, in it, to make a crimp that'll hold. If it doesn't, it's not crimping. It's just squashing. Go out and buy one with a stake.

Even better, buy a *ratcheting crimping tool* (see **Figure 10**). These are preferred because a) they use a compound action mechanism for leverage, applying greater force than you might without the leverage, and b) on the better ones, the ratchet won't release until you've squeezed it all the way down, thus making it less likely a crimp will fail because you didn't squeeze hard enough.

Inexpensive ratcheting crimpers are now available for less than $20. If you want to spend more money, you can buy a hydraulic crimping tool with a set of different sized removable dies (jaws), enabling you to crimp the ends on big 2-gauge battery cables.

Figure 10 A high-quality ratcheting crimping tool (top), and an all-in-one tool with both non-staked and staked crimpers (bottom).

11

How to Crimp a Connector Onto a Wire

- Strip 1/4" off the end of the wire. Remember to gently twist the stripped end clockwise with your thumb and forefinger to make sure there aren't any rogue frayed wire strands sticking out.
- Slide the stripped end of the wire into the plastic barrel of the connector. The wire's insulation should tuck inside the plastic barrel as shown in **Figure 11**. Note the arrow where the red plastic insulator narrows. That's the start of the metal barrel. You should be able to feel where the edge of the wire's insulation catches on the edge of the inside metal part of the barrel, and stop there. If you have any question, you can put the stripped wire side-by-side against the barrel and judge how far up you can push it before the insulation starts going into the metal part.
- If there's bare wire showing at the near end of the barrel, you haven't pushed the wire in far enough. If there's more than just a small amount of wire sticking out the far end, you may have pushed the wire in too far, and you may accidentally make a crimped connection around the insulation, which will leave you scratching your head wondering why it doesn't conduct electricity. See **Figure 11**.

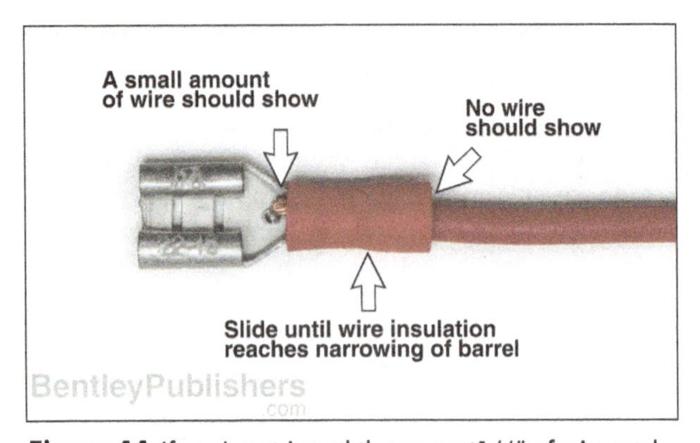

Figure 11 If you've stripped the correct1/4" of wire and pushed the wire into the barrel by the correct amount, no wire should show at the barrel, and only a small amount of stripped wire should protrude through the barrel, as shown.

- Place the wire in the crimper. If you have a non-staked crimper, use the section that's color-coded for your connector size (**Figure 12**). Then squeeze pretty hard. Don't put the entire caloric output of your body on it, however; that'll probably bend the connector.
- If you have a nice ratcheting crimper that won't release until you've squeezed it hard enough, well, squeeze it hard enough so it'll release (**Figure 13**).

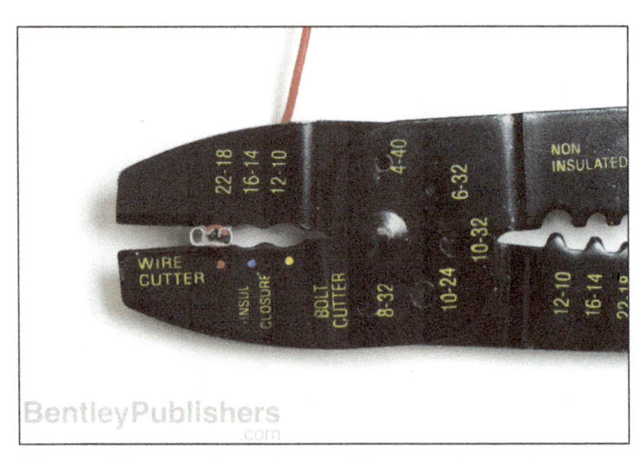

Figure 12 Using the red part of the crimping tool to crimp a red connector onto 20-gauge wire.

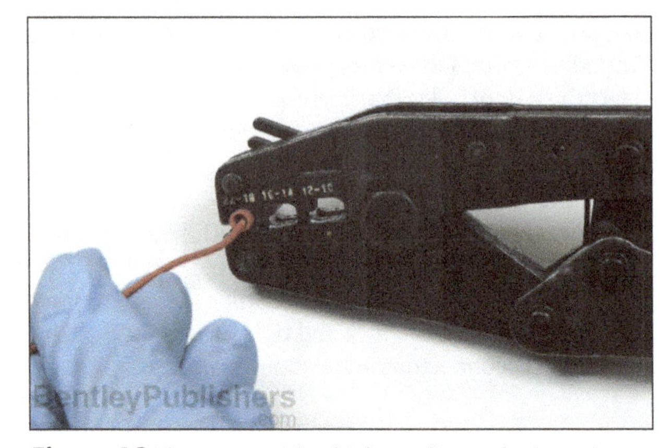

Figure 13 The same, with a high-quality ratcheting crimper.

- If your crimping tool has a stake, though, it's important that you orient the stake on the other side of the connector from its seam. See **Figure 14**. That is, if you look closely at the connector you're about to crimp, you'll see that, where the stripped wire slides in, there's a seam. You should always place the stake on the back side, away from the seam. If you mistakenly do the opposite and "stake the seam," it can widen the seam, and some of the wire strands can escape through it, making a connection that isn't as mechanically or electrically strong.
- When using a staked crimper, squeeze firmly, but not quite as hard as you would with an unstaked connector. It's pretty easy for that pointed stake to bend the connector.

Figure 14 When using a crimping tool with a stake, always position the stake on the side of the connector away from the seam, as pictured.

Note that, in several of the photos above and in the one below, we show a red-handled crimping tool that, at the tip, has color-coded non-staked openings for "insulated closure," but partway down the handle, also has a set of staked openings labeled "non-insulated," meaning they are meant for bare connectors with no plastic base. See **Figure 15**. Unless you have a high-quality ratcheting crimper, a staked crimp is always preferable to an unstaked crimp. So, on a crimper like this, with both staked and unstaked openings, always use the staked opening wherever possible. If you are crimping a connector with a plastic base, you may need to use a larger crimp setting than the connector size indicates. Simply choose the setting that fits.

Figure 15 Using the best-fitting staked "non-insulated" opening to crimp an insulated connector.

Lastly, once you've crimped the connection, hold the connector in one hand and the wire in the other, and try to pull them apart. That crimp should be on there very solidly. Better to have it slide off in your hand than fall off while going down the road. See **Figure 16**.

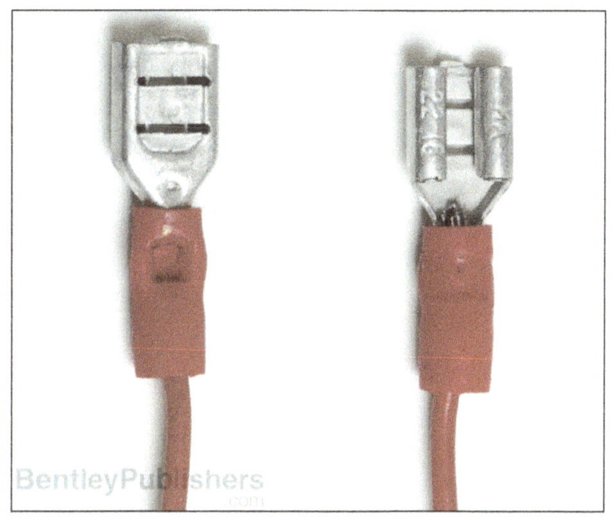

Figure 16 Back and front of the finished crimped product. On the front, notice just a small amount of wire is visible through the terminal end of the barrel, as it should. On the back, note the dimple from the stake.

Making Weather-Resistant Connections with Heat Shrink Tubing

It's one thing to crimp a nice, tight connection for use inside the car. But what about under the hood where the connection is exposed to the elements? First, understand that, unless a connector has a rubber boot on the outside and a rubber seal inside, it is not going to be truly weathertight or waterproof. However, you can try to make it weather resistant by sealing up the connection to keep corrosion-causing water out of it.

One method is to use heat-shrink tubing around the base of a crimped-on connector. Heat-shrink tubing is exactly what it sounds like – when you apply heat, the tubing's diameter shrinks. To use heat-shrink tubing:

- Cut a roughly one-inch-long piece of tubing.
- Slide it over the wire.
- Crimp the connector onto the wire.
- Slide the tubing up over the base of the crimp. See **Figure 17**.

Figure 17 Heat shrink tubing slid over a crimped-on connector.

- Then heat the tubing (there are dedicated heat guns, but a hairdryer on high usually works fine). See **Figure 18**.

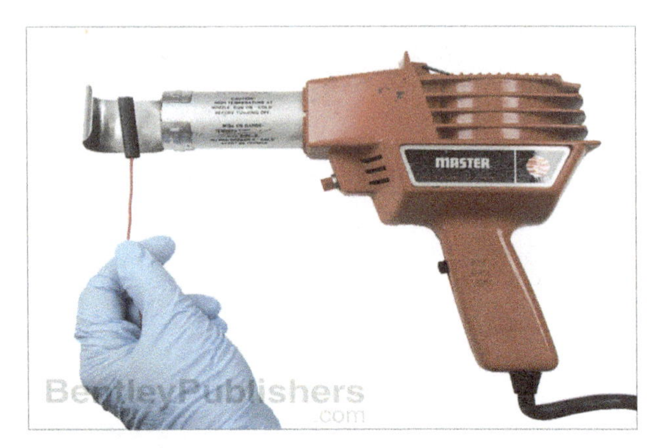

Figure 18 The connector and tubing placed at the business end of the heat gun.

The tubing shrinks in diameter, helping to keep water out of the crimped connection. See **Figure 19**. Just remember – you may need to slide the tubing on the wire before you crimp the connector. With quick connect terminals (the ones pictured in the figure), you can usually slide the tubing over the end of the connector, but with bigger connectors such as ring terminals, the end of the tubing is often too small to slide over the fat end of the connector.

Figure 19 The finished heat-shrunk product. It's not waterproof, but it will help keep moisture out of the crimp.

Instead of buying your own heat shrink tubing and cutting it, there are also "waterproof" crimp connectors you can buy with integrated heat shrink tubing.

Soldered Connections

Crimping Versus Soldering

Soldering used to be the tried-and-true method for making high-strength low-loss electrical connections. Why haven't we mentioned it until now? A few reasons.

- Many vehicle manufacturers now train their dealer technicians that soldering is verboten and crimping the preferred method for repairs.
- Soldering was once the go-to method for many DIYers. But the ratcheting crimping tool produces a strong crimp and is easier for many to do correctly than soldering.

Nonetheless, soldering still has its place, particularly when there are so many wires in such close proximity that using butt splice connectors simply isn't possible due to their physical size. In these cases, soldering and heat-shrinking produce a compact, robust result. Soldering is also recommended for high-amperage connections such as a ring terminal on a battery or alternator cable.

Warming the Wire Joint

The key to making reliable soldering connections is not to cold-solder the joint. That is, don't just heat the solder up with the iron and drip solder on the joint. Instead, heat the joint up with the soldering iron, and use the hot joint to melt the solder. Do this, and your solder connections will outlast your car.

Heat Conductivity

Be aware, though, that soldering does heat up wire, and you don't want that heat traveling to any sensitive electronics. As long as you're soldering connections that are several feet away from any electronics, you're probably fine, but if you have any doubt, don't do it. The nicer soldering irons have an adjustable heat range so you can dial them up to the minimum temperature that heats the solder. The heavier the wire, the more you need to heat it up to get it to melt solder. If you have access to an adjustable one or want to buy one, great, but if not, soldering irons in the 40 to 60 watt range should have ample output to solder wire down to about 12 gauge and can be purchased for less than $20. If you need higher wattage to solder thicker wire, first see if it's possible to simply replace the length of wire instead of soldering. Then be sure that there are no electronics close to the end you're soldering.

Solder Material and Flux

As far as solder itself, there is electrical solder and plumbing solder. Use electrical solder only. Solders have a core that contains a "fluxing agent" that removes oxides from the material being soldered to help make a better bond. Because of the high amount of oxidation typically on plumbing pipes, plumbing solder usually has an acid core to help clean the part and help the solder adhere. To make matters worse, the acid flux is hygroscopic, meaning it absorbs water from the atmosphere.

 **CAUTION —**

Never use solder labeled as "acid core" on an electrical application! Use only "rosin core" solder, which is specifically meant for electrical applications.

How to Solder Wires

- Strip at least 1/2" off the insulation of both wires. The amount you need to strip is a function of the wire gauge. For thin wire, 1/2 is fine. As wire gauge increases to 14, then 12, then 10, you may need to strip over an inch off each end.
- Take each stripped end, one at a time, and with your thumb and forefinger, twist the strands together to make a nice, tight bundle.
- If you are going to heat-shrink the soldered connection (and we strongly recommend that you do), cut at least a one-inch-long section of heat shrink tubing and slip it over the ends of one of the wires. As with the amount of stripping, the length of the heat shrink tubing will depend on the wire gauge. Once the wires are soldered together, if both of the other ends are permanently attached to things, you'll have no way to get the heat shrink tubing on if you don't do it right now.
- Hold one end of each wire in each hand, and cross the two stripped ends at their midpoints as shown in **Figure 20**.

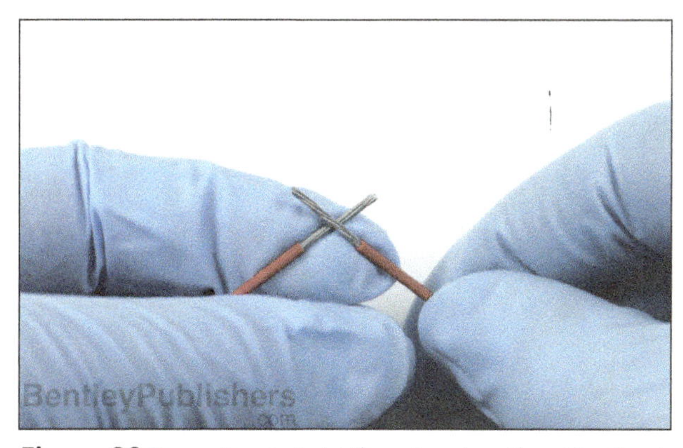

Figure 20 Preparing to twist the wires together. The heat shrink tubing has already been slid onto the wires and is below the gloved hands.

- Using both hands, twist the wires together along their length. See **Figure 21**. There should be enough wire so that the twisted joint will hold temporarily by itself. With thin wire, this is trivial, but the thicker the wire, the more difficult this is to do. You may find you need to strip off additional insulation to get the joint to hold. Another method is to bend both wires into hooks and then twist the hooks in order to increase the mechanical strength of the connection before soldering. With thin gauge wire, that's fine, but as the wire gauge gets fatter, it can make the connection so fat that heat shrink tubing of the appropriate size can't pass over it.

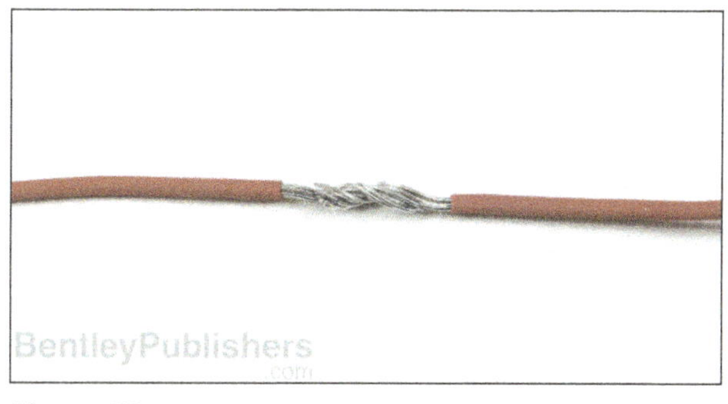

Figure 21 The wires twisted together.

- The wires you're soldering may be in the engine compartment, but also may be in the interior of the car. While it may be possible to rest the wires on a temporary surface like a block of wood while you hold the soldering iron and the solder, it is often convenient, if not outright necessary, to have a device called a "third hand" – basically a pair of alligator clips on an articulating base – to both hold the joint together and keep it above whatever surface the work is occurring over. This is pictured in **Figure 22**. With the wires on the third hand, re-twist the strands at the joint to make them as compact as possible. (Note that we have put heat shrink tubing over the ends of the third hand's alligator clips so they don't bite into the insulation.)
- Now, take the soldering iron, heat up the joint, and touch the solder to it. Keep some space between the soldering iron and the solder, letting the joint melt the solder rather than having the tip of the iron melt the solder.
- ***Do not cold solder the joint by simply dripping solder onto the wires!*** Move the soldering iron and the solder along the joint, watching to verify that solder is flowing into the crevices between the individual wires and strands. See **Figure 23**.

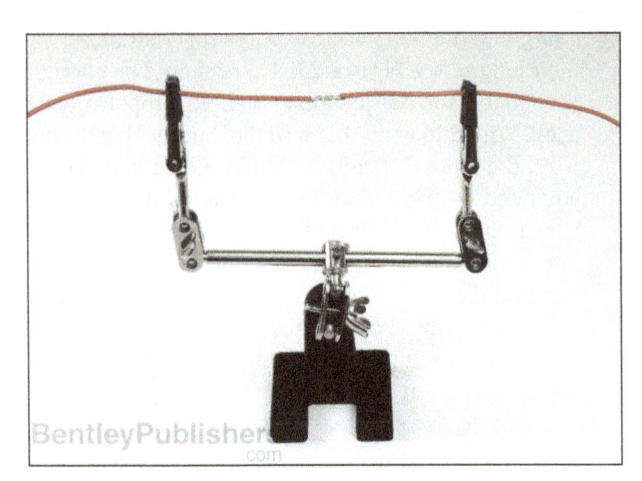

Figure 22 Using a third hand to hold the wires.

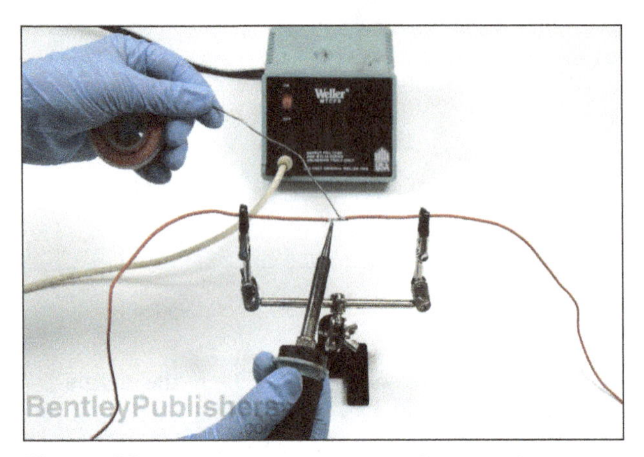

Figure 23 Soldering the connection, keeping the soldering iron and solder apart so the joint itself melts the solder, not the iron.

- The finished joint should have solder flowed completely through the joint. Individual strands of wire shouldn't be visible (since they should be surrounded by solder), and there should be no balls or blobs of solder. See **Figure 24**. If there are, the joint was not heated thoroughly (was cold-soldered). Heat it thoroughly with the soldering iron and add more solder as necessary until the joint is correctly soldered.

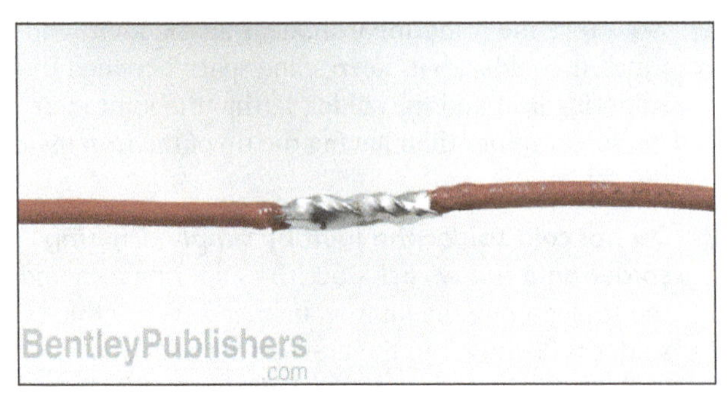

Figure 24 The soldered joint. Note how solder has flowed into all crevices of the twisted pair of wires.

- Slide the heat shrink tubing over the joint (you did remember to slide it over one wire before you soldered them, didn't you?) and heat it with a heat gun or hair dryer. See **Figure 25**. Sit back and admire your work.

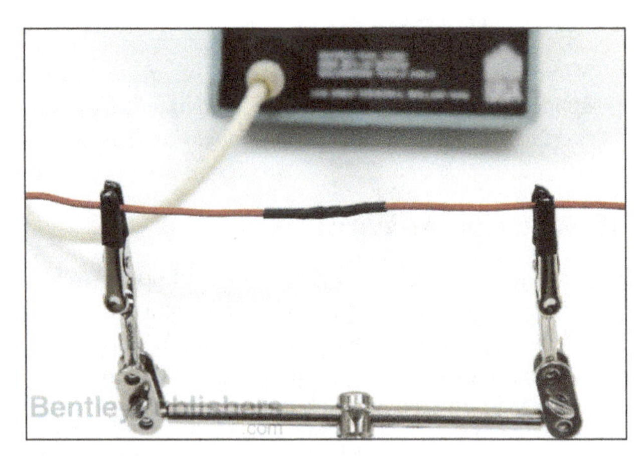

Figure 25 The finished heat-shrunk product. It's not "waterproof," but it will help keep moisture out of the soldered joint.

That's it. Go forth and strip and crimp and solder like a pro. It takes a little time, but it's very satisfying knowing that you've built connections that fix problems, not cause them.

12

Wiring Harnesses

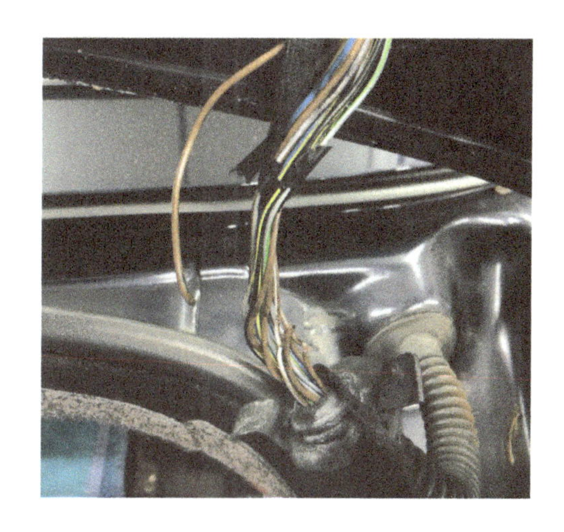

What Is a Wiring Harness?

Nearly all wires in a car are grouped into harnesses – bundles of multiple wires, together with sheathing to enclose them, rubber grommets to seal where they run through body panels, and connectors on the end – that go to a common destination before branching off to individual components. The harnesses have had to keep pace with the exploding complexity of vehicle electronics. A Volkswagen wiring harness repair document from the year 2000 states that there are over 1,400 individual terminals in a typical late-model wiring harness, and that the cost of the harness is more than 26% of the cost of the vehicle, making it the single most expensive component in the car. Things have only gotten more complex since then. BMW states that on one of their 2015 models, there can be almost 2 miles of wiring weighing over 120 pounds.

Through approximately the mid-1970s, most cars had a single monolithic wiring harness that connected nearly everything front to back. With cars' increasing complexity, sub-harnesses appeared that wired specific sections such as the engine compartment, doors, or rear of the car, or specific components such as an optional navigation system. As such, the term "harness" began to change in connotation from "the giant single squid-like entity with tentacles everywhere in your car" (see **Figure 1**) to "a collection of wires and connectors that can't easily be separated." Even the connector-and-wire bundles used to install aftermarket radios into cars are sometimes referred to as "harnesses." In this chapter, though, when we say "harness," we mean something that is original, integral to the car, and not easily removed.

Figure 1 Main wiring harness from a 1983 Porsche 911SC, laid out like a giant squid in the sun. Even on a 25-year-old car, this is nothing to be trifled with. *Photo courtesy Weekend Rides LLC*

When you see a thick section of wiring harness, it needs to terminate somewhere. One end goes out to the individual components, but the other end runs directly to a fuse box or to some intermediate connector. The point is that the physical size causes big harnesses to either have big connectors or many small individual connectors.

When you're troubleshooting an electrical problem, harnesses often seem like they're simply in the way. It'd be much easier to trace the wire and find the break or the short circuit, you think, if the wire were right there out in the open instead of buried inside this harness. That may be, but harnesses serve important functions in terms of wire protection and vibration resistance. Something needs to provide mechanical structure to bundle wires together.

European cars typically don't use the split loom or spiral wrap coverings that American and Japanese cars use. Interior sections rely mostly on special cloth tape that sticks only to itself. See **Figure 2**. (You ever wonder why you leave a sticky mess when you use cheap electrical tape to re-wrap a wiring harness? That's why.) Certain under-hood sections of a harness may be covered with an outer PVC sleeve. Yes, if you need to, for some catastrophic reason, dig out an individual wire, it's an enormous challenge, but harnesses are necessary for the reasons stated above.

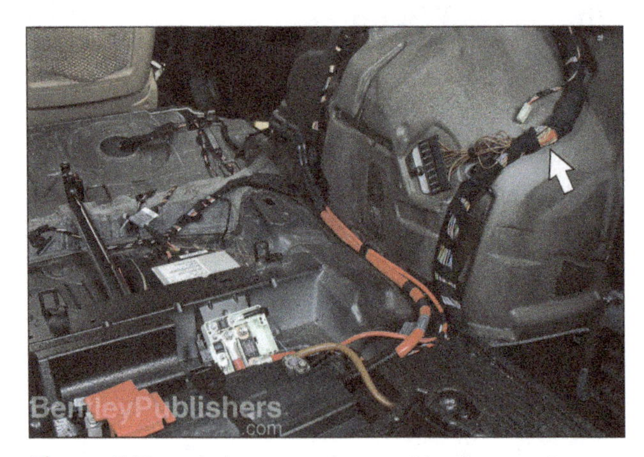

Figure 2 That cloth-wrapped rear wiring harness in a 2008 BMW X3 (**arrow**) is almost an inch in diameter.

What Goes Wrong

Wiring harness issues tend to fall into the following categories:

- Accident damage
- Damaged connectors
- Chafing
- Breaks from repeated bending
- Rodent damage
- Spontaneous systematic degradation
- Burning from an electrical short circuit
- Light harness repair and rejuvenation
- Wiring harness replacement

Accident Damage

The heavily increased use of electronics in cars has lead not only to increased complexity in wiring harnesses, but also to the ubiquity of wiring throughout the car. Simply put, this means that if a car is involved in an accident, there is an increasing likelihood that the accident may damage a portion of a wiring harness. As with other accident damage, it may take less wiring damage than you think for an insurance company to declare the car a total loss.

For example, if a car is hit in the nose, and the electrical connections to the headlights and cooling fan are severed or damaged, and if there is no identifiable line item in the insurance adjuster's worksheet for obtaining and replacing those connectors, the insurance adjuster may call for replacement of the car's harness or sub-harness. If the list price of the harness is $3,000, and if the repair estimate to replace that section of the harness is 10 hours at $125/hour, that's a $4,250 cost that, together with the other accident damage, may well total the car. For this reason, there are companies who specialize in providing connector splice kits for parts commonly involved in impact damage that allow the damaged connector ends to be replaced without replacing the entire harness.

If an accident has completely severed a portion of a harness, at least the extent of the damage is readily apparent. If the harness is small, like a two-wire door harness that only wires a light and a buzzer, splicing it back together is not terribly difficult. The problem is that multiple splices in harnesses containing multiple wires need to be staggered along the length of the wires (see **Figure 3**). If they're not staggered, they'll bunch up together, making the harness look like a snake that swallowed a pig. Not only may this cause clearance problems getting the harness to fit where it's supposed to, but also the "ball of splices" may create electrical interference in other sections of the car.

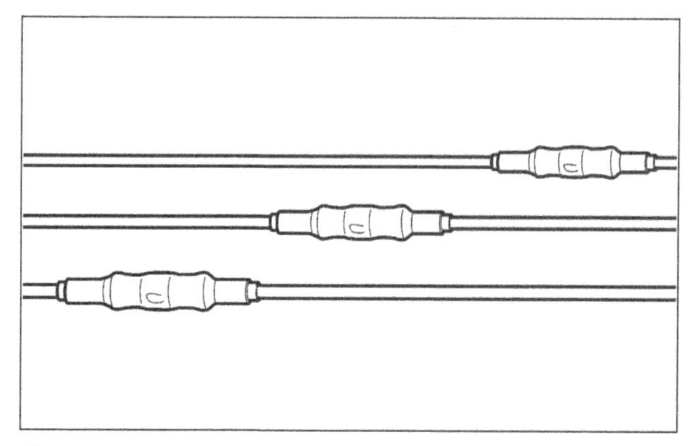

Figure 3 Any splices to a wiring harness should be staggered as shown so they do not "ball up."

The worse situation is where an accident has caused damage that is *not* readily apparent, perhaps partially severing wires that short against each other internally at a later time, creating damage whose cause is unclear. It's a troubleshooter's worst nightmare. If harness damage is more involved than a few isolated wires, you might consider leaving it to one of the companies who specialize in wiring harness repair.

Damaged Connectors

A wiring harness's integrated construction means that connectors are integral with the harness. Thus, replacement of a connector whose pins have broken off, or whose housing is cracked, or whose wires have been degraded by impact or age, is a wiring harness repair. Volkswagen's official "Wiring Harness Inspection and Repair" training document is, in fact, overwhelmingly dedicated to showing different kinds of connectors, how to disassemble them, crimp on new pins, replace rubber seals, and re-assemble.

The sequence of illustrations in **Figure 4** shows Volkswagen's procedure for correctly prying open a particular style of connector, using the correct tool to release the barbs securing the a terminal, and reinserting a wire with a newly crimped-on terminal into the back of the connector. The prying and crimping are well within the capabilities of most do-it-yourselfers because this particular connector is easily pried open and uses wires with OEM terminals which can be crimped on with a standard crimping tool. Note, however, that releasing the barbs securing the terminal requires a specific tool. With practice, it is usually possible to use a jeweler's screwdriver or a small flat piece of metal to release the barbs.

Note, however, that other connectors may be more difficult to pry apart, and that connectors using circular pins require special crimping tools to attach them to wires and special barb-retracting tools to release them from connectors. Thus, the degree to which this sort of connector repair is possible for the do-it-yourselfer varies with the specifics of the connector.

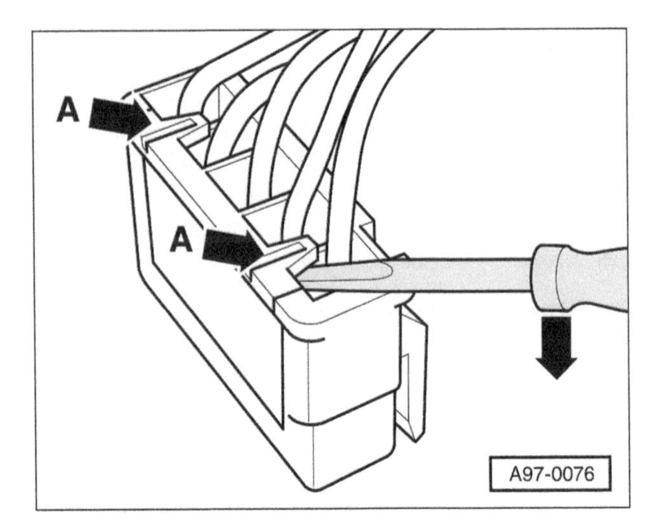

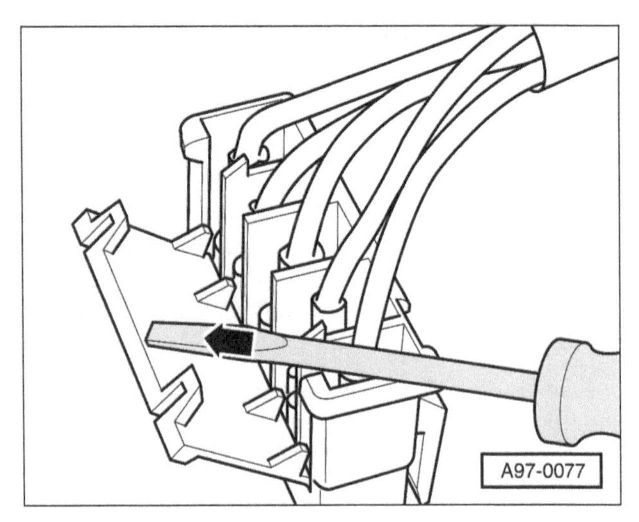

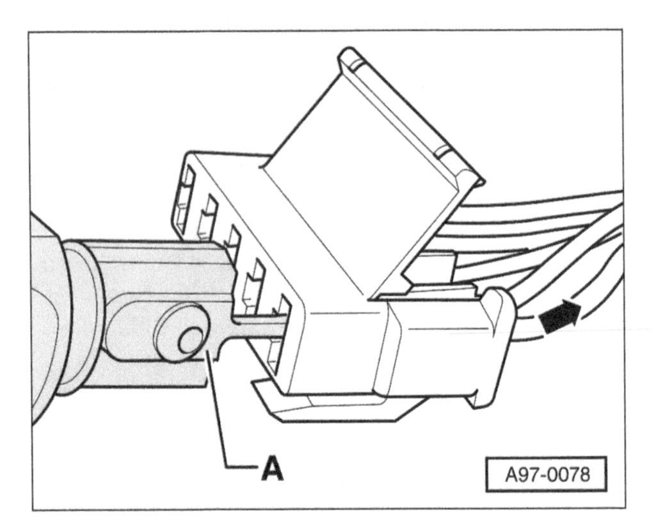

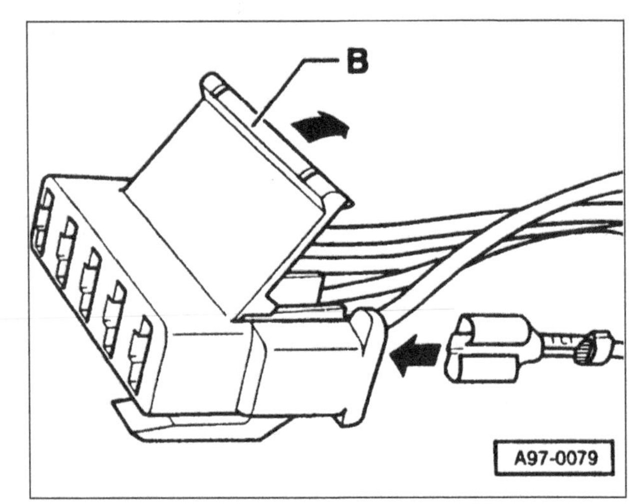

Figure 4 Proper repair of a Volkswagen relay connector. In the first two panels, the connector is pried open. In the third, a special tool is used to release the barbs holding the crimped-on terminal inside the connector and the wire is withdrawn from the back. In the fourth, a wire with a newly crimped-on terminal is inserted into the back of the connector.

If a car is a computerized daily driver used during wet weather, repairs to connectors under the hood need to be made carefully, in a fashion that keeps things as close to original specification as possible. Connector disassembly and repair may involve not only separating inner and outer shells, replacing pins, and using special crimping tools normally available only to dealerships, but also ensuring that weathertight rubber seals inside the connector are intact.

As an alternative to disassembly and repair, as we said when describing accident damage, it may be possible to obtain a cut-off connector with a section of wired pigtail from junkyard wiring harnesses or other vendors in the aftermarket and splice it in, but if the connector is used under the hood, watertight splices like the aircraft-quality ones shown in the chapter on **Wire Repairs** must be used, and the splices should be staggered. Whether this is feasible depends on the number of wires in the connector. It's one thing to splice on a three-wire fan connector, and quite another to attempt it with an ECM connector with dozens of wires.

Chafing

Sections of the wiring harness are usually secured to the body of the car using cable ties, metal or plastic screwed-down loop holders, troughs, or other mechanical structures intended to reduce or eliminate the effects of vibration. Nonetheless, with age and mileage, it is possible for vibration to cause the harness to move against the car and create wear that penetrates through the protective outer layer of the harness. From there, the wear can continue through a wire's insulation and allow the copper conductor to ground out against the body of the car.

If the wire is a passive ground wire, there may be little or no effect. If the wire is an active ground wire, such as the grounding leg of the horn relay, the effect will be to unintentionally complete the circuit (e.g., sounding the horn). The worse case is when the wire is carrying battery voltage, as this will create a short circuit to ground.

If the wire is fused, the result should be that the fuse blows, perhaps intermittently as ground contact is made and broken with continued vibration. If the wire is not fused, though, the wire's insulation and possibly the wire itself will likely burn if it is laid to ground for anything other than extremely short instances.

Systematic detailed physical inspection (meaning, looking at every inch of the wiring harness in situ) usually reveals the location of the chafing point and the damaged wire. At the chafed point, the wire can be cut, spliced, and protected with watertight heat-shrink tubing, and the wiring harness can be re-wrapped.

Breaks from Repeated Bending

Although, in general, wires are unlikely to break in the middle of a wiring harness, breakage from repeated bending motion is an exception. The wire harness section that feeds power seats is susceptible to both bending and chafing, since, as the seat goes back and forth, the wire must bend to follow it, and can chafe as it is dragged along the floor, along the seat rails, or along the bottom of the seat.

For example, reports on user forums and frequency-of-repair sites indicate that BMW E46 3 Series cars often suffer from wire breakage inside the portion of the harness that stretches to the trunk lid. See **Figure 5** and **Figure 6**. The driver's door of the Volkswagen A4 platform (Audi A3, VW Golf Mk4) reportedly exhibits similar wire breakage problems. Again, web forums are excellent places to check if there are commonly reported bending/breaking problems on your car.

Figure 5 Due to repeated bending, the trunk lid harness is a known problem area for broken wires.

In any case, when intra-harness breakage is suspected, the harness must be opened, and the broken wires located and examined. They then must be spliced in a staggered fashion and protected with watertight heat-shrink tubing, and the harness closed.

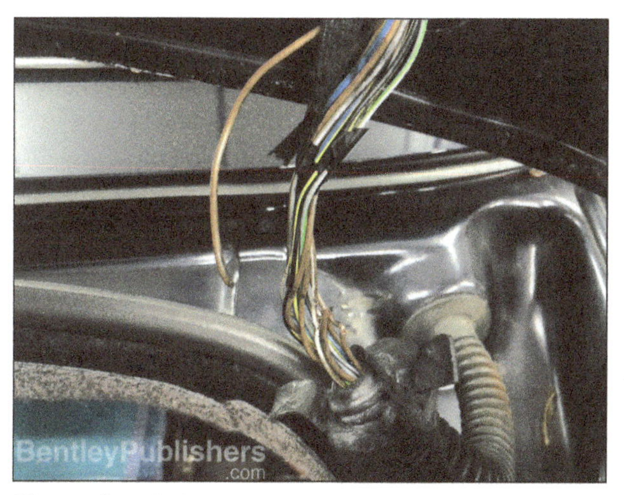

Figure 6 With the sheathe pulled back, broken wires are visible.

Rodent Damage

Reports of wiring being damaged by rodents is nothing new, particularly with cars that are stored outside. Although one school of thought is that rodents simply gnaw on convenient objects to keep their teeth sharp, the insulating materials themselves may be leading rodents to mistake wire for an actual food source. An increasing number of manufacturers appear to be switching from PVC-based to biodegradeable soy-based insulation. Preventive measures include keeping the car garaged, letting a cat patrol the space, and using a variety of products from mothballs to coyote urine (seriously). The damage itself, though, must be addressed like any other harness and connector damage by splicing wherever possible.

Systematic Degradation of Insulation

Mercedes appears to have been an early adopter of biodegradeable soy-based insulation, with use coming in, depending on the platform, around 1991 and continuing through about 1997. The biodegradable insulation appears to have been used throughout the car, but the results appear to have been disastrous in the under-hood areas due to the increased heat. User forums are full of complaints of systemic failures of the insulation literally crumbling off the upper and lower engine wiring harnesses and the harness for the Electronic Throttle Actuator (ETA), particularly on cars with LH-Jetronic injection. Because the problem is systemic, replacement with upgraded PVC-based harnesses is the only viable option.

Burning from a Short Circuit

This is the electrical nightmare scenario. You've smelled "the smell," that acrid scent of burned insulation. Something has toasted. Maybe a fuse should've popped but didn't. Maybe the circuit was unfused. Hopefully you shut it off, or you detached the battery, or the wire whose insulation burned eventually melted completely through, causing current to mercifully stop flowing. But now, here you are, looking at the train wreck.

Why Did It Burn? Assessing the Damage

There are two steps that are equally important: Figuring out what the cause was, and assessing the damage. These steps are, of necessity, intertwined.

As we discussed in the **Circuits** chapter, wires get hot and insulation burns when a wire is called on to carry more current than it's designed for. This can be divided into two classes.

There's "above rated current but not insanely so," where the wire carried more current than it was rated because the device at the other end either was modified or malfunctioned. For example, if the bearings in an electric motor wear out, or if the motor seizes, the current it draws may increase substantially. Perhaps this had been going on for a while, and a previous owner replaced a 20-amp fuse with a 30-amp fuse. When this happens, it's possible for the insulation on a wire to get hot enough to melt but not actually burn, or to burn, but in a localized, less than catastrophic way.

But then there's "insanely above rated current," which is almost always due to a wire carrying 12 volts getting short circuited to ground, with no fuse to interrupt the current flow. This can cause hundreds, perhaps thousands, of amps to flow, instantly turning the wire into a light bulb filament, and burning everything it touches from end to end.

The "end to end" part is key. You need to be able to bound the problem, to be able to know what portions of the harness are affected. Even if you can't directly see the wire, burning will leave telltale marks. Look for a line of discoloration on the outside of the harness and on any plastic connectors.

Exposing the Damaged Section

Having assessed the damage and bounded it physically by finding its endpoints, you need to expose the damaged wiring. On a car where the harness is wrapped in cloth tape, you need to unwrap or cut the tape. But if the section of the harness has multiple wires inside an outer PVC sleeve or jacket, you need to cut it open, gutting it like a fish, preferably using small scissors like a surgical instrument that are angled upward, minimizing the chances of cutting or damaging other wires inside the harness.

Tales From the Hack Mechanic: Burned Wiring Harness

Many years ago, I was driving a beater BMW 2002, little more than a running parts car. I parked it to run an errand. While I was inside, I heard a horn honk and not stop. What a jerk, I thought, just leaning on his horn like that. But after 30 seconds, I thought something was clearly wrong, like someone in a parking lot having a heart attack. I went outside, and found my 2002 filling up with smoke. I detached the battery before the car caught fire. I later figured out that I'd left the headlights on. They were aftermarket lights that drew more current than the original ones, and appear to have been jury-rigged without the benefit of a fuse or a relay.

Why did it happen at this time and not before? Don't know. But the burned wires went from the headlights, up the main part of the harness, through the firewall, and all the way up to the high beam stalk on the steering column. The horn honking was just a by-product of the damage, as somewhere its insulation got burned off as well, and the bare wire laid to ground, causing the honking.

These days, rust-free round tail light BMW 2002s are valuable cars and this was damage that would be worth repairing, but in 1984, it was simply a tired dented old car, and this was the event that cemented its fate as a parts car. But, if I were to have repaired the damage, I would've needed to complete the "end to end" assessment. Did the burning, for example, continue to the ignition switch?

In either case, you then need to use your fingers to separate the wires making up the harness, find the primary burned wire, and see what other damage it caused during its fiery journey. If a wire was "insanely above rated current," expect to find that it took a few others with it (that it melted the insulation of several adjacent wires).

Light Harness Repair and Rejuvenation

If you own a classic car and have restored the outer body and interior, the next aesthetic challenge is often prettying up the engine compartment. Typically the under-hood section of the wiring harness requires attention to match the appearance of other restored areas. We have described connector damage from accidents, but degradation can be more gradual. The under-hood section of the wiring harness suffers more than any other section from exposure due to weather, engine heat, and the presence of automotive fluids. With age and mileage, it's not uncommon to find connectors that have broken off or are about to. Due to engine vibration and heat, this is particularly common on starter, alternator, and temperature sensor connectors mounted directly on the engine.

Fortunately, on a car built before the 1990s, many of the under-hood connections are common single wire quick-disconnect connectors that can be easily replaced using the techniques described in the **Wire Repairs** chapter. On an enthusiast car that's only driven lightly during fair weather, a sheath of heat shrink tubing around the newly crimped-on connector provides a reasonable level of both aesthetics and weather protection.

Unless a harness has suffered a catastrophic event such as burning, many folks find that a systematic in-place rejuvenation can yield satisfying results. Rejuvenation typically involves the following steps:

- Label and photograph the locations of all connectors.
- Disconnect one branch of the harness at a time, undoing all connectors, then popping out or cutting the cable ties that hold the harness in place against the body.
- Inspect all connectors.
- Replace any connectors that have broken off or are about to.
- Unwrap the old cloth tape.
- Clean the wires with a de-greasing agent.
- Inspect each wire for cracks in the insulation.
- If the stranded copper conductors themselves are not damaged, heat shrink tubing can be used to protect cracked or split insulation.
- If the damage is more than superficial, cut out the damaged section of wire and splice in a new section of the same wire gauge.
- Re-wrap the section with new self-adhesive cloth tape.
- Replace the cable ties that hold the harness in place, either with new OEM cable ties if available, or with zip ties of a similar color.

Wiring Harness Replacement

First of all, unless the car is, roughly speaking, a pre-1975 car with a fairly simple wiring harness, we don't recommend harness replacement at all unless the car is highly valuable and/or sentimental and there is absolutely no other option. The newer the car, the more challenging wiring harness replacement is. As you get into post-1995 OBD-II computer-laden cars, removing and replacing an entire wiring harness is likely to stretch into months of nights and weekends. If you're looking at buying a car cheaply because it had an electrical fire that burned the entire harness, you'd be wise to seriously consider finding a different car.

Aside from the general instructions to photograph everything thoroughly and remove and label one connector at a time, the steps needed to replace a complete wiring harness are specific to the particular make and model of car, and thus are beyond the scope of this book. Searching a web forum for your particular car and finding a detailed post from someone who has blazed this path is highly advised.

Availability

If the harness is a relatively simple one for a relatively simple car, and if a combination of poor appearance and lack of functionality make you think that you have no choice but to replace it, the next question is whether a replacement harness is even available. As with many things, it depends on the market. For example, in the BMW 2002 world, the availability of replacement wiring harnesses appears to be in doubt. Folks report that the rear harness appears to still be available from BMW, but the front one is NLA.

Before the advent of modern plastic (PVC-based) wire insulation, automotive wiring, like household wiring, was insulated with braided cloth. The date of the changeover varies with manufacturer. Volkswagen appears to have used cloth wiring through 1952, Porsche about 1954. For safety reasons, harnesses using cloth-braided wire are no longer manufactured. In the pre-1975 British car world, you can buy harnesses where cloth is woven over modern PVC insulation to provide both safety and original appearance. Clearly there is value in full harness replacement if the original harness uses cloth insulation and thus is unsafe.

The manufacturing of reproduction wiring harnesses appears to be a niche business, with companies specializing in muscle cars, or British cars, or hot rods. That's a good thing. You want someone with the deep domain knowledge to ask the question "is it a late '74 with the branch off the main harness for the control of the carburetor dashpots?" Some companies say they can build anything to spec. The problem, of course, is knowing what "spec" is. If you have an original wiring harness to send, there are companies who can reproduce it. But if you have the original harness, rebuilding it may be a more pragmatic road than full replacement.

Buying a Used or Reconditioned Wiring Harness

If there's no possibility other than replacement, with the sky-high cost of new wiring harnesses, it's tempting to purchase a used harness out of a wrecked or otherwise parted-out car. Be very careful. There's a good possibility you may be buying someone else's problems. The harness may have already suffered accident or flood damage. However, note that there are places that sell reconditioned/rebuilt harnesses (the phrase "reprocess the harness" is sometimes used).

Correctness

Depending on the car, there may be several options available for harnesses. In the vintage British car world, for example, you can choose between less expensive harnesses that wear their PVC insulation for the world to see, and more expensive ones that cover the insulation in color-correct cloth braids. And in the muscle car world, you can find decidedly non-original but easy to use harness kits where the destination is actually printed directly on each wire (e.g., "left low beam"). Only you can decide what level of originality is appropriate for your goals and your budget.

Grommets and Connectors

Since a wiring harness inherently connects things to other things, it needs to have the connectors on the ends. On the one hand, anything that calls itself a replacement wiring harness should have the connectors on it, but you need to check that this is in fact the case, as certain connectors may be NLA, and you may need to cut the connectors off your old harness and splice them onto the new one.

At each point where a wiring harness goes through the firewall, there is a rubber grommet. Thus, removal and replacement of the harness requires dislodging the grommets from of the holes and pulling the wiring through them. On 40-year-old cars, these grommets have typically become rock hard and are very difficult to remove without cutting them. Pulling the connectors though the grommets may also be quite difficult. Be certain to verify that new rubber grommets are available. If they're not, an enthusiast web forum may be helpful in identifying "close enough" replacements from other cars.

Figure 7 The complex door wiring harness from a 2006 BMW X3.

Replacing Sub-Harnesses

In the past, a harness for your year/make/model car was no different from any other part. However, with the myriad of electrical option packages available for a given model, coupled with the possibility of mid-year model changes, some manufacturers now report that each wiring harness is *specific to the VIN of the car.* Therefore, if mechanical or body parts containing sub-harnesses are replaced, care must be taken to reduce the possibility of incompatibility. For example, if you purchase a door from a junkyard to replace your dented one, unless you have incontrovertible evidence that the wiring harnesses (**Figure 7**) of the original and replacement doors are identical, you would be wise to remove the original wiring harness from your original door and install it in the new door.

13

Adding a New Circuit

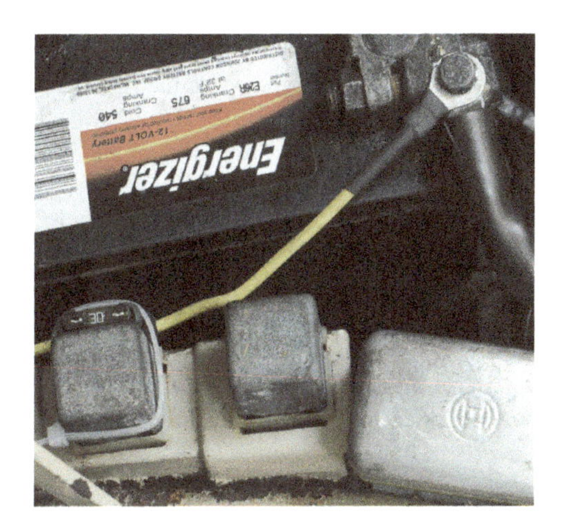

Adding a New Circuit

How exactly do you add a new circuit? It depends on the amperage, where the new device is, and your desire to retain the car's originality. Let's take a look at some the various ways to add a new circuit.

Using an Insulation-Piercing Wire Tap

Pros: Easy installation with no cutting, stripping, or splicing.

Cons: Small physical contact area. Puts fuse in unexpected location. Cuts insulation, weakening wire. Can overload existing circuit. Not a great idea.

During the go-go 1980s, when the stock sound systems in cars were awful and a new high-end stereo drew maybe 50 watts, aftermarket stereos were installed by the thousands using insulation-piercing wire taps, a kind of connector that, as its name implies, allows you to tap into a wire by piercing through a small section of insulation without cutting all the way through the wire. The connector itself does all the work, which is what made it so easy – just buy the red, blue, or yellow connector that's the right size for the AWG range of your wire, put it in place, squeeze the two halves together with pliers, and a small U-shaped cutting edge slices through the insulation and lies against the conductor. There are variants where a second wire is passed through the connector and the two wires are electrically joined when the insulation is cut (see **Figure 1**), and others that present a male push-on terminal onto which you can slide a female quick-disconnect connector.

Figure 1 The frowned-upon insulation-piercing wire tap.

The tap is usually connected to a stand-alone barrel-shaped twist-lock fuse holder typically hanging in space, not secured to anything, free to vibrate itself into non-functionality, behind an under-dash panel where no one knows it's there. Not good.

Using an insulation-piercing wire tap is certainly less invasive than, say, cutting an existing wire and splicing it and a third wire together with a butt splice connector.

However, these days, use of the wire tap is frowned upon for several reasons. Because the U-shaped cutting edge touches the wire along a narrow circumference, its use should really be limited to low-amperage applications (note that this is the case with adding *anything* to an existing circuit). And on 40-year-old cars, wire is brittle enough without cutting notches in its insulation.

Using a Piggyback Connector

Pros: Easy installation with no cutting, stripping, or splicing.

Cons: Can wiggle free. Physical size can pose risk of short circuit. Puts fuse in unexpected location. Can overload existing circuit.

Figure 2 The frowned-upon piggyback adapter.

The ignition switch on old European cars has a set of male quick-disconnect terminals on the back. One of these is for the ignition switch's accessories setting. If you crimp a piggyback connector (a female quick-disconnect terminal that has two male terminals hanging off to the side), you can slide the wire off the accessories tab, slide on your new piggyback connector (see **Figure 2**), then slide the connector from the original wire back onto the piggyback's adjunct tab. There are also piggyback adapters, basically Y-adapters that produce two male terminals where there previously was only one. This is not a bad solution for very low-current electrical devices like an air fuel gauge, but like the wire tap, use of piggyback quick-disconnects is somewhat frowned upon. It is important to be very certain you're not overloading an existing circuit. If you are using the piggyback connector in very limited space, make sure it isn't creating a clearance problem and increasing the likelihood of a short to ground.

Connecting at the Fuse Box

Pros: Puts fuse in centralized expected location.

Cons: Difficulty in accessing back of fuse box. Can overload existing circuit.

The underside of the fuse box may have additional open tabs. It depends on what the amperage is of the existing circuit and whether it has any room for growth. For example, let's say fuse #8 on your car is labeled as general accessories, and uses a 25-amp fuse. Using the 80% rule, it can support up to 20 amps of continuous draw. If you add up the draw of the other circuits already connected to the fuse and they total only 15 amps, it means that you can connect another circuit to this fuse if it doesn't draw more than 5 amps. The cigarette lighter fuse is often a good one to use, since a) it's usually a 16-amp fuse, b) fewer and fewer people smoke in their cars and actually use the lighter, and c) no other circuit is usually on the fuse. If the car is old enough to have a small fuse box held in with one or two screws, this is nearly trivial to do. Just be certain you understand which terminal is on the output side of the fuse.

Connecting at the Battery

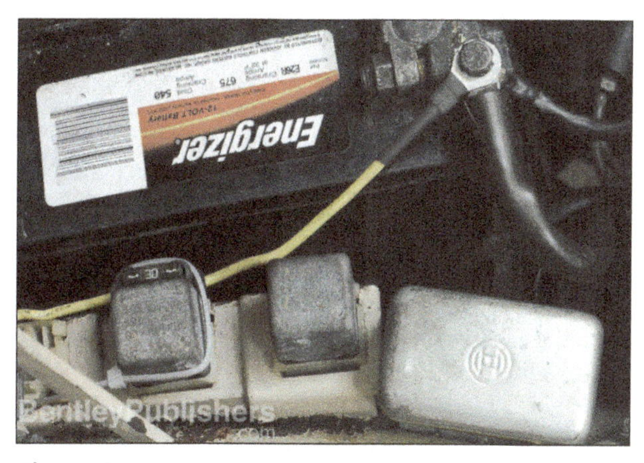

Figure 3 A regular blade fuse wired directly to the battery. Because this is under the hood, ideally a weatherproof fuse holder with a rubber hood should be used.

Pros: Puts fuse in centralized expected location.

Cons: Length of wire. Need to go through the firewall.

Wiring a new circuit directly to the battery (see **Figure 3**) is, without question, the safest, lowest risk, most reliable, most transparent method for adding a circuit. You can determine the fuse and wire sizes yourself and be certain they're correct for your device. You don't run the risk of overloading an existing circuit. You ensure reliability by wiring directly to the battery using a brand-new connector. And, with a new fuse located at the battery itself, the fuse isn't forgotten behind some random trim or access panel.

The basic steps to add a new circuit at the battery are:

- Determine the amperage of the new device by consulting the documentation that comes with it.
- Determine the length of wire required. This can be done by using a piece of rope to simulate the bends that the wire needs to take, then measuring it. If the device requires its own dedicated ground wire, remember to include that as part of the wire length.
- Use the Voltage Drop Index (VDI) technique detailed in the **Wire Gauge** chapter to determine the wire size.
- Procure a ring terminal, butt-splice connectors, and a weatherproof fuse holder for that wire size.
- Size a fuse so that the average circuit load is 80% of the fuse rating (e.g., a 25-amp fuse protects a 20 amp circuit).
- Disconnect the negative battery terminal.
- Crimp a ring connector onto the end of one of the wires from the fuse holder. Use heat shrink tubing to make it weather-resistant.
- Run the length of wire from the new device to the battery. If the wire is passing through a hole in a metal bulkhead, be certain a rubber grommet is in the hole so the hole doesn't cut the insulation. Ensure that the wire is not hard up against any sharp metal edges that might rub through the insulation.
- Secure the length of wire with zip ties to prevent movement from vibration.
- With the wire in place, crimp it onto the short lead coming out of the fuse holder with a butt splice connector. Heat shrink tubing can be used around the butt splice to make it weather-resistant.
- Secure the fuse holder itself against the body of the car with a screw or a zip tie.

Wiring a Circuit Through the Firewall

One of the biggest challenges in running a new circuit is in dealing with the firewall (the metal barrier between the engine and the passenger compartment). The older a car gets, the more originality is valued, and the more you have to make a lawyer's case for drilling or cutting into the body. However, if you opt to power the circuit off the battery, you may have no other choice but to go through the firewall.

There are a few tricks to this. One is to make use of existing points where things such as the main wiring harness, the speedometer cable, and the AC hoses pass through the firewall, and running the wires through their existing rubber grommets (see **Figure 4**). One of the problems, though, is that the older the car is, the less flexible these rubber grommets become. After 40 years, they can be hard as rocks, and if you force a wire through, it can create a gap in the grommet that can allow water in.

Figure 4 Passing a single small wire through the firewall via the rubber grommet of an existing piece of the wiring harness. Note the small gap around the new wire due to the hardening of the old rubber.

If you have a bunch of wires with a big connector on them and need to pass it through the firewall, you can disassemble the connector (as shown in the **Wiring Harnesses** chapter). The technique varies depending on the kind of connector, but if the connector utilizes barbed pins in a plastic housing, you can usually slide a very thin piece of metal like a pin or a jeweler's screwdriver in through the front, on the sides of the pins, to release the barbs and slide the pins out. See **Figure 5** and **Figure 6**. It is highly recommended that you first photograph the connector and wires and take careful notes so you reassemble it correctly. And care must be taken when passing small barbed pins through an existing grommet.

Figure 5 To pass a large connector through the firewall, first completely document the connector. In this case, the wire colors have been written right on the connector. Or even better, take a picture with your phone.

Figure 6 Next, carefully slide the barbed pins out of the connector. These can be released with a special pin sleeve tool, or with a jeweler's screwdriver.

Tales From the Hack Mechanic: Passing Wires Through the Firewall

After owning BMW 2002s for 30 years and going out of my tree trying to figure out how to pass a bunch of wires through the firewall, I learned from the enthusiast forum bmw2002faq.com that there is a hole, sealed up with a rubber plug, that was used by the manual choke cables on early 2002s. This hole was too small, but I also saw references to a nearly one-inch foam-filled hole on the right side, below the glove box. I looked, and surprise turned to joy. I removed the foam plug, installed a rubber grommet, and ran six wires through it.

Figure 7 The super-secret hole in the firewall below the glove box in a BMW 2002.

14

The Charging System (Alternator)

Alternator Basics

The battery, alternator, and voltage regulator together form the backbone of the car's electrical system. The battery supplies the electricity needed to start the car. But once the car has been started, the alternator and regulator take over, supplying the electricity needed to satisfy the demands of all of the electrical users on the car, as well as keeping the battery fully charged. The battery at that point acts mainly as a filter, smoothing out spikes in voltage that come from rapid changes in electrical demand.

Alternators are driven by a serpentine belt (or a V-belt on older cars) run from the crankshaft pulley. The bigger the load on the alternator, the greater the amount of heat it produces. The alternator housing is designed with large cooling holes, and there is often a cooling fan on the back of the alternator pulley. See **Figure 1**.

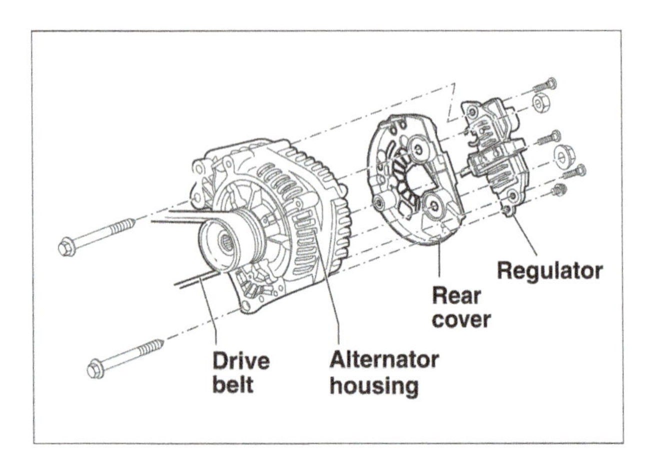

Figure 1 An A4-generation VW belt driven alternator. Notice the holes on the body of the alternator for cooling.

Alternator output increases in response to the electrical load on the charging system and engine speed. Output is low at idle and increases with RPM. Maximum output is achieved at speeds above 2,500 RPM.

The regulator bypasses or enables the alternator tens to hundreds of times per second to create an output in the 13.5 to 14.5 volt range – high enough to satisfy the car's electrical demands and keep the battery charged. Early (pre-1975) voltage regulators were stand-alone mechanical devices, while later cars use solid-state regulators that are integral with the alternator. When we say "the charging system," we mean "the alternator and the regulator," but because most regulators are part of the alternator, the terms "alternator" and "charging system" are sometimes used interchangeably.

What's the Least I Need To Know?

- Alternators/regulators are designed to keep good batteries charged. They are *not* designed to recharge dead batteries. If you are experiencing charging system problems, start the diagnosis using a good-fully charged battery. If you have to buy a new battery, or pull a known-good one out of another car, do it. Then perform the **Charging System Basic Health Test** in this chapter. This tests the battery, alternator, and regulator together.
- If you have a post-1990-ish car with electronic control modules, and the charging system is not functioning correctly, the control modules won't be fed the correct voltage. This can cause a whole host of electrical gremlins that often disappear once the charging system is functioning correctly.
- In the collective experience of the editors at Bentley Publishers, which includes owning and repairing hundreds of cars of a wide variety of makes and vintages, we find it rare that it's worthwhile cracking open an alternator. So we are not going to detail its inner workings. For the most part, alternators can be diagnosed as black boxes and replaced. But you do need to understand the alternator's purpose and what can go wrong.

A Note on Generators

Before there were alternators, there were generators. Generators produce direct current (DC), whereas alternators produce alternating current (AC) that they then convert to DC. Alternators have enough advantages over generators that generators went the way of the dinosaur in the 1960s. Detailed information on generators is outside the scope of this chapter, but most of the general troubleshooting information applies to generator-equipped cars as well.

Also, note that, for newer cars, the Society of Automotive Engineers (SAE) J1930 literature refers to alternators as "generators," but they are, in fact, alternators, and in this book we refer to them as such.

Alternator DIN Terminal Numbers

Let's whip out our DIN 72552 card and look up the connections that are germane to the charging system on a European car. These will come into play later in the chapter. They are:

DIN Terminal Designations for Alternators	
61	Alternator charge indicator
B+	Battery positive terminal
B–	Battery negative terminal
D+	Alternator positive terminal
D–	Alternator negative terminal
DF	Alternator field winding

How the Battery Warning Light Works

Most cars have a battery/alternator warning light, and it's actually quite important. When it is lit, it usually indicates the battery is not being charged.

If this light comes on while you're driving, you should stop immediately. But not because of the battery. Yes, the battery *will* run down if the alternator isn't charging it, but there's a more important reason. Most water-cooled cars have a mechanically driven water pump (note that some late model BMWs use an electric water pump) turned by a so-called fan belt shared with—guess what?—the alternator. If the fan belt breaks, or the tensioner holding it tight fails, both the alternator and the water pump stop spinning. See **Figure 2**.

While the alternator is important, continued operation of the water pump is far more important. Without it, your engine will rapidly turn into an expensive glowing Euopean lump. That's the real reason to immediately stop and turn the engine off. Seriously.

And on air-cooled Porsches and VWs (you know who you are), this is even more important because the alternator is integral with the fan that performs the air cooling. There, the fan belt really is *the fan belt*. If that light is on, you want to be REALLY SURE the belt hasn't been tossed and the fan is still turning.

Figure 2 On most cars, the alternator belt also turns the water pump (**arrow**).

Unfortunately, the converse ("alternator light is off, therefore alternator is fine") is not always true. Trust the **Charging System Basic Health Test** listed later in this chapter more than you trust the light.

However, if you own a pre-computerized European car, where one of the alternator wires is a D+ wire that runs directly to the light and not to the car's ECU, you really want to be certain that the alternator light does come on when you turn the key to the ignition setting, and not only for the reason above. Here's why.

On most European cars built before the 1990s, the alternator will not charge the battery unless the warning light is working. It is sometimes reported that the warning light is wired "in series" with the alternator. Depending on the model, this is true, but not the whole story. What's really happening is that the light itself creates the alternator's excitation current. The warning light on most old European cars is a 2-volt bulb connected between the ignition switch and the alternator's D+ terminal. Neither leg of the bulb is permanently grounded. See **Figure 3**.

The bulb will light whenever there is a difference of at least 2 volts between the ignition switch and D+. When you turn the ignition key to the run setting, one leg of the bulb is fed resting voltage from the ignition switch. The other leg (D+) should be near ground because the alternator windings are not yet energized. Thus, the difference should be about 12 volts, and the light should come on. The cool thing is that, because the other leg is connected to D+, the same current energizing the light now flows through the alternator field winds, producing the magnetic field the alternator needs to start up.

If the bulb is missing or burned out, or if the wattage of the bulb is too low, the alternator doesn't experience the load it needs to fire up. Once the alternator starts up, it creates current, which is then present at D+. The voltage at D+ should be the same as what's feeding the ignition switch (both B+ and D+ should be at charging voltage rather than resting voltage). With the voltages on both legs of the light the same, there is no potential difference, and the light should go out. Thus, the simple design of the "idiot light" is actually quite clever, and having it "in series with the alternator" isn't the act of idiocy it initially sounds like.

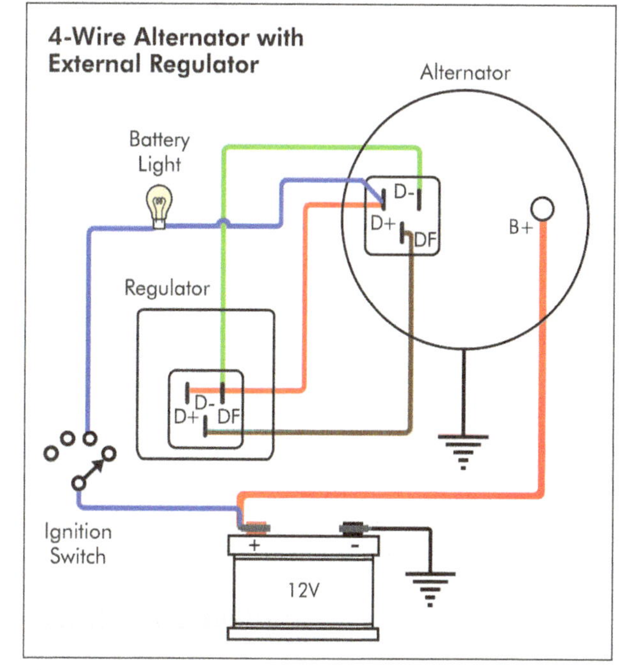

Figure 3 Basic wiring of a four-wire alternator with external regulator.

Testing the Battery Warning Light
(Pre-Computer-Controlled Alternators)

The following test is applicable to pre-computer-controlled alternators only. For a rough idea, computer-controlled alternators were used starting around the early 2000s. To identify a computer-controlled alternator, check the electrical wiring diagram for your vehicle. If the "D+" wire from the alternator runs directly to the engine control module, it is a computer-controlled alternator and this test does not apply to your car.

- Turn the ignition key to run.
- The battery warning light should come on.
- If the light is not lit, turn the ignition key back off. The problem could be the voltage to the bulb from the ignition switch, the ground path through the alternator, or the bulb itself.
- To check the voltage to the bulb from the ignition switch:
  - Set a multimeter to measure voltage.
  - Connect the meter between the D+ line and ground. On a car with an internal voltage regulator, the D+ line is usually a post on the back of the alternator.
  - On a car with an external regulator, pull the plug out of the regulator to check D+ (blue wire).
  - Turn the key on.
  - Verify there is resting voltage (12.6V) present at D+. If there is not, either the bulb is burned out, or the wire from the ignition switch to the bulb has been disconnected. See **Figure 4**.

Figure 4 Testing the voltage at the D+ terminal on connector pulled out from the external voltage regulator.

- If there is voltage at D+, try completing the D+ circuit to ground.
  - Under most circumstances, you wouldn't intentionally ground a 12-volt signal. However, in this case, you are merely completing the light bulb's path to ground. But just to be on the safe side, connect one end of a jumper with an in-line fuse to D+, and the other end to ground, as pictured in **Figure 5**.
  - Turn the ignition key to run. The battery light should illuminate.

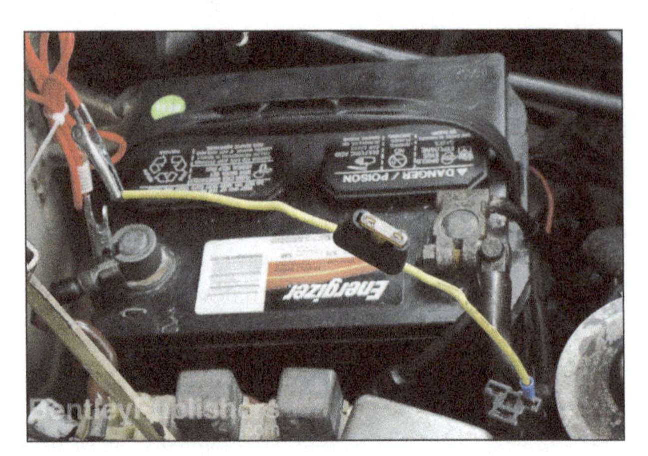

Figure 5 Testing the battery light by grounding the D+ terminal with a fused jumper wire.

  - If the battery light lit when grounding D+ but did not when it was connected through the alternator, it is possible that the alternator is not well grounded. On some early cars, a separate ground strap was used between the alternator and the engine block. Check the integrity and cleanliness of the connectors on the ground strap and retest.
- Once you have gotten the battery warning light to come on, start the car.
- If the battery warning light is now working, the light should go off once the engine is running. If it does not, there is likely a problem in the charging system. Proceed to the **Charging System Basic Health Test** below.

Charging System Testing

Are Your Lights Dim?

First, we offer not quite a test, but a canary in a coal mine. Your headlights and other light bulbs may provide a tip-off that your charging system isn't working correctly. If your headlights have become noticeably dimmer and don't brighten up at all when you increase engine RPMs, odds are your alternator isn't charging the battery at all. If your lights are dim at idle but get much brighter as you increase engine RPM, and if you're burning out small bulbs, it's likely your voltage regulator isn't properly limiting the alternator's output voltage. These seat-of-the-pants observations can give you early warning that something is wrong; the tests below will tell you for certain.

> ⚠ **CAUTION —**
>
> ***NEVER Test an Alternator by Disconnecting the Battery with the Engine Running!***
>
> *Let's deal with this one right now. Grizzled old men and impatient young ones used to conduct this brute-force test – simply disconnecting the negative battery terminal with the engine running – to see if the alternator was working. The idea was that, since this took the battery out of the loop, if the car continued to run, it was operating entirely off the current generated by the alternator, therefore the alternator must be working.* **Do not ever, under any circumstances, do this!** *In the first place, it's likely to blow the diodes in the alternator. But more importantly, the battery acts as a buffer, filtering out voltage spikes, keeping them out of the car's electronics. Without the battery in place, that filter is gone. And the biggest spike of all is the one created by the act of disconnecting the battery itself as the alternator reacts to the changing electrical load.*
>
> *Back in the day, when cars had primitive electrical systems without computers, and alternators that created much more modest output, you might have been able to get away with this primitive alternator test, but with any car with an ECM (meaning any electronically fuel-injected car, which takes us back to Bosch D-Jetronic starting in 1967), it is a recipe for disaster. Incredibly, you still see this test described on web sites today. Plus, there's just no need – you don't learn anything you can't tell from the* **Charging System Basic Health Test** *described later in this chapter.*
>
> *Perhaps the idea was, if your battery was dead, you could easily tell if it was a dead alternator that killed it. But, when you unplug the battery, the voltage spike is likely to pop your alternator's diodes, so if it was good, it probably ain't now. Just take out your multimeter and check for charging voltage of 13.5 to 14.5 volts instead. You did buy several $5.99 multimeters and keep one in the glove box of each of your cars, right?*

Charging System Basic Health Test

Basic health testing of the alternator and regulator is very straightforward. In the **Battery** chapter, we describe the quickie test where you set your multimeter to measure voltage, put its probes across the positive and negative battery terminals while the engine is running, and make sure you're reading charging voltage (approximately 13.5 to 14.5 volts). To make the test more thorough, you should:

- Begin with a fully charged battery and the ignition off.
- Turn the headlights on for two minutes to dissipate any surface charge on the plates, then turn them off.
- Start the car.
- Turn on a heavy electrical load (e.g., headlights, wipers, and front windshield defrost).
- Gently rev the engine to 2,000 rpm and hold it there.
- Set the multimeter to measure voltage, and put the probes across the positive and negative battery terminals. If the battery is not easily accessible, you can instead put the positive probe on the positive post under the hood used for jump starting, and the negative probe on any chassis ground.
- You should see "charging voltage" in the 13.5 to 14.5 volt range.
- Turn your lights, wipers, and defrost off. You may see the voltage increase momentarily, then stabilize at charging voltage.
- Turn the lights, wipers, and defrost back on one at a time. When you turn each one on, you should see the voltage dip slightly, then stabilize at charging voltage.

This test is worth its weight in, well, alternators. If the multimeter shows charging voltage throughout the test, there is nothing obviously wrong with the alternator or regulator, and you should only entertain further diagnosis if there continues to be an unexplained problem. (This assumes in this that there are no apparent mechanical issues – that the alternator spins freely without any bearing noise, and that the belt is intact and sufficiently tensioned that it isn't squealing.) But if the multimeter reads only the battery's resting voltage (12.6V) the whole time, or abruptly drops to resting voltage when the defroster is turned on, you've got a problem.

If you want to do more, you can easily check the voltage while varying both the load and the engine RPM under actual conditions by wiring your multimeter to a cigarette lighter plug and driving the car. See **Figure 6**. Even easier, you can buy one of the sub-$10 cigarette lighter digital voltmeters sold for exactly this purpose. Be aware, though, some variation in voltage is normal during this test, but if it falls abruptly to 12.6 and stays there when you turn on every accessory, you've definitely caught it in the act; there is a problem with the alternator or regulator.

Just Replace the Damned Alternator and Regulator

The **Charging System Basic Health Test** is easy because all you need to do is access the battery or the jump-start posts. However, further tests require you to access the alternator. That's trivial to do in a 30- or 40-year-old car where the alternator is usually fairly high up in the engine compartment and its electrical connections

Figure 6 A multimeter connected through the cigarette lighter is a great way to verify the presence of charging voltage while driving.

are readily accessible. However, this is not the case in many later cars where the alternator can be buried beneath the air filter box and mass flow sensor.

In a world where we like to repair our cars ourselves in an expeditious fashion, there's a temptation to jump to the most probable answer, order the part, install it, and be done with it. So if your alternator light is on, and you fully charge your battery, and the alternator/regulator fails the **Charging System Basic Health Test** above, and the car has an alternator with an internal voltage regulator, and you want to fix the car in one whack because you need to use it Monday morning, should you simply procure a rebuilt alternator/regulator and swap the whole unit in?

In a word, yes.

In a perfect world, if you can do things on a more leisurely schedule, it's certainly preferable to first do the voltage drop test listed immediately below, and if that doesn't show up a problem, remove the internal voltage regulator and examine the brushes. But if you need the car up and running the next day and can't afford to be wrong about the diagnosis, unless you have a receipt showing the alternator's just been replaced, throwing in a rebuilt alt/reg is, in fact, a reasonable approach.

Note that many of the same auto parts stores who will test batteries for free will also test alternators, though it's not a slam-dunk that they'll have the connector assemblies needed to test every single variant. However, as with bringing in a battery to be tested, it's good etiquette to buy a rebuilt alternator from them if yours tests as bad.

Alternator Voltage Drop Test

As we described in the **Circuits** chapter, high electrical resistance caused by corroded connections can reduce voltage. Therefore, before you decide your alternator is bad, you should perform a voltage drop test to check for a loss of voltage between the alternator and the battery caused by a high-resistance electrical connection.

- Start the engine and run it at about 2,000 rpm.
- Turn on the headlights to ensure there's a good electrical load.
- Conduct the Voltage Drop test as described in the **Common Multimeter Tests** chapter. Set your multimeter to DC volts, and connect the two probes along the leg of the circuit you are testing for voltage drop. For this test, that means connecting the red probe to the alternator's B+ terminal, and the black probe to the positive battery terminal (yes, positive, not negative). See See **Figure 7**.
- Ideally, the voltage at both two places should be the same, so the voltage difference measured across them should be zero. If the reading is greater than 0.4 volts, it means that excessive resistance in the alternator to battery cable is sapping voltage.
- If the multimeter reads the battery's resting voltage (12.6V) or close to it, there's likely a detached connector or a break in the wire.

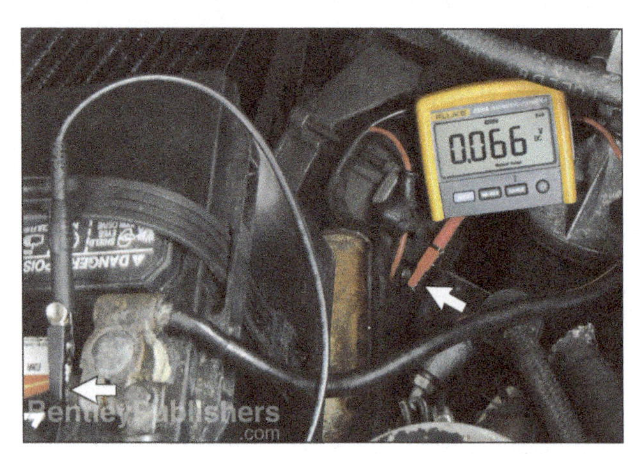

Figure 7 A multimeter connected between the alternator's B+ terminal and battery positive (arrows), measuring voltage drop. Here, the drop is very small – 0.066 mV – indicating that voltage drop is not a problem.

- Perform the same test on the negative side by connecting the multimeter between the alternator's case and the negative battery terminal.
- If the reading exceeds 0.2 volts, check all of the wires in the ground path to the alternator. There may be a ground strap between the body of the alternator and the engine block. And because the engine is vibration-isolated on rubber mounts, there is a large ground strap between the engine and either the battery or the body of the car. Verify that all these ground connections are clean and tight.

If you read this and think "can't I just connect the alternator directly to the battery and see if my charging problem goes away," you are exactly right; you're just thinking about it a different way. Using two heavy-duty test leads (like 8 gauge, with beefy alligator clips), you can connect the alternator's B+ terminal directly to the positive battery terminal, and B– (the case of the alternator) to the negative battery terminal, and then measure the voltage across the battery with the engine running. If you were reading the battery's resting voltage before but are reading charging voltage now, the problem isn't in the alternator—it's in the wiring.

If none of these tests show a problem, the B+ and B– wiring is probably fine, and you are correct to suspect the alternator and regulator.

Alternator Amperage Measurement Test

You might ask "isn't it possible to directly measure the alternator's output by putting an ammeter in series between the alternator's B+ terminal and the battery?" Yes, it is, but you can't use your multimeter in ammeter mode. When running, the current produced by the alternator will likely exceed the multimeter's 10-amp limit and will fry it. Later in the book we'll talk about troubleshooting a parasitic battery drain by connecting a multimeter in ammeter mode between the negative terminal and ground, but for this test, because you need to start and run the car, you'd need to use an old-school ammeter, and it would need to be connected between the positive battery terminal and everything that isn't the starter wire, using beefy wires secured by ring terminals and clamps that won't slip off and short to ground. See **Figure 8**.

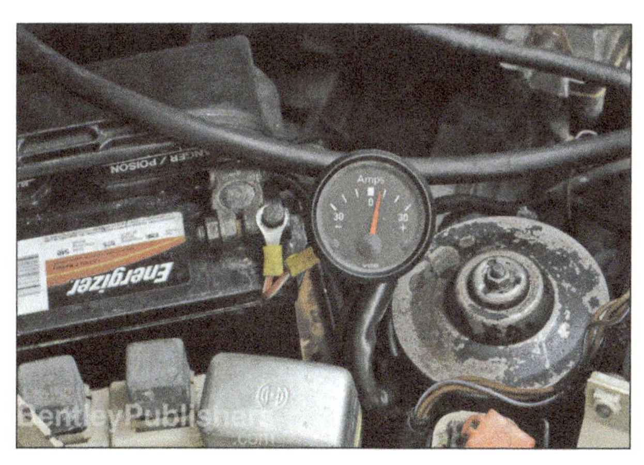

Figure 8 An old-school amp gauge connected between the alternator B+ terminal and the positive battery terminal, showing about 8 amps of charging at idle. And that's before turning on any lights or motors.

If you need an actual amperage measurement to, for example, know what the worst-case load is from your killer 1,000-watt stereo and your new Xenon headlamps while your wipers are on and the rear windshield is defrosting so you can know if you actually need that new 150-amp alternator, another way to do it is to purchase a clamp-on multimeter that measures amperage inductively, saving you from having to disconnect wires and actually splice in an ammeter in series. We discuss these in the **Multimeters and Related Tools** chapter. However, you must be absolutely certain that your clamp-on multimeter measures DC amps, as many of the cheaper ones measure only AC amperage.

Alternator AC Ripple Current Test

The alternator produces alternating current (AC), which is converted to direct current (DC) by a six-diode rectifier, which is built into the alternator. Diodes pass current in one direction only, which is how AC is converted to DC. Three positive diodes control the positive side of the AC sine wave, while three negative diodes control the negative side.

The AC ripple current test checks for bad diodes. If there's more than 0.5 volts, it's an indication that one of the diodes in the alternator could be blown. In addition to causing charging system issues, AC current can result in strange electrical issues and even damage to electronic control units.

To conduct the ripple current test:

- Set the multimeter to measure AC voltage as described in the **Common Multimeter Tests** chapter.
- With the engine running, carefully connect the red multimeter probe to the alternator's B+ terminal, and the black probe to ground.
- Rev the engine from idle up through about 3,000 RPM, and observe the reading on the meter.
- If the reading never exceeds 0.5 volts, the diodes in the alternator are likely fine. If the reading slightly exceeds 0.5 volts, you'd be advised to have the alternator tested. If the reading substantially exceeds 0.5 volts, you'd be advised to replace the alternator.

Bad Diodes?

If you own an oscilloscope or an expensive multimeter that plots waveforms (graphs the voltage over time), you can read up on what the waveforms look like when different diodes are blown. It's true that bad diodes can be a source of parasitic drain on the battery, can reduce the alternator's charging output, and can cause the alternator light to glow dimly rather than go completely out. But you don't need to go looking specifically for bad diodes. Instead, test the overall health of the charging system. If the alternator doesn't pass muster, and if the cause of that is bad diodes, it gets handled by rebuild or replacement.

Alternator Testing

In the next sections, we'll describe how to test the basic types of alternators, from oldest to newest.

Determining Which Type of Alternator You Have

Examine the back of your alternator. All alternators have a threaded post on the back labeled "B+" where a thick wire and a ring terminal connect to the battery. But there will be at least one other cable connected to the back of the alternator. That cable may contain one, two, or three wires. Thus, the alternator will have, in total, two, three, or four wires.

- If the car was built in the mid-1970s or earlier, it most likely is an **externally regulated four-wire alternator.** This is indicated by a three-wire plug as shown in **Figure 9** and **Figure 10**, connecting to an external regulator shown in **Figure 11**. Note that only these early alternators have external regulators; all later alternators are internally regulated.
- If the car was built between the mid-1970s and approximately the mid-1990s, it is most likely an **internally regulated two-wire alternator.** This configuration is indicated by the presence of a large red (B+) cable and a small blue (D+) wire, and by the bolt-on regulator as shown in **Figure 14**. The "D+" connection may be a threaded post or a push-on connector, but the label "D+" should be plainly visible next to the connector.
- If the car was built between approximately the mid-1990s and the mid-2000s, it probably has a **internally regulated alternator with an ignition wire.** In addition to B+, there's typically a molded plug on the back of the alternator with two or three wires in it. These are typically 15 (voltage from ignition), 61 (warning light, essentially the same as D+), and possibly a multifunction regulator (MFR) and a third terminal for analog control by the DME. Though the molded plug with more than one wire is a tip-off, you may need to consult a wiring diagram to see if any of the terminals in the plug carry voltage that is switched on with the ignition key.
- If the car was built after approximately the mid-2000s, it may have a **computer-controlled alternator.** The internal regulator on these alternators is controlled by the car's DME. A molded plug with a single wire in it carrying the signal from the DME is the tip-off.

On 2000 and later cars, it is recommended that you consult a wiring diagram to determine if your alternator is computer controlled. If any of the alternator's wires goes directly to the ECM, it is computer-controlled.

Testing an Externally Regulated 4-Wire Alternator

Classic European cars such as BMW 2002s, pre-'74 Beetles, and pre-1982 Porsche 911s have an alternator that uses an external regulator. There's a post on the back of the alternator for the thick B+ line to the battery as there is on every alternator, but in addition, the D+, D–, and DF lines listed in the DIN table are each present on both the alternator and the regulator; see **Figure 9**, **Figure 10**, and **Figure 11**. In spite of its period-correct charm, this design has two inherent sources of unreliability.

Broken Wires and Bad Connectors

If a car with an integral regulator fails the **Charging System Basic Health Test**, most often the alternator and/or regulator are at fault, but on cars with externally regulated alternators, the three extra wires and six extra connectors carrying the D+, D–, and DF lines between the alternator and regulator provide extra opportunity for malfunction. The failure to charge is often caused by a bad electrical connection created by age, corrosion, and the flexing of those wires.

Specifically, there are three male quick-disconnect terminals on the back of the alternator, as well as on the bottom of the voltage regulator, for the D+, D–, and DF lines. A section of wiring harness with a three-prong connector plug at both ends connects the two units. Each plug holds three female quick-disconnect terminals.

A thin metal spring clip the diameter of a paper clip is often employed to hold the plug on the back of the alternator. See **Figure 9**. In a 40-year-old car, this clip is often missing, allowing the plug to vibrate out. The connector plug bodies are plastic that can crack and crumble with age, allowing the female connectors to push out of the back of the plug and not slide firmly into their counterparts in the alternator and regulator, even when the plugs appear to be solidly in place.

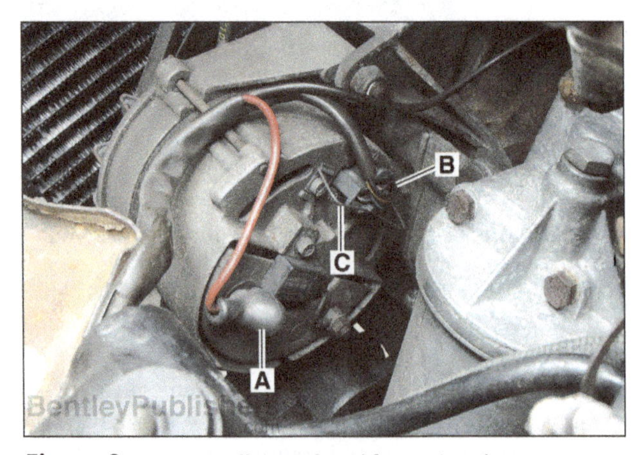

Figure 9 An externally regulated four-wire alternator on a 1972 BMW Bavaria. Note the thick red B+ wire (A), the three-pronged plug (B), and the thin spring clip holding the plug to the back of the alternator (C).

When either the plug body or the quick-disconnect terminals fail, you have three choices: 1) cut the plug off and individually crimp three new female connectors onto the ends of the wires, 2) cut the plug off and splice on a new plug assembly available through the aftermarket, or 3) retrofit a two-wire alternator with an internal regulator (described at the end of this section).

Figure 10 Blowup of the D+, DF, and D– terminals on the back of an externally regulated alternator.

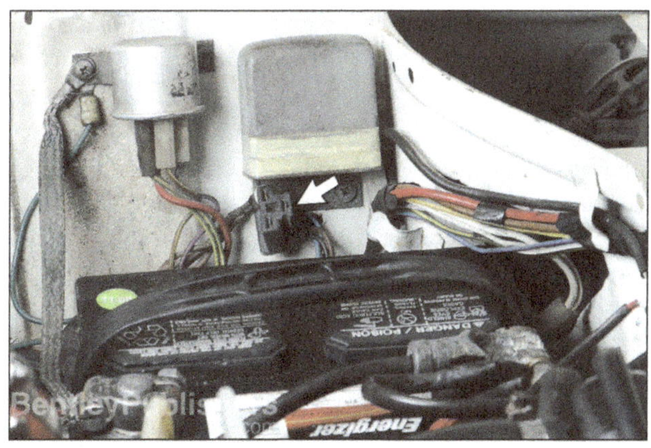

Figure 11 An old-school external mechanical regulator (silver box with yellow tape) on a 1972 BMW 2002tii. The three-pronged plug (**arrow**) has been disconnected to show the connection.

Mechanical Regulators

The second source of unreliability is the construction of the external voltage regulator itself (**Figure 11**). They are mechanical in nature, sort of a combination of an electric relay and the points used in 40-year-old ignition systems, using little electromagnets and springs to pull contact points open and closed, at tens of times per second. These are now regarded as antique voltage regulators. Age, mechanical wear, internal sparking, and exposure to weather eventually cause failure. The voltage regulators in newer types of alternators are solid state with no moving parts to fail and are integrated into the regulator itself.

Interestingly, with new or new-old-stock antique mechanical regulators getting quite pricey, many owners of vintage cars switch to equivalent plug-compatible external solid state regulators, sometimes even contained within an old mechanical regulator housing to retain a vintage look and feel.

Although it's not strictly a reliability issue, the construction of an alternator with an external regulator is typically such that the brushes can't be changed without disassembling the alternator. In contrast, newer internally regulated alternators typically have the brush pack integrated with the regulator as an easily replaceable unit.

Upgrading an Externally Regulated Alternator to an Internally Regulated Two-Wire Alternator

For all the reasons stated, many owners of 40-year-old cars who want them to be reliable for long trips update the alternator and external regulator to an internally regulated two-wire unit. To do this:

- As we will see in the next section, internally regulated two-wire alternators require two wires, B+ and D+.
- The B+ wire is already present. You can either use the existing B+ wire, or, if the new alternator has a higher amperage rating than the old one, run a new B+ wire directly to the battery using 8-gauge wire.
- For the D+ connection, you need to tap into the D+ wire that's part of the car's wiring harness between the old alternator and the external regulator. This is because that wire also goes to the wire that runs to the battery warning light, which is needed to initiate the field excitation process, as per the section **How the Battery Warning Light Works** at the beginning of this chapter,.
- You can run a new D+ wire through the firewall directly to the warning light, cut the old plug off that went into the old alternator, and connect the D+ wire directly to the alternator, but instead, it's easier to make a short jumper wire with a ring terminal at one end and a male connector at the other that connects the D+ post on the alternator to the D+ female connector in the old alternator plug (see **Figure 12**). This also preserves the possibility of reinstalling the original alternator and regulator at a later date.

Because the regulator on a four-wire alternator is external and separate from the alternator's brushes, it does have the advantage that it can be run with the regulator unhooked, allowing additional tests.

Figure 12 An alternator with an internal voltage regulator installed in a 1972 BMW 2002tii. Note the blue wire jumpering the D+ terminal on the back of the alternator to the terminal on the original three-pronged plug (**arrow**) to allow the warning light on the dashboard to function.

Alternator Internal Short Test

- First perform the "Troubleshoot Battery Warning Light" test described earlier in this chapter. If the test is not successful, fix the light or any wiring.
- Disconnect the plug from the back of the alternator. This completely removes the regulator from the circuit.
- Start the car.
- Set the multimeter to measure DC voltage, and connect it between alternator D+ and battery ground.
- With no connection between D+ and DF, there shouldn't be anything telling the alternator to make juice, so it shouldn't output anything at all, so B+ should read the battery's resting voltage (12.6V).
- If B+ is higher than resting voltage, it is likely the alternator is internally shorted.
- Shut off the car.

D+ and DF Wiring Test

- Reconnect the plug to the back of the alternator but disconnect it from the regulator.
- Start the car and repeat the test.
- If the alternator is outputting anything higher than resting voltage at B+, there is likely a short in the connector or cable between the D+ and DF lines.
- Shut off the car.

Bypassing the External Regulator Test

 **CAUTION —**

The following test to bypass the voltage regulator is described in factory repair manuals for older cars with external voltage regulators. Because the test bypasses the regulator, the test may generate voltage levels above 14.5 volts for the several seconds the test is run. Perform this test for only a few seconds maximum to avoid an overvoltage situation.

- Take a short piece of wire and jumper between D+ and DF on the connector that's unplugged from the regulator. This bypasses the voltage regulator.
- Turn the ignition key to the run setting. The battery warning light should go out.
- Set the multimeter to measure voltage and connect it across the positive and negative battery terminals.
- Start the car and run it at about 1,000 RPM **for just a few seconds** while observing the voltage on the meter. If the meter reads close to charging voltage (14.2V) with the regulator bypassed, then the alternator is outputting charging voltage, and it is likely the regulator is bad and should be replaced.

Testing a 2-Wire Alternator

Most European cars built from the late 1970s through the mid-1990s use a Bosch internally regulated alternator with two electrical connections (B+ and D+) on the back. See **Figure 13** and **Figure 14**. The regulator can easily be removed to visually examine the brushes. The alternator should not be run with the regulator removed.

As with all alternators, the "B+" connection to the car's positive battery terminal is the most important. There's a heavy-gauge wire connecting the "B+" post on the back of the alternator. This wire may go directly to the positive battery cable, or may disappear into the wiring harness.

- Inspect this wire as it comes off the B+ post. The ring connector that attaches to the post may simply have broken off the wire.
- If the ring connector is intact but looks corroded or dirty, disconnect the battery, then unbolt the connector from the B+ post, clean it with a small file, and retest.
- If there's still no charging voltage at the battery, carefully use the multimeter to test the voltage at the B+ connector itself.

- If there's charging voltage at B+ but not at the battery, the wire connected to B+ may be internally broken. Disconnect the battery and the connector at the B+ post again and use the multimeter to check the continuity between the connector and the positive battery terminal.

The second electrical connection is the D+ terminal. This is for the "exciter" wire that's connected to the warning light. It should show resting voltage when the ignition is turned on but the car hasn't been started yet. Its checkout procedure is described in the warning light section earlier in this chapter.

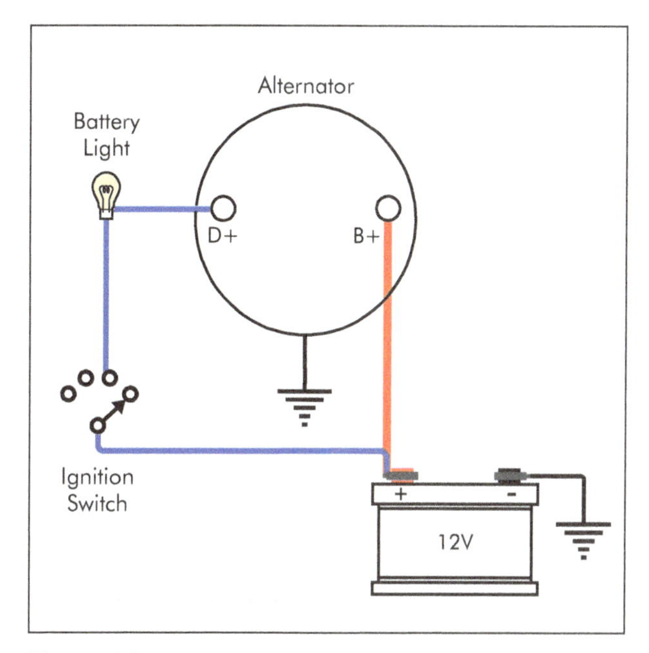

Figure 13 The wiring of a two-wire internally regulated alternator is very straightforward.

Figure 14 A two-wire (B+ and D+) alternator with an internal regulator on a 1979 BMW 635CSi. The thick red wire (1) is B+. The thin blue wire (2) is D+. The third wire at top left is a ground wire.

Lastly, if the alternator isn't charging the battery, the brushes that are integral with the internal regulator pack may be cracked or worn down. To inspect the brushes:

- Disconnect the battery and remove the regulator/brush pack from the alternator by undoing the two screws holding the pack to the back of the alternator. Depending on the car, you may be able to remove the regulator without removing the alternator by using a short screwdriver.
- Inspect the brushes.
- When new, the brushes should protrude about 3/4" from the regulator and should spring back when pushed in. If the brushes don't spring back, or are cracked, broken or worn down to the size of pencil erasers (see **Figure 15**), you've likely found your problem.

Internal regulators are fairly inexpensive. If there aren't other obvious problems with the alternator, throwing in a new or known good regulator and brush pack is usually worth the trouble.

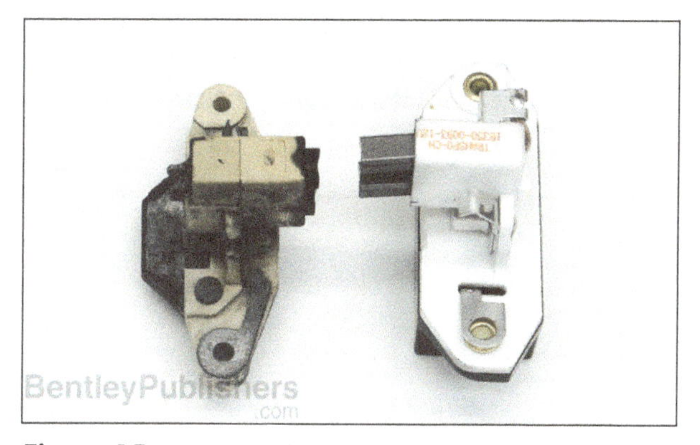

Figure 15 Two internal voltage regulators with integral brush packs. The one on the right is good. The one on the left has its brushes worn down to nubs and no longer making reliable electrical contact with the alternator.

Testing an Alternator with an Ignition Wire

A third wave of alternators came in in the late 1990s. On BMWs, these were used on E46 3 Series and E39 5 Series cars. Like all alternators, they have a large B+ post for the battery wire, but instead of a D+ post, there's a molded plug with a two- or three-pronged socket. These include DIN terminals 15 and 61E (see **Figure 16**).

Terminal 15 provides the alternator with voltage whenever the ignition is on. This appears to have been used for two different things. Initially, the voltage at "15" was a sensing wire, allowing the alternator to know the voltage present at the ignition, and to compensate if there's high resistance in the cabling.

However, many of these alternators feature a Multifunction Controller or Multifunction Regulator (MFR) that allows some level of internal alternator fault detection, as well as certain control functions such as "start load response," where the alternator holds off charging during engine start to avoid putting a brake torque on the engine. After the first few seconds, charging amperage is allowed to increase at a rate of 10A per second. In alternators with MFR, the "15" connection serves to provide power to the integrated controller.

Terminal 61E takes the place of D+ to control, as the DIN spec implies, the warning light, which functions in essentially the same way as in the externally-regulated alternators described above – if there's a voltage difference of more than 2V across the legs of the warning light, the light comes on, but as the alternator spins up and there's voltage on both legs of the light, it goes out. However, the lamp itself is no longer providing the excitation current, so the alternator will still charge the battery even if the bulb is dead or disconnected.

Later versions of MFR include direct communication with the DME via a fourth wire acting as an analog control line. In this case, communication with the warning light is no longer direct; it is passed through the ECM, which then tells the instrument cluster when to illuminate the light.

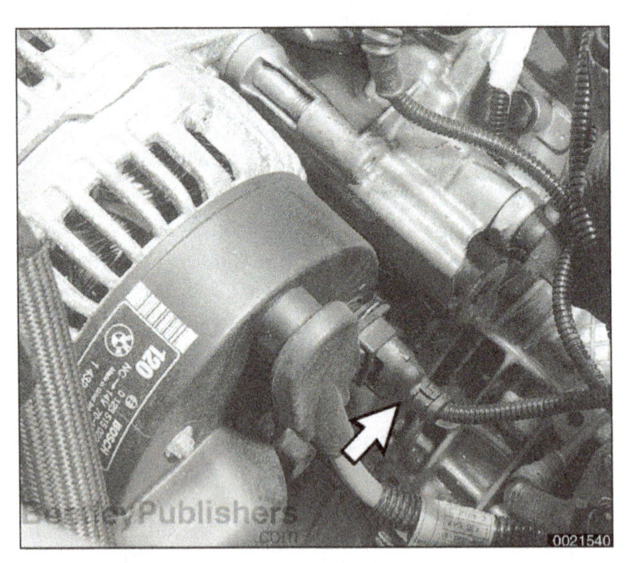

Figure 16 Molded plug (**arrow**) on an internally regulated alternator with an ignition wire on a BMW E46 3 Series.

The **Charging System Basic Health Test** is still valid for these alternators, but if they have some level of DME control and if they fail, there's a possibility that the failure is due to the control signal. Testing by a specialty shop with the proper equipment may be necessary.

To test an internally regulated alternator with an ignition wire:

- Conduct "The Charging System Basic Health Test" as described earlier in this chapter.
- If the alternator fails to output charging voltage, check the B+ line as described in the **Testing a 2-Wire Alternator** section on the previous page.
- Disconnect the harness connector from the alternator. Turn the ignition on and check for battery voltage between terminal 15 and ground. If voltage is not present, check wiring and fuses for faults.
- Reconnect the connector to the alternator.
- Locate the the blue wire coming from terminal 61E. Either back-probe the connector or carefully pierce the insulation with a wire-piercing probe.
- Turn the ignition key on and check for voltage on 61E. Voltage should be less than 1.5V and the charge indicator lamp should be lit.
- Making sure no wires are near the fan, start the engine. The voltage on 61E should increase to about 8V or higher, and the charge indicator light should go out. If the voltage does not increase and the system is not charging, the alternator is probably defective.
- If the voltage does not increase but the system *is* charging, interrogate the system for faults with a scan tool.

Air and Water Cooled Alternators

To keep pace with the demands of electronics-laden cars, the newer the car, the higher the current generated by the alternator. Unfortunately, higher charging current generates more alternator heat. For this reason, more and more cars are employing alternators with either supplemental air cooling (an air duct on the alternator housing is connected to a source of fresh air ducted in from outside the engine compartment) or by water cooling (sealed alternator housing sits in the engine's cooling system to carry off heat). Water cooling in particular makes the alternator less trivial to remove for simple testing. Generally speaking, many alternators with ignition wires are air-ducted, and some computer-controlled alternators are water-cooled.

Testing a Computer-Controlled Alternator

Cars built since the late 1990s, with their myriad of computers, electric motors, entertainment and navigation systems, and high-output lighting, make enormous demands of their electrical systems. To keep pace, many cars now utilize digitally controlled alternators, where the car's ECM reads a variety of conditions including ambient temperature, warm-up status, and electrical demand, and dictates to the alternator what amperage and voltage it should produce. For example, the alternators on most BMWs built since the early 2000s employ a Bit Serial Data (BSD) interface with a single-wire connection to the DME control module. See **Figure 17**.

Understand that this means that the alternator is now a "module" on a network. In this architecture, the alternator warning light is no longer the quaint analog function described earlier. Instead, alternator status is sent by the alternator to the ECM, and if there's a problem, a command is sent by the ECM to the instrument cluster to illuminate the light in response to certain fault conditions. Newer BMWs also use this BSD interface to have the ECM tailor the alternator's charging of the battery based on the stage the battery is at in its life cycle.

Figure 17 Digitally controlled alternator on a 2008 BMW 335i.

The **Charging System Basic Health Test** is still valid for these alternators, but if they fail to charge, there's a possibility that the failure is due to the digital control signal or the interface path that carries it. Fortunately, because the alternator is a module on a bus, it should be visible via a proper scan tool.

To test a computer-controlled alternator:

- Conduct the "Charging System Basic Health Test" as described above.
- If the alternator fails to output charging voltage, check the B+ line as described in the **Testing a 2-Wire Alternator** section earlier in this chapter.
- Interrogate the system for faults with a scan tool.

NOTE —

In addition to computer control, overrunning alternator pulleys have been phased into production as of the late 2,000s to help increase fuel economy and provide smoother engine operation. These pulleys allow the alternator to decouple or free-wheel when the engine decelerates. The pulley is a wear item and special tools are required to replace it. To quickly check for a worn overrunning pulley, run the engine at idle and look closely at the belt tensioner. If the tensioner has excessive movement, the pulley may be going bad. Another check is to rev the engine to about 2,500 rpm and then turn the engine off. If you can hear a brief buzzing sound, it's another sign of a seizing pulley bearing.

Failure Modes of the Regulator

Let's look specifically at the two primary failure modes of the regulator: overvoltage and undervoltage.

Testing for Overvoltage and Undervoltage

Most charging system problems are binary. That is, most of the time, the system is either charging the battery or it's not. The exceptions are overvoltage and undervoltage.

Overvoltage is the condition where the voltage output by the alternator/regulator is too high. It is always the result of the regulator malfunctioning, either being electrically or mechanically stuck in the closed or bypassed condition, or the wiring being shorted, creating the same condition.

WARNING —

Sustained overvoltage in excess of 15 volts can literally cause the battery acid to boil and produce hydrogen sulfide gas – a dangerous, messy, and fragrant occurrence. If you smell rotten eggs, stop! You may see the battery wet with acid. At your earliest convenience, put on gloves, eye protection, and a Tyvek suit, and wash the entire area down with water and baking soda. A boiled battery is toast, so it has to be removed anyway. While it's out, be sure to wash the tray beneath it.

In contrast, undervoltage is the condition where the alternator is functioning correctly and outputting something that's higher than resting voltage, but the regulator is bypassing the alternator too much of the time, resulting in a voltage that is lower than normal 14.2V charging voltage.

To check for both overvoltage and undervoltage:

- Start the car.
- Set the multimeter to measure voltage, and put the test probes across the positive and negative battery terminals. If the battery is not easily accessible, you can instead put the positive probe on the positive post under the hood used for jump-starting, and the negative probe on any chassis ground.
- Gently rev the engine to 2,000 rpm and hold it there.
- You should see "charging voltage" in the 13.5 to 14.5 volt range. If the alternator output voltage is too low, make sure that the alternator is solidly grounded as described earlier in this chapter. If the car has an external regulator, the regulator should be grounded through its case, so verify that the contact point between the case, screw or bolt, and body of the car are clean. Then retest.
- If the charging voltage is too high, the regulator is almost certainly at fault. Replace the regulator and retest.
- If the charging voltage is higher than resting voltage (12.6V) but substantially lower than charging voltage (14.2V), it is possible that an internal alternator diode has failed. Replace the alternator and retest.

Most of the time, fixing charging issues aren't that hard. Install a new battery and a rebuilt alternator with a new regulator, and you're usually good for years. *Charge!*

Tales From the Hack Mechanic: Rebuild or Replace?

When the alternator light came on in my BMW E46 3 Series wagon, the voltage test showed only resting voltage. I needed the car back up ASAP, as it's my daily driver, so I pulled the alternator. I then removed and inspected the internal voltage regulator, and did not see anything wrong with it. "Must be the alternator," I shrugged. I went to a national chain auto parts store. They did not have the connector needed to test the alternator on their machine, but, hey, it's-Sunday-need-the-car-Monday, so I bought their "Authorized Rebuild" alternator and installed it. It solved the problem.

The next day, I read a hail of negative posts online regarding these alternators. I thought, damn, I should've ordered one with a better pedigree from one of my usual BMW-specific parts suppliers. But when I looked on a BMW forum, I found that these had the same bad reputation.

Not wanting the alternator to strand me, the next weekend I pulled it out and took it into a local auto electric shop I sometimes use for a bench test. It passed, but the seen-it-all owner chewed me out. "You're really taking your chance with these rebuilds," he said. "You really have no idea who did them or what they mean by 'rebuilt.' " He then walked me through his Rebuilt Alternator Hall of Shame, showing me examples where a vendor had done nothing but change the brush pack and spray-paint the outside. "Next time," he said, "bring your old one to me if you want it done right. Of course, I'd have to charge you more than you paid for this one." As with most things in life, what you don't pay for, you don't get.

15

Starter

The Starter Motor

It can be helpful to think of the starter motor and the alternator as financial inverses of each other. That is, the alternator is the responsible party making continuous deposits in the bank, so to speak, using the spinning engine to generate the electricity needed to power the car and keep the battery charged. In contrast, the starter is the mad impetuous youth that makes withdrawals, furiously burning through the electricity that the alternator has so diligently been hoarding in the battery. You can also think of the demands placed on the alternator and starter as reciprocal. The alternator is continuously generating electricity whenever the car is running, whereas the demands placed on the starter are dramatic and episodic (as Thomas Hobbes said about the life of man – nasty, brutish, and short).

The starter motor is part of the cranking system consisting of the battery, ignition switch, starter motor, and safety interlocks. As a user, it's pretty simple – turn the key (or, in newer cars, push the button), and the starter fires up and spins the engine. There are, however, two sleights of hand in the process, and to understand them, we must have…

Starter and Solenoid

The starter uses a special kind of relay called a *solenoid* where, rather than an electromagnet pulling two electrical contacts together, the core of the electromagnet is a plunger which slides forward when the field is applied, and springs back when it's shut off. See **Figure 1**.

The movement of the plunger forward and back is what makes and breaks a high-current electrical connection. Physically, the solenoid is a cylinder about the size of a can of tomato paste, and is mounted on the side of the starter.

Figure 1 An old rusty starter.

A solenoid is a perfect choice for a starter motor because the second sleight of hand is that the starter not only has to spin and stop, it has to mechanically engage the engine, then disengage so the now-running engine doesn't force the starter to keep spinning.

Here's how it works. The car's engine has a toothed flywheel gear connected to its crankshaft. The starter motor also has a toothed gear, but it needs some way to connect and disconnect it from the flywheel. So the starter uses an intermediate toothed gear called a pinion gear that is moved by the solenoid's plunger. When you turn the ignition key to start, electrical current is supplied to the solenoid, which energizes it and causes its plunger to move forward.

This does two things: 1) it makes the electrical connection between the starter and the battery, which causes the starter to spin, and 2) it moves the pinion gear forward, connecting the starter gear with the flywheel, causing the spinning starter to be mechanically engaged with the engine, thus allowing the engine to turn. When you release the ignition key, it cuts the current to the solenoid, which causes the plunger to retract, which makes the starter stop spinning and mechanically decouples it from the engine at the same time. It's actually pretty neat, and utterly taken for granted.

We can see that, for a starter to operate correctly:

- The starter must have a solid electrical connection to the positive battery terminal (and to the negative battery terminal through the grounding of the engine). These connections must be capable of carrying the hundreds of amps the starter may draw.
- The solenoid must receive its turn-on voltage through the ignition switch (which is at a much lower current than the starter's direct battery connection).
- The plunger in the solenoid must move the pinion gear forward to engage the teeth on the flywheel.
- The starter must spin with sufficient torque to turn the engine quickly enough for it to start.
- Once the engine has started, the plunger must retract to disengage the pinion gear and shut off the starter.

Starter DIN Terminal Numbers

The DIN lists many potentially starter-related connections, but the ones present on all cars are:

DIN Terminal Numbers for Starter	
30	Thick cable direct to battery
50	Ignition switch to solenoid
15 or 16	Starting voltage to ignition (older cars only)

See **Figure 2** and **Figure 3**.

Note that an explicit ground connection is not listed on the DIN because the starter is grounded through its case to the engine block, which in turn is grounded to the body of the car via a ground strap.

We'll discuss terminals 30 and 50 at length below. Terminal 15 or 16, if present, is used only on old cars. It supplies 12 volts while the starter is cranking, and shuts it off when cranking ceases. On pre-1990s cars with a ballast (current-limiting) resistor installed in series with the ignition coil, terminal 15 may be used to bypass the ballast resistor and supply the ignition coil directly with a full 12 volts to aid in starting while the engine is cold. We'll discuss this in the **Ignition** chapter.

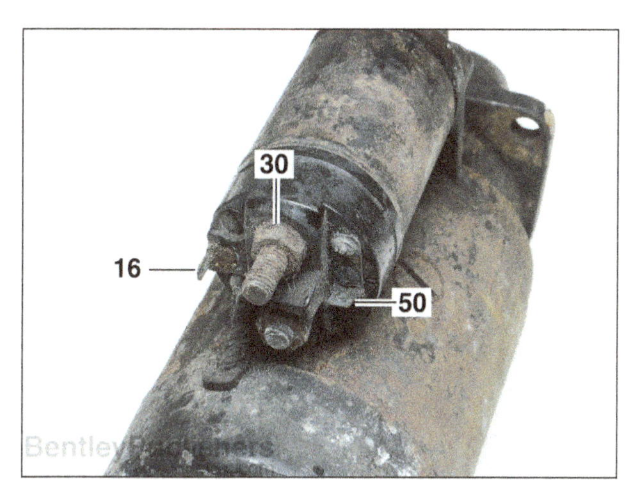

Figure 2 A closeup of the starter.
DIN terminal 50 is from the ignition switch.
DIN terminal 30 is a direct connection to the battery.
DIN terminal 16 used to bypass an ignition ballast resistor at startup (older cars only).

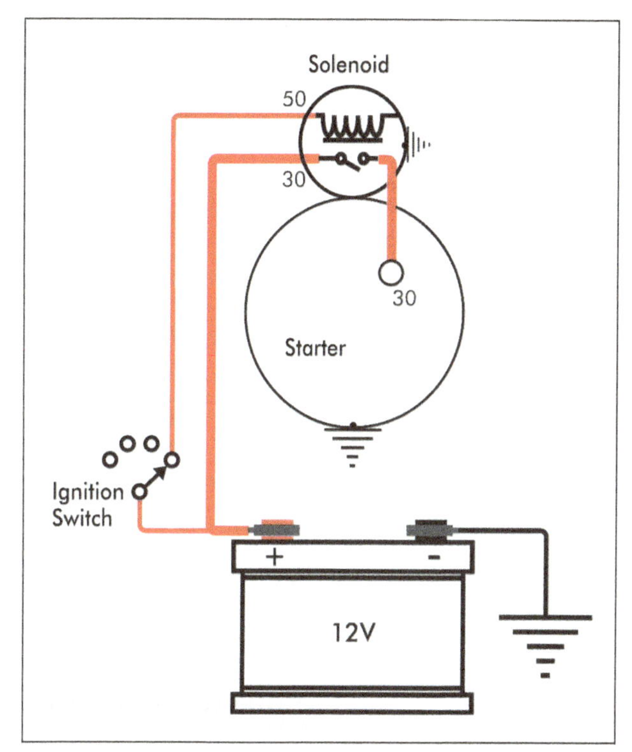

Figure 3 Basic wiring of starter and solenoid. Ignition switch energizes the electromagnet in the solenoid, which pulls the switch closed, making the starter spin, while keeping high-current wiring out of the ignition switch. Both the solenoid and starter ground through the starter's case.

The American Car Starter

If you have experience with the way the starter terminals are labeled on American cars, you'll recognize that "30" is "B" from the battery, and "50" is "S" from the ignition switch. Interestingly, the terminal labeled "M" on American starters for the actual starter motor field windings – the hot connection on the starter motor itself – does not have a corresponding DIN label.

Complete Starter Testing

This is a series of tests that we recommend approaching in the order given below, since it takes you from the easy and obvious to the slightly more involved.

First, understand the following:

- If you turn the key and it goes *click*, suspect the battery first, then the battery cables, then the interlocks that disable the starter (if your car has them).
- If you turn the key and you hear nothing, it may be the key itself. For example, since 1994, all BMWs use a computer chip in the key to prevent unauthorized starting. Most other European manufacturers use similar immobilizer systems. Try another key first if the starter does not engage and you're sure the battery is good. If you suspect the immobilizer system is the source of your cranking problem, a brand-specific scan tool should be connected to interrogate the system.
- The starting circuit (battery-ignition switch-starter motor) on late model cars is often wired through various power modules, junction boxes, and connectors. It is therefore a good idea to know what you're up against before diving in with the multimeter. One of the best ways to understand the circuit is to study the wiring diagram for your vehicle. If necessary, see the chapter **Reading Wiring Diagrams**.
- Suspect your starter only when you've troubleshot your way past the other usual suspects. They can and do die, but as people in medical school say, when you hear hoof beats, suspect horses, not zebras.

1. Begin with a Good Battery

As with alternator testing, starter testing *must* begin with a fully charged battery. When checked directly across the battery terminals with a multimeter, you should see 12.6 volts. If you have a trunk-mounted battery hidden beneath an access panel, expose it and measure the voltage across the terminals. *Do not skip this step and test the voltage under the hood at the battery jump points*.There's no reason to waste time chasing starter problems if the cause is simply a discharged or bad battery.

If the battery doesn't read damned close to 12.6 volts (12.4, but no lower), go back to the **Battery** chapter for information on battery recharging or replacement. Fully inspect and clean the battery cables. The inner surface of the cable clamps must be free of corrosion, and the clamps must be sufficiently tight that they can't be rotated by hand. The ground straps from the battery to the body and the engine block to the body must be intact, with connections that are clean and tight.

If the battery does read 12.6V, for good measure, turn on the headlights and verify they're nice and bright, then turn them off, then turn on the blower motor and verify that the fan spins good and fast, then turn it off. These are further indications of good battery health.

2. Dashboard Lights Coming On?

- If you turn the key to "run" and the dashboard lights are on but very dim, we'd wager you have a discharged battery and haven't followed the instructions to test and remedy that first. Bad reader. Go back to step 1.
- If you have no dashboard lights on at all, the battery is effectively disconnected from the car due to a loose or badly corroded battery cable or, there's a blown fusible link or battery safety terminal.
- Closely inspect the battery cables. Clean the terminals as described in the **Battery** chapter. Use a multimeter to check the continuity between the negative battery terminal and the body of the car. If it's infinite (no continuity), there's a failure in the cable.
- Do the same test on the positive side to verify continuity of the positive battery cable, testing for continuity between the positive battery terminal and the B+ post on the alternator. *The alternator*? Yes. The positive battery cable goes two places. The fat part goes directly to the starter motor, but the thin part *that carries the current for everything in your car except the starter motor* ultimately is connected to the B+ post on the alternator. If you twist the key to "run" and have no dashboard lights, you've probably got nothing at B+.
- On cars with a front-mounted engine and a trunk-mounted battery, directly checking the continuity between alternator B+ and the positive battery terminal may not be easy – you'll need long leads for the multimeter – but, hey, you have no dashboard lights; you have to solve the problem.
- If the battery cables check out good and still have no continuity to the alternator, odds are your car has a fusible link or a Battery Safety Terminal (BST), and it's blown. A fusible link is usually between the battery and the rest of the electrical system *except the starter motor*, so check for continuity of the thick positive cable to the "30" post on the starter.
- In contrast, if the car has a battery safety terminal that's blown, it completely disconnects *everything* at the positive terminal. (Well, almost everything; it typically allows the hazard lights and door locks to work.)

3. What Do You Hear When You Turn the Key?

Back at the beginning of the **Battery** chapter, we said "Click ... It's the moment we all dread ... Or, even worse, you turn the key and hear nothing." Now that we know the basic workings of the starter, we can, in fact, say what that dreaded "click" is.

"The Click"

The "click" is the sound of the plunger inside the solenoid moving forward and making electrical contact. Knowing whether or not you hear the "click" enables you to quickly go to the correct step in the troubleshooting tree and triage which problems are likely due to the battery cables versus which are likely due to the starter.

- Set the car's handbrake.
- Put your right foot on the brake pedal.
- If the car has automatic transmission, put it in "park." If it's a standard, depress the clutch with your left foot.
- Now, turn the key to "start." What do you hear?
- If you hear the normal rapid "RRRRrrrrr RRRRrrrrr" sound of the starter motor spinning the engine, but the car doesn't start, there is absolutely nothing wrong with your battery, cables, or starter, and the failure of the car to start lies elsewhere (for example, in the fuel or ignition systems).

- If you hear "click" or a few clicks in rapid succession, and then nothing, or hear a "click" followed by a labored, much slower-than-normal "RRRRrrrrr," this means that the solenoid is being energized and is trying to fire up the starter, but the starter motor is not able to turn or doesn't have the strength to turn the engine quickly enough to allow the engine to start. This may be because the connections are dirty, the cables are bad, the starter itself is bad, or there's some other mechanical obstruction. Go to step 6.
- If you hear a grinding sound, the solenoid is making electrical contact and the starter is engaging, but it is likely that the pinion gear is either spinning in place, or it is trying to move forward and engage the flywheel but it can't because either its teeth or the flywheel's teeth are damaged. Remove the starter and inspect the teeth on the pinion gear and the flywheel. See **Figure 4**.

Figure 4 Close-up of the starter's pinion gear (**arrow**). If you hear the starter spin but then hear a grinding sound, this gear, or the one on the engine's flywheel that it engages, may be damaged.

- If you hear absolutely nothing (to be clear, THAT'S NOTHING – NO CLICK FROM THE SOLENOID), then the solenoid isn't being energized. Among the possibilities are:

• The wire from the ignition to the starter solenoid has fallen off.

• The connection to the ignition switch, or the ignition switch itself, is bad.

• The solenoid itself is bad.

• The safety interlock from the automatic transmission or clutch is bad.

• The key is bad, or the immobilizer key in the chip is bad. Try another key.

- To verify which of these is the cause, you need to check for voltage at terminal 50 on the starter. Access the starter by whatever method is necessary on your car (see below), then check for voltage at terminal 50 (see below).

Accessing the Starter

Accessing the terminals on a starter motor is far easier in theory than in practice. Water-cooled engines have the starter motor bolted to the engine block, so the terminals are with the engine, often hidden beneath the intake manifold. On a '60s or '70s European car, the starter motor is easily accessible, but on a post-1996 fuel-injected car, the starter is usually buried beneath the intake manifold and is obscured by the brake master cylinder, the power assist booster, and an army of cooling hoses and electrical connections. It may be easier to reach the starter's terminals from underneath by jacking up the car and putting it securely on jack stands. See **Figure 5**.

Figure 5 Starter motor (**arrow**) on the engine block of a 2008 BMW 335i is difficult to access, even with the intake manifold removed.

In contrast, because an air-cooled engine has a "case" instead of a block, and the case is typically smaller than a block on a water-cooled engine, the starter on an air-cooled engine is usually bolted to the transaxle instead of the case, so the terminals are with the transaxle. In both instances, the starter is generally held in place by the very bolts that attach the engine to the transmission or transaxle. Sometimes there's an additional bracket on the starter that bolts directly to the side of the block or transaxle.

To do anything with the starter, you need to access terminal 50. It's not the big post with the fat cable from the battery – that's terminal 30. Terminal 50 is usually a small push-on male quick-disconnect terminal (also sometimes referred to as a "spade" terminal). On old cars, it may be a ring connector on a small threaded post instead.

4. Check for Terminal 50 Voltage

When, on a 40-year-old car, you turn the key and hear nothing, the connector at terminal 50 is a prime suspect.

- Look closely at the connection to terminal 50. If the connector has fallen off the terminal, you've likely found your problem. Female quick-disconnect terminals can loosen up over time. If the connection feels loose and sloppy, take the connector off its post and gently squeeze it with needle-nosed pliers across its short edge to make it seat tighter.
- Or you may see that the wire into the base of the connector is about to break off. In that case, cut the connector off, crimp on a new one, and proceed with the test. See **Figure 6**.

Figure 6 It's quite common on an older car to find the wire on a spade connector hanging by strands (**arrow**), or completely broken off.

- If the connector is good, set your multimeter to measure voltage. We're looking for voltage going *into* terminal 50, so put the positive probe of the multimeter on the wire going to terminal 50.
- If it's a wire with a female quick-disconnect terminal and you need to pull it off the male connector on the starter, that's fine, but put the probe on the wire side, not the starter side.
- Attach the negative probe of the multimeter to any convenient ground. Put the multimeter where you can see it from inside the car and turn the key to "start," or have a friend turn it.

If the multimeter reads zero, you've found the problem – there needs to be voltage at terminal 50 to cause the solenoid to close. On a pre-1980s car with a manual transmission, this wire comes directly down from the ignition switch, but the newer the car, the greater the number of interlocks are between the ignition switch and the solenoid. Virtually every car with an automatic transmission has a transmission interlock switch, preventing the car from starting unless it's in park or neutral. Some automatic cars also require you step on the brake pedal during starting. See **Figure 7**.

At some point, even cars with standard transmissions began using interlock switches to make sure the clutch pedal is depressed before passing voltage to terminal 50. Certain anti-theft systems also suppress the voltage at 50 if the correct key isn't in the ignition.

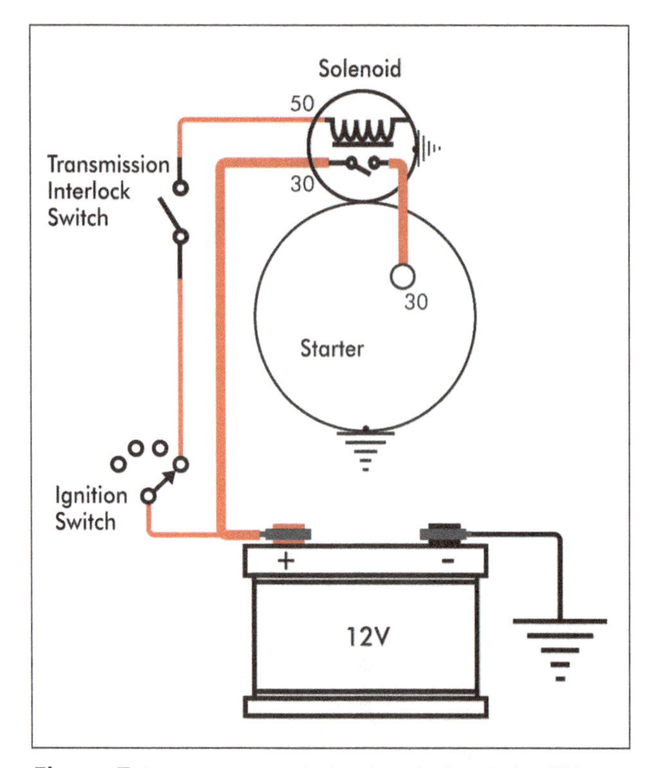

Figure 7 An open transmission interlock switch will keep the car from starting.

5. Hot-Wire the Starter

If there is no voltage at terminal 50, it's a good idea to verify that this is the *only* problem – that the starter will indeed roar to life if there is voltage at 50. There's an old-school way of doing this by using a screwdriver to jumper across from 50 to 30, *but don't do it*. It's not safe. It's much too easy to accidentally short the body of the screwdriver to ground.

The safe way to do it is to use a manual remote start switch designed for exactly this purpose. A manual remote start switch is a momentary-on switch with two leads, each with a small alligator clip.

- Temporarily disconnect the push-on spade connector to starter terminal 50.
- Clip one alligator clip from the remote starter switch to starter terminal 50, being very careful to ensure it cannot swing around and short out against any other metal on the body or engine.

- Clip the other end of the remote start switch to the positive battery terminal or to the big solenoid post that's connected to the positive battery terminal. Because it is difficult to photograph the remote starter switch when it's connected with the starter in the car, below we show it connected on the bench in **Figure 8**.

Figure 8 Connection of one alligator clip of a remote starter switch to starter terminal 50 (**arrow**). The other clip from the remote starter switch (on the left) goes to the positive battery terminal. Note that the positive battery cable is normally connected to terminal 30, the big threaded post in the center.

CAUTION —

Understand that, when you do this, you are bypassing a safety interlock and making it possible to accidentally start the car in gear. Be absolutely certain not to do that. Set the handbrake firmly. If the car has an automatic transmission, be absolutely certain it is in park. If the car has a standard transmission, be absolutely certain it is in neutral.

From a wiring diagram standpoint, **Figure 9** shows what you're doing when you hot-wire the starter.

- Leave the ignition key in the off position (we're testing the starter, but we don't actually want the car to start).
- Now hit the button on the remote start switch.
- If, using the manual remote start switch, you hear the car try to start with a normally fast "RRRRrrrr RRRRrrrr" but it did *not* try to start when you turned the ignition key, then there is nothing wrong with the starter or solenoid, and the problem lies in the connection to the 50 terminal.

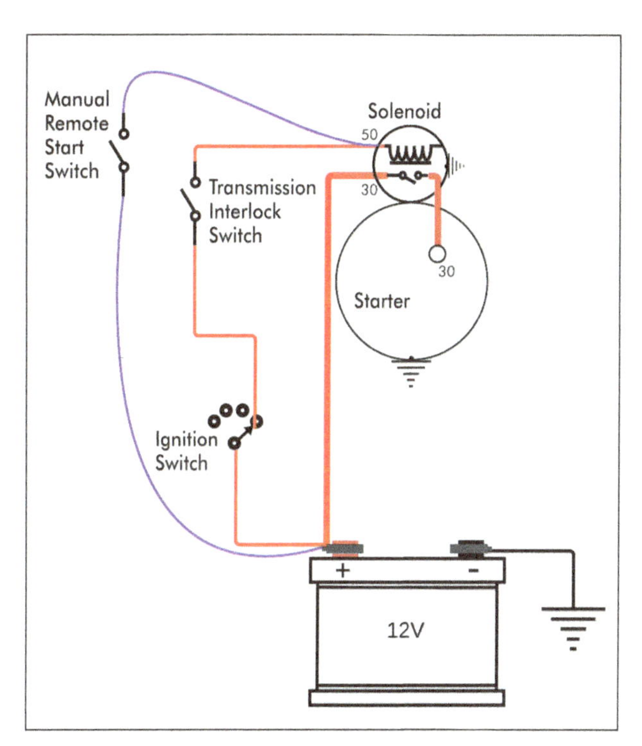

Figure 9 What you're doing when you use a manual remote start switch.

- Consult a wiring diagram for your specific model, first checking the transmission or clutch interlocks, then the feed from the ignition switch itself. Note that, on a computerized car, the transmission interlock is just the beginning; there are typically other gating functions governing whether voltage is actually supplied to terminal 50. For example, BMWs with EWS (the German acronym for electronic drive-away protection) check for the presence of a key with a chip and a radio transponder in it, and disable the starter and ignition if the chip's code is incorrect.

6. Check for Voltage Drop

If you hear "click," then either a labored slow starting sound or nothing, and you're absolutely certain you have a good, fully charged battery and battery cables with clean terminals, then your path has led you inexorably to the legendary Voltage Drop Test.

In the **Circuits** chapter, we discussed, from a theoretical standpoint, how high resistance from corrosion can cause a circuit

to fail, and how you can use a Voltage Drop Test to locate the cause of the failure. In the **Common Multimeter Tests** chapter, we stepped through a generic voltage drop test. Here, at the starter motor, is where theory meets application. We're going to go through it all again, because it's really important.

Revisiting Why You Need to Measure Voltage Drop to Find Resistance

When you turn the key and crank the starter motor, you are generating the biggest electrical load that your car ever sees. How big? Depending on the size of the engine and how cold it is outside, it could be hundreds of amps. The battery cables need to deliver that much current to the starter. Are they delivering it, or is high resistance (old cables, broken strands, corroded connectors, loose crimps) sapping the current? And how do you know?

In theory, you could directly measure the actual current drawn by the starter by placing an ammeter in series between the battery and the starter, similar to what we discussed in the **Battery** and **Charging System (Alternator)** chapters. Except that, well, you can't. Remember that, in the **Common Multimeter Tests** chapter, we were talking about a small current (less than 10 amps) that we could directly measure with the multimeter. And in the **Charging System (Alternator)** chapter, we were talking *hypothetically* about measuring the 50-ish amp charging output from the alternator using an inexpensive standalone amp gauge But you don't own an ammeter capable of directly measuring the hundreds of amps drawn by the starter.

You could, instead, measure the starter's cranking current inductively by using the DC clamp-on ammeter or DC current clamp that plugs into a multimeter, as shown in other auto electrical books, but most folks don't own those tools either. So, from a practical standpoint, most people can't directly measure the current in the starter cables.

So if we're so concerned about the *resistance* of the battery cables, can't we just directly measure *that*? Well, no. You can't. At least not accurately enough. Let's say that your battery cable has 100 individual strands of wire in it, and 99 of them are broken but one is still connected. Can your multimeter tell the difference? See **Figure 10**.

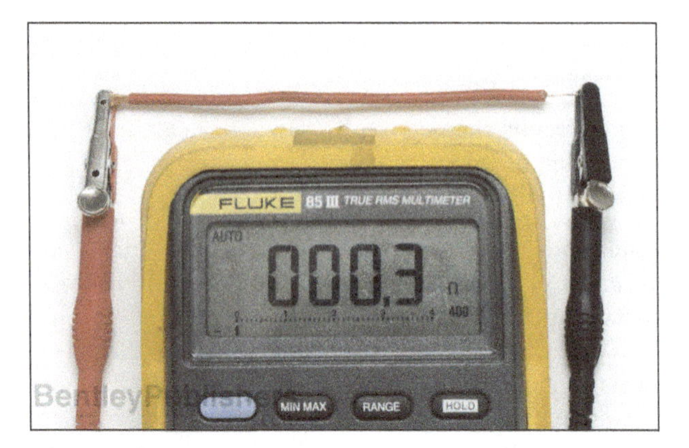

Figure 10 Even with only one strand of wire remaining on the right side of this wire, a resistance measurement still shows continuity, but you can't pull a lot of amperage through one skinny strand of wire.

An inexpensive multimeter has a resolution of about 0.1 ohm; an expensive one perhaps 0.01 ohm. You can look up resistances of 3 feet of heavy 0-gauge wire (about 0.0003 ohms) and skinny 20-gauge wire (about 0.03 ohms) and see that your multimeter doesn't have the resolution to tell them apart. (Yes, you pros who own Fluke 87s and good probe sets can heckle like those two Muppets in the balcony, but for everyone else, it's true.)

When you use a multimeter to measure the end-to-end resistance of a cable, all you're really doing is performing a continuity check. You're looking to see if the resistance is infinite, which tells you there's a break in the cable. But if the resistance is *not* infinite, what *should* it be? What is the "correct" resistance for a starter cable and its connectors?

Many people find the answer surprising: There isn't a "correct" resistance. Yes, you can look up an expected range of resistances for a given wire size in a table, but it's the wrong way to think about it. And this isn't Hack Mechanic Zen or hand waving. Here's why.

As current increases, temperature increases. As temperature increases, resistance increases. Thus, *the resistance varies with the amount of current*. And with a starter motor, the amount of current is so large that the resistance of the cables is going to vary *substantially* when the starter is cranking compared to when it isn't.

So... can't you just measure the resistance in the cables while the starter is cranking, if that's what you're so concerned about? Well, no. A multimeter can only measure resistance on a portion of a circuit that isn't active. When you use a multimeter to measure resistance, what's actually going on is that its internal battery sends a very small current through the multimeter's probe leads, then measures the resulting voltage across them and uses that to calculate the resistance. That resistance may be very different from the resistance encountered when the starter motor is trying to pull 200 amps. Our hypothetical battery cable with 99 of its 100 strands broken will easily transmit the multimeter's small signal through its one unbroken strand, and the multimeter will dutifully report that the cable has continuity (has "low resistance"), but the starter won't be able to pull much current through it. It would be like trying to run a fire engine's water pump through a soda straw.

Because you can't easily measure high starter current (and you don't really know what it should be anyway), and because you can't accurately measure resistance, what you do instead is to measure *the drop in voltage* along each portion of the battery cables while the starter is cranking. This is called a *voltage drop test*, and it is an indispensable tool for troubleshooting *anything* that generates a high electrical load, not just the starter. Unlike measuring resistance, which can't be done while current is flowing, a voltage drop test is *always* done while current is flowing. This gives it the advantage of being performed in place, without removing any cables.

When you do the voltage drop test, you set the multimeter to measure voltage, and place both multimeter probes *along the same positive or negative path*. We're so used to using the multimeter by placing one probe somewhere we expect voltage, and the other on ground, that placing the probes along a 12V or a ground cable is counterintuitive at first, but it's highly effective.

Summary of Basic Voltage Drop Test

- Set the multimeter to measure voltage.
- Instead of putting the red probe where you expect voltage and the black probe on ground, put both probes at different points along an electrical path where you expect voltage and test the circuit while it is under load.
- If the voltage is the same at those points, the multimeter will read zero. But if it reads something other than zero, there is a voltage drop between the points, which means it is losing current due to high resistance.

Another intuitive way to think about the voltage drop test is this. It's often said that electricity follows the path of least resistance. Large multi-stranded copper battery cables should have very low resistance, but broken internal wires or corrosion on the connectors can increase their resistance. By putting the multimeter's probes at the beginning and end of a battery cable, if there's high resistance somewhere in the battery cable, some of the electricity will follow the path of least resistance and flow through the multimeter instead.

There are two basic ways to conduct the voltage drop test. The first is to start by testing the entire length of a cable, and only hone in on the individual sections if the first test shows an unacceptable drop. The second is to start with the smallest section (for example, measuring the voltage drop from the positive battery post to the positive battery clamp) and work outward. Which method you use is up to you; just be consistent.

It's true that accessing the cable connections on the starter is not easy on a post-1996 car. But remember that you're doing this because your starter is turning slowly or not at all, and if you don't find a voltage drop, it likely means you're going to need to remove and replace the starter. So be thorough.

Voltage Drop Test for the Starter

- Disable the car's engine from starting, either by removing the fuel pump fuse or relay, or, on an old-school ignition system, temporarily grounding the wire coming out of the center of the coil with a cable with alligator clips at both ends. See **Figure 11**.
- Set the multimeter to measure voltage.

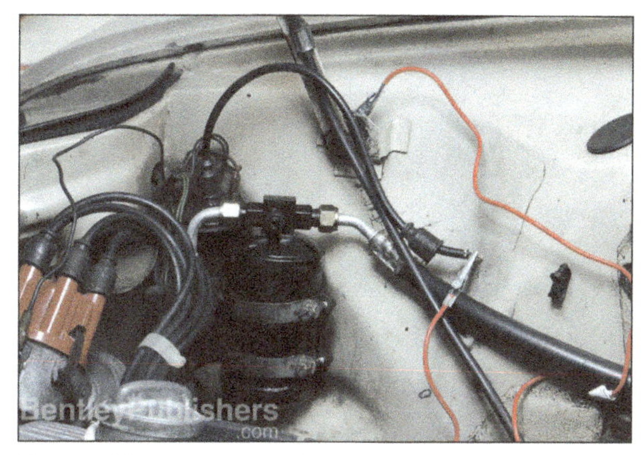

Figure 11 Disabling an old-school ignition system by grounding the center wire from the coil.

- Connect one lead of the multimeter to the 30 post on the starter where the positive battery cable is connected. It doesn't matter if it's the red or black lead. In **Figure 12**, we've connected the red lead for visibility.

Figure 12 One multimeter lead connected to starter terminal 30. (Vacuum hose removed for clarity.)

- Hold the other lead of the multimeter to the center of the post of the positive battery terminal.
- Observe the multimeter reading. It should be near zero, since, with no current flowing, there should be no voltage drop across the cable.
- While continuing to observe the multimeter, have a friend turn the key to "start" and hold it there for about 10 seconds. See **Figure 13**.

What is the voltage drop across the battery cable while the starter was cranking? Unless there's manufacturer-specific information to the contrary, it is common to assume that the drop across any one connection shouldn't be more than 0.2 volts, and the drop across an entire battery cable shouldn't be more than 0.5 volts.

We recommend that you first drop-test the entire length of the positive cable as described above, then individually test:

Figure 13 Other multimeter lead held to middle of positive battery terminal. Meter is showing an end-to-end voltage drop of about 0.28 volts.

- Positive battery post to positive battery clamp. See **Figure 14**.

Figure 14 Measuring voltage drop from positive battery post to positive battery clamp.

- Positive battery clamp to cable eyelet at solenoid.
- Connector at solenoid to solenoid post.

Similarly, on the negative side, test:

- Entire length of the negative battery cable. See **Figure 15**.
- Negative battery post to negative battery clamp.
- Negative battery clamp to cable eyelet at body ground.
- Cable eyelet to body ground
- And the same for the engine ground strap
- If you find a voltage drop in the middle of a cable (between the end connectors), replace the entire cable.

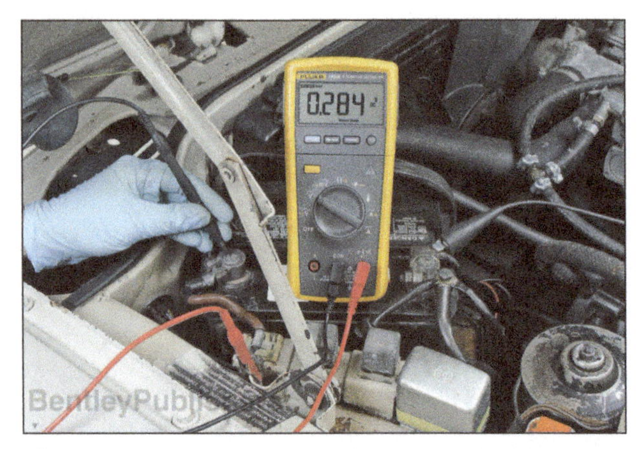

Figure 15 Measuring voltage drop across the length of the negative battery cable with one meter lead on the negative terminal, the other at the cable attachment point on the body of the car.

- If you find a drop where a cable connects to the battery or starter or ground, clean the connection. It doesn't matter if you just cleaned it. Clean it again. Clean it better. Then retest.

It's really pretty simple. You can now walk through the rest of your life exuding voltage drop confidence. Women will want to be with you. Men will want to *be* you. Cars will fix themselves as you walk past. It's pretty freaking awesome.

But if you never find a voltage drop and you're still getting "click," there's one more thing to test before yanking out the starter:

7. Check for Mechanical Restriction

Maybe there's nothing wrong with your starter. Maybe it is trying to spin, but something won't let it. Maybe the engine has a seized piston. Maybe it's thrown a rod. Maybe the air-conditioning compressor is locked up. If you're trying to resurrect a barn-find car, you may have compound problems with internal mechanical friction and an old weak starter *and* corroded battery cables. But to get to the root of the problem, you need to verify that the engine can, in fact, turn.

- Make sure the car is parked somewhere flat, put the handbrake on, put the car in neutral, and do whatever you need to do to manually rotate the engine. Make sure the ignition is in the OFF position.
- If it's a car with a small four-cylinder engine and a directly-coupled fan like a BMW 2002, you can usually rotate the engine just by moving the fan blades.
- Or you can – on a flat area – put the car in gear, release the handbrake, and try to rock the car to get the drive wheels to turn the drivetrain.
- Ultimately, you may need to put a socket and ratchet handle directly on the crankshaft to turn it.

- If you think the compressor, water pump, power steering pump, or any other component is causing a mechanical restriction, remove its belt. But verify that the engine rotates freely.
- If it doesn't, the starter is the least of your problems.

Is It Hot?

If there's no voltage drop, and no mechanical restriction, and it *still* goes "click," try to start it a few times, then reach down and lay your hand on the body of the starter. If it's hot, that's a good indication that current is reaching the starter, it's trying to turn, but some internal malfunction is preventing it. In other words, you have a bad starter. If you've gone through all this and it's *not* hot, it could be that the solenoid isn't passing current to the starter, despite the "click." If that's the case, we'll give it one more chance.

8. Power the Starter Directly

As we keep saying, starters are trivial to pull out of a carbureted car, but can be a bear to remove on a post-1996 car. One or two of the nuts or bolts at the top of the bell housing are typically a total bitch to reach, and, even once it's unbolted, as we said earlier, the intake manifold and brake, cooling, and electrical components may make it difficult to actually pull the thing out. So, before subjecting you to that fate, there is one thin ray of hope – perhaps the solenoid is bad but the starter itself is okay.

There is, in fact, a way to check this without removing the starter. There's a short thick cable with a ring connector coming out of the starter that bolts onto a post on the solenoid, the one that's labeled "M" on American cars but for some reason has no DIN designation. Note that this short cable doesn't bolt to the same post that the positive battery cable attaches to – it bolts to the post between it and the starter. In fact, the electrical function of the solenoid is to connect these two posts together and supply battery voltage to the starter.

For this test, you want to temporarily supply battery voltage to this post and see (more accurately, hear) if the starter spins. If it does, then the starter is most likely okay and the problem is most likely in the solenoid.

The problem is that you need to do this safely. You don't want to be poking around in an obstructed space with a live cable to the battery, trying to touch this post but accidentally touching ground.

So here's how to do it:

- Detach the negative cable from the battery.
- Detach the positive cable from the battery. This ensures that absolutely nothing in the car is electrically connected to the battery during this test.
- Take a 10-gauge wire with alligator clips at both ends. Connect one end to the starter post on the solenoid, and the other end to the positive battery post. See **Figure 16**. Note that this wire doesn't need to be as thick as a standard battery cable because, since the solenoid will be bypassed, the pinion gear won't drive forward and thus the starter won't turn the engine over. Still, don't use anything thinner than 10-gauge wire.

Figure 16 The direct connection to the starter isn't the threaded post with the battery cable on it – it's the shorter threaded post to its left (**arrow**). Use extreme care to not short the hot wire to the grounded starter body.

- Double-check that the clip on the starter post is not touching the body of the starter or the body of the car.
- Now, wearing gloves, briefly put the negative battery cable on the negative battery terminal. This will complete the circuit to the starter, which is grounded through the engine (and you've presumably already verified the integrity of this ground with a voltage drop test). You should hear the starter spin, but because the solenoid is bypassed, the pinion gear won't move outward so it won't turn the engine over.
- If the starter spins in this test but didn't before, the problem is likely in the solenoid.

If the starter *doesn't* spin, there are a couple of possibilities.

- If there's a spark when you touch the negative cable to the battery, the starter is probably mechanically seized.
- If there's no spark at all, the starter is, electrically an open circuit, meaning there's a broken wire somewhere inside it.

If the test shows up that the starter is okay but the solenoid is suspect, it may be possible to replace the solenoid without even pulling the starter off the car. You'll need to read up on your specific car and starter to know whether this is possible, and see if the solenoid is, in fact, available as a separate part.

What If the Starter Really Is Bad?

End of the line. Suck it up. Pull it out. Consult the repair manual specific to your car for the exact procedure.

Can I Get Home Before I Replace It? (The Pop Start)

Just in case you don't know, if your starter has failed somewhere, there's a save-your-butt trick you can do if the car has a manual transmission and doesn't have a clutch interlock. It's called pop-starting.

If you had the tremendous good fortune to park your car on a hill, and there are no parked cars in front of you, you simply turn the key to ignition, put your foot on the clutch and leave it there, put the car in second gear (second seems to give the best compromise between spinning the engine quickly but allowing the car to maintain momentum without the braking effect of the engine slowing it down before it starts), roll down the hill until the car is going between 5 and 10 mph, then quickly let the clutch out.

With the car in gear, the spinning rear wheels will now spin the engine just like the starter would, and if the ignition and fuel system are in good condition, it should start. What'll happen is that the engagement of the engine will slow the car down quite a bit, but hopefully, before it stops completely, the engine will start. This *will* cause the car to lurch forward as if you're a newbie learning to use a clutch, so be prepared.

If the car wasn't parked on a hill (and it never is), you can get your four best friends to push you as fast as they can. Just make sure to have them stop and let the car coast away from them, or else when you let the clutch out, they'll slam into the back of the car.

But if your starter has been failing, and now has finally failed completely, and you do this, when you say, "I *promise* I'll replace the starter when I get home," keep your bargain with your car, okay?

Bench Testing

Okay, the damned thing is out. Can you bench-test it? You sure can. Further, you should be able to triage whether the root cause is in the solenoid or in the starter itself. However, since there's no longer any mechanical load on the starter, it may not fail like it did while it was trying to spin the engine, and the results may not be conclusive. If the starter's failure mode is that it won't spin after repeated attempts, or gets hot to the touch, don't second-guess yourself if you can't recreate the failure on the garage floor; just replace it.

⚠ CAUTION —

To bench-test the starter, you need to immobilize it, either by standing on it or clamping it securely in a good-sized bench vise. Otherwise, if it starts to spin, the torque is going to cause it to jump around like a good-sized bluefish out of water. Seriously. And when it does, it's going to pull the battery cables off, and they're likely to touch together and generate a hail of sparks. So don't do this unless you're certain you've immobilized the starter.

Testing the Starter and Solenoid Together

Note that we already did this above in Starter Test 5 with the starter in the car, using a manual remote start switch. The only real differences are that the starter is out of the car, there's no mechanical load on it, and you're connecting directly to it with jumper cables, lessening the possibility of voltage drop issues.

- Gather together a battery, the starter, a set of jumper cables, and a jumper wire with alligator clip leads at at least one end.
- Connect the positive jumper cable from the positive battery terminal to the big threaded terminal 30 on the solenoid. See **Figure 17**.

Figure 17 Jumper cable connecting positive battery cable to starter terminal 30.

- As per the caution above, immobilize the starter.

 **CAUTION —**

Do not proceed with this test until the starter is immobilized!

- Connect the negative jumper cable from the negative battery terminal to the body of the starter. One of the ears where the starter bolts to the bell housing is a convenient place to clip the jumper cable.
- Connect one end of the alligator clip to terminal 30. Be careful not to touch the other end to the body of the starter, or it will spark.
- Supply battery voltage to terminal 50 by touching the other end of the clip lead wire there. See **Figure 18**. When you do, it will energize the solenoid. Two things should happen:
  - The pinion gear at the end of the solenoid should thrust forward. The starter should spin.

Figure 18 If you touch terminal 50 to battery power while there's power at terminal 30, be certain to stand on the starter so it doesn't jump. A red spade connector has been put on terminal 50 to make it more visible for the photo.

  - If the starter spins but the pinion gear doesn't jump forward, the problem is in the solenoid or in the mechanical connection between the starter and solenoid. Proceed to the next step below.
  - If you hear a good "click" and see the pinion gear jump forward but the starter doesn't spin, the solenoid is probably okay but the starter is probably bad. Proceed to the next test to power the starter directly.
- Remove the power cable from terminal 30. With the starter body grounded, connect power to terminal 50. See **Figure 19**.
  - The pinion gear should jump forward with a loud click. If not, the solenoid is faulty.

Testing the Starter Directly

This is the same test as in Step 8, just with the starter out of the car.

- Connect one negative end of the jumper cable to the negative battery terminal, and the other to one ear of the starter.

Figure 19 Grounding the case of the starter and touching powering terminal 50 will activate the solenoid.

CAUTION —

Do not proceed with this test unless the starter is immobilized!

- Supply power from the positive jumper cable from the positive battery terminal to the threaded post where the solenoid connects to the starter. In the pictures, this is the short threaded post beneath threaded post 50.
- Be certain the cable is touching only the threaded post or the nut on it, not the body of the starter. You will almost certainly need something smaller than the jumper cables to make this happen, such as an alligator clip lead. Because the starter is not connected to the engine, it is not under load, so it won't draw as much amperage as it would if you were actually trying to start the car, So for this test, you can use wire that's thinner than battery cable, but we recommend not using anything thinner than 10-gauge wire.

WARNING —

Do not put your hands directly on the metal alligator clip lead! *Even though the amperage will be much less than if the starter was under load, it is still possible for the starter to pull enough current to heat up the connector and burn your hand.*

- When you touch the alligator clip lead to the positive jumper cable as pictured in **Figure 20**, you should hear the starter spin, but because the solenoid is bypassed, you won't see the pinion gear jump forward.

Figure 20 You can bypass the solenoid and test the starter, but it's hard to get at the post below terminal 50 without a small alligator clip.

- If the starter doesn't spin:
  - If there's a spark when you touch the negative wire, the starter is probably mechanically seized.
  - If there's no spark, the starter is probably electrically open.

Repair, Rebuild, or Replace?

In theory, if the solenoid tests out bad but the starter tests out good, it should be possible to replace only the solenoid. Whether it's worth it, how easy it is to do, and whether the solenoid is available are questions that will be specific to your model.

Considering what a pain it is to get a starter out of and into most cars, you should be very circumspect about repairing it yourself or installing a lowest-cost rebuild of unknown provenance. If your testing has shown that the starter is bad, our advice is to replace it with a new starter.

Tales From the Hack Mechanic: Starter Stuff

I recently helped a friend of mine diagnose a starter problem in a 2007 Honda Civic. The "click" problem with a fully charged battery led me to reading about the car's Denso-brand starter, which, it turns out, has a solenoid whose plunger is a user-replaceable part that can be accessed with the starter in the car. Unfortunately, the Bosch and Valeo starters in European cars don't seem to be designed this way.

Back in the day, it seemed like every second BMW 2002 I owned (and I've had 30 of them) needed the starter replaced. In contrast, I've never had to replace a starter on a newer (say, post-1995) BMW. Whether the newer starters are better, or age and mileage haven't caught up to them yet, I can't say, but I don't read the same level of vitriol about rebuilt starters that I routinely see about rebuilt alternators.

I have never rebuilt a starter, but you can certainly try your hand at it. I've read posts on enthusiast forums describing how problems were found to be due to a general munging-up of the inside from the carbon thrown off as the brushes wore, and how, with a good cleaning, lubing, and replacement of the brushes, the thing gleefully returned to life. Then again, I've also read posts where do-it-yourselfers lost track of the order with which some of the small parts went back together, couldn't figure out how to insert the spring-loaded brushes, threw in the towel, and ordered a rebuilt unit. It's just never been somewhere I've wanted to put my time.

On my 1970s BMWs, when I have had starters or solenoids die, it gives me an opportunity to replace them with a gear reduction starter that's smaller, draws fewer amps, and cranks faster. I did the same when the starter gave up the ghost in my Porsche 911SC. So, for me, these things tipped the decision from rebuild to new purchase.

16

Mechanically Timed Ignition – Theory

Mechanically Timed Ignition

There's an old adage that a car needs gas and spark to run. In these next three chapters, we're going to concentrate on spark. More specifically, we'll talk about the ignition systems used through the mid to late 1970s, where high voltage is created by a single plainly visible coil, triggered by contact points inside a distributor, and the distributor is responsible for sending the spark to the plugs and advancing the spark through mechanical (or mechanical and vacuum) timing. This is called *mechanically timed ignition*.

Mechanically timed ignition is easy to understand and wonderfully straightforward to diagnose, which is good, because it's one of the most likely things on your car to die on you and leave you in the lurch. Suspect the points first and the coil last. As they say, forewarned is forearmed.

This chapter covers the theory of mechanically-timed ignition. The following chapter delves into troubleshooting an ignition system, and the chapter after that talks about electronic (breakerless) ignition.

The Basics: How Ignition Works

Figure 1 illustrates a mechanically timed ignition system for a 4-cylinder engine.

1) An *ignition coil* converts the battery's voltage into the high voltage necessary to ignite the fuel/air mixture in the cylinders.

2) A *distributor* has a rotating shaft that distributes the high voltage to each of the spark plugs.

3) As the distributor shaft spins around, a set of *points* inside the distributor opens and closes. Every time they open, they trigger circuitry that tells the coil to do its thing and turn low voltage into high voltage.

4) The distributor also controls the *timing* of the high-voltage delivery, sending voltage to each spark plug just before its cylinder reaches the top of its compression stroke.

5) The high voltage jumps the small gap in each of the spark plugs to create the spark that ignites the fuel/air mixture.

On post-1990s cars with computer-driven stick coils, there is no distributor, and the second through fourth functions are performed electronically. But on cars with mechanically timed ignition, that distributor is the ticking heart of the whole system.

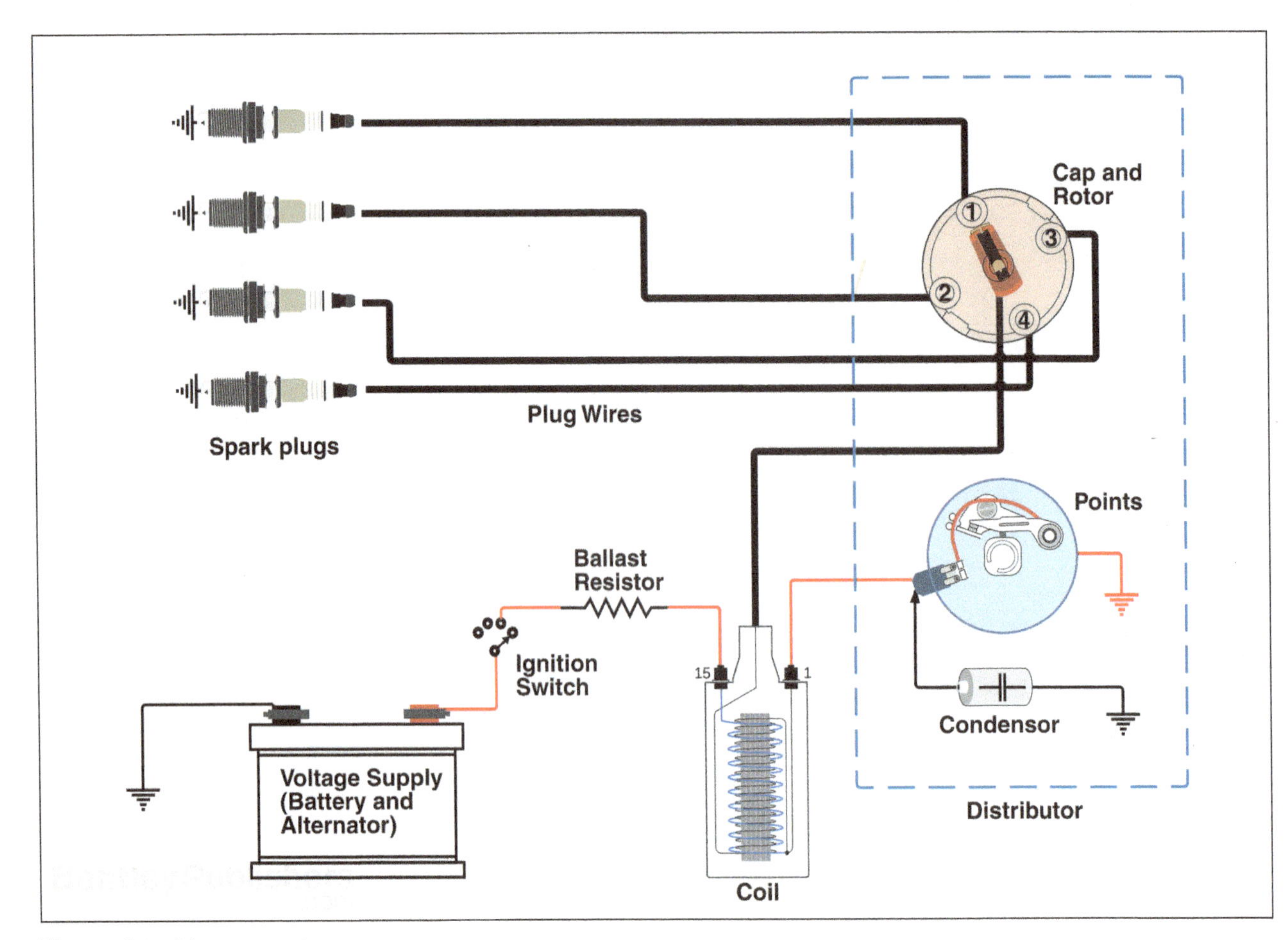

Figure 1 Ignition system layout.

The early history of automotive ignition systems is best understood by realizing that, initially, the production of internal combustion engines predated affordable reliable battery technology. We take for granted that the "spark" in "gas and spark" is an electrical spark, but that was not always the case. A glorious spectacle of ridiculous mechanical contrivances were instead employed in service of this end. This included:

- "Flame ignition" that used a mechanically actuated sliding window to expose each cylinder's combustion gases to what was essentially a Bunsen burner.
- A "hot tube" system where a tube connected a burner to the head, and the length of that tube determined the ignition timing.
- Attempts to use what were basically moving cigarette lighters – files that scraped against flints to generate spark.

In hindsight, using a thermal source or a flint-driven spark was silly, with electrical spark the obvious inexorable solution. But there were problems – how to generate a spark with a high enough voltage that could jump an air gap, and, initially, how to do it without a battery. The no-battery part was addressed by use of a magneto – a permanent magnet rotating past a coil of wire. Magnetos are still widely used in small gas engines that don't have batteries (lawnmowers, leaf blowers, chainsaws, etc.), and as part of redundant ignition systems in small airplanes.

The Amazing Mister Kettering

The person to whom you owe the design and implementation of the mechanically timed ignition system (what we'd oxymoronically call "modern vintage ignition") was Charles F. Kettering. While working at National Cash Register (NCR) in Dayton, Ohio, in 1908, Kettering won a contract from General Motors to improve the troublesome adolescent spark ignition. The application of a single ignition coil, a single set of contact points, a capacitor to prevent arcing across the points, and a rotating arm that "distributes" the spark to the cylinders all dates back to Mr. Kettering's design. (Great piece of trivia: Mr. Kettering then quit his job at NCR and formed a new company, Dayton Engineering Laboratories Company, which you will better know by its initials – Delco.) The Kettering ignition system was first installed in the 1910 Cadillac, and became the blueprint for most ignition systems for nearly 70 years.

So, how does it work? How does an ignition system perform the conjurer's trick of taking 12 volts (well, more like 14.2 volts when the alternator's running) that couldn't hurt a fly, and blasting it up to a voltage high enough to jump a gap in a spark plug and cause a mixture of gasoline and air to burn? We shall explain.

Functions of the Ignition System

From a functional standpoint, your ignition system is actually doing three distinct things:

- It ***creates the spark*** that ignites the fuel/air mixture.
- It ***delivers and distributes the spark*** to the cylinders (unless you own some beasty single-cylindered British motorcycle, the spark needs to go to more than one cylinder).
- It ***controls the timing*** of the spark (when the spark fires relative to the location of the piston in its compression stroke), advancing the timing, and making the spark ignite the mixture sooner at higher engine rpm.

Ignition System Components

Let's address the components that perform these functions.

The Ignition System Component Functions Table lists the ignition components along with which of the above functions they perform. This can be a little confusing because the distributor, as its name implies, distributes the spark, but it also houses components that create and advance the spark. For this reason, we explicitly list which components are inside the distributor and what function they serve. As we describe each component, we'll walk through its portion of the electrical path in the ignition circuit.

Ignition System Component Functions

Component	Function
Voltage Source	Spark Creation
Ignition Coil	Spark Creation
(Distributor) Shaft	Spark Timing
(Distributor) Points	Spark Creation
(Distributor) Condenser	Spark Creation
(Distributor) Rotor	Spark Delivery
(Distributor) Cap	Spark Delivery
Spark Plug Wires	Spark Delivery
Spark Plugs	Spark

16

Ignition DIN Terminal Numbers

Let's quickly review the ignition-related terminals listed in the DIN table, as they'll come up again and again. Clarifying comments in brackets are ours.

DIN Terminal Numbers, Ignition System	
DIN Terminal	**Description**
1	Low voltage [connection on coil to condenser and points in distributor, tachometer; also frequently labeled "–"]
4	High voltage [output from center of coil to center of distributor cap]
15	Switched positive after battery (ignition switch output) [battery voltage input to coil; also frequently labeled "+"]
15a	Output at the series (ballast) resistor to the ignition coil and the starter

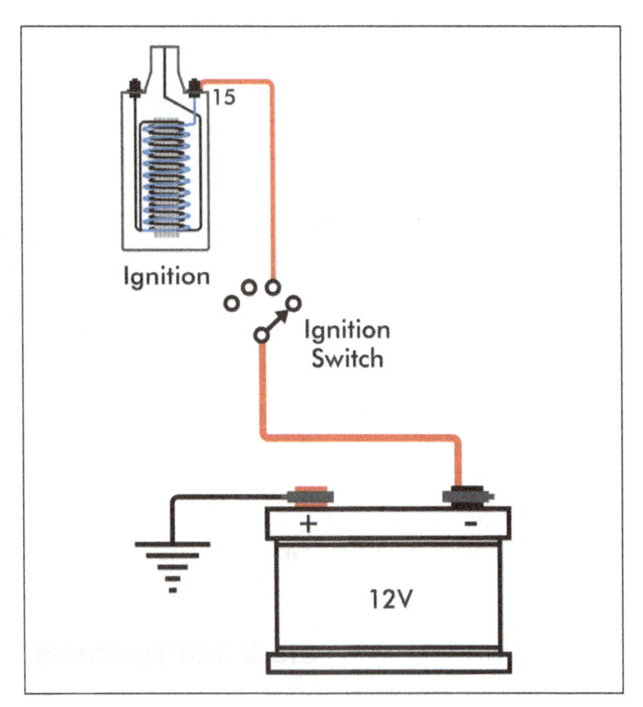

Figure 2 The power path of the ignition system on a car with a distributor and external coil is incredibly straightforward.

Voltage Source

Battery, Alternator, and Ignition Switch

Voltage to power the ignition comes from the battery, through the ignition switch, and out to the 15 terminal on the coil. See **Figure 2**. As with all other electrical components, once the car is up and running, it is really the alternator that is supplying the voltage.

Ignition Relay

An ignition relay is often employed to avoid having the ignition current run through the ignition switch itself. The higher the amount of current that is drawn by the ignition coil, either from the coil itself (you don't think those "high-powered" coils get the energy out of thin air, do you?) or from related accessories that may be powered off the coil's 15 terminal, the less desirable it is to have the entire ignition current run up the wiring harness and pass through the ignition switch. For this reason, a car may be originally fitted with an ignition relay or have had one added. An ignition is wired as follows. Looking at the high and low current sides of the relay:

- ***Relay terminal 30:*** The high-current side of the relay is fed current by a fused ignition-enabled circuit.
- ***Relay terminal 87*** is connected to ***Coil Terminal 15***.
- ***Relay terminal 86:*** The low-current side of the relay is connected to the ignition switch.
- ***Relay terminal 85*** is connected to ***ground***.

An ignition relay is shown in **Figure 3**. In this way, rotating the key to the "ignition" setting energizes the low-current side of the relay, which throws the relay's internal electromagnet. This pulls the switch closed and connects the coil to the battery, and prevents having the high-current connection pass through the ignition switch. For simplicity, fuses are not shown.

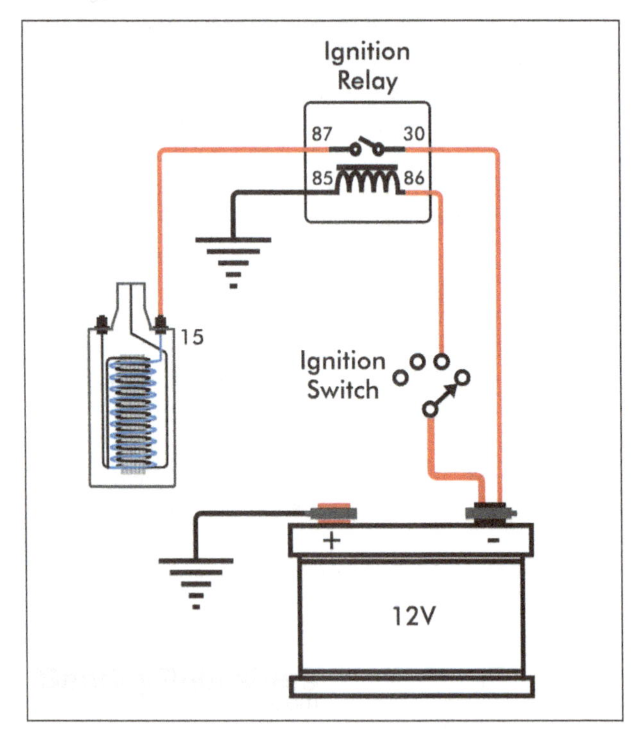

Figure 3 An ignition relay can be used to avoid having ignition current pass through the ignition switch.

Ignition Coil

The ignition coil is the device that turns low voltage into high voltage capable of jumping the gap at the spark plugs. It does this by having two sets of windings. When current flows through the primary winding, a magnetic field is created. When that current is shut off, the magnetic field collapses and a secondary field of a much higher voltage is set up in the secondary winding. This high voltage fires the spark plugs.

The Coil Is Triggered by the Points

It is the opening and closing of the points (also called contact points or breaker points) that is the action that breaks the primary low-voltage circuit, which in turn energizes the secondary high-voltage circuit to create the voltage necessary for the spark plugs to actually spark.

How the Coil Works

In the **Circuits** chapter of this book, we gave an example of building a simple electromagnet. An electric current, when flowing around a ferrous (iron-bearing) core, creates a magnetic field. What you can't see is that, when you shut off the current, that shuts off the magnetic field, and the changing magnetic field actually, for a moment, creates another electric field. This is an example of Faraday's Law of Induction ("a changing electric field creates a magnetic field, and vice versa").

This is what's going on inside an ignition coil. An ignition coil actually has two coils of wire, called the *primary* and *secondary* windings. See **Figure 4**. The primary winding is made of about 200 turns of heavy-gauge wire that surround an iron core. The secondary winding is made up of a far larger number of turns (about 20,000) of very fine wire.

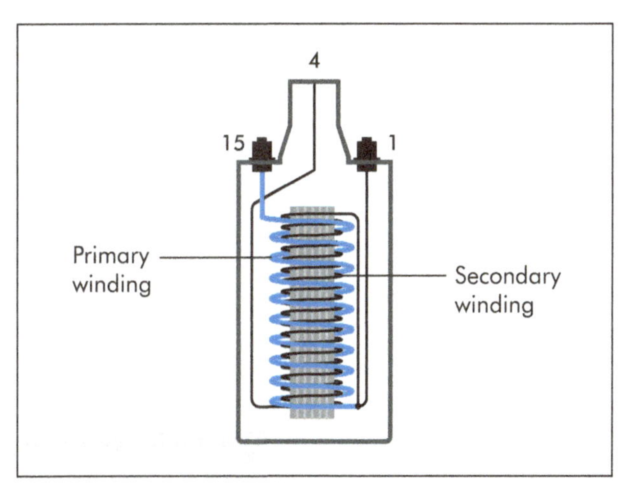

Figure 4 Inner wiring of ignition coil showing connections at: primary winding (terminal 15), secondary winding (terminal 1), output of secondary winding (terminal 4).

The ignition coil on a European car with mechanically timed ignition is nearly always a Bosch coil about the height of a soda can but slightly narrower in diameter. Terminal 4 is the fat spark plug wire that comes out of the center of the coil and goes to the distributor. It's frequently unlabeled, but there's no confusing it with anything else. Terminals 1 and 15 straddle terminal 4 and are either small threaded posts or male quick-disconnect terminals. On a Bosch coil they are physically labeled "1" and "15," though on aftermarket coils they may be labeled "–" and "+" respectively. See **Figure 5**. The DIN may call the ballast resistor output "15a," but it's rarely labeled or referred to that way in manuals.

Figure 5 Bosch coil with standard DIN terminal numbering: 1 (or "–") left, 4 center, and 15 (or "+") right.

Let's look in detail at the circuits made by both sets of windings.

Primary (Low-Voltage) Circuit

The primary circuit creates an electromagnet that can be shut off. Current flows into the primary winding from the voltage supply (battery/alternator/ignition switch) to coil terminal 15, and flows out through coil terminal 1. From there, the primary current takes a slightly circuitous path to ground. A wire is connected to 1 that goes to an insulated (non-grounded) terminal on the outside of the distributor. This terminal is actually one side of the condenser, which is located inside the distributor. The wire tees from there to one side of the points. When the points are closed, the primary current flows through the closed points to the metal body of the distributor, which is grounded by its contact with the engine. This completes the primary circuit. See **Figure 6**.

Note also that, because the 1 terminal is the readily accessible point where the opening and closing of the points can be monitored, it is where you connect a dwell meter with an integrated tachometer. It is also, on most 1970s and 1980s European vehicles, where the car's tachometer is connected, though the connection is usually made on the terminal connected to it on the side of the distributor.

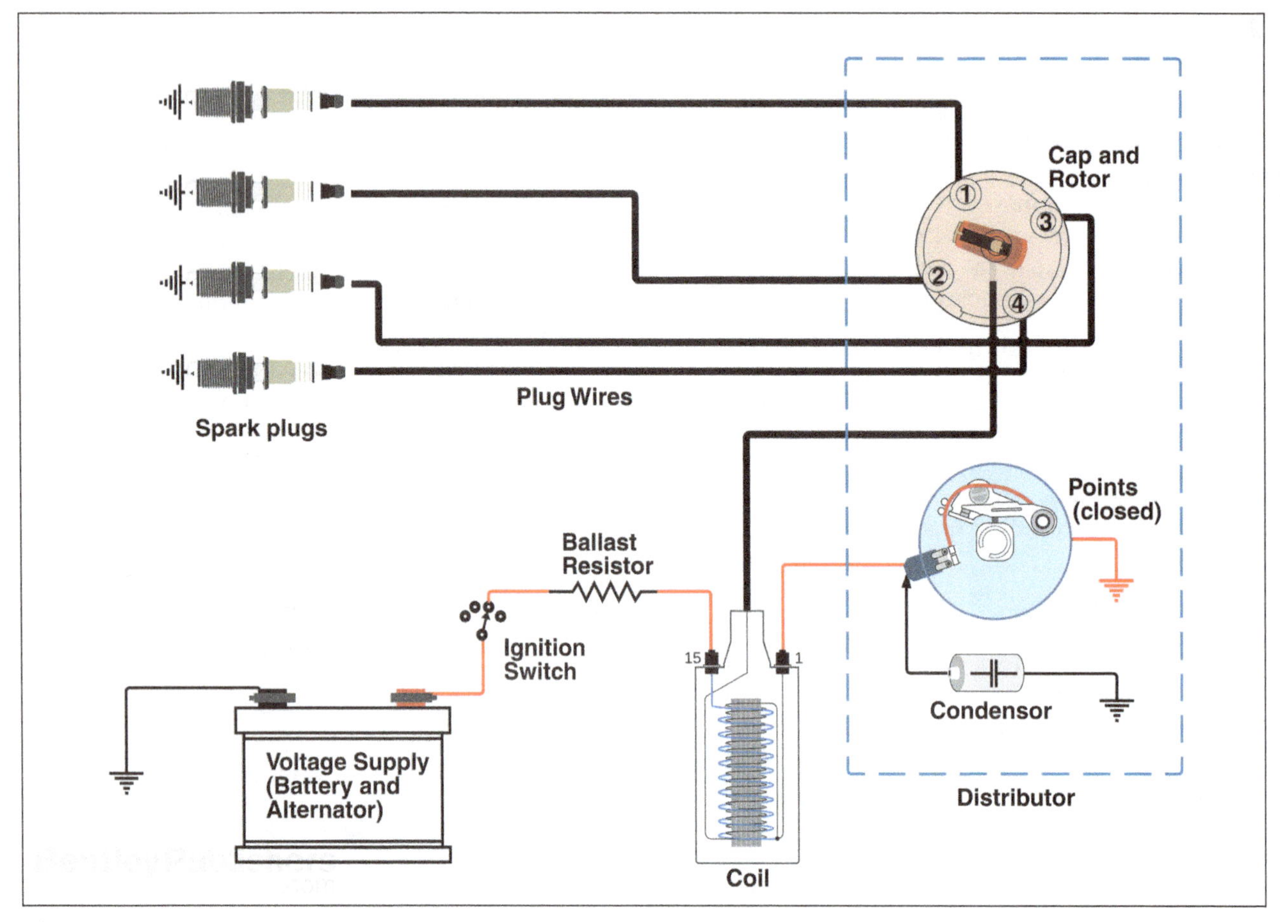

Figure 6 Diagram of ignition current flow through the (low-voltage) primary circuit with points closed.

Secondary (High-Voltage) Circuit

The magic happens when the points open. This breaks the primary circuit. The current stops flowing through the points to ground, and the magnetic field in the coil collapses (shuts off). Recalling Faraday's law that a changing magnetic field creates an electric field, when a magnetic field abruptly shuts off, it is certainly changing very rapidly, so the collapsing magnetic field induces current in both the primary and the secondary winding of the coil. The secondary (high-voltage) circuit is shown in **Figure 7**. It is the high voltage in the secondary circuit that flows to the spark plugs, igniting the fuel-air mixture.

The voltage of the current induced in the primary winding is about 300 volts. Because there are about 100 turns of secondary wire per turns of primary wire, the voltage of the induced current in the secondary winding is much higher than in the primary winding – tens of thousands of volts. When the points were closed, the primary current flowed through the points to ground, but when the points are open, where does that current go? If something clever wasn't done, the primary current would create a spark across the point gap just like the secondary current creates a spark across the spark plug gap. That "something clever" is the condenser, and we'll talk about it in its own sidebar.

Note that, although one end of the secondary winding is attached to terminal 1, terminal 1 should not strictly be thought of as the input to the secondary winding because the high voltage in the secondary winding is created by the coil itself, not by an external source on terminal 1 (we'll get back to this when we talk about the condenser). The output, however, of the secondary winding comes out the big receptacle in the center of the coil. It goes from there, out a fat spark plug wire, to the distributor.

Ballast Resistor

Pity the poor misunderstood ballast resistor. It exists only so it can be bypassed during starting. Why the hell would you want that? Actually, it makes a lot of sense.

Once a car is up and running, the alternator should be supplying the coil with about 14.2 volts. But during the act of starting, the ignition coil gets dealt a triple whammy.

First, from an ignition standpoint, starting is more difficult than sustained running due to the cold engine and the viscous oil. Second, the alternator isn't up and running yet, so the coil doesn't see 14.2 volts; it instead sees whatever battery voltage is – more like 12.6 volts,

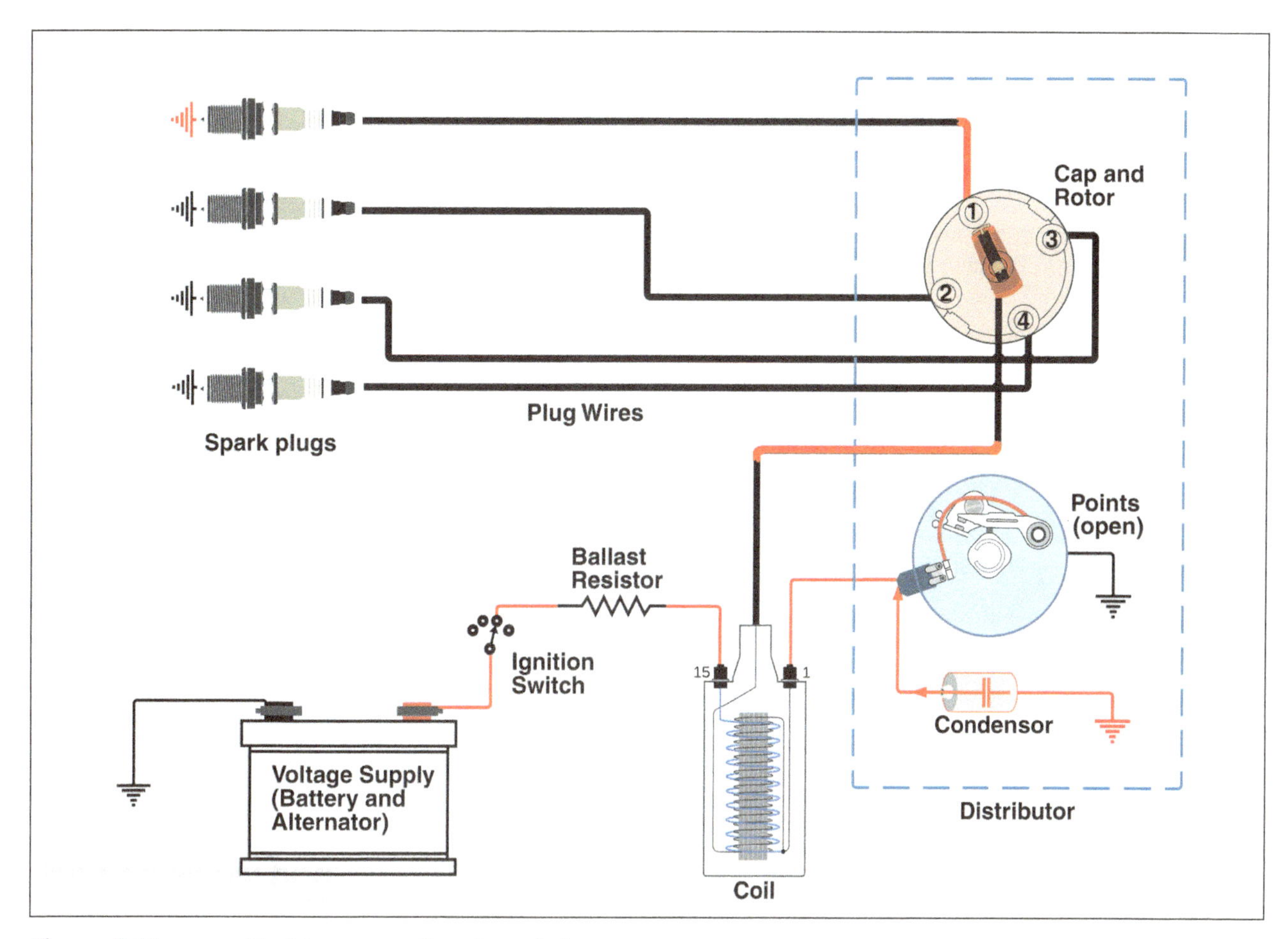

Figure 7 Diagram of ignition current flow through the (high-voltage) secondary circuit with points open.

minus any voltage drops. Third, while the starter motor is cranking, it draws a lot of current from the battery, which lessens the voltage that the coil sees. Because of all these effects, the coil may only see about 9.6 volts during startup.

For these reasons, ignition systems need to give the coil more juice during starting. They achieved this by using a low-resistance coil coupled with a *ballast resistor.* While the engine is starting, the coil is fed battery voltage. The coil's low resistance allows it to draw more current, creating a stronger spark. It can't, however, be run this way for long without overheating the coil, burning the points, or frying the condenser, so once the engine starts, the ballast resistor is brought into the circuit, increasing resistance and thereby limiting current.

Technically, the ballast resistor is never "bypassed" in the sense of being taken out of the circuit. It's more accurate to say that, during starting, the coil is fed full battery voltage through an alternate path. The ballast resistor is still connected, but current is literally following the path of least resistance.

And here it is in the flesh in **Figure 8**.

Figure 8 Coil, ballast resistor, and "ignition relay" on a 1972 BMW 2002tii.

Distributor

The distributor consists of the metal distributor body, the condenser mounted to the side of the distributor, the cap on the top of the distributor, and the rotating shaft and the points inside the distributor. See **Figure 9**.

As we said earlier, the points and condenser create spark, the cap and rotor deliver spark, and the rotating shaft adjusts the timing of the spark.

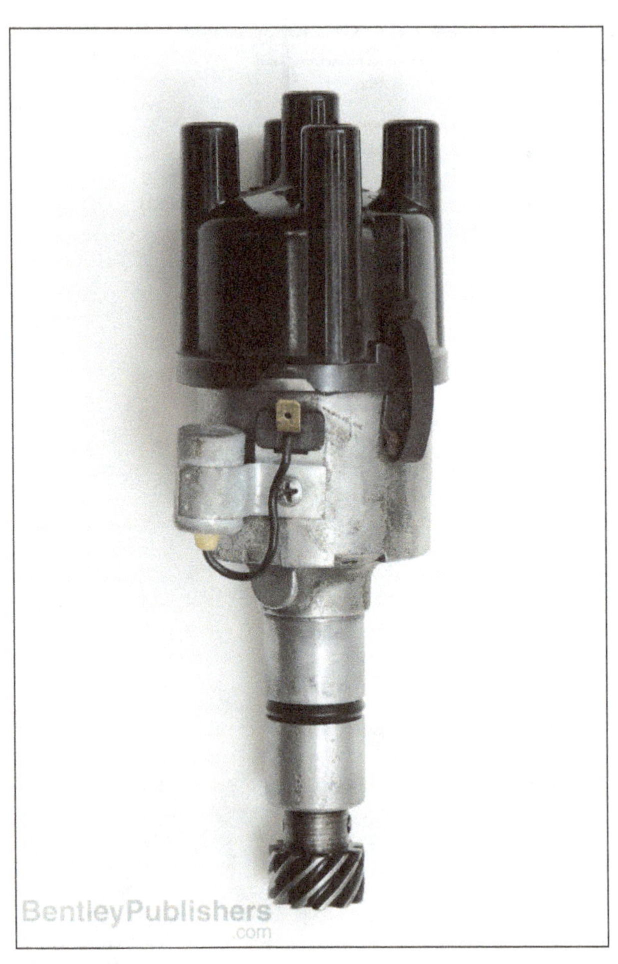

Figure 9 An old-school mechanical advance distributor. We'll describe the pieces in detail in this chapter.

Points

In the **Ignition Coil** section, we discussed how the opening and closing of the points is the triggering action that breaks the primary low-voltage circuit, which in turn energizes the secondary high-voltage circuit to create the spark. But why are the points located inside the distributor? Because something must open and close the points in synchronicity with the motion of the engine, and that something is in fact a set of lobes on the rotating shaft inside the distributor. There are as many lobes on the shaft as there are cylinders in the engine. See **Figure 10**. The points are kept closed by a spring. One side of the points is fixed to the distributor, but the other side of the points has a little nylon block that rides on the distributor shaft lobes, making the points open when the lobe's corresponding cylinder is nearing the top of its compression stroke. (In fact, setting the dwell and setting the timing are the actions needed to adjust the synchronization of piston location to plug firing.)

Figure 10 Inside of distributor for 6-cylinder engine on a 1972 BMW Bavaria showing points (A), nylon block (B), rotating six-lobed shaft (C), and condenser (D). The rotor is usually on top of the shaft, but has been removed for clarity.

Condenser

The coil gets all the attention, but the condenser is really the small piece of brilliance that makes it all work. The short description is that the condenser prevents the spark from arcing across the points. The long description is a bit involved, but it's actually really cool. We have included it as a sidebar.

Cap and Rotor

The high voltage comes from the center of the coil to the center of the distributor cap via a a wire the same thickness as the spark plug wires. See **Figure 11**. From there, a little sprung contact inside the cap touches the center of the rotor. The rotor, as its name implies, spins around, driven by a gear at the base of the distributor, which is turned by the engine. Inside the cap along its inside circumference, there is one tab for each cylinder. The tip of the rotor comes very close to but does not actually touch the tabs. When the tip of the rotor is opposite one of the tabs, the spark jumps a very small gap between the rotor and the tab, sending the high voltage out the corresponding connection on the top of the cap. From there it goes through the spark plug wire to fire the plug.

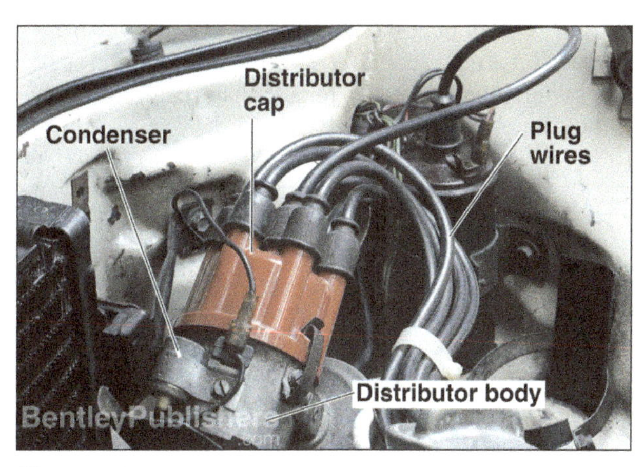

Figure 11 Distributor for 6-cylinder engine on a 1972 BMW Bavaria.

Plug Wires and Spark Plugs

Spark plug wires are unlike any other wires on your car in that they are designed specifically to carry the high voltage from the coil to the distributor and from the distributor to the plugs. They are heavily insulated so the high voltage won't leak out and cause an unintended spark before reaching the spark plug. Plug wires are very flexible; they're designed to be able to flop around, to some extent, with the natural rocking of the engine on its mounts. They typically have high resistance to cut down on radio frequency interference (RFI) that could cause audible noise on onboard audio systems or damage vehicle electronics.

The plug wires must connect the plugs to the distributor cap in the proper firing sequence for the engine. As the rotor spins around and distributes the spark to the tabs on the cap, the high-voltage pulse travels down each plug wire, and reaches each spark plug. The plugs are grounded because they're screwed into the engine, enabling the spark to jump the gap at the end, and ignite the air/fuel mixture.

Spark plugs of the same physical size (meaning thread diameter and the socket required to install them) can have a variety of physical characteristics such as the distance the electrode projects into the combustion chamber, the length of the insulator, and the tip materials. These all affect the plug's heat range. The owner's manual and repair manual will list the recommended spark plugs, but web forums are a good place to go for opinions on whether a plug with a different heat range is appropriate for a 40-year-old car running on unleaded fuel, as well as alternatives if you have a modified engine or ignition. (The BMW 2002 community, for example, is generally fairly negative on the use of platinum plugs in a 40-year-old car that experiences a much wider lean-to-rich mixture swing than a post-1996 computer-controlled car for which they're generally intended.)

The Magic Condenser

Astute readers and former teacher's pets in physics class will have already asked the question "If the collapse of the magnetic field induces a current in the secondary winding, doesn't it also induce a current in the primary winding?" Yes, smarty pants, it does. Recall that terminal 1 — the output of the primary winding—is connected to a tee in the distributor, and from there it's connected to both the points and to the condenser. The condenser is just a capacitor, a device that stores charge. So when the points open, the current that was flowing through the primary winding can no longer flow through the points to ground, and is redirected to charging the condenser.

This serves two purposes. The first is that it gives the primary current somewhere to go so it won't jump the point gap. But it's the second purpose that's genius. While the condenser is charging, the primary current drops, and the magnetic field collapses, inducing a current of about 300 volts in the primary winding. Because the primary winding is still connected to the condenser, this packs the condenser with additional charge until it's at a voltage that's much higher than battery voltage.

When the condenser is fully charged, it discharges. Where? The only place it can: terminal 1, which is connected to both the primary and secondary winding. Because the points are open, the discharge into the primary winding can't go to ground, so it swims upstream, pushing back against the battery current (remember, the other end of the primary winding – terminal 15 – is still being supplied battery voltage).

But the real trick is that, because the direction of current in the primary winding has been reversed, the magnetic field not only collapses, it collapses and reverses. Since the physics behind all this is that a changing magnetic field creates an electric field, you can appreciate that a collapse and a reverse is a much larger change than just a collapse. Therefore, this collapse-and-reverse induces a current at a very high voltage in the secondary winding which then flows from the coil to the distributor to the spark plugs.

In electronics-speak, the use of the condenser forms a "resonant circuit." Some people like to think of the condenser as a spring that stretches out, then snaps back. It's a bit more fun to visualize one of those cartoon characters who absorbs a hail of bullets, then spits them all back out machine gun style from whence they came.

Neat, huh? You can also appreciate how the condenser needs to be matched to the coil and the points, and that not just any capacitor will do.

Everyone say it: Thank you, Mr. Kettering. 1910. Sheesh.

Note that many spark plugs and wire sets each come in resistor and non-resistor versions, and you don't want resistors in both places. Note also that there's a resistor in many rotors, and that the incorporation of all of these resistors was intended to reduce ignition whine when listening to AM radio, and who does that anymore, so the consequence of having too little wire/plug resistance on a pre-computerized car is generally minimal. On a post-1996 car, though, it's best to stick with recommended plugs and wires to reduce the possibility of interference with vehicle electronics.

On 1970s European cars, the connectors on both ends of the plug wires usually screw into the conductor in the middle of the wire using something that looks pretty much like a wood screw that goes into a cross-section of the multi-stranded wire. While this makes it so the connectors can be easily replaced if they're broken, it's a poor electrical and mechanical connection and a common point of failure in old cars. See **Figure 12**.

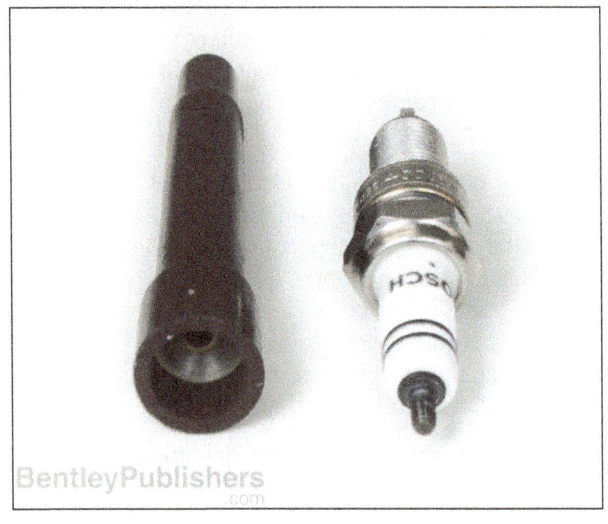

Figure 12 Spark plug with the ferrule removed, and a detachable plug wire connector (left) that must be screwed into the plug wire.

For this reason, plug wires where the connectors are crimped on are preferable. This often means using spark plugs where the removable ferrules have been left on the end of the plug. See **Figure 13**.

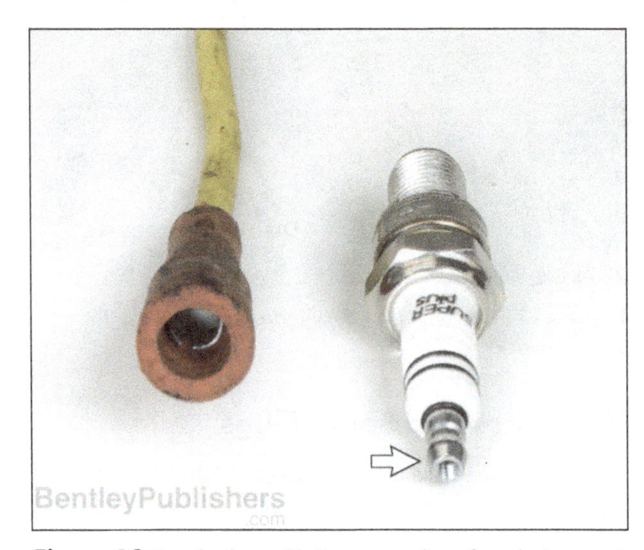

Figure 13 Spark plug with its screwed-on ferrule (**arrow**) attached, and a plug wire (left) with a crimped-on connector that snaps onto the ferrule.

Now that we've given a good overview of the basics of mechanically timed ignition, in the next chapter we'll talk about how to set your ignition timing.

17

Mechanically Timed Ignition – Troubleshooting

Troubleshooting

Damn, we've covered a lot of ground on how mechanically timed ignition works. Of course, you're most concerned when it doesn't work. And, unfortunately, for all its pre-MTV charm, it doesn't work an awful lot of the time (at least the points part of it). In this chapter, we list what you can check for.

Before we begin, a few things are worth pointing out.

 **WARNING —**

We've said in this book that most systems in a car run on 12 volts, and that 12 volts does not present an electrical hazard. The ignition system is an exception. The coil's secondary winding generates tens of thousands of volts to enable the voltage to jump the gap in the spark plug. If you are exposed to this voltage, it generate a very strong, potentially deadly electric shock. **Be absolutely certain to wear rubber gloves and shoes with insulated soles, and be certain the floor is dry when testing for the presence of spark from the coil or the spark plugs!**

The troubleshooting process is presented in a systematic fashion, but there are two quick checks we can do to make sure we're dealing with an ignition issue and not something else.

Checking for Spark

First things first. Checking for spark is the first test to determine if your problem is ignition based. If you find a good spark, chances are good your no-start issue is fuel or maybe even mechanical in nature (no compression, for example) and you get to exit this chapter.

 **WARNING —**

The high voltage from the coil's secondary winding can be tens of thousands of volts. If you are exposed to this voltage, it can generate a very strong, potentially deadly electric shock. **Be absolutely certain to wear rubber gloves and shoes with insulated soles, and be certain the floor is dry when testing for spark at the plugs!**

- Put on a pair of rubber gloves.
- Grab one of the connectors from the spark plugs and pull it off the plug. Take care not pull on the plug wire, as that may pull the wire out of the connector.
- The spark plug contact is often recessed inside the connector, so it can't be touched directly to ground. To address this, insert a spare spark plug into the end of the connector and ...
- Hold the electrode of the spark plug against chassis ground. See **Figure 1**.

Figure 1 Testing for spark at the end of wire and plug. Electrode is held against chassis ground.

- Alternately, take a thin screwdriver, insert it into the end of the connector, and hold the shaft of the screwdriver close to ground. See **Figure 2**.

Figure 2 Testing for spark at the end of the plug wire using a screwdriver held near chassis ground.

- Have a friend operate the starter. With the starter motor cranking, you should see a clear bright bluish-yellow spark jump the gap several times a second. If the engine starts, the frequency of the spark will increase.
- Turn the ignition off.

If there is no spark, let's quickly check for voltage at the ignition coil next. If power is not getting to the coil, the engine will not start.

Checking for Voltage at the Coil

For the ignition to work, the coil must be supplied with battery voltage from the ignition switch.

To test for this:

- Set the multimeter to measure voltage.
- Connect the negative probe to chassis ground.
- Turn the key to ON (ignition on, engine not running).
- If the car has no ballast resistor, touch the positive probe to coil terminal 15.
- If there is a ballast resistor, touch the positive probe to the input side of the ballast resistor (the side not connected to the coil).
- You should see battery voltage (about 12.6V). It may be lower due to the "sag" from not having the alternator running. This is normal. See **Figure 3**.

Figure 3 Checking for the presence of battery voltage at coil terminal 15. The voltage is a little low due to the "sag" from the alternator not running.

- If the meter reads near zero, the ignition switch isn't supplying voltage. Trace the 15 ignition wire back to the ignition switch looking for a broken connector, broken wire, bad ignition switch, bad ignition relay, or other source. For a generic layout of the ignition wiring on a car with mechanically timed ignition, see the **Mechanically Timed Ignition – Theory** chapter.
- If the car has a ballast resistor:
  - Touch the positive probe to coil terminal 15. The ballast resistor will drop some of the voltage across it, so you won't see 12.6 volts at terminal 15. Depending on the resistance, you should see between about 8 and 9 volts.
  - If the meter reads near zero, the ballast resistor has likely burned out. Turn the ignition off and set the multimeter to measure resistance. Check the resistance across the ballast resistor. It should be about one ohm. If the meter reads "OL" (over limit"), the ballast resistor is bad. Replace it.

Note that if there's no voltage at terminal 15, and you want to verify that the rest of the ignition is okay, you can simply connect terminal 15 to the positive battery terminal using a wire with alligator clips. Then try to start the car. If the car has a ballast resistor, however, don't leave it connected this way for long, as the ignition will be running at a higher voltage and drawing more current than it is designed for.

If you don't have a spark but you do have 15 voltage at the coil, the next logical thing do to is check the points.

Testing the Points

Nearly everyone who has owned a car with mechanically timed ignition has had the experience of the car dying, or not starting, because the points aren't opening. Due to pitting of the contact faces of the points (**Figure 4**) and wear of the nylon block, the points gradually open less and less, eventually causing ignition failure. This is by far the most common source of problems on mechanically timed ignition systems.

It is also possible for the nylon block between the points and the distributor cam to break off entirely.

There are several ways to check if the points aren't opening and closing.

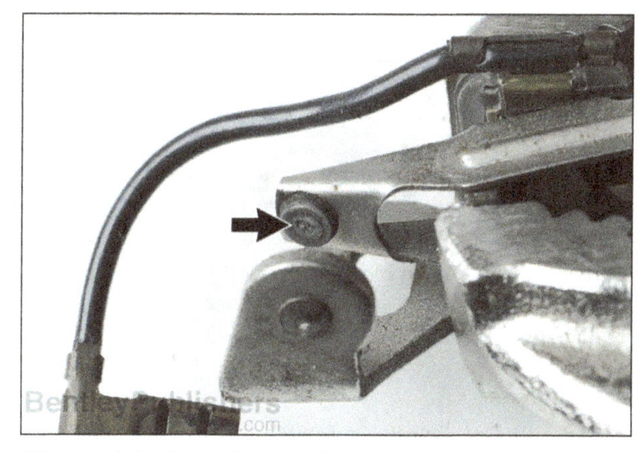

Figure 4 Surface of points showing pitting (**arrow**).

Visual Method

The visual method is very simple.

- Remove the distributor cap.
- Have a friend try to start the car while you watch the points.
- Do you see the points opening and closing?

If you don't, they're not, but if you do, that's not always enough. It's possible that due to the presence of mounds and pitting, the points appear to be opening but are never fully electrically separating, in which case they should be replaced.

Multimeter Method

Remember, when the points are closed, they complete the primary winding's circuit to ground, and when they open, the complete the secondary circuit to ground. So…

- Set the handbrake, chock the wheels, and put the car in neutral.
- Set the multimeter to measure voltage.
- Connect the red lead to terminal 1 on the coil, and the other to ground.
- Turn the key to ignition ON.
- Rotate the engine until the nylon block is on one of the high points on the distributor's lobes and the points open.
- The meter should read battery voltage, or close to it. See **Figure 5**.

Figure 5 With nylon block positioned on high point of distributor cam lobe (**arrow**), points are open, and meter measures battery voltage between coil terminal 1 and ground.

- Continue to rotate the engine until the points close.
- The meter should read zero.

Replace the Points and Re-Gap

If any of the above tests fail, it's highly likely the points are worn or not properly opening. Remove the points and inspect them. In the old days, you'd file any mounds and pits off the point faces, but now you'd only do that if you were in the middle of nowhere and didn't have a replacement set of points.

In either case, install a new set of points, gapping them using a feeler gauge or the trusty matchbook cover if you have nothing else. If you don't have a repair manual handy, search the web for a video of this DIY procedure. If the points were closed, odds are this will get the car to start.

Remember that changing point gap changes the dwell, and changing the dwell changes the timing, so when you can, reset the timing, but this should get you on your way.

Testing the Coil

Coils last a long time. However, you can burn out a coil. One way this happens is by leaving the ignition on for a long time while the car is not running. If the ignition is left on while the points are closed, the primary side of the coil—those fat, low-resistance windings—just sit there, dumping current through the points to ground, and getting hot. They're not designed to be on constantly. They're designed to cycle on and off.

You can also burn out a coil by allowing it to fire but not giving the high voltage a path to ground. For this reason, in any of the above procedures that call for cranking the engine while the distributor cap is off, you really should ground the center wire from the coil by connecting a jumper from it to a convenient engine or chassis ground.

If there's voltage to the coil, and the points are opening and closing, then the coil should be doing its primary/secondary thing and generating spark. There are a few quick ways to check if the coil is firing.

Tachometer Method

On nearly all cars with mechanically timed ignition, the tachometer is connected to coil terminal 1. Therefore, if, while you're cranking the engine to try to start it, the tachometer moves off zero, it's likely the coil is firing, but if the tachometer sits there at zero, it's likely the coil isn't firing.

Visual Method

 ***WARNING —***

The high voltage from the coil's secondary winding can be tens of thousands of volts. If you are exposed to this voltage, it can generate a very strong, potentially deadly electric shock. ***Be absolutely certain to wear rubber gloves, shoes with insulated soles, and be certain the floor is dry when testing for spark from the coil!***

- Put on a pair of rubber gloves.
- Pull the thick ignition wire out of the center of the distributor cap. This wire comes from the ignition coil.
- Position it 1/8" to 1/4" from any nearby engine compartment ground such as an exposed bolt.
- Have a friend try to start the car.
- While the starter motor is cranking, you should see a clear bright bluish-yellow spark jump the gap to ground several times a second. See **Figure 6**.

Figure 6 Testing for spark from the coil by holding the coil center wire close to ground (**arrow**), trying to start the car, and checking for spark.

Timing Light Method

In addition to being used to set the timing, the timing light is an extremely useful troubleshooting tool because you can use it to check for the presence of high voltage in the coil wire and plug wires. This test assumes you have already checked the points, as described earlier.

- Connect the timing light's power leads to the battery.
- Connect the timing light's inductive probe around the thick wire that goes from the ignition coil to the center of the distributor. You don't need to remove the wire.
- Have a friend try to start the car.
- Squeeze the trigger switch to the timing light and hold the end of the light close to any dark surface where you can see it flash.
- You should see the light flash steadily. If not, the problem is likely a bad connection, a bad coil, or a bad condenser; there's really not anything else it could be. And *this*, my friend, is the joy of working on a vintage car.
    - Check the wire between coil terminal 1 and the condenser on the distributor. Be certain each connection is firmly gripping the terminal. Give them a squeeze with a pair of pliers if they feel loose. If no faults are found, continue with the remainder of the tests in this chapter.
    -

NOTE —

Because the tachometer is connected to coil terminal 1, there is the slim possibility that the tach is malfunctioning and causing terminal 1 to short to ground. For this reason, it's good practice to pull the tach wire off coil terminal 1 when rechecking for spark.

- If the light flashes on the coil wire, check that high voltage is reaching the spark plugs.
    - Connect the inductive probe around one of the plug wires. You don't need to remove the wire.
    - Have a friend try to start the car. The timing light should flash steadily. Repeat on the remaining plug wires.

If there's spark at the wire coming out of the center of the coil but there's no spark out to the plugs:

- Check that the distributor cap is correct for the distributor, is seating correctly, and is not cracked.
- Check that the rotor is correct for the distributor and cap and that the electrical contacts on the rotor and inside the cap are not badly pitted or corroded.
- Check that the cap and plug wires are mated together correctly.
- Check that the ground strap from the battery to the engine is connected and is not corroded.

Tales From the Hack Mechanic: Fooled by Plug Wires

I once bought a dead BMW E21 323i from a guy who had just rebuilt the engine and could not get the car to start. I found that it had spark coming out of the coil but no spark at the plugs. On close examination, I found that the plug wires were wildly incorrect. The distributor cap on this car has little electrical posts, like the tips of spark plugs, in the middle of the sockets for each of the plug wires. It thus needed plug wires that had the female versions of these connectors. Instead, he'd bought plug wires for an older model, which did not have these connectors and thus did not make electrical contact with the posts in the cap. When I installed the correct plug wires, the car started right up.

Testing Coil Resistance

Testing Primary Winding Resistance

- Set the multimeter to measure resistance.
- Turn the car's ignition off.
- Disconnect the wires from the coil's 15 and 1 terminals, then connect the multimeter across those terminals. See **Figure 7**.
- If the car has a ballast resistor, do the same across the resistor.
- Read the resistance measured by the multimeter.

Figure 7 Testing the resistance of the coil's primary winding between terminals 15 and 1.

Though you'd need to confirm the exact resistance values from a factory repair manual or user's forum, the resistance of the coil's primary winding should be about 2 or 3 ohms. If the car has a ballast resistor, the resistance of the ballast resistor should be about 1 or 2 ohms. But if you're only reading a fraction of an ohm, the coil may be internally shorted and may draw too much current. And if, instead, you see an open circuit, the primary winding is broken.

Testing Secondary Winding Resistance

To test the resistance of the secondary winding:

- Set the multimeter to measure resistance.
- Turn the car's ignition off.
- Disconnect the wires from all of the coil's terminals, then connect the multimeter across the coil's 1 and 4 (the big one in the middle) terminals.

Figure 8 Testing the resistance of the coil's secondary winding between terminals 4 and 1. Here, it's reading 9.29 kilo ohms, or 9,290 ohms – about 10,000 ohms.

- Read the resistance measured by the multimeter. See **Figure 8**.

Because the secondary winding is longer and thinner, the resistance should be much higher than the primary – in the neighborhood of 10,000 ohms. But as with the primary, if you're only reading a fraction of an ohm, the coil may be internally shorted and may draw way too much current. And if you see an open circuit, the secondary winding is broken.

As with other tests, the absence of a negative is not a conclusive positive. That is, if the coil has obvious shorts or open circuits, it's clearly bad, but if it passes these tests, it still might malfunction while under load. If you continue to have unexplained problems, you can try to expose the coil to mechanical and thermal stress by heating it with a hairdryer or tapping it with the handle of a screwdriver while conducting the above resistance tests. These may expose a broken winding that's making intermittent contact.

But, in general, you should suspect the wiring, points, and condenser before you jump to replacing the coil. While it is possible to bench-test a coil by setting up a condenser, plug wire, and spark plug, we don't recommend this, as it usually involves cobbling together components whose condition is unknown and possibly worse than those in your car. It's generally preferable to swap components, one at a time, with those in the car.

Testing the Condenser

We've learned from the previous two chapters that the condenser performs a crucial role in the tuning of the high-voltage circuit, and also prevents the spark from arcing across the point gap, There are several tests you can perform on the condenser to verify its integrity.

WARNING —

The condenser is a capacitor, which is an electronic component used to store charge. Rubber gloves should be worn when handling a condenser that has been in a running engine.

Tales From the Hack Mechanic: The Vexing Condenser

Condensers can be particularly vexing. Perhaps it is a sign of the times, globalization and so forth, but the condenser is the only electronic component where I have experienced failures with brand-new parts in BMW-logo'd boxes with correct part numbers, as well failures within a few hundred miles. This makes it difficult to troubleshoot an ignition system because it muddles the concept of a "known good" part.

Visual Method

You can perform a basic functionality test on the condenser, with no tools at all, as follows:

- Take the cap off the distributor.
- Have a friend try to start the car while you watch how strong the spark is across the points.
- What do you see?

There should only be the faintest visible spark across the point gap. If instead you see and hear a bright blue and yellow spark arcing across the point gap, this means the ability of the condenser to absorb charge is compromised, and it should be replaced. If this is the case, odds are the points are burned as well. Inspect and replace them as necessary.

Direct Capacitance Measurement

The most direct test on the condenser is to evaluate its value as a capacitor. If you own a high-quality multimeter that measures capacitance, simply follow the instructions that come with the multimeter. The capacitance of the condenser should be in the neighborhood of 0.2 to 0.3 microfarads. Note that the multimeter may read out in nanofarads. In **Figure 9**, the reading of 247 nanofarads is 0.247 microfarads.

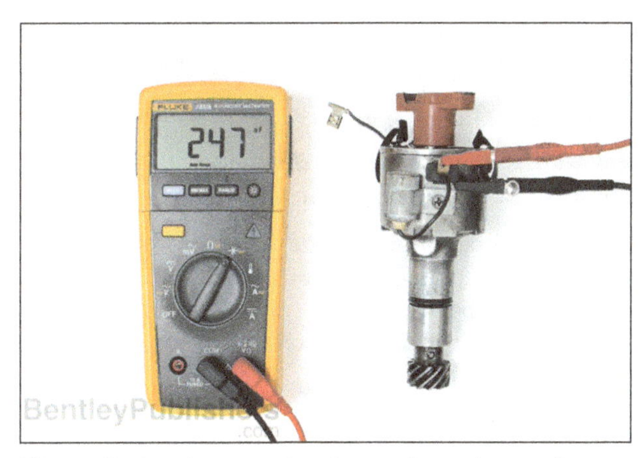

Figure 9 Directly measuring the condenser's capacitance. It should be approximately 0.2 to 0.3 microfarads. The multimeter is reading 247 nanofarads, which is within this range.

Tales From the Hack Mechanic: The Electromagnet, Part 1

I had a maddening problem with my 1972 BMW Bavaria. The car began intermittently stumbling and dying while running, but the spark appeared to test fine at idle. While driving it, I caught it in the act and noticed that, when it stumbled, the rpm shown on the tachometer fell abruptly to zero. This indicated that, during the stumble, the coil wasn't firing. I poked and prodded the wires around the coil and distributor, and found that when I tugged the wire from coil terminal 1 to the condenser on the side of the distributor in one direction, the car barely idled, but when I tugged it the other direction, the car ran fine. It turned out that the little screw holding the condenser to the side of the distributor was loose. The condenser's ground connection is through the body of the distributor; there's no ground path to allow the condenser to discharge. I simply tightened the screw and the problem went away.

Troubleshooting Rough Running

Check for Spark at Each Plug

The old-school method of disconnecting one spark plug wire at a time (pulling each connector off its plug) while the car is running is still a valid technique. The idea is that, if pulling off one of the wires doesn't make the engine run worse, odds are that cylinder isn't producing power. This could be because:

- There's no spark coming into that plug wire.
- The spark plug isn't firing.
- That fuel injector is clogged.
- There's no compression in that cylinder.

WARNING —

The high voltage from the coil's secondary winding can be tens of thousands of volts. If you are exposed to this voltage, it can generate a very strong, potentially deadly electric shock. **Be absolutely certain to wear rubber gloves and shoes with insulated soles, and be certain the floor is dry when testing for spark at the plugs!**

By pulling off one connector, then bringing the connector close to ground and checking for spark, you can verify the first point.

Check for Intermittent Spark

You've probably seen enough photos of new/good/bad plugs to know one from the other. If you're troubleshooting a rough-running condition, you'd be wise to pull the plugs and see where they are on this continuum. Old fouled plugs can certainly generate a weak or intermittent spark. If you have any question, replace them. Spark plugs are cheap.

You can also use a timing light and put it on each spark plug wire one at a time to check for intermittent spark. Rev the engine while watching the flashing of the timing light. If the timing light flashes solidly, the high voltage is traveling through the plug wire and being discharged as spark. But if the light flashes erratically, you've found a fault.

Tales From the Hack Mechanic: Saved by a Timing Light

I once had a rough-running problem in my 2002tii I initially thought was a fuel-related problem, but when I put the timing light on the main coil wire, the rough running was completely correlated with intermittent flashing of the light. I eventually found I had a mismatched coil and ballast resistor that, once it ran for a while, was causing the ignition to overheat.

Don't overlook using the tachometer as a diagnostic tool. As we said earlier in this chapter, the tachometer is connected to coil terminal 1. If the car begins to stumble and the tach abruptly drops to zero, that's a likely indication that the coil isn't firing.

Check the Plug Wires and Connectors

The spark plug wires on distributor-equipped cars often have the connectors screwed into the stranded ignition wire. Age, vibration, and corrosion often weaken the mechanical and electrical bond. It's not unusual to try to pull a connector off a spark plug only to have the wire come off in your hand.

The connectors themselves can be problematic. Missing or torn rubber insulator boots can allow the spark to jump to the head instead of going through the spark plug. With age, the plastic end of a connector that normally shrouds the top of a plug can crack off, exposing the top of the plug and opening up another possible path to ground.

Check for Evidence of Arcing

Due to the high voltage in the secondary circuit, the current will sometimes jump a gap other than the intended ones inside the distributor cap and at the spark plug. If, for example, the rubber insulating boot around the center 4 terminal of the coil is not in place, current may jump from there to the 15 or 1 terminal. When a high-voltage spark jumps across plastic, a carbon-rich track is sometimes deposited in the plastic, making it likely that arcing will continue to occur along the same path. Replace any component that shows visual evidence of arcing.

One good way to check for stray spark is to run the engine in the complete dark with the hood open while you look for sparks indicative of arcing to ground. If your plug wires or their connectors show any of these signs of failure, or if they are just plain old, replace them.

Arcing can also occur inside the distributor cap, either running between the input contact and a plug connection, or between two plug connections. If there is any visual evidence of arcing, replace the cap.

18

Electronic Ignition

Electronic Ignition

As we said in the **Mechanically Timed Ignition** chapter, Charles F. Kettering's basic design for a points-based ignition system that created, distributed, and advanced the spark was the standard for nearly 70 years. However, technology marches on, and in the mid-1970s, both in response to the maintenance requirements of points and the increasingly strict emission control standards, things began to change.

Enter "electronic ignition." The Society of Automotive Engineers (SAE) uses the terms Distributor Ignition (DI) for ignition systems that have a distributor, and Electronic Ignition (EI) for those that don't, but these terms can be misleading, since distributors can use either contact points or electronic triggering. While the term "electronic ignition" certainly implies "something other than points," electronic components made their way into the ignition system in a few important ways that span years of technology. We'll step through them in this chapter.

Electronic Triggering

Other than the simplicity and period-correct quaintness of points, they're really more trouble than they're worth. The mechanical making-and-breaking of contact points and wear of the nylon block that rides on the cam lobe are *the* wear-and-tear issues most likely to cause a pre-1980s car to die. So it's no surprise that electronic triggering to replace points was one of the first incursions of solid-state electronics into a car. Since points are also called "breakers" (recall that the opening of the points breaks the low-voltage primary circuit, causing the high-voltage secondary circuit to fire), electronic triggering is also referred to as *breakerless ignition*.

When it began to be widely adopted in the 1970s, electronic triggering was sometimes referred to as "transistorized ignition," since somewhere in the triggering electronics, a transistor is employed as the low-level switching mechanism that actually takes the signal from the sensor and uses it to stop and start the current flow to the ignition coil. Ironically, for all the British car jokes about Lucas being "the prince of darkness," Lucas Electrical appears to have been the first to introduce a transistorized ignition (in 1955, and employed on Coventry Climax Formula One engines in 1962). Widespread adoption of so-called transistorized ignition occurred in the 1970s. These days, now that you can't spit around a car without hitting a solid-state device packed with transistors, calling an ignition "transistorized" is no longer particularly meaningful.

There are three methods of electronic triggering: VRT, Hall Effect, and Optical.

- The **Variable Reluctance Transducer (VRT, also called a pickup coil)** uses a rotating metal star-shaped disk called a reluctor wheel in the center of the distributor to change the magnetic field detected by a coil of wire wrapped around a magnet. VRTs were used as the ignition triggering mechanism in many European cars during the 1970s. See **Figure 2**.
- The **Hall Effect sensor** is similar to the VRT but uses a different magnetic sensing technique. Both VRTs and Hall Effect sensors are described in more detail in the **Tools for Dynamic Analog Signals** chapter later in the book. Hall Effect sensors are still used in retrofitted electronic triggering systems, as we'll cover at the end of this chapter.
- **Optical triggers** use a light-emitting diode (LED), a photovoltaic sensor, and a rotating wheel with slits in it to alternately block then allow the light through. For the most part, optical triggering systems were early aftermarket replacements for points, and have largely been supplanted by aftermarket Hall Effect systems.

For all three triggering methods, the sensor package, or most of it, lives inside the housing of a conventional mechanical advance distributor. Supporting triggering electronics may reside in a box attached to the firewall of the engine compartment (see **Figure 3**). And there's still a single conventional ignition coil that's fired by the triggering electronics and sends its high voltage to the center of the distributor cap, from where it is distributed to the individual spark plugs...

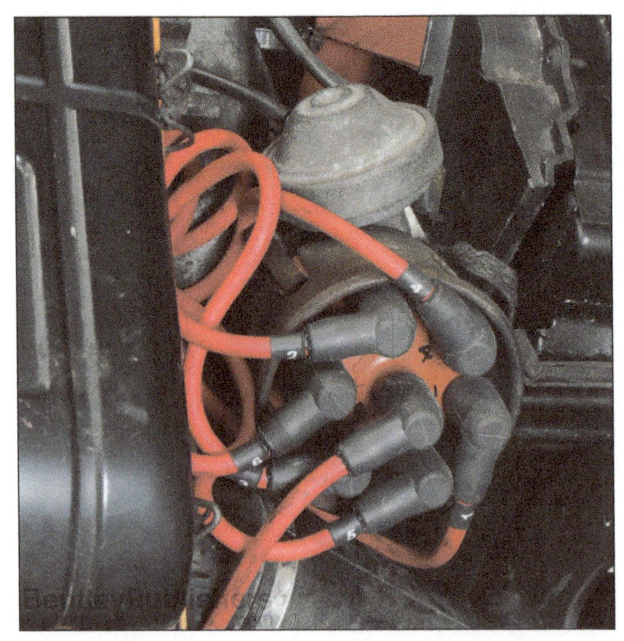

Figure 1 Electronic ignition in a 1979 BMW 635CSi uses a grounded metal shield around the distributor to keep the rotor spark from interfering with vehicle electronics.

Figure 2 Star-shaped reluctor wheel for electronic triggering inside the distributor of a 1979 BMW 635CSi. Note the six fingers, one for each cylinder.

Again, the advantage of electronic triggering is substantial – there are no more points to wear out and no need to adjust the point gap to mechanically set dwell, since it is set electronically. Indeed, aftermarket electronic triggering units do a brisk business in the vintage car world. The downside is that the triggering unit is more of a black box. You can't merely pop the distributor cap and verify with your own eyes that you see the points opening and closing. But given the improved reliability, it's not much of a downside.

Figure 3 Transistorized ignition and resistor pack on a 1979 BMW 635CSi.

Capacitive Discharge Ignition (CDI)

As we've said many times, ignition systems work by setting up a current in the coil's primary winding that creates a magnetic field. When the current is shut off, the field collapses, which induces a much higher voltage in the secondary windings that then discharges through the coil's center terminal.

There is, in fact, a name for this – Inductive Discharge Ignition (IDI). The downside of IDI is that a certain amount of "dwell time" is needed for the magnetic field to build up and fully saturate the coil before collapsing.

The other option to IDI is Capacitive Discharge Ignition (CDI). The 1968 Fiat Dino appears to be the first production car to employ CDI. In the European car world, CDI was installed on late 1960s through 1980s Porsche 914s and 911s. See **Figure 4**. Many BMW models in the late 1970s through early 1980s had a box of a similar size and appearance, but it isn't CDI; it's just the driving circuitry for the electronic triggering.

Figure 4 CDI box used on late 1960s through late 1980s Porsche 911s. *Photo by Rafael Corujo*

CDI bypasses the step of building up and saturating the magnetic field inside the coil. Instead, an external charging circuit is employed that uses a power supply and a transformer to energize a high-voltage capacitor. Upon receiving a trigger, the charging circuit discharges the capacitor, rapidly sending hundreds of volts through the primary winding of the coil. The rapid discharge creates a fast rise in the voltage in the secondary winding, which in turn creates a stronger spark than a traditional inductive system.

Because CDI doesn't need to saturate the coil, its spark tends to be more robust and stable at high RPM than IDI. The stronger CDI-generated spark is also reportedly an advantage in engines that are prone to plug fouling from oil burning. For this reason, CDI is employed ubiquitously on two-stroke engines.

The downside of CDI is that the spark is of shorter duration than a spark created by an IDI system. To mitigate this problem, CDI is often coupled with multiple spark discharge technology to allow each plug to fire multiple times at low- to mid-engine rpm, typically reducing to a single spark above approximately 3,000 rpm. The "stronger spark" retrofit ignition units from the two best-known aftermarket vendors appear to be both CDI as well as multiple spark discharge.

CDI systems are usually electronically triggered, but they don't necessarily have to be, as the triggering mechanism is a separate issue from the discharge method. However, a CDI system needs to be paired with an ignition coil whose windings are specifically designed for the CDI application.

Care must also be taken to use plug wires that can handle CDI's higher voltage output without breaking down or imparting interference into the audio and possibly the other electronic systems. The CDI electronics box itself is usually somewhere between the size of a cigarette pack and a box of Cracker Jacks, covered with finned heat sinks, and is typically bolted to the side of the engine compartment.

Lastly, if you look up CDI online, you may see confusing references to an "exciter coil." These are used to energize the charging circuit as part of a CDI configuration on a two-stroke engine on a motorcycle, ATV, snowblower, or other non-automotive application without a power supply. They are not germane to the automotive CDI application.

Electronic Timing Advance (Motronic)

The traditional mechanical advance distributor performs both the functions of determining the spark advance and distributing the spark to each cylinder. These functions are both performed mechanically. That is, spark distribution is performed using a spinning rotor that, on its way around, contacts the terminals for each of the spark plug wires. Spark advance occurs via a system of weights and springs inside the distributor that move the mounting plate that the points are attached to as engine rpm increases, causing the coil to fire earlier.

Then, in the mid-1970s, a significant change occurred. Instead of mechanically advancing the ignition timing via springs and weights, a "load map" is stored in the ECM's memory, and the amount of ignition timing advance is read from the map as a function of engine rpm, throttle opening, temperature, and other variables. Using the load map, the ECM triggers the coil to fire at the desired time. So not only is adjusting the points no longer necessary, neither is setting the ignition timing. To give credit where credit is due, Chrysler appears to have been the first to implement electronic timing advance with its "lean burn" engines.

The early 1980s brought another step change: the widespread adoption by European automotive manufacturers of the Bosch "Motronic" system. Chrysler may have been the first to offer electronic spark advance, but Motronic's claim to fame was combining electronic spark advance with electronic fuel delivery to give birth to what we now call *digital engine management*, with spark advance and fuel volume both under computer control.

Motronic Cap and Rotor

Note that, with early Motronics, because there is still only one ignition coil, something must still "distribute" the spark to each of the plugs, so there is still a cap, rotor, and plug wires. However, they're no longer part of a separate entity you could call a "distributor." Instead, the rotor is mounted directly to the end of the camshaft, and the cap is typically bolted directly to the engine's front timing cover. See **Figure 5**. A wire from the coil goes to the center of the cap, distributing the spark to the individual plug wires, just like on a conventional distributor. The system has fewer components than points-based ignition, but Motronic rotors and caps do wear, crack, and fail.

Figure 5 Motronic rotor (**arrow**) in a 1987 BMW 325i, mounted at the end of the camshaft. The cap attaches directly to the front timing cover.

Electronic Trigger Signal from the Crankshaft

Simplicity in one place often creates complexity elsewhere. Since there's not a distributor with lobes opening and closing points or causing pickup coil or Hall Effect sensors to spin around and generate triggering signals, the trigger must come from somewhere else. And it does – it comes directly from the ECM. But the ECM has to know where the engine is in its rotation cycle. For this reason, along with Motronics came the incorporation of Crankshaft Position (CKP) sensors.

We discuss these in detail in the **Crankshaft Position Sensors** chapter, but they are either pickup coils or Hall Effect sensors that work in concert with a toothed positioning gear to transmit a sine wave or a square wave to the ECM, allowing it to calculate engine position and trigger the ignition correctly relative to top dead center.

Direct Ignition (Stick Coils)

Once electronic triggering had eliminated points, and Motronics had replaced mechanically advanced timing, the next logical step, in the early 1990s, was to do away with any rotating-contacting electrical components altogether. In the process, plug wires went away as well.

Rather than using a centralized ignition coil and a rotating distributor that are connected to each other and to the spark plugs with long plug wires, direct ignition systems most often use one ignition coil per cylinder. The coils are integrated with the spark plug connectors, placing one coil very close to each spark plug. Each coil is triggered to fire by the car's ECM, which factors in the appropriate amount of spark advance as a function of engine RPM, load, temperature, and other factors.

The coil-plug design was originally referred to as Coil on Plug (COP), which connoted that each coil was positioned directly above its spark plug connector via an internal flexible coupling (usually a conductive spring). This configuration eliminated the long spark plug wires, which can be a source of both voltage loss and electronic interference, but still allowed the COP units to be used on existing engine designs where the spark plugs were easily accessible on the exhaust side of the head. So-called Coil Near Plug (CNP) configurations, with a short plug wire between each coil and plug connector, were also employed for engine designs where the exhaust manifold did not allow sufficient clearance for COP.

Stick coils, also called pencil coils, were introduced in the mid-1990s as a further refinement of the COP configuration. In a stick coil, the conductive spring between the coil and plug connector is eliminated, creating a compact rigid assembly about 4" long and an inch in diameter. Whereas a CNP coil is designed to live out in the open on the exhaust side of the engine, a stick coil is designed specifically to fit into top-of-head spark plug holes located between the camshafts on a double overhead cam engine to be completely hidden beneath a plastic engine cover. See **Figure 6**. The downside is that this places each stick coil in a high heat and vibration environment (which is to say that stick coils do fail).

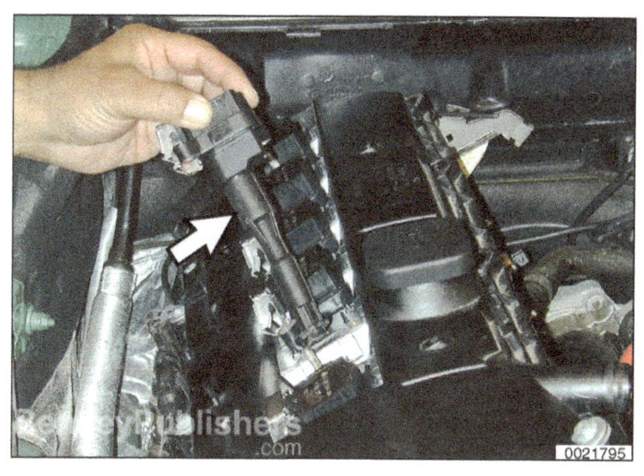

Figure 6 One of six stick coils (**arrow**) in a 2006 BMW X3.

In addition to COPs and stick coils, the term "coil pack" is meant to refer to a set of individual direct ignition coils configured in a single mechanical structure, usually sharing a heat sync and a single connector. However, in general usage, coil on plug, stick coil, and coil pack are often used interchangeably.

The use of individual coils means that each coil fires less often, and that reduces misfires. Here's why. Recall that, in the **Mechanically Timed Ignition – Theory** chapter, we explained how conventional ignition systems have real limitations as the number of cylinders increases. Ignition dwell is expressed as the angle the points are closed for each cylinder's rotation in the engine. Thus, for a 4-cylinder engine, dwell is an angle relative to a maximum of 360/4 = 90 degrees. As you progress upward in cylinder count, for a 12-cylinder engine, dwell is an angle relative to a maximum of 360/12 = 30 degrees.

So, with a mechanical distributor where rotating cam lobes close and open the ignition points, as the number of cylinders increase, the dwell gets smaller because the amount of time available for dwell (available for the points to be closed, which charges the primary winding) becomes less and less. But a direct ignition system doesn't have this limitation. Because there's one coil per cylinder and it's triggered not by rotating lobes forcing open points but by a computer, it can use a duty cycle – a dwell – that's longer. In fact, many stick coils use an internal current limiting sensor to address the opposite problem – the primary winding can overheat from a dwell that's too long.

We cover the troubleshooting of stick coils in detail in the **Stick Coils** chapter.

Troubleshooting Electronic Triggering

The coil troubleshooting steps listed in the **Mechanically Timed Ignition – Troubleshooting** chapter are identical for electronic triggering. Unfortunately, with the three different triggering technologies and the variations in the ways they're integrated into distributors, it's difficult to give a single set of universal troubleshooting instructions for the triggering sensors and electronics. You'll need to consult a repair manual specific to your car.

However, where VRT and Hall Effect sensors are used, the description of these sensors in the **Tools for Dynamic Analog Signals** chapter, and the tests listed in the **Camshaft Position Sensors** and **Crankshaft Position Sensors** chapters at the end of the book may be of help.

Note that, when any triggering mechanism is integrated with a conventional distributor and single ignition coil, that triggering mechanism must supply voltage to the primary winding (DIN terminal 15), have it flow out through DIN terminal 1 to ground, and then cut off the ground path. Thus, you can refer to the **Mechanically Timed Ignition – Troubleshooting** chapter for troubleshooting steps.

In addition, if there is an externally mounted triggering box, its power and ground connections can be verified.

19

Reading Wiring Diagrams

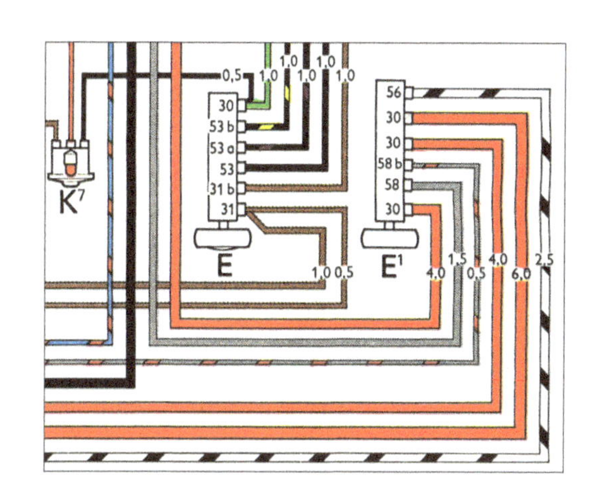

Wiring Diagrams

Ah, the wiring diagram. The Rosetta Stone of the electrical world. Love them or hate them, at some point you have to deal with them. We'll teach you how.

As discussed in the **Circuits** chapter, the electrical diagrams used in this book are basic conceptual illustrations drawn to show the functional relationships between components, but they are not technically wiring diagrams. While they are useful for understanding a circuit, they do not show important things such as which fuse the horn is on, or the color of the wiring.

A factory wiring diagram system will show you all of these details. In addition to detailed circuit information, European car wiring diagrams use the German DIN standard throughout the schematics. For example, if a terminal is labeled 30, it tells you that positive (+) voltage is supplied to that terminal at all times directly from the battery. The other two most commonly used terminal designations are terminal 15 (key-on voltage) and terminal 31 (ground). See **The DIN Numbering Standard** chapter for more details.

When Do You Need a Wiring Diagram?

If a brake light isn't working, and you want to use a multimeter to check if there's power at the connector to know if the problem is the bulb or the wiring, that's pretty straightforward. There are only two or three wires. You probably don't need a wiring diagram to figure out which wire is power and which is ground. In addition, on things like relays, the DIN designations tell you, for example, that anything labeled "30" should have power directly from the battery. So the DIN designations are essentially an on-the-car crib sheet saving you from needing to consult a wiring diagram.

At some point, however, a problem you're trying to fix becomes sufficiently complex that you really have no choice but to consult a wiring diagram. For example, in troubleshooting a turn signal, you often wind up needing to look at the switch that's part of the stalk attached to the steering column. Since these switches often control wipers and washers as well, in order to know which set of pins on the connector you need to test for continuity, there's really little choice but to find a wiring diagram for your car and slug it out.

Where to Find Factory Wiring Diagrams

Early cars (1960s – 1990s)

Wiring diagrams for cars from the 1960s through the 1990s were once available in paper format. Some were part of a factory multi-volume repair manual binder set (vintage Porsches, for example) while others had dedicated ETMs (Electrical Troubleshooting Manuals) for a specific chassis (BMW and Mercedes, for example). See **Figure 1**.

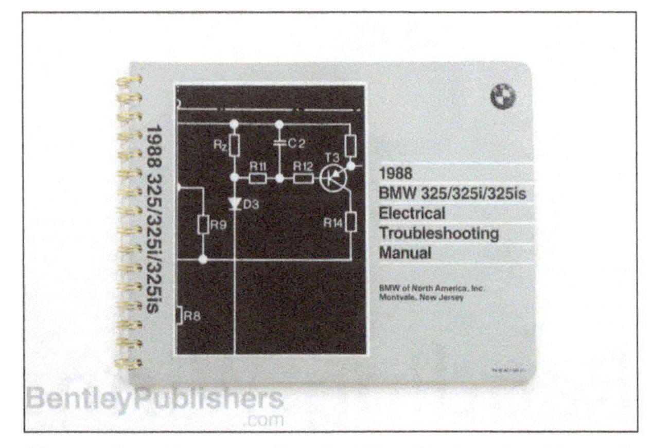

Figure 1 BMW paper Electrical Troubleshooting Manual (ETM) for a 1988 E30 3 Series chassis.

Paper manuals often included lots of hard-to-find electrical documentation, such as component locations, fuse details, connector illustrations, pinout charts, ground locations, and even troubleshooting trees. It may be difficult to find original copies these days, but well worth the money if you can locate one for your car on the Internet or from a used automotive literature reseller. Aftermarket repair manuals are also a great source for paper wiring diagrams. See **Figure 2**.

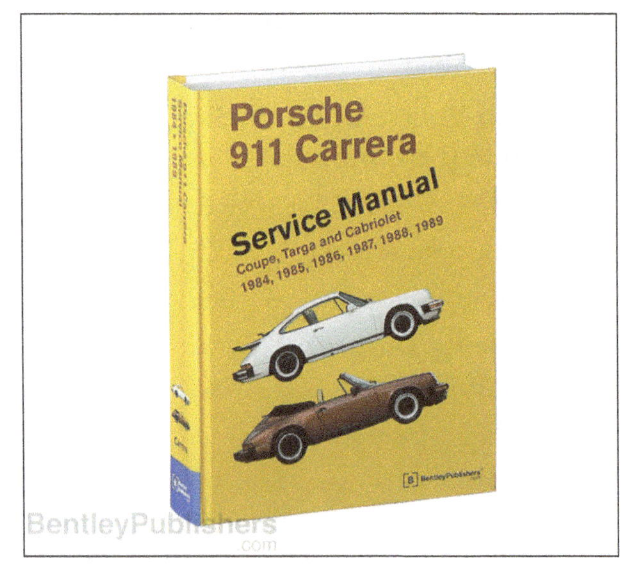

Figure 2 This aftermarket Porsche manual has wiring diagrams, component locations, fuse positions, and more.

Late cars (late 1990s – 2016)

Wiring diagrams and other electrical system technical information (fuse charts, ground locations, connector and component locations and illustrations, pinout charts, nominal values, etc.) for cars built after around 1996 can be sourced through the manufacturer's service information website. Although printed aftermarket repair manuals may contain good usable wiring diagrams applicable to most vehicle configurations, the increasing complexity of automotive electrical systems results in substantial wiring variations for different option packages and nationalities. With a paid subscription, you can probably access the wiring diagrams and other electrical, technical data, and repair information specific to your car's VIN. Manufacturer service sites typically also include factory training documents, service bulletins, repair information, and a lot more.

The service information web sites for most of the European manufacturers are listed in the table below. As an example of cost, a one-day subscription to Mercedes-Benz's STAR TekInfo (www.startekinfo.com) is $19 per day or $100 per week (as of 2016).

German wiring diagrams share certain similarities between brands (for example, the DIN terminals and designations, wire colors, and common symbol sets), but also have aspects that are unique to manufacturers. We suggest that you review the sample diagrams in this chapter to give you a richer understanding of how to decipher German wiring diagrams.

OEM Service Information Web Sites

Alfa Romeo	http://www.techauthority.com
Aston Martin	www.astonmartintechinfo.com/home
Audi	https://erwin.audiusa.com
Bentley	https://erwinusa.bentleymotors.com
BMW	http://www.bmwtechinfo.com
Ferrari	www.ferraritechinfo.com
Fiat	www.techauthority.com
Lamborghini	http://www.lamborghini.com/en/servicing-and-maintenance/independent-operators/
Maserati	www.maseratitechinfo.com
Mercedes-Benz	http://www.startekinfo.com
MINI	http://www.minitechinfo.com
Porsche	https://techinfo2.porsche.com/PAGInfosystem/VFModuleManager?Type=GVOStart
Rolls Royce	http://www.rrtis.com
Saab	http://www.epsiportal.com
Smart	http://www.smarttekinfo.com/SmartTek/
Volkswagen	https://erwin.vw.com
Volvo	http://www.volvotechinfo.com

When working on electrical or electronic circuits, observe the following precautions. Read the **Safety** chapter before beginning any electrical work.

CAUTION —

Airbags and pyrotechnic seat belt reels are deployed by explosive devices. Handled improperly or without adequate safeguards, these components are very dangerous. Use care when working at or near airbags.

Prior to disconnecting the battery, read the battery disconnection cautions in the **Battery** *chapter.*

Connect and disconnect ignition system wires and connectors only while the ignition is switched OFF. Keep clothing, hands, and feet dry if possible.

Switch the ignition off and disconnect the negative (–) battery cable before removing electrical components.

Wiring Diagrams for German Cars (Early Models)

The electrical systems on 1960s and 1970s era cars were simple enough that the entire vehicle electrical plan could be drawn on no more than a few sheets of paper.

For example, the wiring diagram of a pre-1974 BMW 2002 takes up only a single page, whereas that of a post-1974 takes two pages. The BMW E21 320i's wiring diagram spreads over four pages but is still connected page-to-page.

In general terms, the wiring diagrams on early German cars were similar in their layout conventions, symbols, and look and feel between marques. The picture of the alternator looks like an actual alternator, the starter looks like a starter, etc. These concepts can be applied to other German brands of similar vintage.

Below is an example of an early wiring diagram – a 1970-1971 VW Type 1 (Beetle). See **Figure 3**. Each factory wiring diagram has a key containing a list of the symbols used in the diagram. The key for Figure 3 is shown on the following page in **Figure 4**.

A bit of physical reality is imparted by the fact that the headlights are shown at the top of the page, the tail lights at the bottom of the page, and other items proportionally located within the page. The diagram is in color, where the colors of the drawn wires are the actual colors of the actual wires. Ah, those were the days.

This wiring diagram lists wire diameter and DIN terminal numbers. To walk through one section, the battery (**A**) at lower left is connected to the starter (**B**) via a black 25.0 mm^2 (approximately 4-gauge) wire. DIN terminal 50 on the starter motor is connected via a red 4.0 mm^2 (approximately 10-gauge) wire, emerging (via connector T1) as a 2.5 mm^2 red and black striped wire, and connecting to DIN terminal 50 on the ignition/starter switch (**D**).

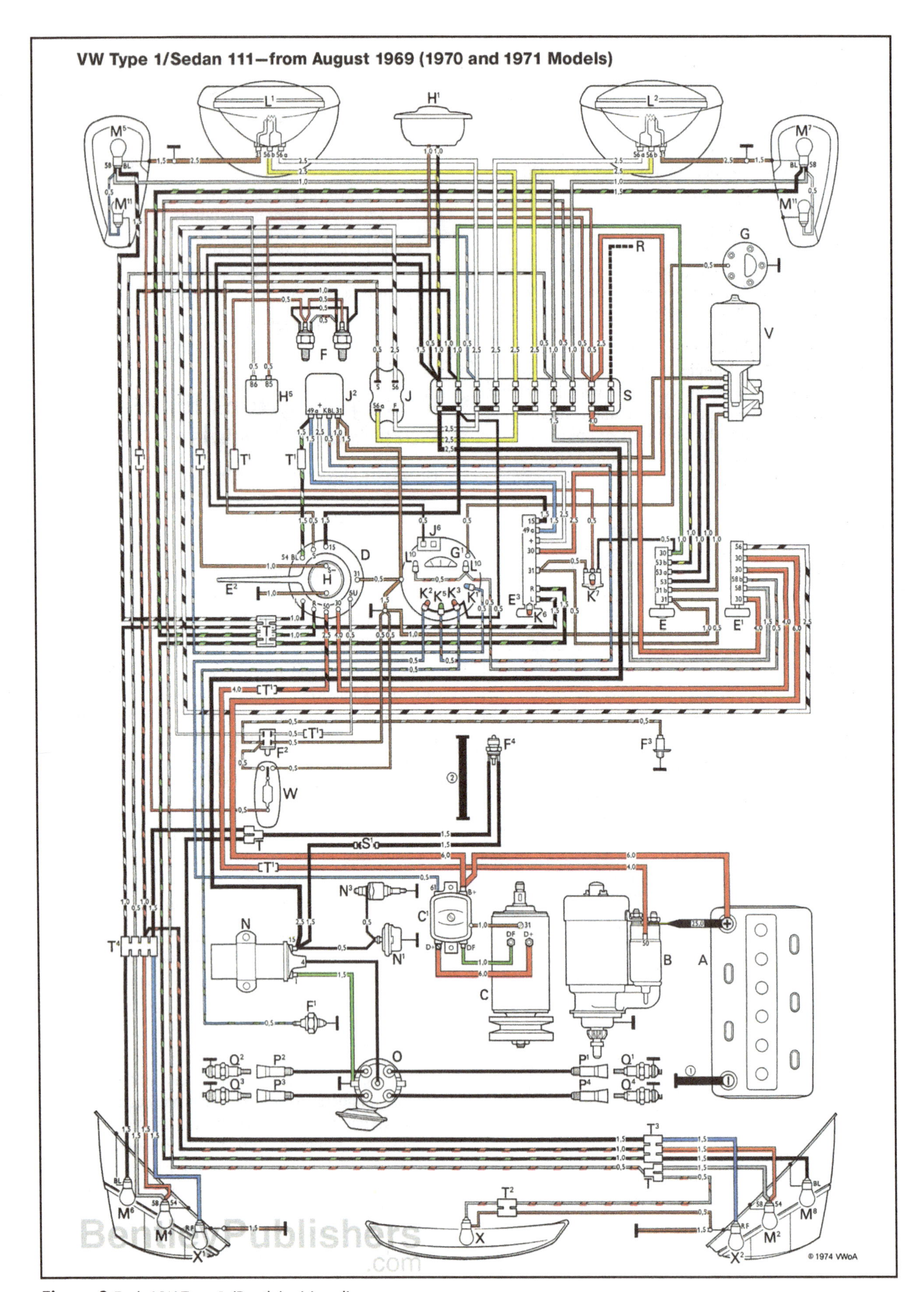

Figure 3 Early VW Type 1 (Beetle) wiring diagram.

B. **VW Type 1/Sedan 111—from August 1969 (1970 and 1971 Models)**

A – Battery
B – Starter
C – Generator
C¹ – Regulator
D – Ignition/starter switch
E – Windshield wiper switch
E¹ – Light switch
E² – Turn signal and headlight dimmer switch
E³ – Emergency flasher switch
F – Brake light switch with warning switch
F¹ – Oil pressure switch
F² – Door contact switch, left, with contact for buzzer H 5
F³ – Door contact switch, right
F⁴ – Back-up light switch
G – Fuel gauge sending unit
G¹ – Fuel gauge
H – Horn button
H¹ – Horn
H⁵ – Ignition key warning buzzer
J – Dimmer relay
J² – Emergency flasher relay
J⁶ – Vibrator for fuel gauge
K¹ – High beam warning light
K² – Generator charging warning light
K³ – Oil pressure warning light
K⁵ – Turn signal warning light
K⁶ – Emergency flasher warning light
K⁷ – Dual circuit brake system warning light
L¹ – Sealed beam unit, left headlight
L² – Sealed beam unit, right headlight
L¹⁰ – Instrument panel light
M² – Tail and brake light, right
M⁴ – Tail and brake light, left
M⁵ – Turn signal and parking light, front, left
M⁶ – Turn signal, rear, left
M⁷ – Turn signal and parking light, front, right
M⁸ – Turn signal, rear, right
M¹¹ – Side marker light, front
N – Ignition coil
N¹ – Automatic choke
N³ – Electro-magnetic pilot jet
O – Ignition distributor
P¹ – Spark plug connector, No. 1 cylinder
P² – Spark plug connector, No. 2 cylinder
P³ – Spark plug connector, No. 3 cylinder
P⁴ – Spark plug connector, No. 4 cylinder
Q¹ – Spark plug, No. 1 cylinder
Q² – Spark plug, No. 2 cylinder
Q³ – Spark plug, No. 3 cylinder
Q⁴ – Spark plug, No. 4 cylinder
R – Radio connection
S – Fuse box
S¹ – Back-up light fuse
T – Cable adapter
T¹ – Cable connector, single
T² – Cable connector, double
T³ – Cable connector, triple
T⁴ – Cable connector (four connections)
V – Windshield wiper motor
W – Interior light
X – License plate light
X¹ – Back-up light, left
X² – Back-up light, right

① – Battery to frame ground strap
② – Transmission to frame ground strap

Figure 4 Key for early VW Type 1 (Beetle) wiring diagram.

Audi and Volkswagen Wiring Diagrams (Late Models)

In late Audi and Volkswagen wiring diagrams, electrical assemblies are represented as rectangles, with the major portions of their relevant electrical circuitry – switches, relays, and so forth—drawn inside the rectangle.

Another concept that is unique to VW and Audi diagrams is that they are known as *current track diagrams*. As the name implies, the circuit depicts the flow of current from power at the top through the device and to ground at the bottom. Current track diagrams have a very different look and feel than early physical wiring diagrams, and they take a bit of getting used to.

2000s-era VW Wiring Diagram Example

Below are two samples of wiring diagram from a 2000s-era Volkswagen that uses the current track format.

Let's step through the first sample wiring diagram in **Figure 5**, starting at the red/black (ro/sw) wire labeled D/50 on the right side.

- The diagram is divided into current tracks from left to right, with power originating at the top of the diagram to ground at the bottom. The D/50 wire, if you follow it down to the bottom, is on track 8, and is the low current control for the starter solenoid from the ignition key.
- The presence of the label D/50 at the end of the wire means that this circuit originates somewhere else.
- Looking at the key in the lower left corner of the figure, we see that "D" is the Ignition/Starter Switch. 50 is the DIN code of the terminal of the starter switch, and as we know, that's the code for the ignition switch to starter wire.
- The notation 2,5 indicates that the wire is 2.5 square mm. From the color code, ro/sw is a red wire with a black stripe.
- Looking at the key, T10a/1 indicates that this is a 10-pin connector on the left side of the engine compartment. The 1 indicates that this wire is pin 1 in the connector.
- T4e/1 indicates that this wire continues through pin 1 on a 4-pin connector attached to the transmission.
- The wire continues to terminal 50 on a square labeled B. From the key, we see that's the starter.
- Looking inside the starter square, we can see, in fact, that 50 is connected to the coil of a relay (the starter solenoid), and 30 (another DIN designation) connects to the solenoid switch, on the other side of which is a small icon for a motor, which is the starter itself.
- Continuing to follow the circuit through 30 takes us onto track number 7, which shows the physical grounding of the starter motor (the thin line does not indicate a physical wire; rather it implies a direct connection to ground).
- There's a connection labeled 25,0, or a 25mm^2 black wire. This is the big heavy wire from the starter to the battery, and indeed, following the wire, it goes to rectangular symbol A, which is the battery.
- Continuing left along the bottom, we get to track number 4. On the ground side of the battery, there is another 25mm^2 wire connected to 1 in a circle. Looking that up in the key, it says ground strap, battery to body. Looking at the other circled number in the key, we see it is also a ground.
- Following track number 4 upward, it connects not only to the starter, but also to the fuse box, showing fuses S162 and S163. Fuse S163 is shown going into the main relay panel to terminal 30 (power at all times).
- Continuing along the bottom, track number 2 shows ground strap, transmission to body.
- Track number 1 shows a physical connection to ground (the engine block) to feed the generator (alternator) and voltage regulator.
- Terminal B+ is represented by a DIN number (battery positive terminal), and is shown feeding the fuse box via a 16mm^2 red wire. Note that B+ does, in fact, connect back to the battery positive terminal after passing through the maxi fuse S162.
- Terminal D+ (also a DIN number) is for the alternator positive terminal. The key shows this blue wire as going to T4e/2 (same connector as from ignition switch, but pin 2).
- 100 (listed at the top of track number 1) indicates that this wire continues onto current track 100, which means you will have to locate track 100 and then look for a 1 in a rectangular box (there you will find that D+ continues on to the battery indicator light in the instrument cluster).

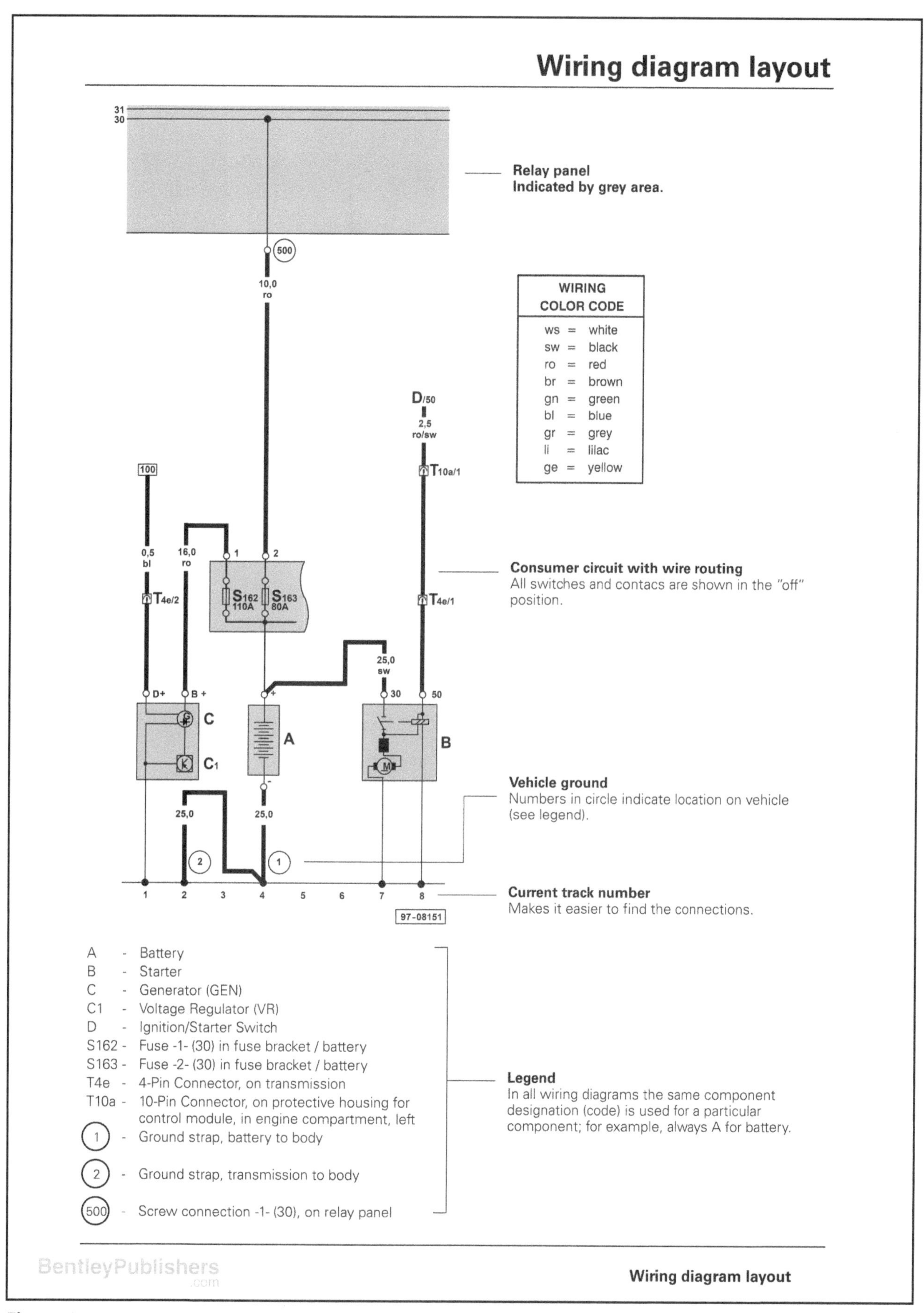

Figure 5 Sample wiring diagram from a 2000-era VW.

How to Read VW Wiring Diagrams)

1. **Relay location number**
   - Indicates location on relay panel
2. **Arrow**
   - Indicates wiring circuit is continued on the previous and/or next page
3. **Connection designation - relay control module on relay panel**
   - Shows the individual terminals in a multi-point connector
   - For example: terminal 24 indicated
4. **Diagram of threaded pin on relay panel**
   - White circle shows a detachable connection
5. **Fuse designation**
   - For example: S228 = Fuse number 228, 15 amps, in fuse holder
6. **Reference of wire continuation (current track number)**
   - Number in frame indicates current track where wire is continued on another wiring diagram page
7. **Wire connection designation in wiring harness**
   - Location of wire connections are indicated in the legend
8. **Terminal designation**
   - Designation which appears on actual component and/or terminal number of a multi-point connector
9. **Ground connection designation in wire harness**
   - Locations of ground connections are indicated in legend
10. **Component designation**
    - Use the legend to identify component code
11. **Component symbols**
    - See **Common symbol set**
12. **Wire cross-section size (in mm^2) and wire colors**
    - Abbreviations are explained in color chart beside the wiring diagram
13. **Component symbol with open drawing side**
    - Indicated component is continued on another wiring diagram page

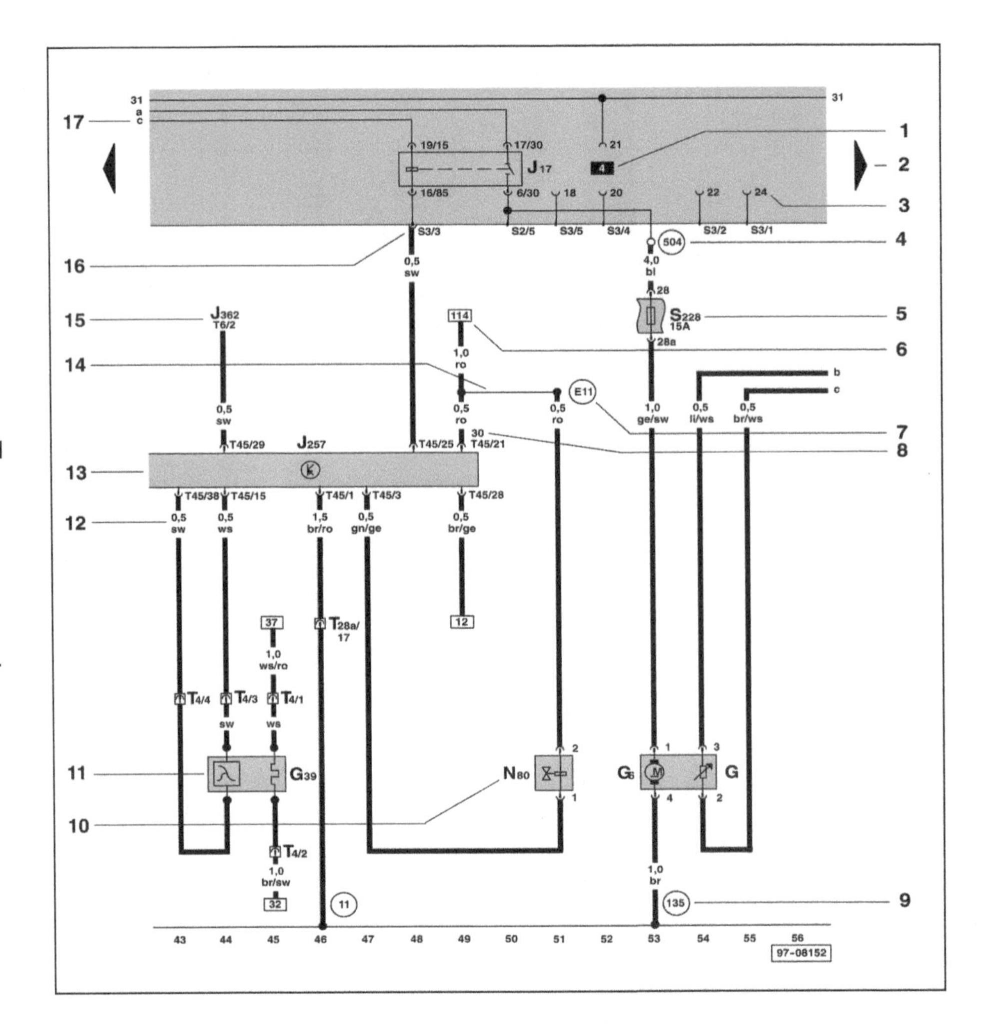

14. **Internal connections (thin lines)**
    - These connections are not wires
    - Internal connections are current carrying and are shown to allow tracing of current flow inside components or wiring harnesses
15. **Reference of continuation of wire to component**
    - For example: Control module for anti-theft immobilizer (J362) on 6-pin connector, terminal 2
16. **Relay panel connectors**
    - Shows wiring of multi-point or single connectors on relay panel
    - For example: S3/3 - Multi-point connector S3, terminal 3
17. **Reference of internal connection continuation**
    - Letters indicate where connection continues on previous and/or next page

Common Symbol Set

Below (**Figure 6**) is a symbol set used in 2000-era Volkswagen and Audi wiring diagrams. There have been a lot of additions to the standard symbol set used throughout the years, but the symbols for power, ground, switches, relays, fuses, lights, variable resistors, and other things have remained constant.

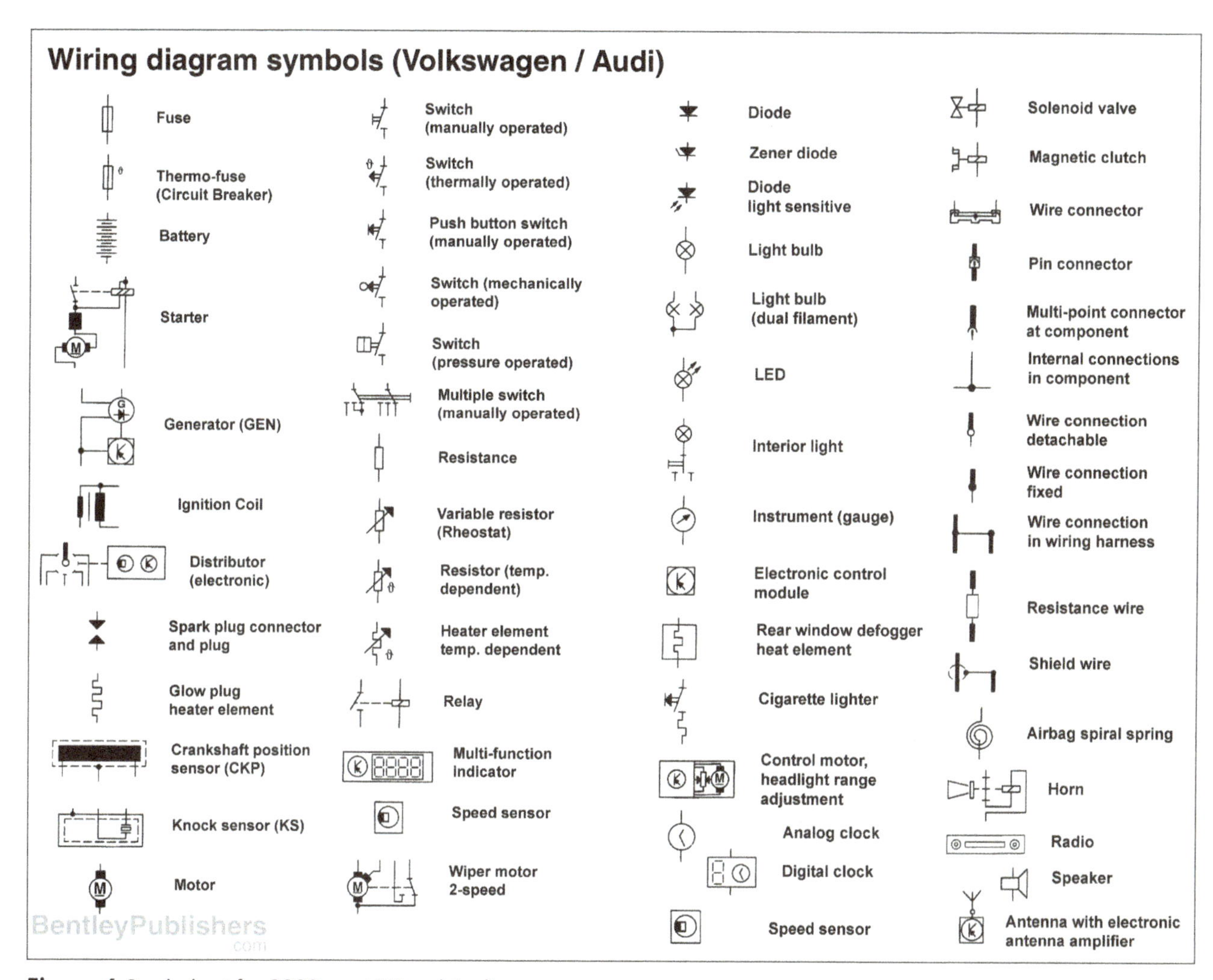

Figure 6 Symbol set for 2000-era VW and Audi.

Wire Colors

Because many wiring diagrams are drawn and printed in black and white, there is a code for wire colors. Often the color abbreviations will be in the native language. For example, in German, red is "rot." For VW and Audi wiring diagrams, this is abbreviated as RO. Note that the abbreviations change with manufacturer. The decoder chart will normally be listed as part of your wiring diagram.

If a wire has a striped tracer, the primary color is listed first, followed by a slash and then the secondary color. For example, a brown wire with a red stripe is listed as ro/br.

Volkswagen Audi Wiring Color Codes		
ws		white
sw		black
ro		red
br		brown
gn		green
bl		blue
gr		grey
li		violet
ge		yellow
or		orange

BMW Wiring Diagrams (Late Models)

BMW schematics (wiring diagrams) divide the vehicle electrical system into individual circuits. Electrical components are represented in such a way that their general layout and function are self-explanatory.

- A component (or connector) which is completely represented in the schematic is shown as a solid box.
- A component (or connector) which has other connectors in addition to the ones shown in the schematic is shown with a dashed line.
- Switches and relays are generally shown in the rest position (OFF).

Electrical assemblies are represented as rectangles, with the major portions of their relevant electrical circuitry – switches, relays, and so forth – drawn inside the rectangle.

BMW Wiring Diagram Example

Below (**Figure 7**) is a wiring diagram from a 2004 BMW 5 Series.

Each component has a unique alpha-numeric designation. For example, G1 is BMW's designation for the battery. On some diagrams, only the alpha code will exist and you will need to search BMW's TIS (Technical Information System) to identify the component.

Let's step through a small part of the diagram, starting at the top left.

- X6042 is the designation for the battery chassis ground cable. G1 is the battery and X6406 is the connection to the battery negative post. X is the designation for a connector (e.g., X6011).
- The positive battery cable continues into the engine compartment to jump start terminal (G6430) and exits as a red (RT) 25mm^2 wire directly to the starter (M6510a) at terminal X6512.
- Looking at the starter, it can be seen that it has two wires and a connection to ground through the engine block. The numbers in the gray box are mapped to the key on the right side and indicate internal starter function.
- The 2.5mm^2 white (WS) wire then leads to pin 22 at the CAS module (connector X10318). It feeds internally to a relay in the CAS module, which is fed to terminal 30 power (power at all times) from fuse F7 via a large red /green (RT/GN) 4.0 mm^2 wire.

Wire Colors

Because many wiring diagrams are drawn and printed in black and white, there is a code for wire colors. Often the color abbreviations will be in the native language. For example, in German, red is "rot." For BMW wiring diagrams, this is abbreviated as RT. Note that the abbreviations change with the manufacturer. The decoder chart will normally be will be listed as part of your wiring diagram.

If a wire has a striped tracer, the primary color is listed first, followed by a slash and then the secondary color. For example, a red wire with a white stripe is listed as RT/WS.

BMW Wiring Color Codes		
WS	□	white
SW	■	black
RT	■	red
BR	■	brown
GN	■	green
BL	■	blue
GR	■	grey
VI	■	violet
GE	■	yellow

Common Symbol Set

Below (**Figure 8**) is a limited symbol set used in BMW wiring diagrams. There have been a lot of additions to the standard symbol set used throughout the years, but the symbols for power, ground, switches, relays, fuses, lights, variable resistors, and other things have remained constant.

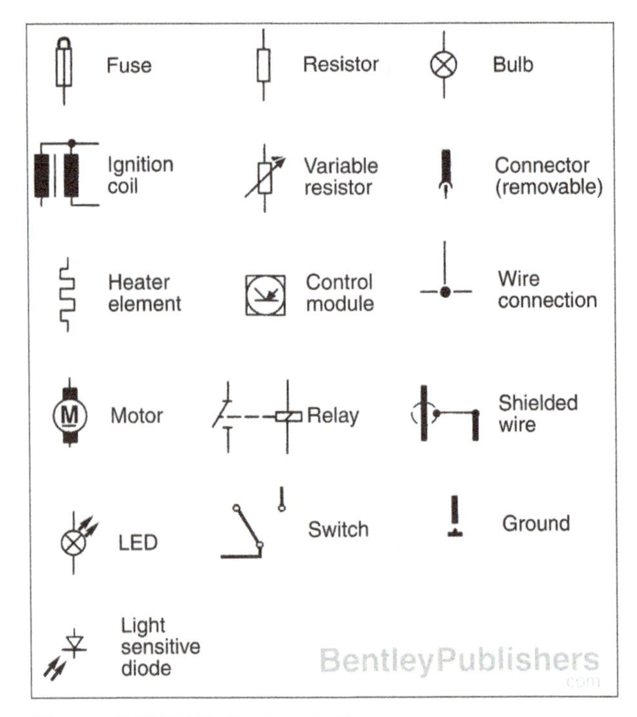

Figure 8 BMW limited symbol set.

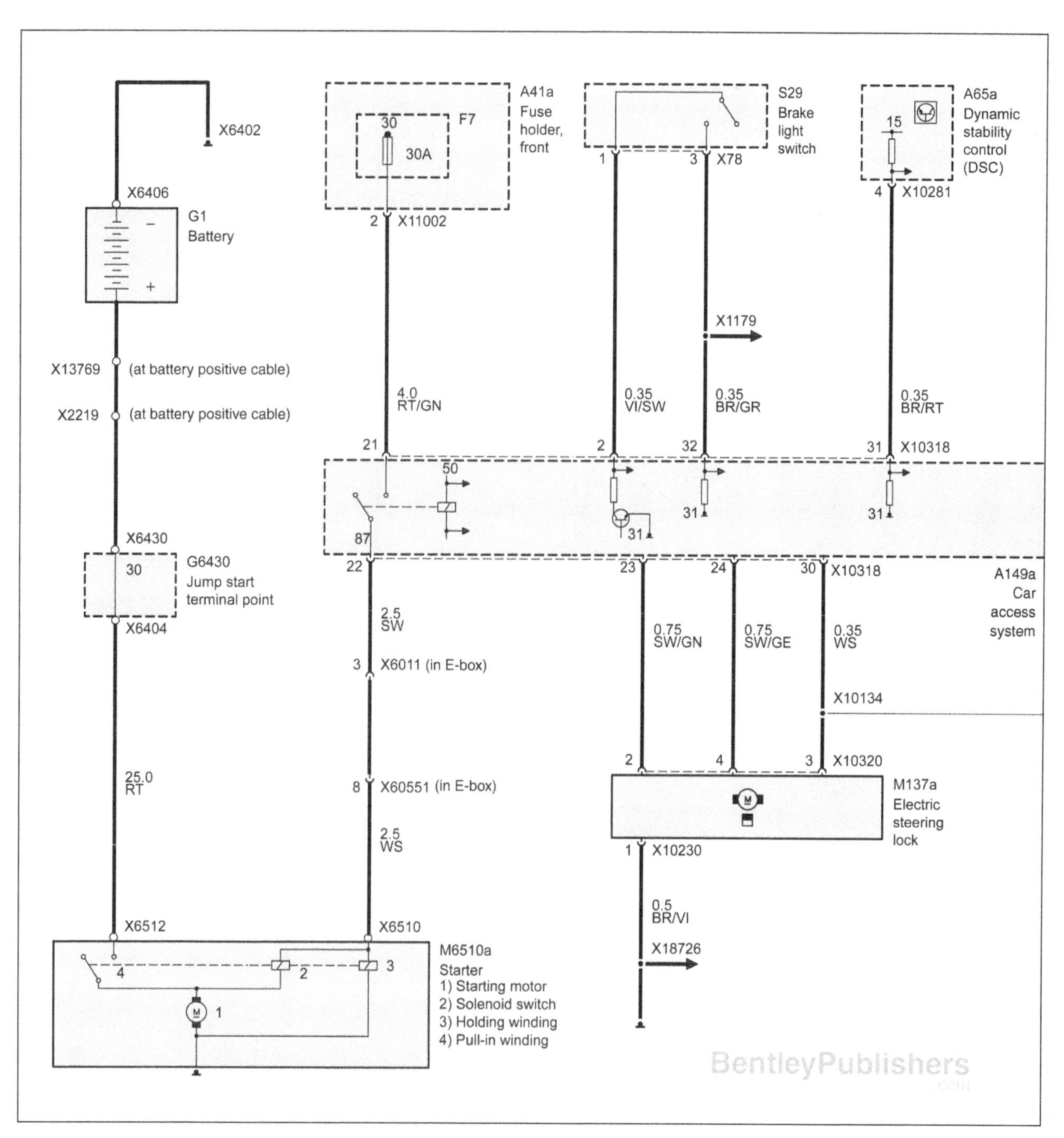

Figure 7 Wiring diagram from a 2004 BMW E60 (5 Series).

Mercedes-Benz Wiring Diagrams (Late Models)

Using Mercedes-Benz wiring diagrams for the first time may require some interpretation. All components are identified using an alpha-numeric code, which are keyed to a list included with each diagram. For example, G1 is the designation for the battery, but "battery" will not be written on the diagram. See **Figure 9**.

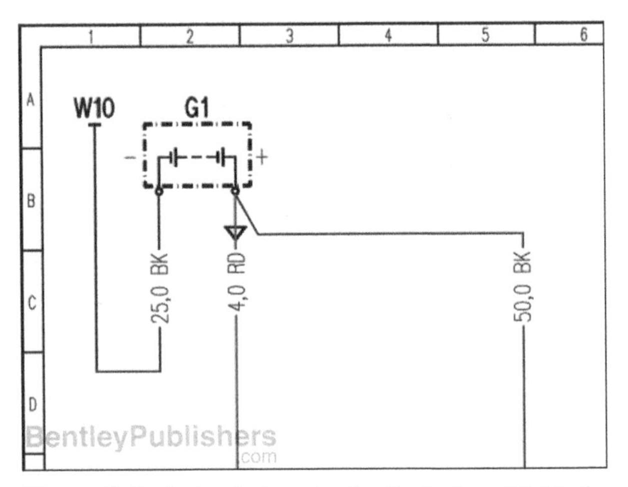

Figure 9 G1 is the designation for the battery. W10 is the designation for the 25mm^2 black (BK) battery ground cable.

Abbreviations, acronyms, and not-so-obvious conventions are used throughout the diagrams.

Components and connectors grouped within a dashed area indicate variations in wiring or components which are model, engine, or equipment specific. Components in parenthesis or with a slash in their description (e.g., N47-1 (N47-2) or ASR/SPS control module) indicate information applicable to either both or just one system. See **Figure 10**.

- A component (or connector) which is completely represented in the schematic is shown as a solid box. Partial component views are wrapped with a dashed line. A component (or connector) which has other connectors in addition to the ones shown in the schematic is shown with a dashed line.
- Switches and relays are shown in rest position (generally OFF).
- **X** indicates a connector (e.g., X18).
- **Z** indicates a connector sleeve/splice (e.g., Z49).
- **W** indicates a ground connection (e.g., W11)
- ☞ **PE** indicates that the component is displayed in more detail on another diagram. Each diagrams has a unique ID based on the MB Workshop Information System (WIS) repair scheme (e.g.,15.00).

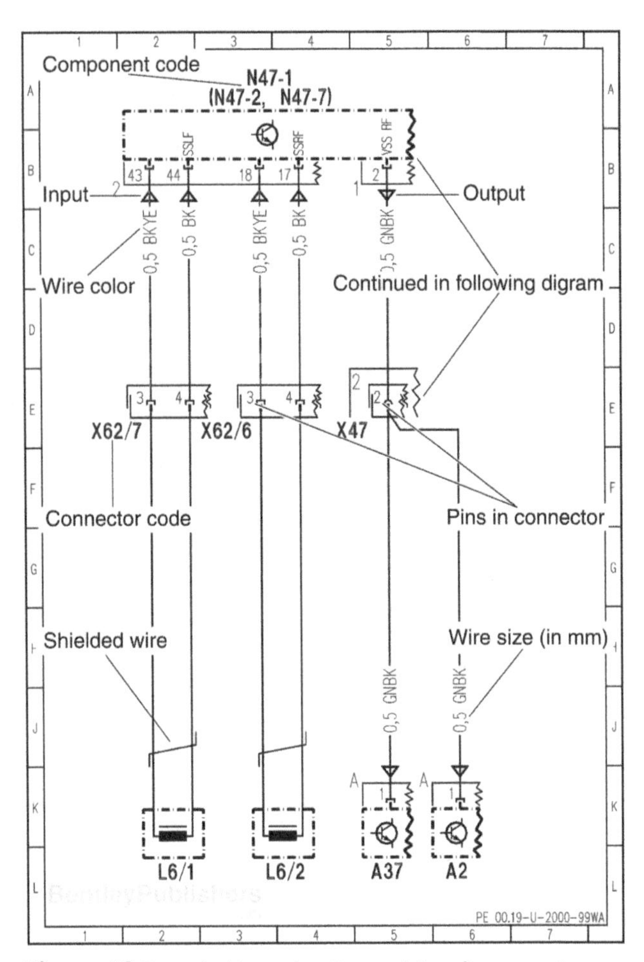

Figure 10 Sample Mercedes-Benz wiring diagram. As you can see, MB diagrams need a decoder list.

Mercedes-Benz Wiring Diagram Example

A sample starter/alternator wiring diagram from 2000 W202 C-Class can be found below. See **Figure 11**.

Let's step through a small part of the diagram, starting at the top left.

- G1 is the battery. W10 is the battery ground cable. The positive cable runs from the battery into X4, which is called the terminal block (circuit 30, left footwell).
- X4 terminal block feeds + power to a number of components, namely F1 (15-amp fuse), S2/1 (ignition/starter switch), and M1 (starter motor).
- The two large red wires (2.5 mm^2 and 4.0 mm^2) go from terminal block X4 to terminal 30 on the ignition/starter switch. Notice there are no fuses between the battery and the ignition/starter switch or in the terminal 30 feed to the starter.
- Terminal 50 on the ignition/starter switch (output for starter control) then travels through pin 1 in X26 (harness connector) to pin 40 on connector C of N3/10 (ME-SFI control module). From there the digram is not complete enough to show how terminal 50 activates the starter motor.
- If you follow terminal 50 off of M1 (starter motor) to the right, you can see it comes from K40/4 (passenger side fuse and relay module box). You can also see the K40/4 is partially shown. You are directed to diagram PE 54.20-2200 for more information on the fuse relay module box. (Group 54 is for Vehicle Electrical – Equipment and Instruments).
- G2 is the generator (alternator).
  A1 is the instrument cluster.
  N7 is the exterior lamp failure monitoring module.
  Z (Z3/26, Z93) components are connector sleeves or welded connections in the harness. All of this information will be included in the decoder list that accompanies each diagram.

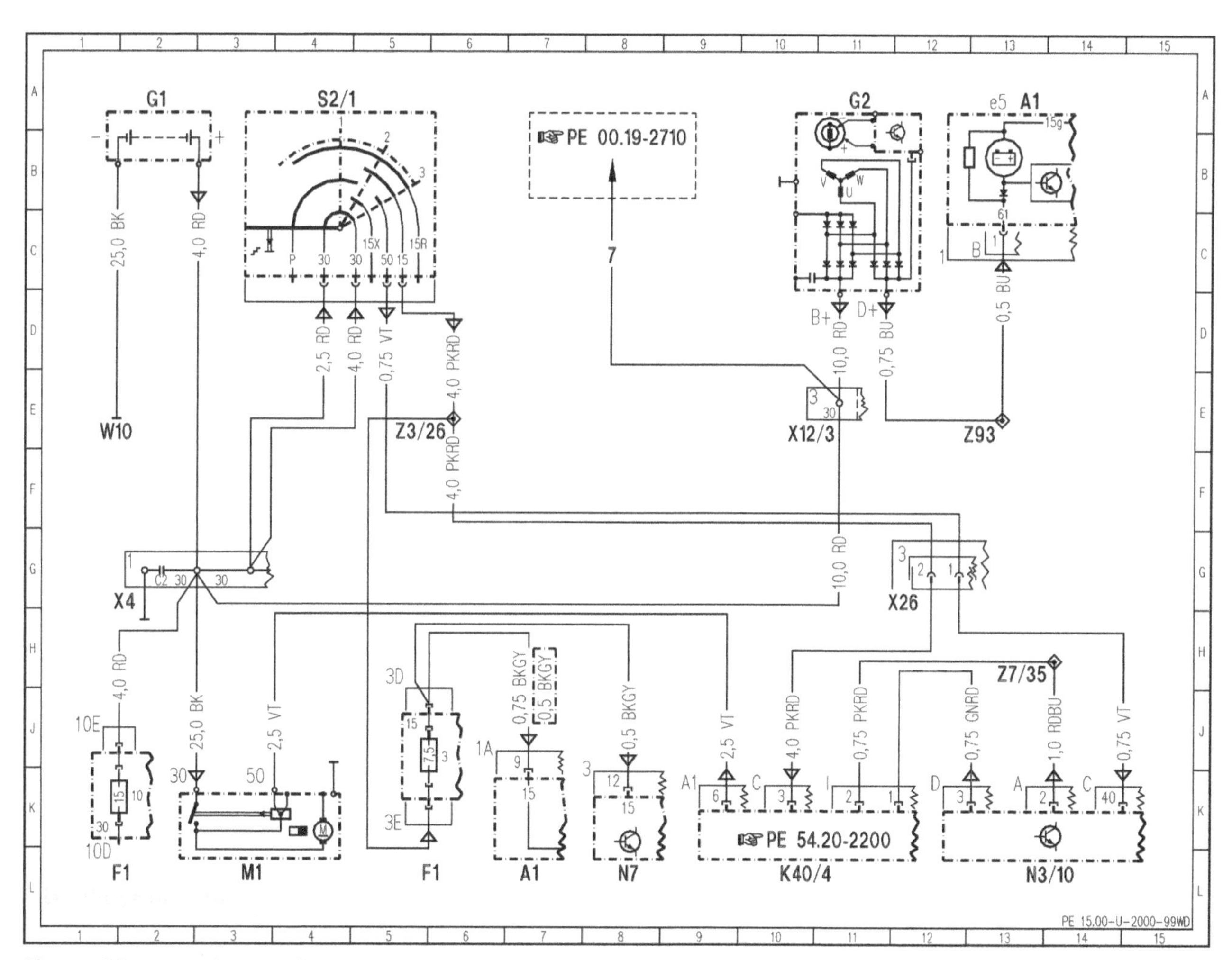

Figure 11 Wiring diagram from a 2002 Mercedes-Benz C-Class (W203).

Wire Colors

Because many wiring diagrams are drawn and printed in black and white, there is a code for wire colors. The color chart will normally be will be listed as part of your wiring diagram.

If a wire has a striped tracer, the primary color is listed first, followed by a slash and then the secondary color. For example, a red wire with a white stripe is listed as RD/WT.

Mercedes Wiring Color Codes	
BK	black
BR	brown
BU	blue
IV	ivory
GN	green
GY	grey
TR	neutral / transparent
OR	orange
PK	pink
RD	red
VI	purple
WT	white
YL	yellow

Common Symbol Set

Below (**Figure 12**) is a limited symbol set used in Mercedes-Benz wiring diagrams. There have been a lot of additions to the standard symbol set used throughout the years, but the symbols for power, ground, switches, relays, fuses, lights, variable resistors, and other things have remained constant.

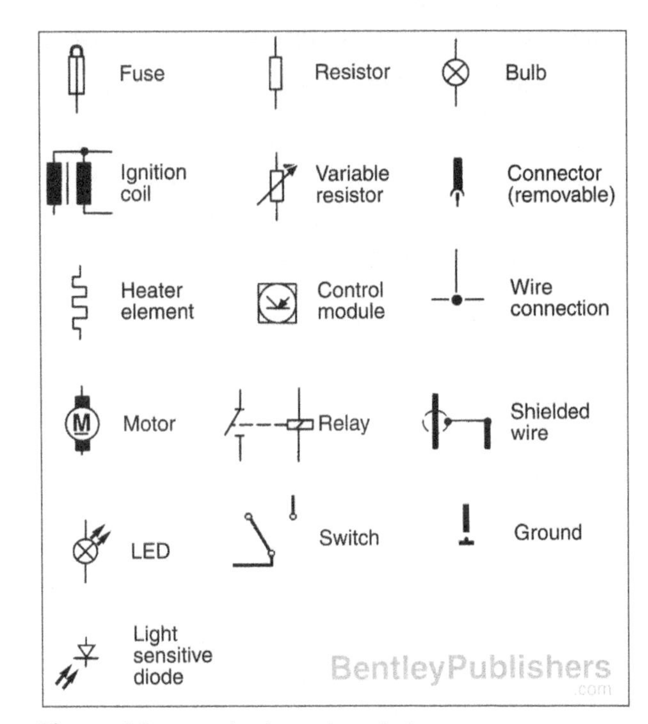

Figure 12 Mercedes limited symbol set.

Porsche Wiring Diagrams (Late Models)

Using Porsche wiring diagrams for the first time may require some interpretation. Abbreviations, acronyms, and not-so-obvious conventions are used throughout the diagrams. Many of the wires are continued on to other pages or "sheets," as Porsche calls them. These conventions are detailed in the sample wiring diagram below.

2000-era Porsche factory schematics are composed of multiple sheets. For example the 2004 Porsche 911 contains as many as 38 separate sheets of wiring. Some sheets are single pages, but most are multiple pages. Each diagram shows the wiring, connectors, terminals, and electrical or electronic components of the circuit. It also identifies the wires by color or terminal coding.

Components and connectors grouped within a dashed area indicate variations in wiring or components that are model, engine, or equipment specific.

- Up to MY2001, the wire continuation link (number-letter combination in a box, e.g., D125) references a grid coordinate. The letter is for the grid position across the top of the diagram. The number is for the grid position along the side. See **Figure 13**.

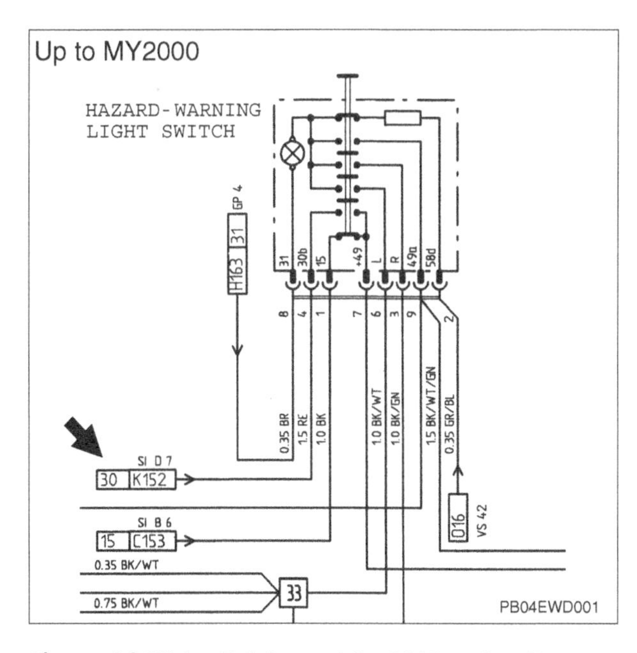

Figure 13 Wiring link (**arrow**) for 2000 and earlier cars. This black wire continues on to grid position K152. SI D7 is a fuse in position D7. 30, or DIN terminal 30 means that positive (+) voltage is supplied to that terminal at all times directly from the battery.

- On MY2001 and later, the wire continuation link (open arrow) references a diagram number followed by a grid coordinate, as shown. For example, 17/D6 indicates diagram Sheet 17, grid position D-6. The sheet numbers are identified on the top of each wiring page. See **Figure 14**.

MY2001 and later

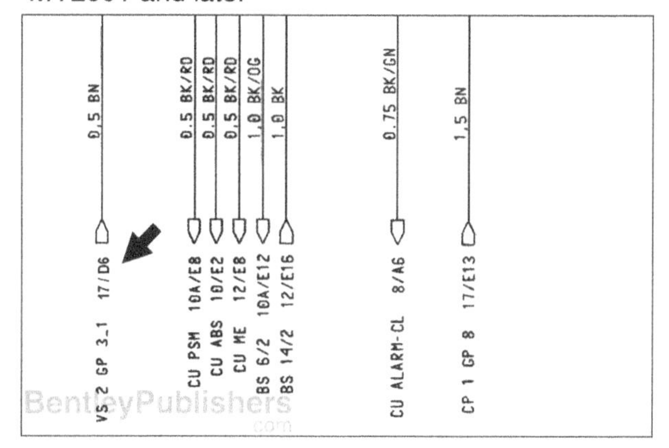

Figure 14 Wiring link (**arrow**) for 2001 and later cars. This brown (BN) wire continues on to Sheet 17, grid position D6. VS 2 is a wire splice. GP 3_1 is a ground point.

Additional Porsche wiring diagram conventions are detailed in **Figure 15**.

- Switches and relays are shown in rest position (generally OFF).
- **X** indicates a connector (e.g., X 2/3).
- **BS** indicates a bridge connector (e.g., BS 10/1).
- **GP** indicates a ground connection (e.g., GP 4).
- **VS** indicates a wire splice (e.g., VS 42).
- **SI** indicates a fuse (e.g., SI D7).

Wiring Color Codes

Wire insulation colors in the wiring diagrams are abbreviated and shown in the table below.

BLACK	BK	BK
BROWN	BR	BN
RED	RE	RD
ORANGE	OR	OG
YELLOW	YE	YE
GREEN	GN	GN
BLUE	BL	BU
VIOLET	VI	VT
GREY	GR	GY
WHITE	WT	WH
PINK	RS	PK
GOLD	-	GD
TURQUOISE	TK	TQ
SILVER	-	SR

PB04EWD005

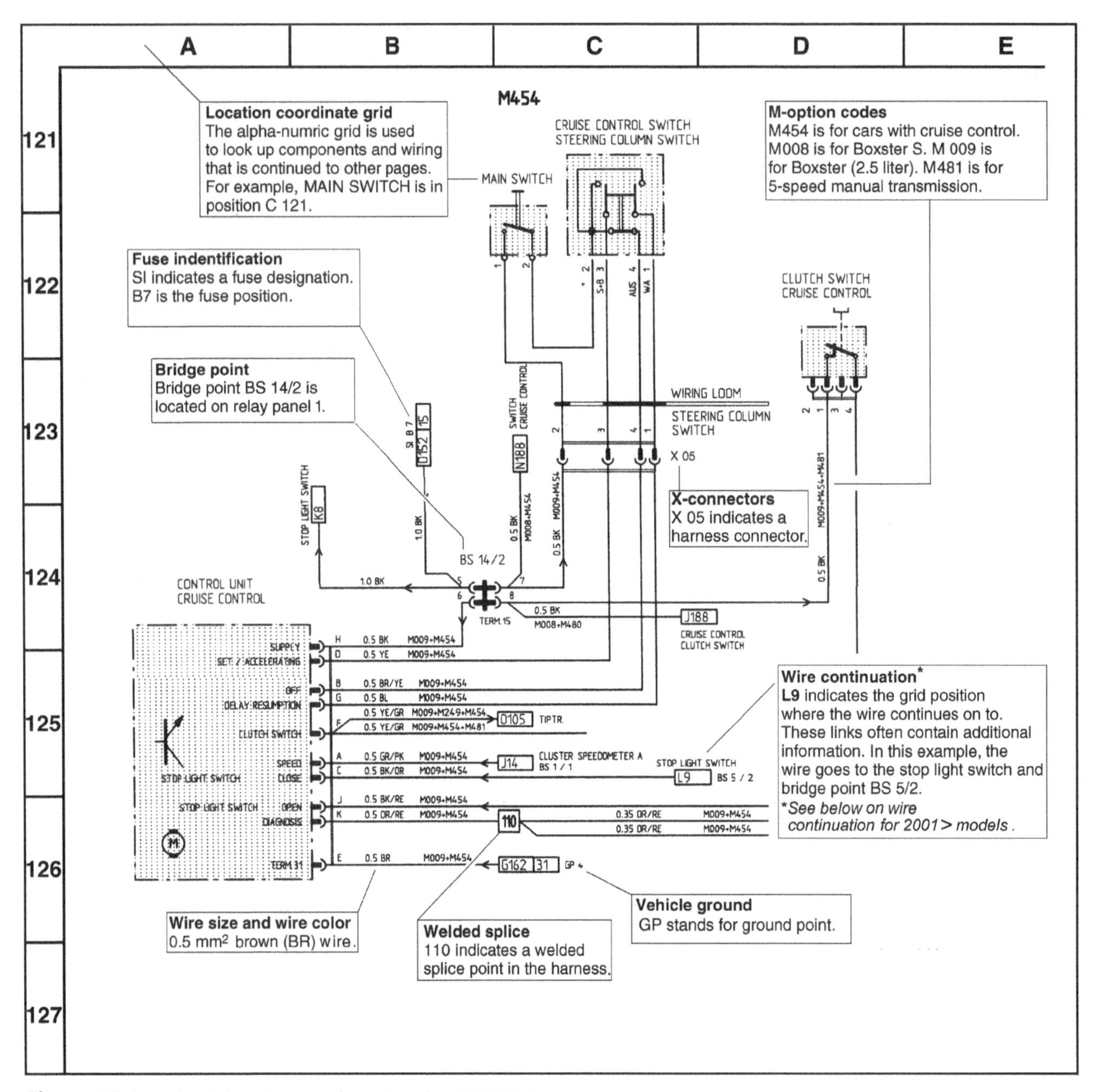

Figure 15 Sample wiring diagram for a Porsche 911 (996).

Porsche Option Codes, Abbreviations, and Acronyms

Porsche option codes, abbreviations, and acronyms are used throughout the factory wiring diagrams. For example, you might see M 476 in bold at the top of a diagram. M 476 is the option code for Porsche Stability Management (PSM). You will also likely see many non-intuitive abbreviations and acronyms used extensively over and over, such as SI (fuse), BS (bridge point), CP (connecting point). Other acronyms will not be used as frequently, which is why having these lists handy can be quite useful. Many enthusiast's websites have such lists and can easily be printed out.

20

Energy Diagnosis and Parasitic Drain

Energy Diagnosis

So. You haven't driven your car in a day or two or three. You hop in it, turn the key, and... click. You know the problem isn't your battery because you just replaced it. You know the problem isn't your charging system because you've done the basic health test described in the **Charging System (Alternator)** chapter.

The likely culprit is a "parasitic drain" – an electrical load that remains on and draws current when the car is off, causing the battery to discharge. It's time for an energy diagnosis on the vehicle's electrical system.

We're going to explain how to do this (measure parasitic drain) and give you some tips on how to find the source of the unwanted current draw. There are risks, however, in the procedure.

 **CAUTION —**

Measuring parasitic drain requires disconnecting the battery. Disconnecting the battery carries some risk of the radio and the car's other electronic modules losing presets. If your radio is an anti-theft radio, be certain to have the anti-theft code on hand before proceeding.

How Much Parasitic Drain Is Too Much?

What you optimally want to see is **less than 50 milliamps (mA)** of current. More than 80 mA is, by industry standards, a problem. The 50 mA number assumes that all electronics have "gone to sleep" and all electrical consumers are switched off. We'll talk more about the sleep mode in a moment. 80 mA may not sound like a lot, but factor in cold weather and a not-so-healthy battery and the car may not start if it hasn't been driven in a week.

Trickle Charger

If, after testing, the draw is within specs, but you either don't drive the car often enough or just make short drives with high loads on the battery, it is normal for the battery to go flat, say in a few weeks. To keep the battery healthy and charged, you may want to consider a battery maintainer (so-called trickle charger).

Sleep Mode

On many late '90s and later cars, control modules and other electrical consumers remain alive after shutdown for a short period of time (16 minutes to an hour is a general rule of thumb). After that, the car enters into what is referred to as sleep mode or rest state. For accurate results, all modules and electrical consumers need to be asleep. It is also important to know that each time a door or trunk is opened, or the key remote is used, the car wakes up, which wakes up multiple electrical consumers.

So when checking parasitic draw, be sure the car can enter sleep mode by simulating a locked car. If any doors or lids need to remain open during the test, roll all latches to the closed position and lock the car. This will enable sleep mode conditions. On a BMW, if the hood must remain open, pull the hood alarm switch up into the service position.

Advanced Power Management

From about the mid-2000s, advanced power management software (and sometimes hardware) became commonplace. For example, the Audi A6 uses a dedicated energy management control module for power monitoring and energy diagnosis. On BMWs, this function is through the DME engine control module via the Intelligent Battery Sensor (IBS) on the negative battery terminal.

On these late cars having battery drain issues, it would make sense to check for power management fault codes using a dedicated scan tool as part of your energy diagnosis test plan.

Disconnecting Battery Precautions

Unless you own a low-current clamp meter, measuring parasitic drain requires disconnecting the battery and connecting a multimeter between it and the car's body. You need to be aware of the consequences of disconnecting a battery. On pre-computer cars, it's a total non-issue. On '90s-era cars, it's a minor annoyance. If you have an anti-theft radio, you'd be wise to make sure you have the radio code and know how to program it in.

However, on a post-1996 electronics-laden car, there may be other issues. Many electronic modules have Keep Alive Memory (KAM) which is lost if the battery is disconnected. These may be comfort or convenience parameters such as radio stations, seat and mirror positions, and other personal settings, but there may also be adaptively learned acceleration and shift maps.

You can buy little 12V devices that plug into a car's cigarette lighter and provide a low amount of power so modules don't lose their KAM settings.

A Few Notes on Measuring Current

Measuring current is fundamentally different from measuring voltage or resistance in that all of the current in the circuit must flow through the meter. In this chapter, we refer to the meter's 10A high-current setting and 300mA low-current setting, but the settings will vary meter-to-meter. For additional information, see the **Common Multimeter Tests** chapter.

The terms "current" and "amperage" are used interchangeably, and while the word "draw" refers to current, and "drain" refers to undesirable current, the two words are used interchangeably also.

There are so-called "clamp meters" that are easier to use than a multimeter, but to be useful for this measurement, you'd need one that can reliably measure faint parasitic drains in the tens of milliamps.

With many meters, if the 10A setting is used, the meter will display in amps (for example, 0.69 amps), whereas if the 300mA setting is used, the meter will display in milliamps (for example, 690 milliamps).

Measuring Parasitic Drain

Connect an External Amp Gauge

To begin this check, we must make sure we don't have a drain so high that we risk damaging our meter. Most meters can handle 10 amps. Over 10 amps and the internal fuse will blow. A 10-amp drain is big but by no means impossible. You can perform an approximate check on the drain either using an inexpensive clamp meter with 1A accuracy, or you can buy an amp gauge on eBay for less than ten bucks. Be certain to buy one that says either "direct connect" or "internal shunt" to ensure that the ammeter is self-powered and doesn't need to be wired to battery voltage in order to measure amperage.

- Buy a pair of 10-gauge wires with big alligator clips at one end like battery chargers have. You can buy these inexpensively online or at an auto parts store.
- Connect the other ends of the wires to the back of the amp gauge. Use ring terminals for secure connections. Consult the **Wire Repairs** chapter for how to crimp connectors.
- Turn absolutely everything in the car off.
- Disconnect the car battery's negative terminal.
- Connect one alligator clip to the negative battery terminal. It doesn't matter which clip. See **Figure 1**.
- Connect the other alligator clip to the negative battery cable (or directly to the body of the car).
- Check the amp gauge. If it reads less than 10A (and it bloody well should), then it's safe to connect your 10-amp multimeter.

 ***CAUTION —***

Only connect the amp gauge between the negative battery post and the body of the car! NEVER connect between the positive cable and positive battery terminal. If any wire touches the body of the car, it will be a direct short circuit to the battery!

Don't start the car with the amp gauge connected! The amp gauge and that 10-gauge wire aren't big enough to carry the load of the starter motor!

Figure 1 Use amp gauge to verify that drain is under 10A. Here, the current reading barely budges from zero, so it's safe to connect the multimeter.

Install a Battery Cutoff Switch

In the test outlined below, you're going to need to switch the multimeter from the 10A setting to the 300mA setting. Installing a cutoff switch makes this easy to do without waking the car and pulling a load higher than 300 mA and blowing the multimeter's fuse. In other words, the cutoff switch allows us to reconfigure meter settings without ever having to disconnect the battery.

These switches come in a few flavors, but the ones that have a rotary knob in the middle and attach between the negative battery post and the clamp end of the negative cable are inexpensive and easy to install. See **Figure 2**.

If clearance around the battery post makes installation of a switch impossible, use an in-line cutoff switch. This will requiring an additional battery cable. In our case, the switch just fits.

While you can accomplish the same thing as a cutoff switch using a jumper wire with alligator clips at both ends, the battery cutoff switch is safest.

In addition, the cutoff switch lets you continue to drive the car with a battery drain problem because you can now throw the switch to disconnect the battery when you park the car and not come back to a dead battery.

These switches work well except on cars with trunk-mounted batteries and those maddening electronic soft release switches on the trunk or rear hatch where, when you disconnect the battery, you can't easily unlock the trunk.

20

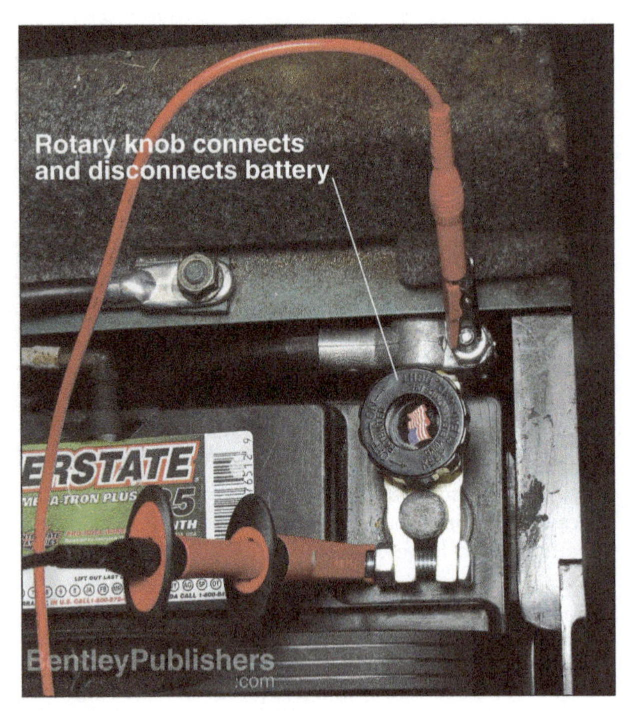

Figure 2 A battery cutoff switch, with the multimeter probes attached on either side with alligator clips.

Configure Multimeter for High Current

You now have some rough idea how big the parasitic drain is from connecting the clamp meter or amp gauge.

- Check to make sure the drain isn't bigger than the current rating of your multimeter. Most multimeters will clearly say something like "10A fused" on one of the input sockets on the front.
- Set the multimeter to measure DC amperage for a high current (e.g., 10A) measurement. Configure the meter's rotary knob and use the correct input sockets for the high current setting. If there's any doubt what the correct configuration is, consult the manual for the meter.

Connect Multimeter to Battery Switch

Next, connect the multimeter across the battery switch. But before you do this, prepare the car for sleep mode, if applicable. See the notes given earlier in the **Sleep Mode** and **Advanced Power Management** sections.

Use multimeter probe tips with alligator clamps integrated into them. You need them to do this test. In addition to spreading the electrical load over a wider physical area, the clamps stay put, allowing you to use your hands for actual work.

 **CAUTION —**

Do not use the pointy multimeter lead for this test! Adapt the pointy leads with alligator clips. The pointy leads are not designed to carry high current.

- Clip red probe to bolt holding switch to negative battery terminal. See **Figure 2**.
- Clip black probe to bolt holding negative battery cable to switch.
- It doesn't matter if positive and negative are reversed. The meter will simply read a negative current value.

Be very aware that you are about to have the car's entire electrical load passing through the multimeter via those skinny wires. So, do not turn on your headlights. Do not turn on your massive power amp. And for heaven's sake, do not crank the starter motor. If you do, it'll try to send hundreds of amps through your multimeter with spectacular results.

Disconnect Battery, Check the Meter

- Twist battery cutoff switch counterclockwise to disconnect battery. All current is now flowing through the meter.
- What does the meter read?

The trunk light is not yet unplugged, so the current is in our case 0.69 amps (690 milliamps). See **Figure 3**. Yes, we should have had you disconnect it first, but we wrote it this way to make a point.

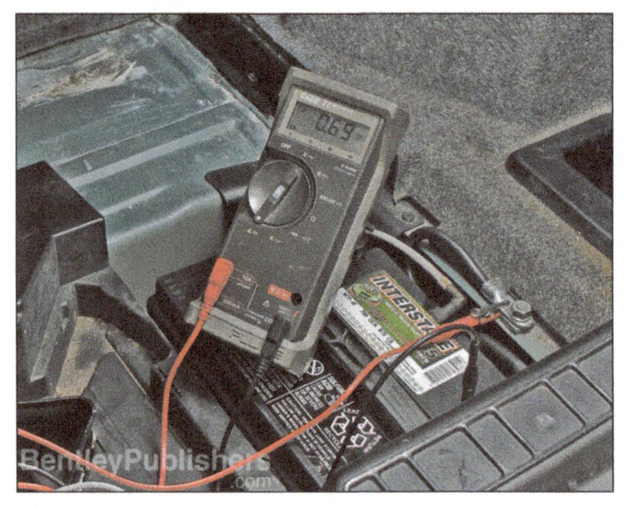

Figure 3 A measured current draw of 0.69 amps (690 mA), most of it coming from the trunk light.

Now find the hood or trunk light switch (a small push-button usually located on the body of the car at the trunk or hood) and manually push it in. If the current drops and rises when the switch is pressed and released, then the light is being switched off. For now, simply remove the bulb and continue troubleshooting.

But if the light isn't going out when you push the switch in and the meter reading doesn't change, you may have found your problem—a bad switch. Test the switch, and replace it if necessary.

With the trunk light disconnected, the current falls dramatically to about 10 milliamps. See **Figure 4**. Note that the meter is still set on the high current setting, so the measurement of a low current is not highly accurate.

Figure 4 Lights out! Current reading drops dramatically to about 0.01 amps (10mA).

The important thing to note is that we've just shown, using the trunk light, what you'll need to do to find a parasitic drain. That is, you saw a current reading on the meter indicating a parasitic drain, disconnected a source of current, and verified that when you did, there was an effect on the meter.

Configure Multimeter for Low Current

If the reading on the meter is now low enough to use the low current setting (under 300 mA), let's take advantage of the cutoff switch and change the red test lead to the low current socket to get a more accurate reading.

NOTE —

If the meter reading is still higher than the meter's low current threshold (e.g., 300 mA), figure out what's causing the high drain before continuing. See the following section called **Finding Parasitic Drain**.

- Twist the knob on the battery cutoff switch clockwise to reconnect the battery. Note that the meter current will drop to zero.
- Set the meter knob to the low current setting and reposition the red test lead to the low current socket. If there's any doubt, consult the manual for the meter.

Disconnect Battery, Check the Meter

- Twist the knob on the battery cutoff switch counterclockwise to disconnect the battery and watch the meter come to life. See **Figure 5**. Note that the reading was 0.01 amps on the 10A setting, and on this more accurate setting, the reading is 6.44 mA.

Figure 5 The reading has become more accurate, changing from 0.01 A to 6.44 mA.

- Now wait. On a late model car, you may need to wait for 16 minutes to an hour for the modules to go to sleep.

The next thing you may see is the reading falling further on the multimeter. It may drop either quickly or in steps. Depending on the era of the car's electrical system, there may be modules that energize, then go to sleep, awaiting commands.

After an hour, see if the amperage stabilizes at no more than 50 milliamps. If it does, you do not have an excessive drain, which isn't to say that you might not have an intermittent problem.

But if the reading is high, like over 100 mA, you have a good-sized drain that requires further diagnosis.

Finding Parasitic Drain

You've done the test and found a drain. What now? You may have read about pulling every fuse one at a time. Hold off on that. It's really the last thing you want to do, not the first.

Check the Easy Things

There are some common sources of parasitic drains that affect many cars.

- Check the alternator, since a shorted diode in the alternator can drain the battery. You'll need to electrically disconnect the alternator.
  - First disconnect the negative terminal to the battery.
  - Detach the main alternator B+ wire and tape it up with electrical tape to prevent it from touching ground, as it is hot to the battery.
  - Disconnect any other wires and/or plugs to the alternator.
  - Reconnect the battery and repeat the parasitic drain test.

- Verify that no interior, hood, and trunk lights are remaining on.
- Check radios and audio power amplifiers, particularly if they're non-original. Disconnect them at their connectors, or pull their fuses.
- If your car has a power antenna, check to see that the motor isn't remaining on. Disconnect it at its connector, or pull its fuse.
- Verify that none of the power window switches are stuck in the on position. Disconnect them at their connectors, or pull their fuses.
- If none of these usual suspects are ringing the bell, consult a web forum. It's likely you're not the first person to experience the problem.

Start Pulling Fuses

As a last resort, yes, you may need to pull fuses one at a time as a way of locating the circuit that is the source of the drain. On pre-1990s-era cars with only 6 or 8 or 12 fuses, this is very straightforward, but the newer the car, the more fuses it has, and the more difficult this procedure is.

- First, know where the fuses are. Are they in a fuse box in the glove compartment? In the trunk? Under the hood? All of the above? A high-end post-2000 European car may have a hundred fuses.
- Next, be aware that, on a post-1996 car that has control modules, you'll need to be able to access these places without disturbing the car's sleep mode. For example, if the fuse box is in the glove compartment, you can't simply open the passenger side door to get at it, because each time you do, it wakes up the car and you have to wait for modules to go back to sleep. To get around this problem, you'll need to roll the door latch to simulate a closed door.
- Removing and installing fuses may also wake up the car, and you'll need to wait for it to go back to sleep. So verify that, when a fuse is re-inserted, the ammeter reading hasn't increased. If it has, wait for it to go back down.
- But once you get through those issues, when you pull a fuse and see the multimeter reading fall, consult a wiring diagram to see what circuits, systems, and devices are on that fuse.
- Then unplug those systems one by one until you've found the cause of the drain.
- If you isolate the problem to a fuse but can't isolate it further, that may be enough to turn up an answer via a search on a good enthusiast forum (e.g., "drain on fuse 39").

Tales From the Hack Mechanic: Current Draw Case Study

Subject: BMW X3 (E83 LCI)
Issue: Battery dead after parked for a week

After three batteries in two years, the owner handed his 2007 BMW X3 over to the Bentley Technical Team for analysis. As we approached the parked truck, we immediately noticed that the orange automatic transmission selector lever light was staying on. This little light is a meaningful telltale; if it's still lit after 16 minutes, it likely means a module is staying awake and drawing current.

We began by plugging in a BMW-dedicated (Autologic) scan tool. There were a lot of under-voltage codes in various systems. Given the battery history, this was to be expected. We also read codes for door mirrors, door windows, and LIN (Local Interconnect Network) bus faults, most of which were pointing to the things in the passenger door. Based on experience with an E46 3 Series, we knew that a bad mirror could affect the LIN and keep the General Module (GM5) awake. We thought we'd found a smoking gun.

We then prepared the vehicle for sleep. We rolled closed the latches on the opened front doors, hood and rear hatch, pulled the hood alarm switch up into the service position, opened the glove box (fuses are there) and disconnected the glove box light. We locked the vehicle using the remote.

After connecting the ammeter, we saw 1.1A for 90 seconds, then 150 mA, and after 10 minutes the current dropped to its final resting place of 55 mA. Not a concern for a daily driver, but a problem for this garage queen.

One-by-one, all mirror and window related fuses were removed. Pulling fuse F49 sent the draw to 7 mA. Bingo. Consulting the Bentley X3 Repair Manual, we learned that F49 was a power feed to the GM5. Not very helpful, but we did find the Bee Gees (*Staying Alive*) module!

Time to remove the passenger door panel and unplug the mirror. No change16 minutes later. Hmmn. How about the window motor? Another 16 minutes, no change. To expedite things and eliminate everything else in the door, we disconnected big harness connector at the A-pillar. No joy16 more minutes. On to the driver's door.

We again disconnected the big connector at the A-pillar to be sure we were on the right path. 16 minutes later, we had our 7mA and no orange light. Using the process of elimination, the source of the drain turned out to be a faulty electric door latch, even though we had no central locking faults, this part is not on the LIN bus or on the mirror/window circuits, and not a common failure part.

We got lucky on this one. There are more than 50 items going through the GM5, from footwell lighting to the Panorama sunroof, including central locking.

Alternate Method: Voltage Drop Across Fuses

The tried-and-true method of pulling fuses has a long and successful history, but is not always the best method with modern vehicles with bus systems and multiple control units. With these, removing fuses one by one can be problematic because some modules use multiple fuses. If you remove one fuse and then reinsert it, you can sometimes cause a module or a bus system to wake up. This can create power draws whose cause is not clear.

There is another method for finding the source of a parasitic drain. Rather than directly measuring the current between the negative battery terminal and chassis ground and watching how the current changes when fuses are removed, the voltage drop across each fuse can be measured. This has the advantage of not requiring fuses to be pulled.

Recall from the discussion on voltage drop in the **Circuits** chapter and the **Common Multimeter Tests** chapter that, when current is flowing through a circuit, every resistance has some voltage drop across it. The resistance across a fuse is small, but it is enough to cause a voltage drop that is measurable with a high-quality multimeter. This measured voltage can then be looked up in a table to find the current causing that drop in voltage through a fuse of that size.

To use the voltage drop method, you need:

- A high-quality multimeter capable of accurately measuring millivolt readings (e.g., Fluke 87 or equivalent).
- A fuse voltage drop chart. These charts are available online for different kinds of fuses (e.g., standard ATO/ATC blade fuses, mini fuses, maxi fuses, etc.).

The charts list the voltage drop produced by different currents flowing through fuses of different ratings.

Below is an excerpt of a fuse voltage drop table for standard ATO/ATC blade fuses. The different fuse ratings are listed across the first row. In this excerpt, voltage drops from 3.0mV to 3.5mV are listed in the first column. The rest of the cells in the table are the currents in milliAmps that would produce that voltage drop for each fuse rating from 1A through 40A.

The steps for the voltage drop method are similar to what we described for locating a parasitic drain by looking at the current through the battery.

- Gain access to the fuses. This usually means opening both the hood and the glove box. If there's a glove box light, disconnect it so you don't create another draw.
- With the doors and hood open, roll the door and hood latches closed so the vehicle's ECU thinks the car is ready to be driven.
- Start the vehicle, let it idle, and cycle all electrical devices. Then shut the vehicle off. This allows the vehicle to fire up all the electrical systems and go to sleep normally.
- With the key off, let the vehicle go to sleep normally for 18 minutes. This should allow all the control units and bus systems to go to sleep.
- Set the multimeter to the DC mV scale and turn autoranging off.
- Zero the multimeter leads (touch the leads together and select the meter's option to subtract off the residual reading) to improve the accuracy of the readings.

Excerpt of Fuse Voltage Drop Chart Across Standard ATO/ATC Blade Fuses (milliAmps)

mV	1A	2A	3A	4A	5A	7.5A	10A	15A	20A	25A	30A	35A	40A
3.0	24	56	96	132	168	275	390	625	888	1190	1523	1863	2083
3.1	25	58	100	136	174	284	403	646	917	1230	1574	1925	2153
3.2	26	60	103	140	179	293	416	667	947	1270	1624	1988	2222
3.3	27	62	106	145	185	302	429	688	976	1310	1675	2050	2292
3.4	28	64	109	149	190	312	442	708	1006	1349	1726	212	2361
3.5	28	65	113	154	196	321	455	729	1036	1389	1777	2174	2431

- Using pointed test leads, measure the voltage drop across each fuse by pressing the test leads on the exposed test points on the fuse as shown in **Figure 6**.

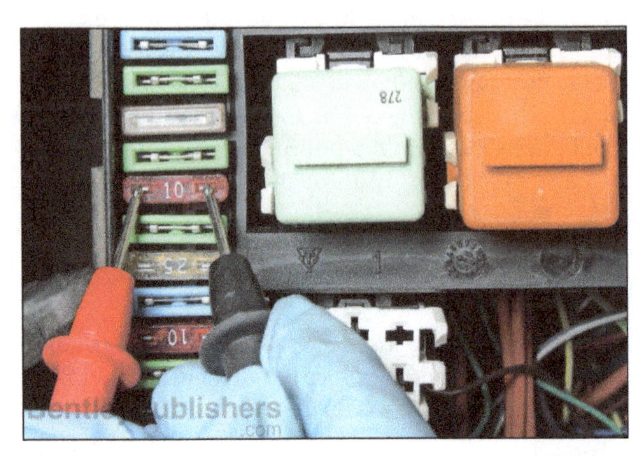

Figure 6 Measuring the voltage drop across each fuse to help locate the source of a parasitic drain.

- The voltage drops across the fuses are very small and may fluctuate a bit, but with a high-quality meter, the readings should be relatively accurate from fuse to fuse. You can verify the test method by checking the glove box light fuse with and without the light connected.

- Take the measurement of the voltage drop across each fuse. Using the measurement of the voltage drop and the rating of the fuse, look up the level of current in the table.

- If, for example, you measured a voltage drop of 3.0mV across a 35A fuse, you would look in the left column of the table for 3.0mV, then look across that row to find the cell under the 35A column. Thus, a 3.0mV voltage drop would correspond to 1863 milliAmps, or 1.863 amps, of current, which, if you're hunting for a parasitic drain, is a sizable amount.

21

Audio Head Unit

Head Unit

It used to be that there was one box in the dashboard that contained all aspects of a car's sound system except the speakers. As audio functionality expanded to things like external CD changers and power amplifiers, satellite mounting points became common, and hence the in-dash box got a new term: Head Unit.

A head unit can be as simple as a radio, or as complicated as the nerve center of an audio and video infotainment system. Detailed information on the latter is beyond the scope of this chapter, but we'll try to give a useful sweep of head unit issues before things got incredibly complicated.

Head Unit Removal

The first thing you need to do in order to upgrade a head unit is remove it, and that can be vexing if you don't know the trick. In the 1970s, most head units were supported by a bracket that needed to be unbolted at the back, typically requiring you to remove the side of the console to gain access.

In the mid-1980s, most original and aftermarket radio manufacturers used a head unit slid into a mechanical support sleeve and snapped into place with sprung clips. Special tools are required to slide into the head unit to release spring clips. See **Figure 1** and **Figure 2**.

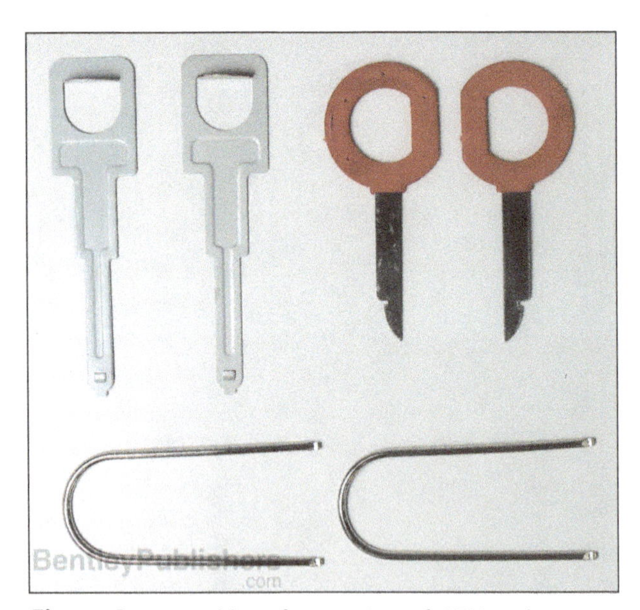

Figure 1 Removal keys for a variety of OEM and aftermarket head units.

Other head units, such as those in most BMWs from the mid-1980s through the mid-2000s, are held in by two small retaining screws accessed by flipping open small hatches on the front of the unit. See **Figure 3**. Some use screws with a 2.5mm hex head that can be removed with a standard Allen key, but others employ screws with non-standard five-sided heads, necessitating a special tool to loosen them without damage.

Figure 2 Use of removal keys for an aftermarket radio in a 1999 BMW Z3.

Figure 3 Small hidden retaining screws holding in a BMW head unit.

Whether clips or screws are used, the removal tool is usually not expensive. Don't try to pry the thing out with a screwdriver; just buy the tool. Consult a manual or enthusiast forum for your car to learn what the exact removal tool is.

Antenna Connections

The antenna in 1970s cars employed a connector still widely used in the aftermarket today, commonly known as a Motorola connector (also sometimes referred to as a "DIN plug"). At around the mid-1990s, most stock European radios stopped using the Motorola-style antenna connectors and began using another form factor sometimes referred to as a "flat connector." See **Figure 4**.

With the advent of antenna diversity (a method for using more than one antenna) in the mid-'80s, even more types of aerial connections were employed.

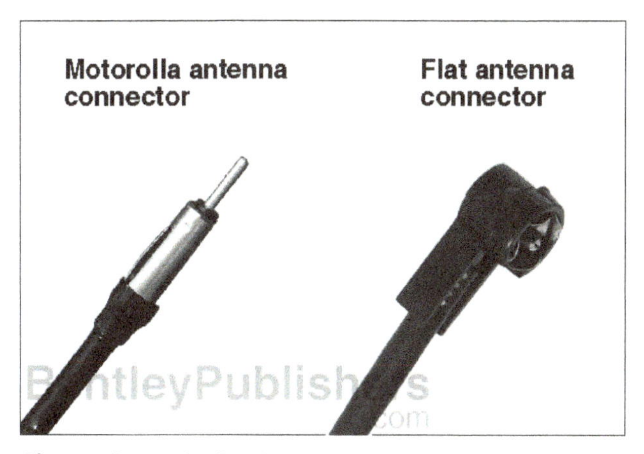

Figure 4 Standard radio antenna connectors.

Connections by Era

When replacing or upgrading sound systems in a car, it's necessary to understand what sound system wiring infrastructure was present when the car was built, as that dictates what's involved in connecting something to it. There's a rough breakdown by era.

First, let's get the terminology straight.

- *Head unit wiring harness* refers to an adapter that lets you connect a head unit to your car. Unfortunately, this same term is also used to describe the car's original wiring that connects to the original head unit, so to avoid confusion...
- *Integral wiring harness* is the term we are going to use in this chapter to refer to the car's integral wiring that plugs directly into the back of the original head unit.
- *Pigtail* is a connector with loose wires that needs to be spliced onto something. Be aware when you're shopping that vendors sometimes also use "head unit wiring harness" to refer to a "pigtail."

1970s

Anyone who has installed radios into 1970s-era cars knows that there was no integral wiring harness for the radio – that is, there's no bundle of wires leading to a single unified connector with all the needed terminals on it. With four wires for a monophonic system (power, ground, and speaker plus and minus), plus the antenna cable, who needed a harness? The separate wires were connected to the back of the radio in a variety of ways.

Early Blaupunkt radios typically presented a male spade connector on the back for power, onto which a female spade needed to be plugged, with the wire stretching in whatever way was convenient to a nearby fuse or hot wire. The ground connection was typically a threaded stud with a nut that held a ring terminal. Two speaker wires typically plugged into the back of the radio with a mini banana plug. The move from mono to stereo introduced left and right speakers, increasing the number of wires from four to six.

1980s

As car stereos began to include things such as front and rear speakers, power antennas, external power amplifiers, and the ability to control the illumination of the display with the dashboard dimmer, the number of wires going to the head unit multiplied. Instead of a head unit having a hard-wired bundle on the back, they began having a single multi-pin socket into which a single plug was connected.

This centralization of connections solved several problems. One was that it allowed head units, which had become frequent theft targets, to be put into removable slide mounts so they could be taken with the owner when the car was parked. The most popular of these were called "Benzi Boxes."

Eventually, detachable face plates took the place of wholly removable head units. But, more generally, the change from hard-wired cables to multi-pin connectors dramatically reduced the number of physical connections to the head unit.

Single and Double DIN

The 1980s brought with it what are known as DIN (Deutsches Institut für Normung, the German Institute for Standardization) radio sizes. One of the DIN standards is the International Standards Organization (ISO) 7736, which governs standards for the physical dimensions of automotive head units. ISO 7736 was adopted as an international standard in 1984 and is still in use today.

Most head units are single DIN, but there are also "double DIN" units which are twice as tall. Depth is, surprisingly, not part of the DIN standard, but depths are typically slightly shallower than the width. These dimensions are shown in the **Head Unit Dimensions Table**.

Head Unit Dimensions		
	Single DIN	Double DIN
Height	50 mm	100 mm
Width	180 mm	180 mm
Depth	Not specified, typically less than width	Not specified, typically less than width

Standard Aftermarket Wire Colors

As the car stereo aftermarket matured, a standard set of wire colors evolved. Sometimes called "EIA (Electronic Industries Association) wire colors," these are shown in the **Standard Aftermarket Wire Colors Table**. Note that this is the standard used by the *aftermarket*. It may bear no resemblance whatsoever to the wire colors originally used in the car.

Standard Aftermarket Wire Colors	
Color	**Function**
Yellow	12V constant for station preset memory
Red	12V switch on from ignition accessory setting
Black	Ground
Blue	Power antenna remote up/down
Blue/White	Amplifier remote turn-on
Orange	Headlight/parking light dimmer
Orange/White	Dash light adjustable dimmer
Green	Left rear speaker +
Green/Black	Left rear speaker –
White	Left front speaker +
White/Black	Left front speaker –
Purple	Right rear speaker +
Purple/Black	Right rear speaker –
Gray	Right front speaker +
Gray/Black	Right front speaker –
Brown	Telephone mute

Non-Standard 16-Pin Connector

This standard set of wires colors totals 16, and thus is usually presented on a 16 pin connector. Although the colors are fairly standardized across aftermarket stereo manufacturers, the form factor, the size of the connector, and the pinouts on the plug were not standardized until the use of so-called "ISO connectors," which we will cover later in the chapter.

If you buy a new head unit, it should come with a pigtail, which as we said, is a plug that connects to the socket on the back of the head unit, along with about 9" of wire with stripped ends. A sample 16-pin pigtail for a Sony head unit is shown in **Figure 5**. In addition to the pigtail using the standard EIA colors, the purpose of the wire may be printed on each wire of the pigtail as well.

The 16 pins include neither the antenna connector nor the standard RCA connectors for line-level signals to a power amplifier. Both of those use shielded cables to suppress noise, and thus often require separate connectors.

For installation, the 16 wires need to be connected to the corresponding power, ground, speaker, and other wires that are present on the car. As we said, in the 1970s, cars didn't really have integral wiring harnesses for the sound system, so "connected" means either splicing into the wires that were present from a previous installation, or doing a from-scratch installation and running wires for power, ground, speakers, raising of the power antenna, instrument illumination, etc.

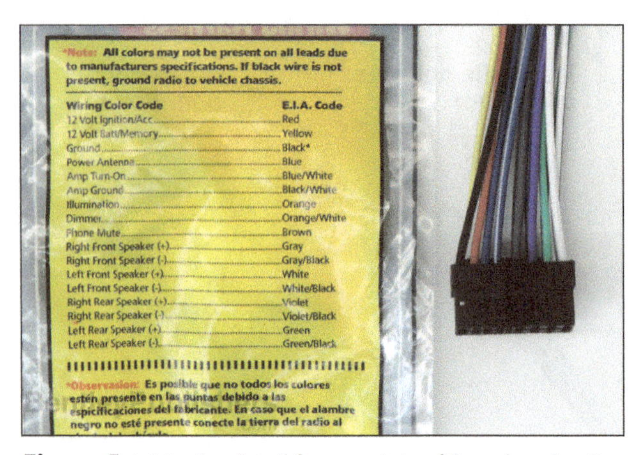

Figure 5 A 16-pin pigtail for a variety of Sony head units, using the standard EIA wire colors.

1990s

Let's take a step away for a moment from the aftermarket and look at the way that a typical stock head unit from the 1990s is connected. The rear of a stock "BMW Business" head unit is shown in **Figure 6**. What initially appears to be a permanently attached wiring bundle on the right is actually a large detachable connector from the car's integral wiring plugging into the back of the head unit.

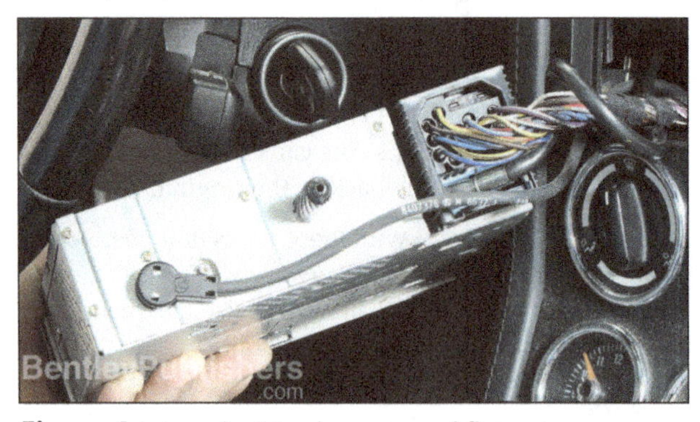

Figure 6 Integral wiring harness and flat antenna connector on the back of the head unit from a 1999 BMW.

When we pull the integral harness connector off the back of the head unit, we can see that the connector is a large male plug with female pins in it. The integral harness connector mates with a corresponding connector on the back of the head unit, a female socket with male pins in it. See **Figure 7**.

This is just one BMW-specific example, but if you're going to replace your head unit, it is crucial to understand what the connectors on the end of the car's integral wiring harness and the head unit both look like. Then, when we start to talk below about changing the head unit and buying or building an adapter for it, you will understand which end is which.

Figure 7 Connector with female pins on 1999 BMW Z3 M Coupe's integral wiring harness. Inset (top left) shows mating connection at back of head unit.

Building a Simple Harness

Now that you've seen the above example, you can appreciate how, if you wish to upgrade the sound system and replace the head unit, you do not have to chop the plug off the car's integral wiring harness and splice a new pigtail onto it. What you do instead is procure a wiring harness that plugs into the existing integral wiring harness and adapts it to plug into the new head unit.

You can usually buy such a harness, but for a 1990s-era head unit, it's simple to build the harness yourself. Procure the pigtails (one that plugs into the car's integral wiring harness, and the other that plugs into the back of the new head unit) and splice them together.

Using the 1999 BMW Z3 M Coupe's wiring as an example, the connector at the end of the integral wiring harness is male with female pins, so to connect to it, you'd need to procure a pigtail that's female with male pins, just like the original head unit. Using a vendor who specializes in these things is your safest bet. Such a pigtail is shown in **Figure 8**. If you look at it closely, you can see that, in addition to using standard EIA colors, the function of each wire is printed on the wire.

A new head unit will almost certainly come with the pigtail that plugs into it. Let's say you're installing a Sony head unit into the BMW. You'd use the 16-pin Sony pigtail with the standard EIA colors. You'll then have two new pigtails, one male with female pins, the other female with male pins, both with standard wiring colors. See **Figure 9**. You simply take one wire at a time of the same color from each pigtail, double-check the wire functions that are likely printed directly on each wire, then splice the wires together with butt-splice or other appropriate connectors (see the **Wire Repairs** chapter). If one pigtail or the other doesn't have a certain wire (such as orange for dimming the display when lights are on), it means the feature isn't supported by the head or by the car, whichever wire is missing.

Figure 8 A pigtail that plugs into the wiring harness of the 1999 BMW. Inset shows wire labels.

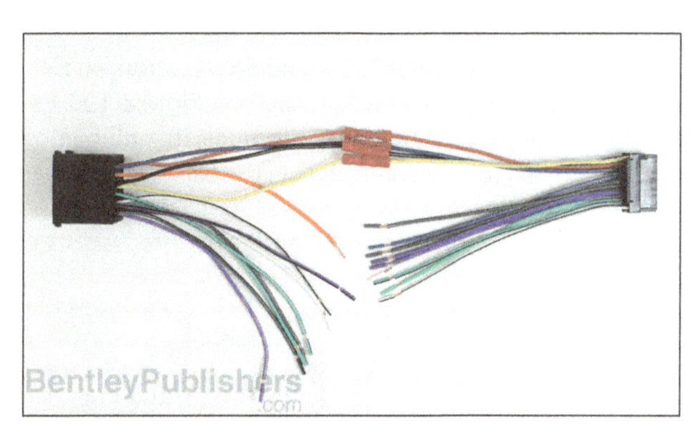

Figure 9 Splicing together two pigtails.

2000s

ISO 10487 Connectors

A standard for audio head unit connectors was developed in the mid-1990s. It's called ISO 10487 and was adopted by BMW, Volkswagen, Audi, Mercedes, and others. It consists of a set of four connectors, two of which (A and B) are always present, and two of which (C and D) are optional. See **Figure 10**.

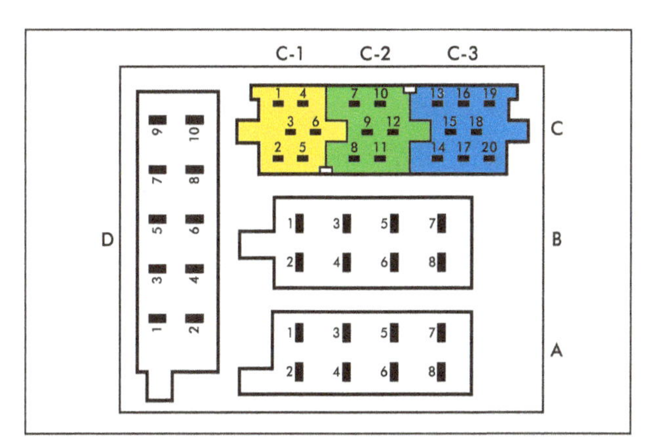

Figure 10 Pinout of ISO 10487 connectors A, B, C, and D.

All four of the ISO connectors as they appear on the back of a head unit are shown in **Figure 11**. The connectors are female sockets with male terminals.

Figure 11 Female ISO 10487 connectors on the back of a head unit.

The two 8-pin rectangular ISO A and B connectors on the car's integral wiring harness are shown in **Figure 12**. They are male plugs with female terminals that plug into the A and B ISO female connectors in the back of the head.

Figure 12 Male ISO 10487 connectors A and B with female sockets on a car's wiring harness.

The pinouts of all of the ISO connectors are shown in the following tables. Connector A is usually black and contains the power and control connections as listed in the Connector A table.

The speed signal on A1 is used by some head units to boost the volume at higher speeds. To work, it needs to be connected to output from a wheel speed sensor. Note that if the speedometer stops working when this pin is connected, the head unit is probably using this pin for some other function, and grounding it.

Connector B is usually brown and contains the speaker connections as listed in the Connector B table.

Together, A and B comprise the set of standard power, control, and speaker signals,. The list of pinouts from

ISO Connector A Pinouts	
Pin #	**Function**
1	Speed signal
2	Phone mute
3	Optional
4	+12V always on
5	Output to power antenna
6	Illumination
7	+12V switched
8	Ground

ISO Connector B Pinouts	
Pin #	**Function**
1	Right rear +
2	Right rear –
3	Right front +
4	Right front –
5	Left front +
6	Left front –
7	Left rear +
8	Left rear –

these two 8-pin connectors, not surprisingly, has a nearly one-to-one correspondence with what's on the not-quite-a-standard 16-pin connector described above. So, if you're interfacing a new head unit with a 16-pin connector to a car with ISO connectors (or vice versa), you simply buy or build the adapter.

Connector C is optional. The pin spacing in connector C is smaller than A and B, so it is sometimes referred to as a mini-ISO. It may be one single 20 pin connector or it may be three smaller sub-connectors that snap together to make up the larger 20-pin connector (see **Figure 10**). When C is made up of three small connectors, C1 is usually yellow, C2 green, and C3 blue. C1 is the connection to an optional external amplifier, with the following pin numbers:

ISO Connector C1 Pinouts	
Pin #	**Function**
1	Line out left rear
2	Line out right rear
3	Line out common ground
4	Line out left front
5	Line out right front
6	+12V switched

C2 is the connection to an optional remote control (meaning steering wheel buttons), with the following terminals:

ISO Connector C2 Pinouts	
Pin #	**Function**
7	Receive data
8	Transmit data
9	Chassis ground
10	+12V switched
11	Remote control in
12	Remote control ground

C3 is the connection to an optional CD changer, with the following terminals. The CD changer is obsolete, but the C3 connector has become a de facto retrofit interface for the iPod and the smartphone via a small interface box.

ISO Connector C3 Pinouts	
Pin #	**Function**
13	Data in (bus)
14	Data out
15	+12V always on
16	+12V switched
17	Data ground
18	Audio ground
19	Audio left
20	Audio right

Connector D is a 10-pin connector for the navigation system to provide an interface to the CD player for the data holding the maps needed by the nav system.

It's worth noting several things about ISO 10487:

- It's absolutely essential to be clear on whether you're talking about the connectors on the car's wiring harness or the connectors on the head unit itself.
- If both the car and the head unit have ISO connectors, then, in theory, you just plug the ISO connectors into the back of the head unit and go. But if one is ISO and the other isn't, you'll need to buy or build adapters.
- If you see an "ISO adapter" with ISO female sockets with male terminals at one end and some other connector on the other end, it's likely meant to allow you to connect a new non-ISO head unit to the car's integral wiring harness. Conversely, if you see an "ISO adapter" with ISO male plugs with female terminals at one end and some other connector on the other end (or just bare wires), it's likely meant to allow you to retrofit a head unit that has ISO connectors.

Quadlock Connectors

Starting in the year 2000, many European manufacturers began using a single 40-pin *Quadlock* connector. As the name implies, Quadlock replaces the four individual ISO connectors with a single large connector with a lock lever. As with ISO, the wiring harness side is a male plug with female terminals, and the head unit side is a female socket with male terminals. The wiring harness side of a Quadlock connector is shown in **Figure 13**. The head unit side and its pin numbering is shown in **Figure 14**.

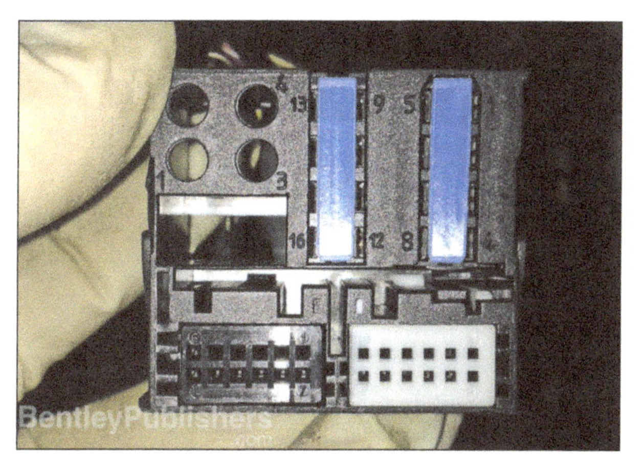

Figure 13 Wiring harness side of Quadlock connector showing snap-in C and D connector blocks.

Figure 14 Pin numbering of head unit end of Quadlock connector from a 2003 BMW 5 Series.

Quadlock shares some of the quirks listed under ISO 10487 and adds some of its own. The labeling of the Quadlock connector blocks does not appear to be standardized. Some sources label them A through D as we've shown. Others label the large square connector block in the upper right as A and the two smaller blocks as B and C. Some refer to them as blocks 1 through 4. And there may be additional manufacturer-specific connector blocks present on the Quadlock connector.

The pinouts on the connectors are not guaranteed to be identical across all Quadlock connectors. Below we list the pinouts for late-2000-era Volkswagens.

Quadlock block A is an 8-pin connector containing the speaker outputs. It is in fact identical to ISO connector B. It is, however, the *only* Quadlock connector block whose pinouts are the same as ISO.

Quadlock A Pinouts	
Pin #	**Function**
1	Right rear +
2	Right rear –
3	Right front +
4	Right front –
5	Left front +
6	Left front –
7	Left rear +
8	Left rear –

Quadlock block B is an 8-pin connector containing the voltage supply and CAN bus lines.

Quadlock B Pinouts	
Pin #	**Function**
9	CAN bus high
10	CAN bus low
11	Display voltage supply, positive
12	Voltage supply, negative
13	Display HV CAN bus low
14	Display HV CAN bus high
15	Voltage supply, positive
16	Anti-theft coding control signal

Quadlock block C is a 12-pin connector containing the telephone and microphone signals.

Quadlock C Pinouts	
Pin #	**Function**
1	Microphone input, –
2	AUX output, audio, right
3	AUX output, common signal
4	Microphone output, –
5	Telephone audio input, left, –
6	Telephone audio input, right, –
7	Microphone input, +
8	AUX output, audio, left
9	Microphone output, +
10	Telephone mute switch for radio
11	Telephone audio input, left, +
12	Telephone audio input, right, +

Quadlock block D is a 12-pin connector containing the CD changer and CD audio input signals.

Quadlock D Pinouts	
Pin #	**Function**
1	AUX signal input, left
2	AUX signal earth
3	CD changer, audio signal earth
4	CD changer, voltage supply, +
5	Not assigned
6	CD changer, DATA OUT
7	AUX signal input, right
8	CD changer, left audio channel
9	CD changer, right audio channel
10	CD changer, control line
11	CD changer, DATA IN
12	CD changer, CLOCK

Buying an ISO or Quadlock Harness

Building a harness to adapt a new head unit to your car has become more involved than simply splicing two pigtails together, as it was for 1990s-era cars. There are often multiple wires and connectors that come off the main portion of the harness and need to be spliced into or connected with other wiring. See **Figure 15**.

The safest way to do this is to go to an online vendor and enter the year, make, and model of the car, and what sound-related options the car has, and the brand and model of the head unit you're installing. You will be presented with your options for a premade harness.

Figure 15 Harnesses with ISO or Quadlock (pictured) connectors can be complex.

22

Dynamic Analog Signals – Theory

Introduction

Thus far, most of the content in this book has centered around static (unchanging) analog signals. That is, when we've used a multimeter, we've done one of two things with it:

- Measured a static voltage
- Measured a static resistance

Actually, most of our efforts have been even more limited than that. Not only have the values of interest been static, they've usually existed in only two states. That is, with voltage, in most cases we've simply wanted to know if voltage is present or not (is it 12V or 0V?). Similarly, when we've been looking at resistance, most of the time it's been in order to find out if the wire has continuity or not (is the resistance near zero or is it infinite?). Yes, there were exceptions (telling 14.2V "charging voltage" from 12.6V "battery voltage," or performing a voltage drop test, or checking if an electrical component had a certain resistance – for example, 50 to 150 ohms for the coil of a relay), but even in those examples, the values we were looking at were not changing while we were measuring them.

We've been living in Static Land.

That's all about to change.

In a computerized car, many devices either output signals or are controlled by signals that are *not* constant and instead change in a very specific way over time. We are going to explain what these signals are. It is going to add *enormously* to your knowledge of how your car works.

What Are Dynamic Analog Signals?

When we say that something is an *analog signal*, we mean that the measurement we take with the multimeter or other tool is the actual data. This is distinct from a *digital signal*, where the data are decomposed into bits and bytes inside a communications packet which is part of a message transmitted over a bus (we'll discuss this further in the chapter on **Modules, Buses, and Digital Data**).

With digital signals, you might be able to see the changing voltage levels on a multimeter or an oscilloscope, but those voltage levels are only a representation of communications activity. They are not the actual data. In contrast, the sine waves and square waves we'll be discussing in this chapter are the actual signals, not messages containing the signals.

The *dynamic* part means that the signals are changing. And that immediately raises questions on how you measure these signals.

Measuring a Dynamic Signal

In most cases, when we say "measurement," we imagine a single number. If we need to measure something that isn't changing, or is changing very slowly, then we simply take the measurement, and we have what we need. For example, the width of the opening in the kitchen for the refrigerator may be 36". You know because you measured it with a ruler. The resting voltage of your battery is 12.6 volts. You know because you measured it with a multimeter set to measure voltage. Done.

But if something is changing, how do you measure it in a way that provides you all the information you need?

Let's begin with a very simple non-automotive example. Think about using a thermometer to measure the temperature inside your house during the winter. You probably set the thermostat to 68 degrees during the day and lower it to 60 at night. So if you took measurements with a thermometer once every twelve hours for three days starting at midnight, you'd see the measurements toggle between 60 and 68 degrees.

But if you only took measurements during the day, they'd always read 68 degrees, and you'd conclude—incorrectly—that "the house is always at 68 degrees." Likewise, if you only measured at night, you'd always see a temperature of 60 degrees. And neither of those sets of measurements convey the whole picture.

You can see from this example that, if something is changing, in order for a set of measurements to be any good to you, you need to measure it often enough to be certain that measurements occur during the changes. Fortunately, the multimeter measures quickly enough that this is generally not a problem. It's your eye that's the problem, as it may not be able to see the changes on the meter's display quickly enough. We'll explain in this chapter how a meter's use of *parameters* gets around this problem.

Terminology

We're going to lay out the terminology that is used to describe dynamic analog signals. Then we'll use these terms to describe some of the most common signals in a car.

Periodic

Periodic signals are those that repeat after a certain amount of time. In the automotive world, periodic signals usually have the shape of either a sine wave or a square wave. We'll discuss both of these in this chapter.

Non-Periodic

Non-periodic signals do not repeat.

Parameters

Parameters are numbers that are used to describe a signal. They're usually inputs into a mathematical equation governing how the signal changes, though you rarely need to worry about the specifics of that.

Amplitude

The *amplitude* of a periodic or non-periodic signal is its maximum value minus its minimum value. This is also referred to as the signal's dynamic range.

Period

The *period* of a periodic signal is the amount of time it takes the signal to repeat.

Frequency

Frequency is usually defined as "1 divided by the period." But rather than just plugging in a formula, a better way to think of it is that period is the time between repeats of the signal, but frequency, as its name implies, is how often (how frequently) the signal repeats. It is the number of repeats that happen per unit time. For more information on what we mean by "unit time," see the sidebar at the end of this chapter.

Sine Waves

In **Figure 1**, we show a sine wave. If you hated math in school, you might hear the term "sine wave" and run screaming from the building. Do not fear. A sine wave is simply a smoothly varying up-and-down wave with a particular shape. Mathematically, the shape comes from choosing one point on a rotating circle and plotting its up-and-down motion. Imagine painting a dot on a tire and following its path as the car it's on drives down the street. It would literally trace out a sine wave. Use a bigger tire, and the sine wave will be taller. Drive faster, and the sine wave will oscillate faster. That's what it is. That's all it is. Put down the pitchforks.

We're going to use this graph of a sine wave to introduce some powerful concepts about periodic signals.

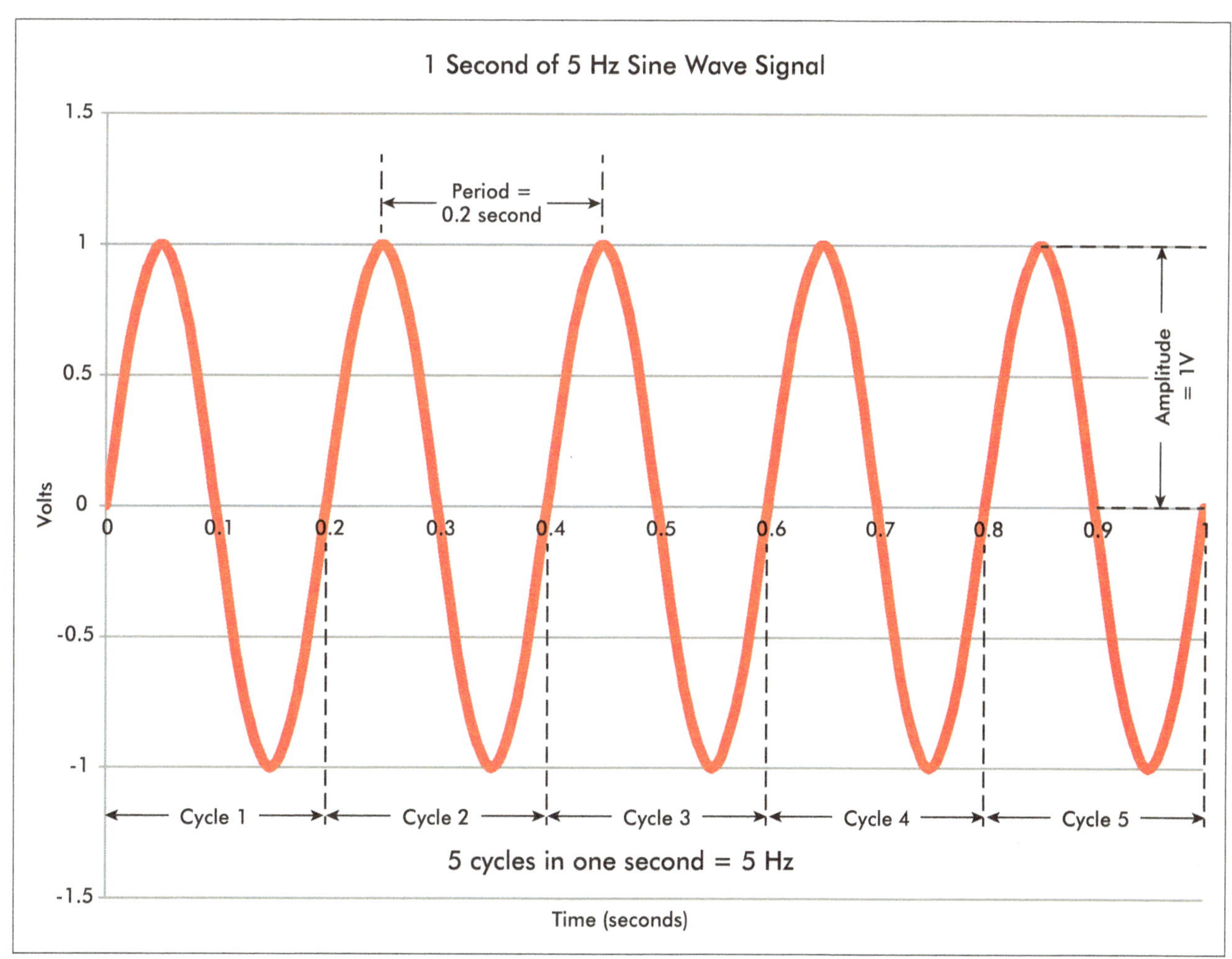

Figure 1 One-second section of a sine wave with amplitude of 1 volt, and five cycles of the signal resulting in a frequency of 5 Hz.

First, you only have to look at this graph to see that **the signal repeats**. That is, a sine wave is an example of a periodic signal.

Second, it's obvious that if you knew two things (that is, two parameters), you'd know everything about this graph. Those two parameters are:

- The dynamic range of the signal (the amplitude).
- How long it takes the signal to repeat, to change from minimum to maximum (the period).

Think about that for a moment. Two parameters completely describe this signal. If a signal is periodic – if it repeats – we can "boil down" the amount of information we need to assess the signal by representing it with one or two parameters. If we have those parameters, we don't even have to look at the graph of the signal.

Coming attraction: This is exactly what automotive multimeters do – rather than display a graph, they boil down periodic signals and show you a handful of parameters that contain the relevant information in the graph.

Amplitude and Frequency of a Sine Wave

Let's look a little closer at the graph of the sine wave in **Figure 1**. By convention, the amplitude of a sine wave is half the vertical distance between its peaks and valleys (that is, if the signal goes from 1 volt to –1 volt as pictured, it has an amplitude of 1 volt). The frequency of the sine wave is the number of full positive-negative cycles per unit time. When the unit time is per second, the frequency is represented by the unit hertz, abbreviated as Hz.

Musical Pitch Example

You might think you don't know about sine waves and their amplitude and frequency, but you actually do, because any sound that you perceive as having a musical pitch has a sine wave-like structure with peaks and valleys. The louder the sound, the bigger the amplitude. And the more of those peaks and valleys there are every second, the higher the frequency (the higher the pitch). If you play a musical instrument, you're probably familiar with concert pitch A above middle C, which has a frequency of 440 Hz. If you plotted its sine wave, you would see it cycle high and then low 440 times per second. If you played another note an octave above this, it would be at a pitch (would vibrate at a frequency) twice as high, at 880 Hz.

House AC Electric Example

House electricity is 120 volts of alternating current (AC). The "alternating" part means it goes up and down in the shape of a sine wave. (Digression: *Why* is an AC signal shaped like a sine wave? Because AC power is created in power plants by large rotating generators, with magnets exciting current in coils of wire. The fact that generators spin around, that sine waves are created by plotting a point on a circle as the circle spins around, and that an AC voltage signal is a sine wave, is not a coincidence.)

When we talk about a 120-volt AC (also written as "120 VAC") signal, it is the amplitude that is 120 volts. Well, it's the peak amplitude – how far the signal is above and below zero. If you could plot your 120 VAC signal, you'd see a sine wave going up and down with a 120-volt amplitude. And if you counted the peaks, you'd find it goes up and down 60 times per second. In other words, the frequency of AC wall power is 60 Hz.

If that sounds familiar, it's because you may have heard that in North America and most of South America, AC power is 120 VAC at 60 Hz, but in most of the rest of the world, it's 230 VAC at 50 Hz. In both cases, they're talking about the amplitude and frequency of the AC signal's sine wave.

In **Figure 1** we showed one second's worth of a 5 Hz signal in which you could clearly count the five peaks and valleys. However, if we showed one second's worth of a 60Hz signal, the peaks and valleys would be so close together that you couldn't count them, so instead in **Figure 2** we show 0.2 seconds, or 1/5 of a second, worth of a signal. If the signal has a frequency of 60Hz, then there are 60 peaks and valleys per second. So if we show 1/5 of a second, we'd expect to see 60/5 = 12 peaks and valleys, and indeed that's what we see in **Figure 2**.

If you actually used your multimeter to measure the AC voltage of the electricity in your house, the meter would read approximately 120 VAC. That's the amplitude of the sine wave. And if your meter also has a frequency setting, it would read about 60 Hz, the frequency of the sine wave. So, even though the sine wave signal itself is alternating up and down, the two parameters of the signal – amplitude (120 VAC) and frequency (60Hz) – are essentially constant.

But here's the most important part. Now that you know that the amplitude is 120 VAC and the frequency is 60Hz, do you need to actually view a graph of the sine wave signal? Probably not. Unless you were interested in the purity of the signal – how closely it actually resembles a sine wave, how much noise it has, etc – all the information you need is represented in the "boiled-down" parameters of amplitude and frequency. This shows the power of using parameters.

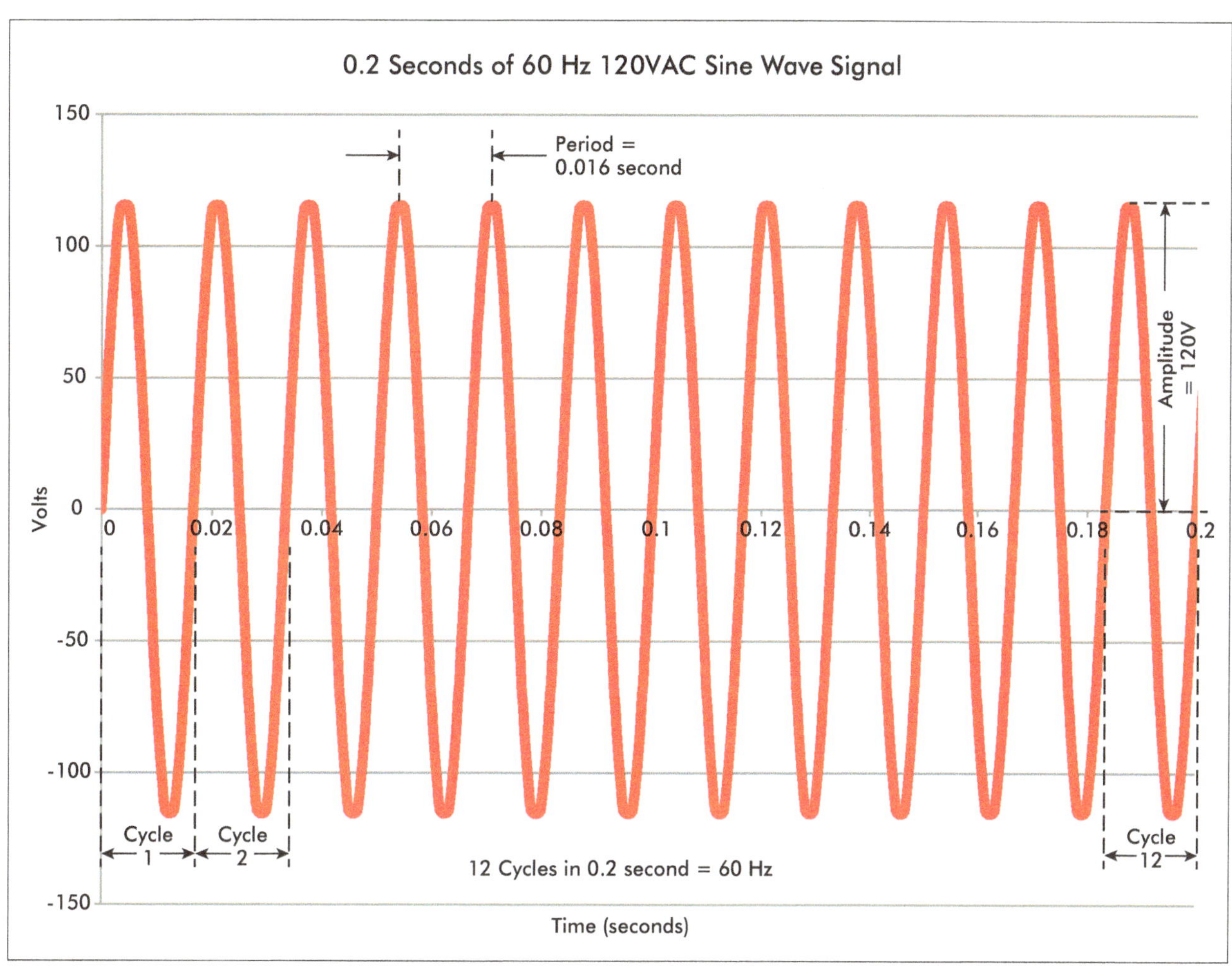

Figure 2 0.2-second section of a typical 60 Hz 120 VAC signal from household electrical current, 12 cycles of the signal resulting in a frequency of 60 Hz.

Devices That Use Sine Waves

We've been talking about sine waves for two reasons. The first is that a sine wave is a simple, fairly easy to understand signal that is useful when introducing the concept of dynamic signals that can be boiled down to one or two parameters. But the bigger reason is that there are actual devices in cars that output sine wave signals. Variable Reluctance Transducers (VRTs) output a sine wave and are employed as crankshaft positioning sensors, ABS sensors, and other sensors on pre-2005-ish cars. More information on VRTs is in the **Introduction to Sensors** chapter.

Square Waves (Pulse Trains)

If a sine wave is a naturally created smoothly varying phenomenon, a square wave (also sometimes called a "pulse train") is the opposite. It is a series of right angle on-off-on-off signals, or "pulses." There is nothing "natural" about a square wave. Its presence is the result of electronics that have taken some other phenomenon, processed it, and created a square wave output. This almost always means that a sensor outputting a square wave has digital electronics inside it and requires power.

Amplitude and Frequency of a Square Wave

The amplitude – the height of the square wave (5 volts in **Figure 3** and **Figure 4**) – is rarely important in the automotive realm. The device receiving the square wave signal typically has some minimum voltage it'll react to, but it typically won't react differently to pulses of different amplitudes. For example, if a square wave signal is used to control a fuel injector, increasing the amplitude won't cause more fuel to be sprayed.

As is the case with a sine wave or any other periodic signal, the frequency of a square wave is how often the square pulses occur per unit time. In our sample square wave shown in **Figure 3**, you can count that there are five pulses in the one second of data shown, so the frequency is five times per second, or 5 Hz.

As with any other periodic signal, the frequency can also be calculated as one divided by the period, and vice versa. It can be seen from **Figure 3** that the rising edge of the first square wave is at 0.1 seconds, the next at 0.3 seconds. Therefore the period is 0.2 seconds. This means that the frequency is 1/(0.2 seconds) = 5 Hz, which checks with how we calculated it by counting pulses.

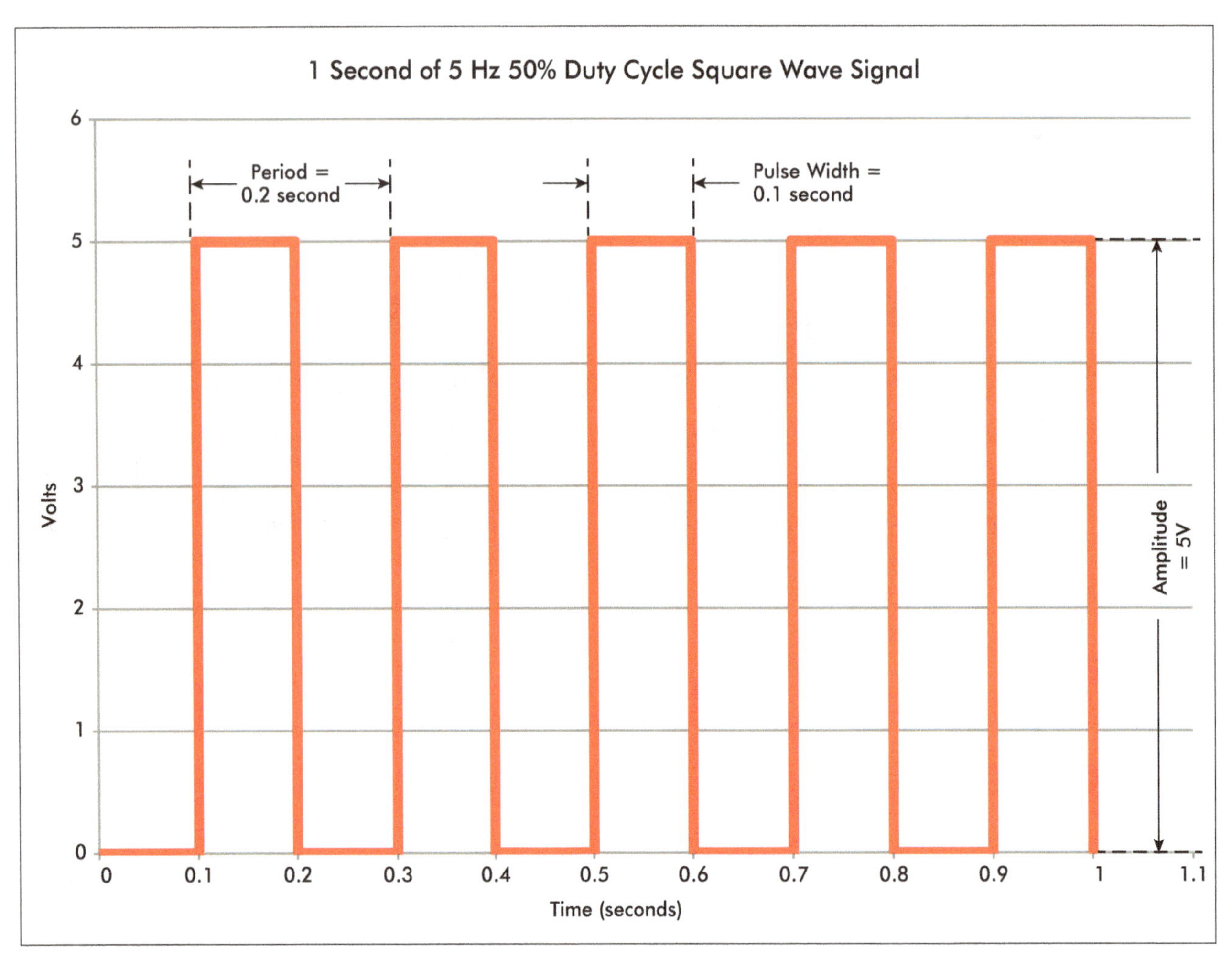

Figure 3 Sample square wave with 5V amplitude, 0.1-second pulse width, 0.2-second period, 50% duty cycle, and 5 Hz frequency

For many sensors such as wheel speed sensors and camshaft positioning sensors, frequency is the parameter that the ECM is looking for. As the wheel or engine spins faster, the frequency increases.

Pulse Width

The pulse width of a square wave is the length of time that the square wave is high (that the pulse is on). In the sample plot in **Figure 3**, we have labeled that the third pulse turns on at 0.5 seconds and turns off at 0.6 seconds. Thus the pulse width is 0.1 seconds.

Duty Cycle

The duty cycle is the ratio of the pulse width (the amount of time the square wave is high) to the period (the amount of time from one pulse to the next), expressed as a percentage. In **Figure 3**, the pulse width is 0.1 seconds, and the period is 0.2 seconds. Therefore the duty cycle is 0.1/0.2, or 50%. Put another way, the pulse is on for as long as it is off. This is plainly visible by simply looking at **Figure 3**, where the square peaks and the square valleys are the same size. Generally speaking, if a sensor outputs a square wave signal where frequency is the important parameter, that square wave looks like the one in **Figure 3**, with a fixed 50% duty cycle. Note that as the frequency changes, the actual pulse width changes, but the duty cycle can still remain at 50%.

In **Figure 4**, we've decreased the pulse width from 0.1 to 0.05 seconds, but left the period as 0.2 seconds. This results in a 25% duty cycle. Put another way, the pulse is on for 25% of the time, and off for 75% of the time. This is also plainly visible from the figure, where the square peaks are 1/4 as long as the period.

Note that, if we keep the period/frequency the same but alter the pulse width, we alter the duty cycle. Thus, *if frequency is constant, pulse width and duty cycle are two different ways of specifying the same thing.*
This brings us to...

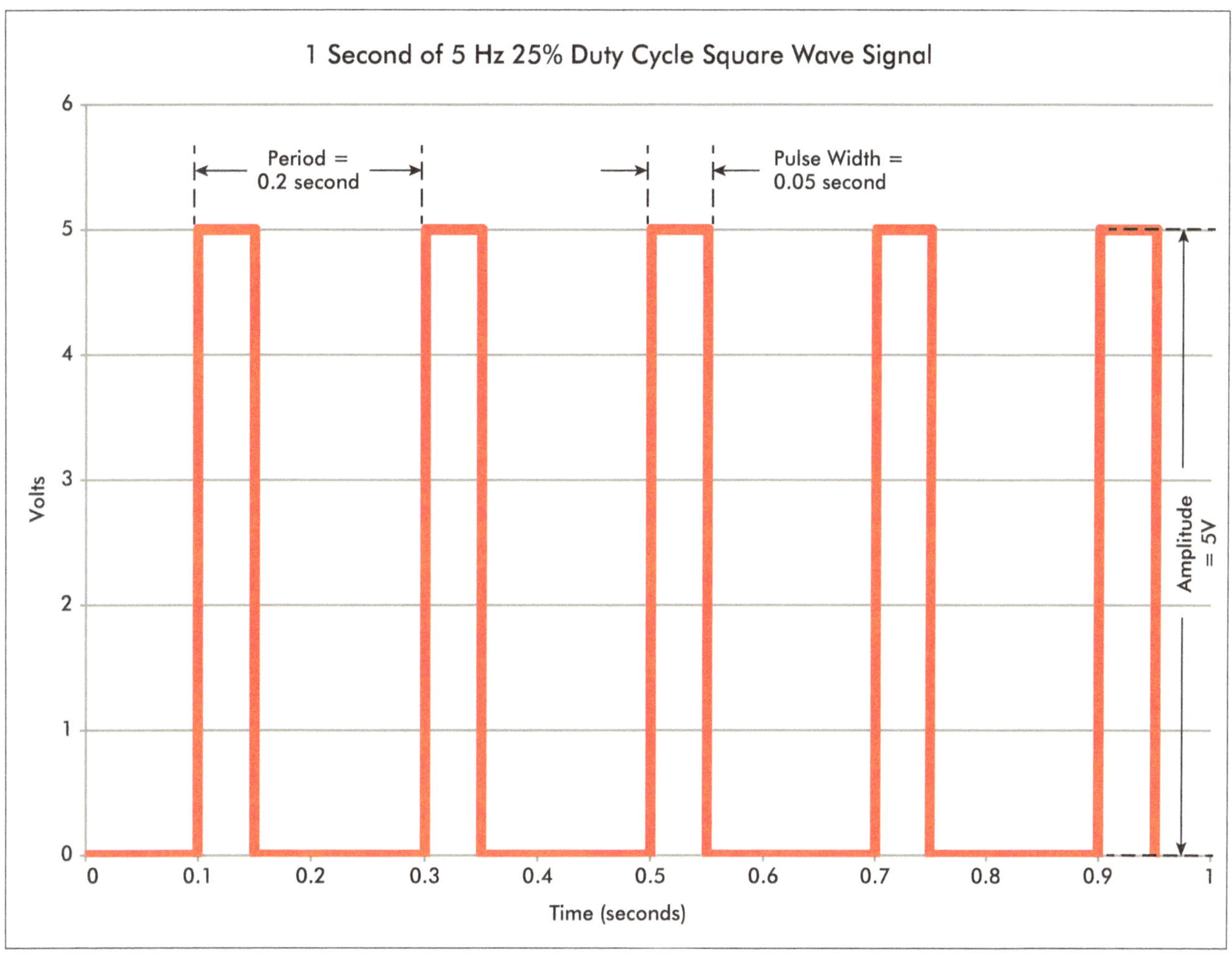

Figure 4 Sample square wave with 0.05-second pulse width, 25% duty cycle, and 5 Hz frequency.

Pulse Width Modulation (PWM)

Pulse Width Modulation (PWM) is, as its name implies, using alteration of the pulse width as a way of sending a signal to control something. What its name doesn't tell you is that, in a PWM signal, frequency is usually kept constant. Therefore, changing pulse width in PWM is exactly the same thing as changing the duty cycle; it's just calling it by a different name. If a device functions by supplying or receiving its signal via a constant frequency but a changing duty cycle, it is often referred to as a PWM device.

Devices That Use Square Waves

There are many automotive components that either output, or are controlled by, square waves. These can be broken down into sensors (devices that output a square wave to the car's ECM) and actuators (devices that are controlled by a square wave sent by the ECM).

On the sensor side, Hall Effect sensors output a square wave and are the technology inside post-2005-ish crankshaft, camshaft, and wheel speed sensors. In addition, post 2005-ish mass airflow sensors output a variable frequency square wave.

On the actuator side, fuel injectors and certain fuel pumps are controlled by a square wave-shaped pulse. But fuel injectors are special, as they respond to changes in both the frequency *and* the duty cycle (pulse width) of the square wave. The higher the frequency, the more often the injectors squirt fuel, but the bigger the duty cycle (the longer the pulse width), the longer they spray during each squirt.

Square Wave Parameter Table

Here's what you need to know in order to understand the bigger picture of how the different square wave parameters are used for different devices on cars.

- If a square wave is created by a sensor that is used to *measure something*, it usually has a *variable frequency* but a *constant duty cycle*, and the parameter of interest is usually the *frequency*. An example is the crankshaft positioning sensor.

Square Wave Perameter Table				
Square Wave Function	**Input or Output**	**Frequency**	**Duty Cycle**	**Parameter of interest**
Measure something	Input to the ECM (sensor)	Variable	Constant	Frequency
Control something that isn't tied to the spinning engine	Output from the ECM ("actuator") such as a fuel pump	Constant	Variable (PWM)	Duty cycle (pulse width)
Control something tied to the spinning engine	Output from the ECM ("actuator") such as a stick coil or injector	Variable	Variable (PWM)	Frequency and duty cycle (pulse width

- If a square wave is created by the car's ECM and is used to *control something that isn't spinning*, it usually has a *constant frequency* but a *variable duty cycle*. This is also known as *Pulse Width Modulation (PWM)*. The parameter of interest is usually the pulse width, which is effectively the same thing as the duty cycle. An example is a fuel pump whose fuel volume is controlled by the pulse width.
- But if a square wave is created by the car's ECM and is used to *control something that IS spinning, or is related to something that's spinning*, it may have both a *variable frequency* and a *variable duty cycle*. This is the case with fuel injectors, where the frequency of the pulse train increases as engine rpm increases (fuel must be sprayed in more quickly), but the duty cycle (pulse width) increases with engine load (each spray must last longer).

This is so important in understanding what you're going to be looking at that we've set it off in a table.

Non-Periodic Signals

Hopefully we've made clear the concept that the periodic nature of sine waves and square waves allows both the car's electronics and you the mechanic to boil down those numbers into one or two parameters – usually frequency and pulse width or duty cycle – in order to judge if a device is functioning correctly without having to actually look at a waveform. This is good, since you *can't* look at a graph of a waveform on a multimeter. However, there are other important sensors on your car that output signals that change but whose change is not periodic. How slowly or how quickly they change is the driving factor behind how they should be measured.

Slowly Changing Signals

Certain sensors – temperature sensors, throttle positioning sensors, and mass airflow sensors – continuously offer up a reading for the ECM to read, but the reading is changing slowly enough that it can simply be examined with a multimeter or plotted with any graphing meter, even one with a slow 1 Hz. sample rate. Then, if possible, the sensor should be exercised to its full range, or near – the radiator warmed up, the throttle opened, the engine revved – to verify that the voltage is changing by the appropriate amount.

Note that some devices such as temperature sensors and throttle positioning sensors naturally output a resistance value, but they use a reference voltage and some circuitry to convert that resistance into a time-varying voltage. While the resistance can be measured directly, it is preferable to measure the value after it has been converted into voltage, as the voltage reading is typically more stable, and is what the ECM is reading anyway.

Rapidly Changing Signals

The digital engine management system on nearly every electronically fuel-injected car built since the late 1970s drives the engine from lean to rich and back again several times per second. The oxygen sensor measures the oxygen content of the exhaust and reports it back to the ECM so the ECM can adjust the mixture accordingly.

On a traditional narrow-band oxygen sensor, signals from the oxygen sensor should be centered at about 0.45 volts and oscillate between 0.1 volts and 0.9 volts between one and five times per second, with the transition taking about 0.1 seconds.

You may read that the signal from an oxygen sensor has a sine wave pattern, but that's not technically correct – it is not periodic, and there is no information contained in its frequency. This update rate of 1 Hz to 5 Hz falls into a funny area. The signal is changing just fast enough that it's difficult to really see what's going on with a multimeter, and too fast to capture on a graphing device that captures data at only 1 Hz.

We'll talk about the oxygen sensor in detail in the **Testing Oxygen Sensors** chapter later in this book.

A Final Note on Frequency

The question "what is the unit time for frequency" is crucially important because, when you see that electronic frequencies are usually referred to in hertz, which is "per second," that means things are usually changing pretty darned fast—too fast to see a pattern with your own eyes or count as numbers scroll by on the screen of a multimeter. We'll come back to it in the next chapter when we talk about **Tools for Dynamic Analog Signals**.

Let's use the simplest possible example of something in your car that is dynamically changing—engine rpm, which is a frequency. Can you imagine if, in order to know when to shift gears, you had to look at a graph of the engine's ignition pulses, count the number of peaks within a given unit time, and divide it by the number of cylinders? You probably never think about it, but inside your car are electronics that do exactly that for you. Not only are they detecting engine rotation, but they then reduce it to the rpm number (the frequency) that you need and present it to you in a familiar form – a needle on a round gauge.

We've covered an enormous amount of ground in this chapter, and hopefully left you with a working understanding of dynamic analog signals, which ones are periodic and which are not, and the parameters of interest. In the next chapter we'll discuss the tools needed to measure these signals and extract these parameters.

Why We Measure RPM Instead of RPS

If frequency is "events per unit time," the next question is "what is the unit time?" Is it a second? A minute? An hour? A day? It's not a foolish question.

There's a guideline that the time scale – the "unit time" – for frequency should be such that the frequency isn't less than one. For example, if you said the frequency was "1/24 per hour," that's completely non-intuitive and unhelpful. Events are quantized. Fractions of events are not meaningful. If something happened every 10 minutes, you wouldn't say "a tenth of it happened per minute," you would say "it happened every ten minutes." If you wanted to express it in terms of a standard unit of time, you'd say "it happened six times per hour."

Engine rpm, as its name clearly states, is measured in "revolutions per minute." It's a frequency. Its scale is "per minute." But anyone who has looked at a tachometer knows an engine spins between about 1,000 and about 6,000 times per minute. So why isn't it expressed as RPS – revolutions per second? The numbers would make perfect sense, since 3,600 rpm is 60 rps. What's more, rpm could then simply be specified in hertz, which has a direct tie into the audible exhaust note of the car.

So why is it measured in rpm?

The answer may be apocryphal, but is very revealing. Use of the term rpm apparently dates back to the Industrial Revolution. The earliest engines were very large, driven by waterwheels or steam. They rotated much more slowly than a car engine. And it was difficult to measure anything accurately to within a second when portable hourglasses ("egg timers") were used as timekeepers. They needed to select a longer period over which to count. Thus, revolutions per minute. True? Who knows? But it's a great story.

23

Tools for Dynamic Analog Signals

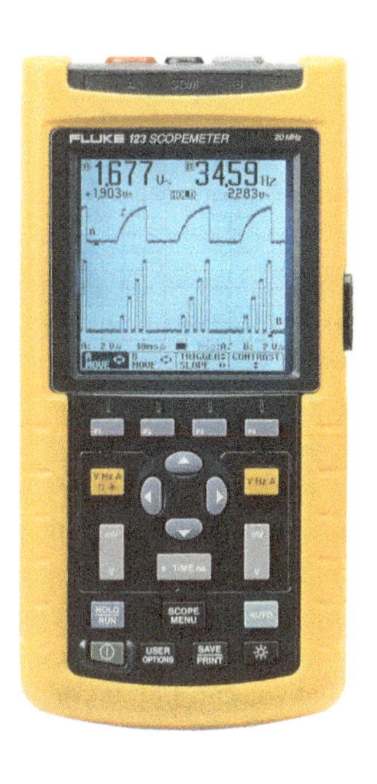

Introduction

In the last chapter, we stepped through a discussion of dynamic analog signals. You should now understand that the signals on your car can be divided into the following broad categories:

- Static signals that basically don't change (e.g., the voltage at your tail lights should be very much the same now versus five minutes from now).
- Those that change, but do so slowly enough that you can still measure them with a conventional multimeter (e.g., a temperature sensor registering the slow rise in coolant temperature).
- Those that change quickly enough that you won't know what you're seeing if you look at them with a conventional multimeter, but are changing slowly enough that you can still sort of do it (e.g., an oxygen sensor whose reading changes one to five times per second).
- Those that change very quickly but are periodic (e.g., square waves and sine waves from wheel speed or crankshaft position sensors changing hundreds of times per second), making it so you need to look not at the reading on the meter, but instead at an important parameter such as frequency, pulse width, or duty cycle.
- Those that change extremely quickly and are not periodic (e.g., the high-speed waveforms from the coil and plugs firing in an ignition system).

In this chapter, we're going to concentrate on the tools needed to detect and measure these things.

What's the Least I Need to Know?

Unless you purchase a pedigreed automotive multimeter meter such as a Fluke 88V, you really don't know if a specific multimeter is going to detect a specific square wave and measure its frequency until you try it. So you can spend a lot of money and buy a Fluke 88V, or spend a much smaller amount and buy an inexpensive hand-held scope meter or a pocket oscilloscope and actually be able to see the graph of the signal itself in real time.

Why is this is so Confusing? Back in the **Multimeters and Related Tools** chapter, we touched on the "Hz" button that some meters have, allowing them to measure frequency, but we lacked sufficient context to fully understand the issue, so we deferred any detailed discussion. Now, having covered dynamic analog signals in detail, we can be much more specific about which tools are needed to measure which signals. We cannot stress strongly enough that you need to know what a signal is in order to select and use a tool to measure and interpret it correctly.

But even if you do, the central confusing issue is that multimeters do not advertise themselves as "square wave detectors." And when you dig into it, you find that not all meters are capable of doing it.

Measuring Sine Waves

In the previous chapter, we said that, if you plugged a multimeter set to measure AC voltage into a household electrical outlet, you'd see the voltage's amplitude (120V) and its frequency (60 Hz) displayed on the meter. But those numbers are fixed; you couldn't make the voltage or frequency change. However, on an automotive sensor, the frequency, and possibly the amplitude, *will* change. They're supposed to.

You should be able to measure a VRT's sine wave signal easily with a multimeter's AC voltage setting. **This works because an AC signal is, in fact, a sine wave**, and vice versa (that little squiggle above the "V" for AC voltage is literally a graphic of a sine wave).

In the **Common Multimeter Tests** chapter, we described how to measure AC voltage, but we didn't really talk about AC frequency. A typical sine wave frequency measurement is conducted as follows. Additional detail is contained in the last eleven sensor-specific chapters of this book.

- Set the meter to measure AC voltage. This means turning the meter's knob to the V with the squiggle above it, and inserting the probes in the same sockets that are used for a DC voltage measurement (red probe in the VΩ socket, black probe in the "COM" socket).
- Put the probes on the terminals on the device whose frequency you're measuring.
- Verify that, when you spin whatever needs to be spun (e.g., for an ABS sensor, spin the wheel), the meter displays a voltage.
- Set the meter to measure frequency (Hz). It's usually a button press, but it may instead be a setting on the dial.
- Verify that the frequency increases when the wheel is spun faster, and decreases when it's spun slower.

How a Meter Calculates Frequency

Before we get into square wave detection, you need to understand how a meter is measuring the frequency of a sine wave.

When a multimeter performs a frequency measurement, it is using a *frequency counter.* The exact mechanism of the frequency counter varies, but it is likely counting the number of times the signal crosses the *x-axis* during a certain time interval. This is known as *counting zero crossings*. It's an old technique in the electronics world. But it requires the signal to cross zero – to go from positive to negative and back. Since that's exactly what an AC voltage signal does, counting zero crossings works perfectly for sine wave signals from sensors. This is why using a multimeter set to measure AC voltage should detect a sine wave signal from a sensor and, if the meter has an "Hz" button, report its frequency.

Measuring Square Waves

So. You now understand measuring the frequency of signals that are sine waves. You own or are considering buying a multimeter and you want to know: If I set it to measure AC voltage, will it detect a square wave and measure its frequency?

Unfortunately, the answer is: Maybe it will, maybe it won't.

There are several issues. First, unlike voltage, resistance, and current measurements, whose settings vary little from meter to meter, the settings for frequency and duty cycle are highly meter-dependent. You need to consult a user manual to be certain that a meter is capable of these measurements. And even then, you may not know until you try it.

Second, many meters are optimized for performing electrical measurements on household electricity, which is at a much higher voltage and current than the periodic signals output by sensors in your car.

But here's the main problem. **A square wave is not, strictly speaking, an AC signal.** Yes, a square wave is "alternating," but not in the AC current sense. An AC signal goes both positive and negative, but a square wave is usually only positive, toggling between zero and some voltage level. It may or may not cross zero. In fact, some automotive square waves don't touch zero at all, and instead toggle between two non-zero values (for example, between 0.3 and 0.6 volts).

For this reason, the frequency counting that occurs on the AC voltage setting – the one that likely relies on counting zero crossings – may or may not detect a square wave, and thus may or may not measure the frequency of a square wave. It is signal-dependent and meter-dependent.

If you need to verify the presence of a square wave for an automotive application – and you *do* need this for many things (see the sensor-specific chapters at the end of this book) – we recommend that you look for a meter capable of performing a pulse width measurement. The ability to measure pulse width is almost always paired with a duty cycle measurement (usually represented by a "%" setting), but the presence of a duty cycle measurement alone isn't sufficient, as some meters have a duty cycle measurement as part of the AC voltage setting to provide a metric on the quality of household electrical power.

As we keep saying, although most meters have a little sine wave squiggle icon over the AC V setting, meters *do not* have little icons for square waves. See **Figure 1**. You just need to know that pulse width and percent duty cycle imply square waves. Which is why we're explaining it in such great and gory detail.

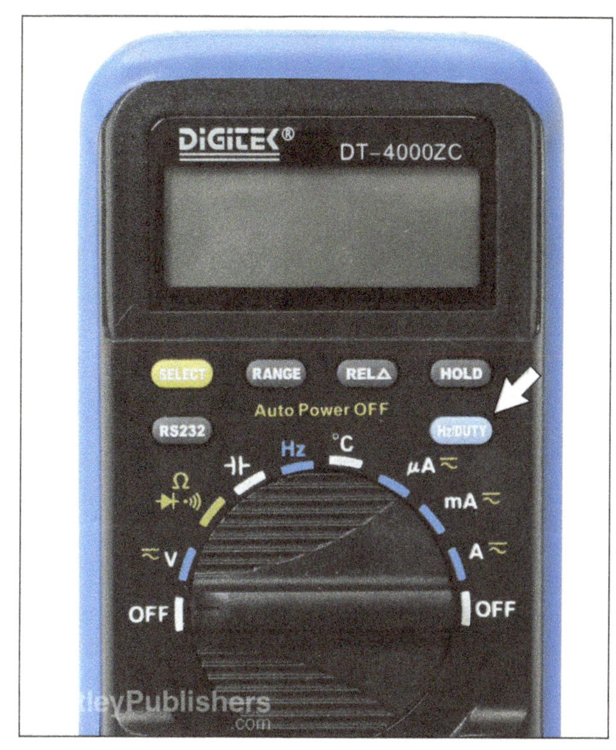

Figure 1 This inexpensive meter has a plainly labeled Hz/Duty switch (**arrow**), but no ms pulse width or triggering settings, and may or may not be able to detect square waves and measure their frequency. Note also the squiggle next to the V, indicating AC voltage.

But even with pulse width measurement capability, you still need to know *how* the meter is measuring frequency in order to be certain it'll work on *all* automotive square waves. Fluke has detailed user manuals posted on their web site that explain how their meters perform frequency and pulse width measurements. In contrast, with an inexpensive meter that comes with just one or two pages of documentation, it's nearly impossible to get that level of reassurance in advance; you just need to try it and see if it works.

Automotive Multimeters

The term "automotive multimeter" implies that a meter performs frequency, duty cycle, and pulse width measurements along with the more old-school rpm and dwell measurements. But you can't assume anything.

For example, the Fluke 88V, the gold standard of automotive meters, has "Automotive Multimeter" stamped right at the top of the meter, and shows labels for Hz % and trigger adjustment right on its function keys. See **Figure 2**. It is indeed capable of these measurements.

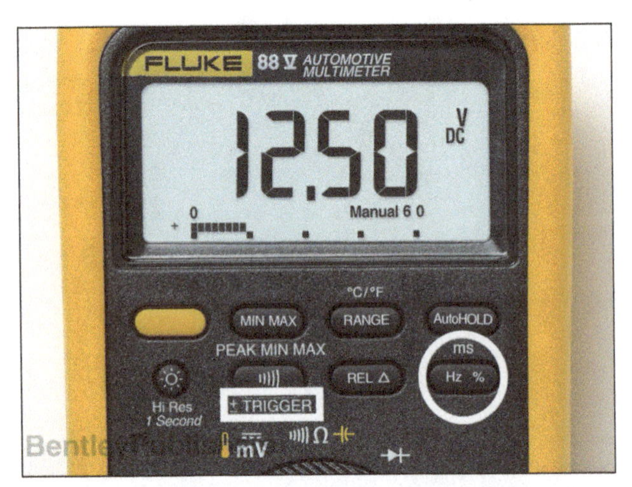

Figure 2 The Fluke 88V has a clearly labeled Hz, % (duty cycle) button.

However, the newer Fluke 233 is also stamped "Automotive Multimeter"" on its front, but shows only a "Hz" label on the front. See **Figure 3**. Its frequency setting is associated with AC voltage and current measurements only, and it does not appear to perform pulse width and % duty cycle measurements.

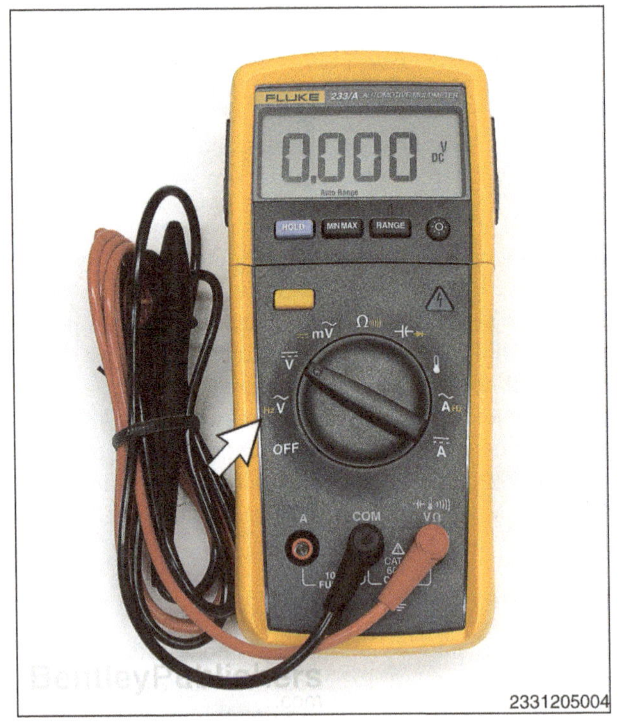

Figure 3 The Fluke 233 Automotive Multimeter can measure the frequency of AC signals (**arrow** on "Hz" setting), but it lacks % duty cycle, pulse width, and triggering adjustment capability.

Triggering

In order for a meter to measure a dynamic quantity such as frequency, it must have a reference point. As we said, a meter usually measures frequency by counting how often the signal crosses zero. Every time the meter sees the signal cross zero, it internally says, "I've got one." You can imagine a similar method for calculating % duty cycle or pulse width, where the meter figures out which parts of the square wave are the peaks and which are the valleys, and then counts how many of them there are per second, or how long each one is, so it can display the frequency or pulse width parameter on the screen.

The more general term for this internal referencing is *triggering*. And the issue of whether a meter can detect a square wave and measure its frequency comes down to the question of whether it can trigger on that square wave. Inside the multimeter, the triggering is usually based on three parameters:

- A threshold value
- Whether the meter is looking for a rising edge or a falling edge
- The steepness of the slope on the rising or falling edge

Depending on the meter, all of these trigging parameters may be adjustable, or some may be, or none may be. For example, when you set the meter to perform a frequency measurement on an AC signal, you usually specify nothing; the meter usually sets the threshold at zero and counts the number of times the signal crosses it.

How to Use a Fluke 88V to Measure Frequency

See Test Set-Up on page 23-6

How do we know for certain that the Fluke 88V actually detects square waves and measures their frequency? Because the 88V's 74-page illustrated manual tells you so. Although the manual never uses the terms "square wave" or "pulse train," it explicitly shows you how to use the meter to measure frequency, pulse width, and duty cycle for fuel injectors, MAP sensors, and other devices. It also gives the accuracy specs for the meter's internal frequency counter, and includes a table of the meter's internal trigger levels for the various measurements. This high level of documentation is one of the things you're paying for when you buy a Fluke product.

The Fluke 88V manual has a detailed section on triggering and frequency counting. In short form, what it says is this:

- In its frequency mode, the 88V is not counting zero crossings. Instead, it's counting *threshold crossings*.
- That threshold isn't set by typing a number into the meter. It's determined by the meter's rotary dial and range settings (even though it's an autoranging meter).
- The button on the 88V for "±trigger" adjusts the trigger slope, not the trigger threshold.
- **From a practical standpoint, if you don't know the voltage levels of the top and bottom of the square wave, you may need to try both the AC V and DC V settings on the rotary knob, and manually alter the range between the 600 mV, 6V, and 60V scales, until the meter triggers on the signal and displays a frequency.**

With all that background, we can finally explain how to use a Fluke 88V to detect a square wave and measure its frequency, pulse width, and % duty cycle. See **Set-Up 23-A** on the following page.

- Set the rotary switch to AC V.
- Insert the probes in the same sockets that are used for a DC voltage measurement (red probe in the VΩ socket, black probe in the "COM" socket).
- Connect the test leads to the signal and ground terminals of the device whose signal you wish to measure.
- Press the "Hz %" button once to measure frequency. "Hz" will appear on the display.
- If the frequency reads 0 Hz, manually set the meter's range to 600 mV, then 6V, and if necessary, 60V.
    - Press the "Range" button to select Manual Range.
    - Press the "Range" button repeatedly to cycle to the desired range.
- If the frequency still reads 0 Hz, change the rotary knob from AC V to DC V, and check both ranges as per the previous step.
- Press the "% Hz" button again to show % duty cycle.
- Press the button a third time to show pulse width in milliseconds.

Using a Fluke 88V to Measure Frequency

Test Set-Up 23-A

See page 23-5 for step-by-step procedure

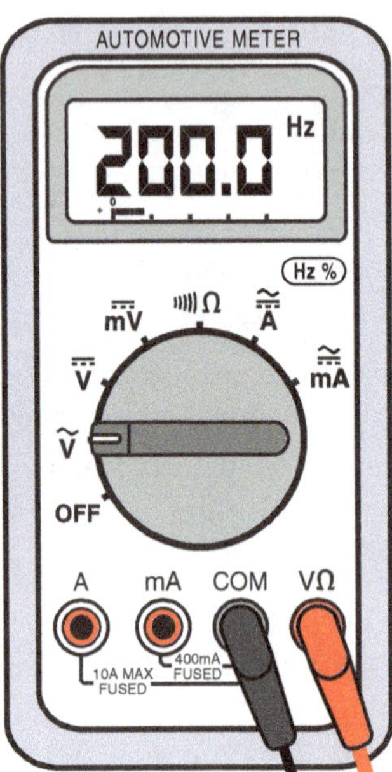

Test Conditions

Dial set to:	Ṽ AC Volts
Buttons:	(Range) (Hz %)
Red input terminal:	VΩ
Black input terminal:	COM
Red lead:	where you want to measure signal (+)
Black lead:	chassis ground (–)

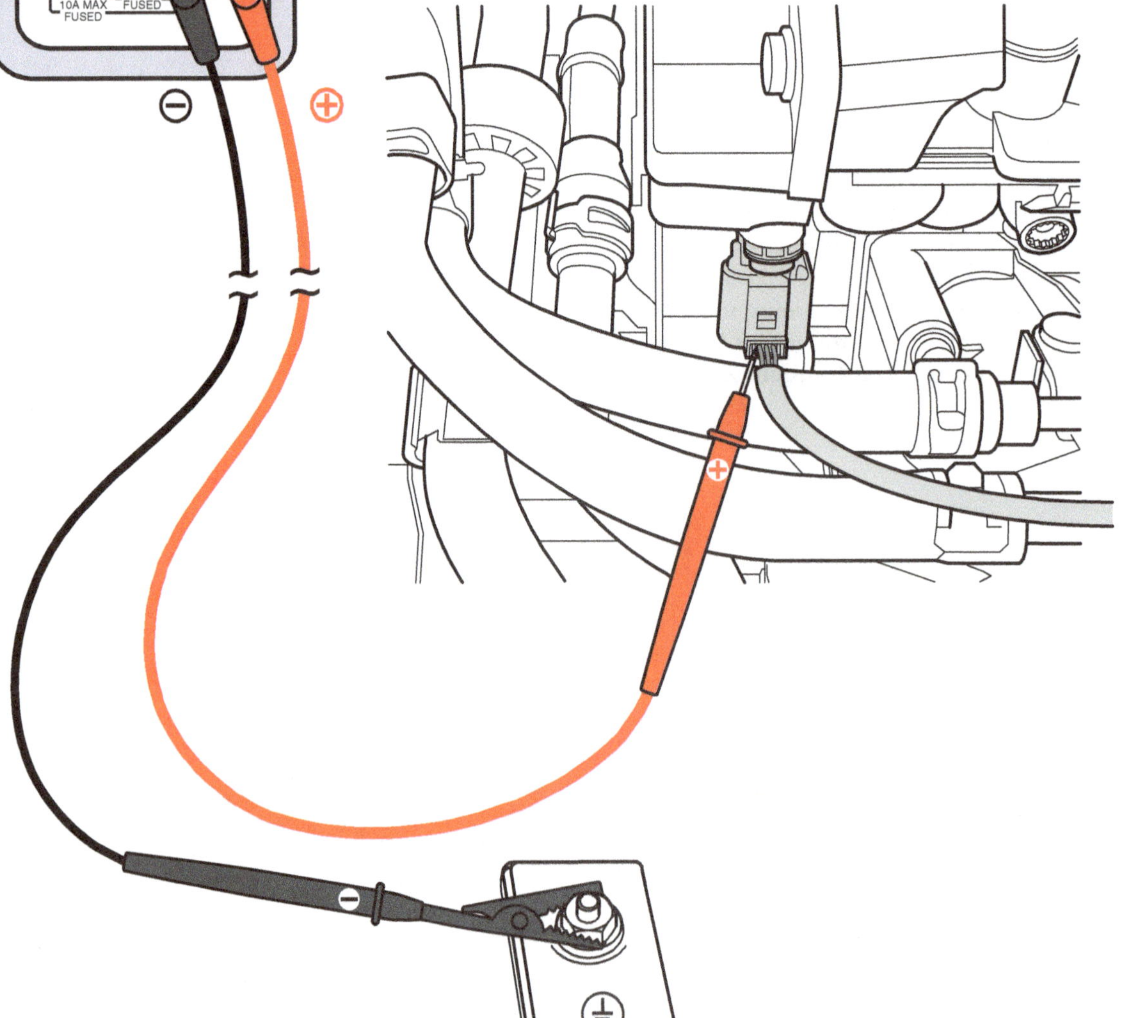

Capturing and Graphing Data

We're going to say a few words about logging and graphing meters so you can understand that a) you shouldn't confuse them with portable oscilloscopes and b) they're not terribly useful in the automotive realm.

Logging Meters

Some multimeters have data logging capability that allows data to be stored and then downloaded to a PC where it can be plotted and viewed. It's typically a three-step process. First, you log the data on the multimeter. Second, you transfer the data to a computer via a serial or USB interface. Third, you run a piece of software (supplied with the meter) on the computer to read in and display the downloaded data. It is appealing to think that you could buy an inexpensive logging meter and, the handful of times you need to graph something (for example, seeing the waveform from the ABS sensors whose functionality you want to validate), you could log it, download it, and graph it.

However, **the logging meters we've looked at do not log data any faster than once per second.** They appear to be designed for catching long-duration intermittent problems (the literature uses terminology like "event logging" and "trend capture"). You need to look at the specifications carefully. So, while it may be possible for you to spin a wheel slowly enough to capture meaningful data from an ABS sensor on a logging multimeter, you shouldn't assume a logging meter will help you look at the output of something that's changing rapidly, like a camshaft positioning sensor on a running engine.

Figure 4 The Fluke 289 graphing meter is playback only with a recording rate of once per second. *Reproduced by permission, Fluke Corp.*

Graphing Meters

The next step up from logging meters are graphing meters. Be careful! Graphing multimeters look like they're plotting data in real time (see **Figure 4**), but they generally do not have real time displays – they have on-meter playback displays. Put another way, **a hand-held graphing meter is not a portable oscilloscope.** Further, as is the case with logging meters, the sample interval is very coarse. You should not assume a graphing multimeter logs and displays data faster than once per second unless there is explicit documentation saying that it does.

Oscilloscopes

If you really want to know if a dynamic signal is present, use an oscilloscope. These days, they're available smartphone-sized and cheap. No "automotive meter" is going to be as good. You should just get one. We did. It works.

There. We said it.

Now, let us justify this surprising statement.

In the tutorial on periodic data in the previous chapter, we said that if we had a tool to graph a sensor's signal in real time, we could simply look at the graph and know that the sensor was or was not operating correctly by the presence or absence of the signal and by the quality of the signal. The tool for that kind of real-time display is an oscilloscope. However, most people don't own one.

Thus, we presented the next best thing: If we had a tool that detected the presence of both sine waves and square waves, calculated their important parameters (frequency, and in the case of square waves, pulse width and duty cycle), and displayed these parameters, it would largely eliminate the need to actually look at a real-time graph of the data. That tool is, in theory, the automotive meter.

In truth, however, most of the time, we don't care what the actual frequency or % duty cycle or pulse width values are. We almost never need to, for example, rev the engine until we see a certain frequency on a certain sensor, and then take a screwdriver and adjust something. Most of the time we simply want to know whether a sensor or device is working. Because most people don't have oscilloscopes, you try to use an automotive meter so you can know that a signal is present by seeing that the parameters change when you spin the wheel faster or rev the motor.

However, there's a Catch-22 here. Because of the uncertain pedigree of most automotive multimeters, you can easily find yourself in a situation where you hook up the meter, don't detect a signal, and aren't sure if the problem is the signal, the meter, or you.

At the end of this chapter we have a table in which we make recommendations on which tools are appropriate to measure which classes of signals, and recommend that you use an "automotive meter" to detect square waves, but nothing is going to be as good as an oscilloscope. They're much less expensive than they used to be. If you're routinely troubleshooting sensors that output dynamic analog signals, you should buy one.

Having said that, be careful what you wish for. Oscilloscopes are enormously flexible and powerful tools, and using one to simply confirm the presence of a square wave barely scratches the surface of what they can do. However, they have a non-trivial learning curve in terms of triggering, thresholds, sampling rate, time base (the display scale of the *x-axis*), volts per division (the display scale of the *y-axis*), and other things. And just as you can connect an automotive meter to a signal and see a zero reading because it is not triggering on the signal, it is quite possible to hook up an oscilloscope and see nothing because the configuration parameters are wildly off.

Fortunately, most modern scopes have "auto" modes where they try and sense the update rate and range of the data and configure themselves accordingly. And even if the scope is improperly triggering, the data are usually still plainly visible – they're just scrolling by very quickly on the screen. If you reach this point of seeing rapidly scrolling data, a simple trigger threshold adjustment usually gives you what you need and locks the data in view.

Oscilloscopes fall into five broad categories.

Analog Bench Oscilloscopes

Old-school cathode ray tube (CRT)-based bench oscilloscopes with the green phosphor screen can be had at yard sales, electronics surplus stores, and eBay for very short money, typically less than $50. They are, however, big and bulky and obsolete, and require all inputs to be set manually, which can be more than a bit daunting for the first-time user. Unless you're already facile with using one, don't get one.

Digital Bench Oscilloscopes

A digital bench oscilloscope intended for general electronic testing will likely do anything you need in the automotive realm. They're more compact and lighter than the old analog CRT-based units, but require wall power. And their cost is high. If you're hardware-savvy, need to have multiple channels of input, and troubleshoot other electronics, great, but if you're looking for something specifically for the garage, there are better choices.

PC-Based Oscilloscopes

There are USB plug-in electronics modules that turn your PC into a digital oscilloscope (see **Figure 5**). Some of these are sub-$100 units for the do-it-yourselfer, but some are high-end units with all the capability of an expensive bench oscilloscope and are specifically aimed at the automotive diagnostic market, offering high-speed sampling and multiple input channels. These higher-end units allow you to do things like look at graphs of both voltage and current simultaneously from an ignition waveform.

The advantage of a PC-based scope is that, because the module is relying on the ability to piggyback the capability on an existing PC, you pay for only the electronics module and the software. The downside is that you're tethered to a PC, and that creates extra bulk and cabling. Unless you have enough room in your garage for a roll-around cart with a dedicated computer on it, and need to use it a lot, it may be more trouble than it's worth.

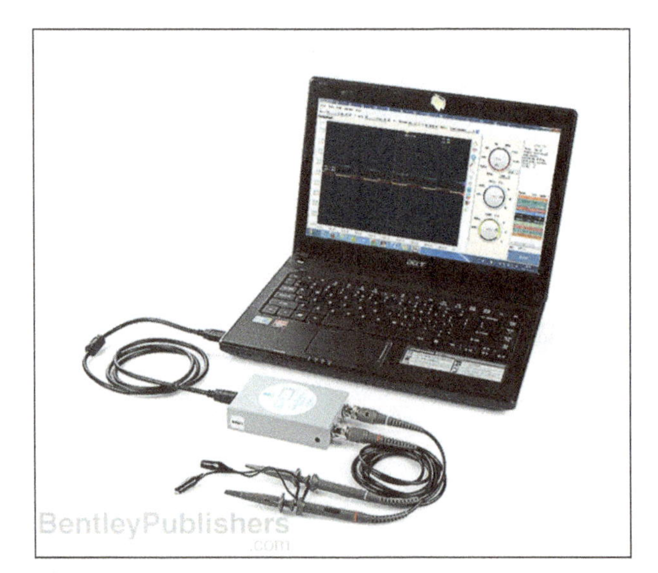

Figure 5 **An inexpensive PC-based USB plug-in oscilloscope module.** *Photo, Sain Smart*

Hand-Held Scope Meters

Snap-on and other automotive tool manufacturers have long made large cart-sized automotive scopes marketed to independent garages. As electronics became smaller and lighter, hand-held scope meters from both Fluke and Snap-on, specifically aimed at the automotive market, began appearing around the mid -1990s. Compact (just a little larger than the size of a multimeter) and rugged, these provide the technician with what is needed for hand-held real-time measurements. See **Figure 6**. The downside is that the cost can stretch to four figures.

However, hand-held scope meters without the Fluke or Snap-On pedigree are now available in the low hundreds

Figure 6 Fluke's 123 ScopeMeter® is an excellent hand-held scope, but the cost is beyond the reach of most DIYers. *Reproduced by permission, Fluke Corp.*

of dollars. As with all the other tools listed above, there is a trade-off of cost against performance (don't expect the $150 scope meter to have the performance, sensitivity, features, and documentation of the Fluke). In fact, the cheapest of these is unlikely to have a frequency counter, so it won't actually give you a numeric display of a waveform's frequency or pulse width or duty cycle. If you want those numbers, you'd need to actually go old school and measure the graph on the screen against the number of divisions on the graph's axis.

But if all you need to do is verify the presence of a square wave from a common automotive sensor and watch the waveform change as you rev the engine, an inexpensive scope meter will likely be more than adequate.

Pocket Oscilloscopes

In addition to the multimeter-sized scope meters shown above, there are shockingly small and inexpensive "pocket oscilloscopes" that will display, in real time, the square waves and sine waves we've discussed. The downside is flimsy construction, questionable reliability, scant documentation, and reduced functionality as compared with a full-sized scope (like the cheapest scope meters, the cheapest pocket scopes may not have frequency counters).

Figure 7 These days, you can buy a pocket oscilloscope and see waveforms in real time. We did.

Still, we purchased the one shown in **Figure 7** – a DS0201 Nano – for $60. It's literally smaller than a smartphone, making it easy to hold in one hand while you use the other hand to affix the probes. We updated its firmware to a better freeware package that we read about online, and have been using it to look at square waves and sine waves from a variety of sensors and devices. In our opinion, it's well worth the $60. If you enjoy doing this sort of diagnostic work yourself, you should just buy one. We did. Why *guess* what the signal looks like when you can *know*?

Tales From the Hack Mechanic: My Oscilloscope Moment

What if, when looking at the output of some sensor, you don't get a valid frequency or a pulse width measurement on a meter? Unless you have a lot of experience in this area, you don't know if you're doing it wrong, or if the meter isn't capable of detecting the signal, or if the signal really isn't there. So what do you do?

Well, if you're me, you spend sixty bucks, buy an inexpensive pocket oscilloscope, and hook it up.

Here at the shop, we were trying to test a wheel speed (ABS) sensor on a 2006 BMW E60 5 Series car. It's a two-wire Hall Effect sensor. The Bentley manual described it as outputting a "pulse" every 0.75 seconds while at rest, and outputting a "pulse train" when the wheel is spinning. We hooked up our Fluke 233 automotive meter, and could not detect an AC frequency and our Fluke 88V wasn't handy. We tried two other meters – a Digitec DT-40000ZC, and an old Craftsman Professional, both with Hz and % duty cycle settings – and still could not detect a pulse train.

We plugged all three meters into the wall and made sure all three saw 120 VAC and 60 Hz. On the Digitec and Craftsman, we also saw a 50% duty cycle, as one would expect of a sine wave (it's high as often as it is low). We went back to the wheel speed sensor and still saw nothing.

So I spent $60 and bought a pocket oscilloscope.

I first wanted to test the oscilloscope with a wave whose parameters I could adjust. I found the web site "onlinetonegenerator.com" in which you can specify the wave shape (e.g., sine wave or square wave) and enter the frequency. I selected a 400 Hz sine wave and verified that I heard a tone in my headphones. I then connected the pocket oscilloscope to the computer's headphone jack and adjusted the triggering until I could see the signal on the scope. There it was, a 400 Hz sine wave, about 150 millivolts peak-to-peak. So I knew without question that there was a 150 mV 400 Hz sine wave signal at the headphone jack.

I then changed the sine wave to a square wave, verified that I could see it on the scope, and tried to detect it with the three multimeters. Only the venerable Craftsman detected the square wave and reported the 400 Hz frequency. The other two meters did not.

I took the pocket oscilloscope down to the shop, connected it to the wheel speed sensor, and we saw the "resting pulse" square wave output as plain as day, as shown in **Figure 7**. When we spun the wheel, we saw the "pulse train" square waves, increasing in frequency as we spun the wheel.

So, yes, to check the tool we selected because we can't usually look at the signal in real time, we need to, uh, look at the signal in real time.

Choosing the Best Tool Based on Signal Type

The correct tool for measuring dynamic analog signals depends on how quickly the signal is changing. There are certain things such as the full high-frequency graph of the voltage or the current in the spark plug as it fires that you can *only* see on a rapid-sampling oscilloscope, but for many applications, several tools will work. Some are a better fit than others. Below, we parse the choice of tool by the type of signal.

Non-Periodic Signal, Changing Very Slowly

If the signal is changing slowly (for example, slower than once per second), you can simply connect a multimeter, set it to measure DC voltage, look at the numerical voltage reading, and watch it change. If you can do something to effect a change in the reading – rotate a throttle positioning sensor, warm up the engine to increase the temperature, for example – then go ahead and do it and make sure the changes in the reading are within the device's acceptable range.

Periodic Signal, Changing Very Slowly

If the device uses periodic signals, you may be able to coax it into generating signals that change slowly enough that you can simply use a meter. For example, if the device is a Hall Effect ABS or camshaft positioning sensor, it should output a square wave of varying frequency regardless of how slowly the wheel or engine is spun. So it should be possible to connect the sensor to the multimeter, display the voltage on the multimeter, spin the wheel or the engine slowly by hand, and see the reading on the meter change from zero to whatever the square wave amplitude is (for example, 5 volts), then back to zero as the top of the square wave passes.

However, note that, in the case of signals from a VRT-based ABS or crankshaft positioning sensor, the signals are sine waves, but there is a physical principle that makes the signal's amplitude very low if the toothed wheel is turned slowly. In this case, it may not be possible to turn the wheel or the engine quickly and smoothly enough to enable the magnetic effect while also turning it slowly enough to see the reading on the meter change in a meaningful and definitive way.

Sine Wave Signal

If the device is a VRT-based ABS or crankshaft positioning sensor that outputs sinusoidal signals, and you can't manually rotate the wheel or engine and get it to output a slowly varying sine wave whose DC reading you can actually watch on a meter, then you can instead set the meter to measure AC voltage. If the meter also reports frequency (Hz), so much the better; you should be able to hit the frequency button and see the frequency increase as the VRT's wheel spins faster. You won't be able to see the actual time-changing reading, but you may not need to in order to verify functionality – seeing the AC voltage and frequency, and seeing both of them increase as rotation speed increases, should be sufficient to verify device functionality.

Square Wave Signal

If you need to verify the presence of a square wave and you don't have an automotive meter, you can try to judge if the device is functioning by measuring DC voltage. Your multimeter is actually taking a bunch of readings, averaging them together, and displaying the averaged number on the screen. So, for example, if the square wave has a 5V amplitude and a 50% duty cycle, you should be able to set the meter to measure voltage and, in theory, see 50% of 5V, or about 2.5 volts. If the device is a PWM device and the duty cycle increases with, for example, engine speed, you should be able to see the average reading increase. If, on the other, hand, the device is not a PWM device but a variable frequency device, the change will be much less clear-cut.

The above is exactly the situation an automotive meter is made for – for sampling a square wave signal and being able to tell you the boiled-down parameters of frequency, pulse width, and duty cycle. You should, however, know whether you are measuring a constant duty cycle at a variable frequency (like a Hall Effect wheel speed sensor or a crankshaft position sensor) or a PWM signal with a variable duty cycle (like a mass airflow sensor) to know if the reading makes sense.

Noid Lights for Injectors and Solenoids

If you are testing an injector or solenoid and need to verify that the car's ECM is outputting a signal, there is a device called a "Noid" light. Some sources state that "Noid" stands for Neon Organic Iodine Diode, but most Noid lights appear to be a simple incandescent bulb. It's more likely that "Noid" originated from "soleNOID," as that's what they are used to test. Noid lights typically come in a set with different Noid lights that have pins for different injector harness connectors. You simply unplug the connector from the injector and plug the Noid light into the injector harness. If, when you try and crank over the car, the Noid light flashes, then you know there's a signal present. If the light doesn't flash, there's no signal. If the light stays on, the signal isn't being toggled.

The appeal of the Noid light is that it is designed for very low current draw. This is critical because injectors and PWM solenoids are driven by the ECM or other control modules. If you plugged in a regular test light, it's possible the light might draw more current than the module can supply and damage the module.

Frequency, Alternate Tool Option

As we noted in the previous chapter, engine rpm was something of an outlier in Static Land – it was an example of a frequency-based parameter that the car synthesized and calculated for you. So, as it happens, if you need to measure frequency, and you don't have an automotive multimeter capable of doing it, and you have an old school tachometer that's part of an engine analyzer (which usually combines a tachometer and a dwell meter), there's a neat little hack you can do. Set the analyzer to measure engine rpm for whatever number of cylinders you want, clip the analyzer's probe to wherever your positive signal is, and use the formula:

Frequency = rpm * number of cylinders * 2 / 60

As a shortcut, set the analyzer to four cylinders. The frequency in Hz is then represented by rpm / 30.

Duty Cycle, Alternate Tool Option

There is one other outlier in Static Land. If you've had to adjust the points in an old-school ignition system, you know that there's this odd thing called "dwell" that is a measure of the amount of time the points "dwell" together. Dwell isn't actually a time measurement – it is oddly expressed as an angle. It turns out that, both conceptually and electronically, dwell is fairly closely related to duty cycle. So, if you need to measure duty cycle and you don't have an automotive multimeter capable of doing it, similarly to using a tachometer to measure frequency, you can actually use a dwell meter to measure duty cycle if you multiply the measurement by a scaling parameter. Since, on the four-cylinder scale, dwell is a number out of 90 degrees, and since duty cycle is out of 100%, not 90, we just need to scale the dwell reading by 1.1. So use the engine analyzer set to measure four-cylinder dwell, connect the probe to whatever your positive square wave signal is, multiply the dwell measurement by 1.1, and that's the duty cycle.

Ignition Pulses

If you're looking to display and examine the full waveform of the ignition, including the charging of the coil's primary winding, the discharge into the secondary winding, the firing of the plug, and the subsequent ring-down, the only thing that will work is an oscilloscope with a high enough sample rate to capture the features of interest.

Table of Tools for Common Sensors

The last eleven chapters of this book provide detailed information on troubleshooting several common sensors and actuators. We list them below, along with the signal they output and the kind of tool you need in order to look at the signal. To be clear, when we say "automotive meter," we mean a meter that is capable of measuring the frequency and pulse width of square waves in an automotive electrical environment. You need to verify that the meter you select can actually do this; the words "automotive meter" on the front of the meter do not in and of themselves constitute proof.

Tools for Common Sensors and Actuators

Device	Signal	Tool
Inputs (Sensors)		
Camshaft	Fast, sine or square wave	Automotive multimeter, scope
Crankshaft	Fast, sine or square wave	Automotive multimeter, scope
Mass Airflow	Slow non-periodic or square wave	Multimeter, automotive multimeter scope
Oxygen	1 to 5 Hz non-periodic	Multimeter, scope
Temperature	Slow non-periodic	Multimeter
Throttle	Slow non-periodic	Multimeter
Wheel Speed (ABS)	Fast, sine or square wave	Automotive multimeter, scope
Outputs (Actuators)		
Solenoid	Static or fast square wave	Multimeter, automotive multimeter, Noid light, scope
Fuel pump	Static or fast square wave	Multimeter, automotive multimeter, Noid light, scope
Fuel injector	Fast square wave	Automotive multimeter, Noid light, scope
Stick coil	Fast square wave	Automotive multimeter, scope

24

Modules, Buses, and Digital Data

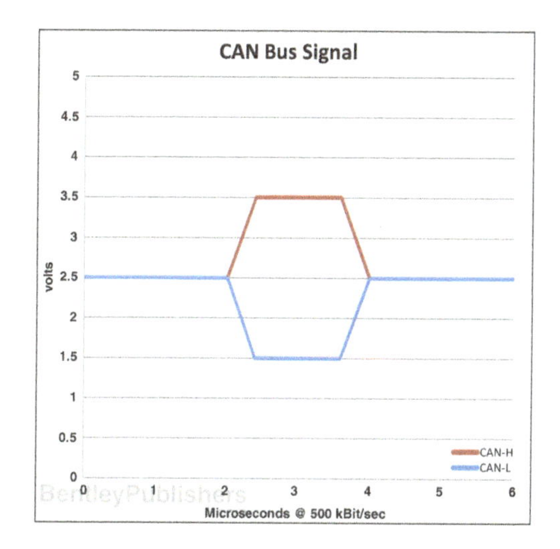

Computers in Cars

In the next three chapters, we're going to discuss the relationship between computers in cars, on-board diagnostics (OBD-II and OBD), and scan tools. In this chapter, we'll concentrate on computers (modules, buses, and the digital message data that are exchanged).

What's the Least I Need to Know?

- Most cars built since 1996 have way more than just one electronic control module.
- The proliferation of modules required a bus – a data network – over which the modules can communicate with the sensors and actuators on the vehicle and with each other.
- Bus communication takes the form of sending and receiving messages. The messages are digital data that can't be read with a multimeter or understood by looking at them on an oscilloscope. Reading them requires an external device specifically designed to decode the messages. That external device is known as a code reader or a scan tool.
- Digital message data are a third kind of electrical signal in the car, fundamentally different from the static signals described in the first half of this book, and the dynamic analog signals (square waves and sine waves) described in the **Dynamic Analog Signals** chapter.

Computers and Emission Controls

To understand computers in cars, it really helps to appreciate that emission controls and computers in cars evolved in parallel. By the late 1970s, in order to meet the increasingly stringent federal emission requirements, virtually every car had electronic fuel injection and an oxygen sensor monitoring the richness of the exhaust mixture. The injection was controlled by an Engine Control Module (ECM). Early ECMs contained no digital computer chips and were in fact analog circuits, collections of resistors and capacitors on printed circuit boards (though they did have a couple of transistors to create the basic injector pulse), but eventually they contained solid state chips, becoming what we think of as "computers."

Soon, the car's ECM began to be used not only for controlling the fuel mixture, but also to perform rudimentary On-Board Diagnostics (OBD). In 1996, a uniform set of standards known as OBD-II was enacted. We will discuss OBD-II in detail in the following chapter, but we make this brief introduction here because many people are familiar with plugging an OBD-II scanner or code reader into the car's OBD-II port. You should understand that this is just one of a number of examples of a computer communicating with the computers inside your car by exchanging messages over a bus.

And to understand those things, well, let's just hold hands and enter the funhouse.

Electronic Modules and Networks

The term "Electronic Control Module" – the thing which performed the rudiments of what we now call digital engine management – was coined back when your car had just one computer. Now, so-called computers have sprouted in cars like wildflowers. ABS, automatic transmission, climate control, vehicle lighting, supplemental restraint, telematics, infotainment, and active suspension are all controlled by their own dedicated processing unit. All with microprocessors inside. All sending and receiving messages. All communicating with other sensors in the car, and many communicating with each other.

First, what do you call them? You can't call them all ECMs because that term still connotes the computer that's performing the digital engine management. They are sometimes called *controllers* (and we'll get back to that very shortly), but the more popular term is *module*.

So the mid-1990s really heralded The Age of the Module. Prior to that, when you hit the brakes in a car, the brake pedal switch would send 12 volts to the brake lights to turn them on. But in The Age of the Module, the brake pedal information probably is sent to a brake control unit, which is receiving wheel speed, vehicle speed, brake pedal pressure and acceleration, and other information, and sends a "stop" command which is received by some sort of a lighting control module, which in turn sends a command to the brake lights to turn on.

When you have computers connected and operating in this way in your house or business, there's a word for that – a network. And when you have modules connected in your car, that's exactly what it is.

There are tremendous advantages to networking modules in a vehicle. First, from a processor load and software complexity standpoint, if you're going to introduce this level of computer control, it make sense to distribute the load.

Second, the information supplied by a sensor is often needed by different modules. Wheel speed, for example, may be used by the ECM, ABS, and transmission control modules. If you think about it, how would you connect a sensor's output to three modules? You'd have three sets of wires going from the sensor to each of the three modules. Imagine extrapolating this situation to include all the sensors and modules in a car, and you begin to see the problem. A networked architecture makes it so there can be a single set of wires linking sensors and modules.

Third, by establishing a network and sending the data as digital messages, the current required is low, so the wires connecting the sensors and modules together can be thin and lightweight.

Network architecture is complex, but the actual wiring itself is simpler – not simpler than a premodule car, but simpler than it would be if the signals weren't presented as digital message data on a network.

Digital Data

We just said "the signals are presented as digital message data on a network." What does the term *digital message data* even mean?

There are really three kinds of electrical signals in a car. We outlined the first two in the **Dynamic Analog Signals** chapter, but we'll summarize them below. The thing to appreciate is that signals containing digital message data comprise a third kind of electrical signal.

1) Static analog signals are those that don't change during the time frame you are taking a measurement. Most electrical troubleshooting on cars involves static analog signals. You simply connect a multimeter and read it to, for example, verify whether or not a device is receiving 12 volts. There's not a question of whether you're catching it at a time when the signal is varying between 12 volts and some other value because it's not varying.

2) Dynamic analog signals are those that are continuously available to be measured but change during the time of the measurement. As we described in the previous two chapters, you need to know what a signal is supposed to be – how it varies – in order to know what tool to use to measure it (multimeter, automotive multimeter, scope) and make sense of it. Some time-varying analog signals like those from oxygen sensors vary non-periodically, but others are periodic – usually sine waves or square waves – in which case you usually need a tool that measures frequency, pulse width, duty cycle, or some other salient parameter. A module on the car is usually receiving that signal, extracting that parameter (for example, frequency), and using it.

3) Signals containing digital message data are fundamentally different from the previous two types of signals. With digital message data, there is no "measurement" because the signal is not continuously available. Instead, there are discrete packets of information. For example, in the "turn the brake lights on" example we presented earlier, the message is not continuously present. It's a discrete event. The signals in digital message data are voltage levels that vary rapidly to transmit the bits and bytes that make up the message.

You can plot the changing voltage levels in a signal containing digital message data on a high-speed oscilloscope, and doing so will tell you if data are being transmitted over the bus, but you can't read a message on a scope. These aren't sine waves or square waves you can look at and visually extract a parameter like frequency or pulse width. Whatever is receiving the messages needs hardware and software to assemble the packets in order to decode and read the messages.

So what tool do you use to read the code in a digital message? This requires a scan tool or code reader to decode and read the messages and feed it back to you in the form of a trouble code or readable text (as we'll cover in the **Scan Tools** chapter).

This information is summarized in the **Types of Electrical Signals in a Car Table.**

Types of Electrical Signals in a Car			
Signal	**Era**	**Data**	**Tool**
Static Analog	All	Fixed numerical reading	Multimeter
Dynamic Analog	Approx. 1975 to present	Waveform	Automotive meter or scope
Digital	Approx. 1996 to present	Message	Scan tool to read and decode messages

Hopefully you now understand why you're never going to be able to connect a multimeter to an OBD-II port to read out a trouble code. That would be like trying to use a record turntable needle to read a DVD.

Buses and Protocols

There are two other computer terms that get thrown around in the automotive world – bus and protocol.

For our purposes, the *bus* is the data communications system, both hardware and software, responsible for transferring data to and from the multiple modules (also called controllers) on the network.

For computers to communicate with one another, standard methods of information transfer and processing are used. These are referred to as *protocols.* What are the rules for transmitting bits and bytes around on the bus? How do you go from information to signals on wires and back to information again? How are the data packed into messages? How are those messages sent? How do different modules know if a message is intended for them? How fast are data transmitted? How are errors in transmission handled?

All of those questions are addressed and answered in the specification of the protocol. Like *bus*, *protocol* includes both hardware and software. Whether the data are carried by optical cables or by wires, how many wires, and what the voltage levels are, are all part of the protocol.

In the automotive world, CAN (Controller Area Network), LIN (Local Interconnect Network), MOST (Media Oriented Systems Transport), FlexRay, and Ethernet are the accepted standards in bus and protocol design.

Let's get back to that other word for module – controller. We said that, when the modules/controllers are interconnected and communicating, they form a network. In fact, a specific implementation of this is called a CAN (Controller Area Network). And the whole thing, the controllers and the data pathways connecting them, is called the *CAN Bus*. Many people have heard of the CAN bus, and understanding the meaning is often a eureka moment.

Now, the CAN bus is only one of a number of different vehicle buses, but in the world of post-OBD-II European cars, it is probably the most commonly used. And its very name carries its context: The need to distribute the processing load created the need for modules (controllers), which created the need for the controllers to be connected, which created the CAN bus.

Physical Layer

When Bosch created the CAN bus, it did not, in fact, specify what's known as the physical layer, meaning Bosch was not specific as to whether the bus had to be implemented as a single wire, two-wire, twisted pair, shielded twisted pair, optical cable, or other wiring. In fact, the three classes of CAN buses, shown in the **CAN Bus Classes Table**, do not all operate on the same physical layers. When we talk about a CAN bus on a European car, we're most often talking about CAN C, implemented over a two-wire twisted pair.

CAN Bus Classes		
Class	**Description**	**Speed**
CAN A	Single wire (uses vehicle ground)	33.3 Kbs to 83.33 Kbs
CAN B	Two-wire twisted pair	95.2 Kbs
CAN C	Two-wire twisted pair	Up to 500 Kbs

CAN Bus Signals

CAN bus signals over twisted pair wiring use differential signals labeled CAN-H and CAN-L. The signal voltages go in opposite directions so they can be constantly compared against each other and error-checked. The nominal signal level is 2.5 volts. During communication, CAN-H goes high to 3.5 volts, and CAN-L goes low to 1.5 volts. See **Figure 1**.

Thus, the total voltage in the twisted pair remains constant, and, with the noise cancellation quality of the wiring's twisted pair configuration, noise is rejected.

Keep in mind that this is high-speed digital communication of single bits. In order to actually see a graph like this, you'd need a high-speed digital oscilloscope. You're certainly not going to catch a voltage transition like this on a multimeter. However, if the CAN-H or CAN-L voltage levels are stuck at zero or pegged at 5V, you *should* be able to see that on a meter, indicating that something is wrong.

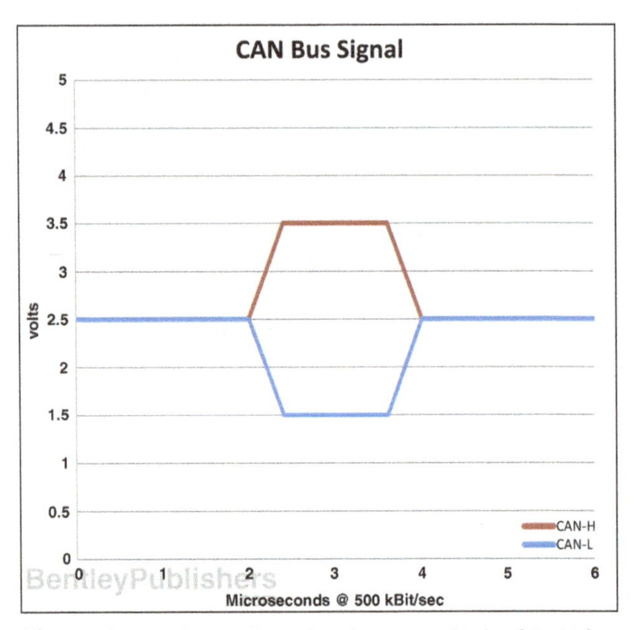

Figure 1 CAN bus voltage levels go to a high of 3.5V for CAN-H, and a low of 1.5V for CAN-L.

Network Topology

You'll sometimes read references to the *network topology* that goes along with an automotive bus. This refers to how the modules are connected.

There are three basic network topologies: ring, star, and bus. See **Figure 2**.

In a ring topology, all modules are connected like beads on a bracelet. If a wire breaks or a module goes down, communication is still possible, but expansion of the network is difficult.

In a star topology, the module in the center of the star is a "master," relaying commands to the "slave" modules at the points of the star. If the master module goes down, communication ceases.

A bus topology (CAN bus), requires each module to be a master capable of initiating communication, helping to ensure bus reliability while providing flexibility in adding additional modules.

Figure 3 is an example of various network topologies for a 2002 BMW 7 Series.

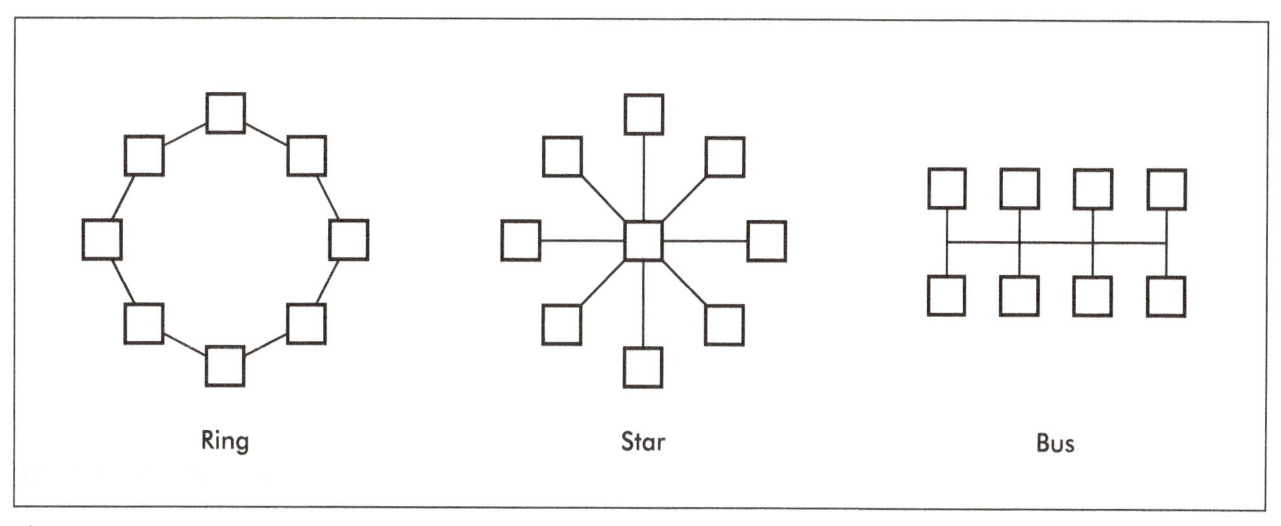

Figure 2 Examples of ring, star, and bus network topologies.

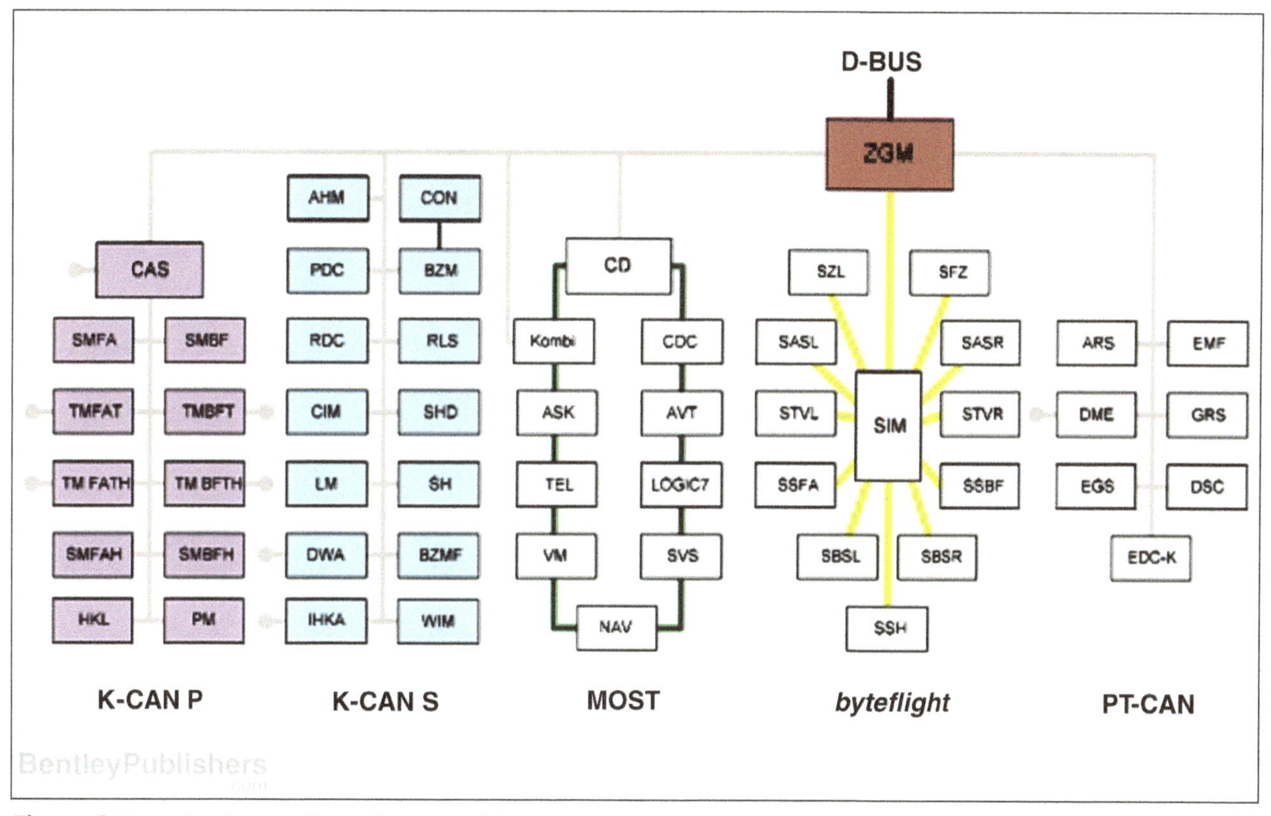

Figure 3 Example of network topology on a 2002 BMW 7 Series (with notorious BMW acronyms in all their glory).

At this point you may be asking, why do I care about networks, buses and protocols used in my car? You probably don't, any more than you care about how your home computer connects to the Internet (unless your OBD-II code reader spits a Diagnostic Trouble Code [DTC] of "DE87 CAN Bus failure / CAN bus communication error".

While the scanner or code reader insulates you from digital bits and bytes, a basic understanding of network topology can't be a bad thing if you have bus-related faults. It's helpful to know just enough to be able to understand why your OBD-II reader runs really slowly but your friend Bob's is much faster; maybe his is running over the CAN bus and yours isn't.

Let's now move beyond the ivory tower of bits and bytes and onto the nuts and bolts of OBD-II and the Check Engine light.

25

OBD-II

Introduction

OBD-II is a rich topic with multiple books and websites dedicated to the subject. However, most people interact with OBD-II the same way they interact with their doctor – only when there's something wrong. Often, the "something's wrong" event is that the Check Engine Light is on and the car won't pass the annual state inspection until it's turned off.

After reading this chapter, you should have a good grasp of the OBD-II philosophy, why the Check Engine Light comes on, what you need to do to fix it so the light goes off and stays off, and why, even after that, you may need to drive the car quite a way before it'll pass inspection.

What's the Least I Need to Know?

- OBD-II is used on 1996 and later U.S.-spec cars and is for emissions-related faults only.
- OBD-II uses a set of self-tests that are always monitoring the health of the emissions-related components of the car. If a problem is detected, the Check Engine Light (CEL) is illuminated.
- There is an OBD-II connector on the car that allows you to connect an OBD-II code reader or scan tool to obtain more information about why the CEL is lit and whether your car is ready to pass inspection.
- Using an OBD-II code reader or scan tool, you can clear the fault codes and make the CEL go out, but unless you fix the underlying problem that threw the codes in the first place, the light will come back on in short order.
- There are three key OBD-II concepts: **Readiness**, **Monitors**, and **Drive Cycle**. When all is good, the OBD-II system is set to "ready." To be ready, all the self-tests (monitors) must run successfully. To successfully run the monitors, a specific driving trip (drive cycle) is required. If you don't have readiness, you will not pass inspection. We'll discuss each of these key concepts in detail in this chapter.

What Is OBD-II?

The following brief description is taken directly from the EPA web site: "The OBD-II system monitors virtually every component that can affect the emission performance of the vehicle to ensure that the vehicle remains as clean as possible over its entire life, and assists repair technicians in diagnosing and fixing problems with the computerized engine controls. If a problem is detected, the OBD-II system illuminates a warning lamp on the vehicle instrument panel to alert the driver. This warning lamp typically contains the phrase Check Engine or Service Engine Soon. The system will also store important information about the detected malfunction so that a repair technician can accurately find and fix the problem."

That's fairly clear, but if you ask different people what OBD-II is, there are perspectives other than just that of the EPA:

- To the vehicle owner, OBD-II is that pesky Check Engine Light. When it comes on, you often need to pay someone money to make it go out.
- To the technician (who may also be the vehicle owner), OBD-II is a physical interface – a plug – to which you can connect a code reader to find out which Diagnostic Trouble Codes (DTCs) are responsible for setting the CEL. The codes can aid in the diagnosis of what caused the code that lit the light and what you need to fix in order to make the light go out.
- To the vehicle manufacturer, OBD-II is diagnostic software and hardware that is an integral part of the vehicle build spec, from the fuel tank gas cap to variable valve actuators and much more. It overlaps with, but is not the same thing as, other manufacturer-specific onboard diagnostics.
- To the feds and the State of California, OBD-II is a self-diagnostic system to ensure that 1996 and later vehicles operate cleanly and efficiently throughout the entire service lifetime of the vehicle, and to help any technician diagnose and repair the emission system without requiring access to proprietary dealer-only tools.
- To your state, OBD-II is part of an annual vehicle inspection program. In many states, if the CEL is illuminated, the car will not pass inspection.
- To the engineer, OBD-II is a set of sensors, computer hardware and software, and messages that comprise an emissions monitoring system that has self-diagnostic, real time, and data logging capabilities.

So, which is it?

All of them. OBD-II is all of these things. How you see it depends largely on how much work you perform on your car yourself.

OBD-II encompasses the following pieces. Some are physical hardware, some are software, but they are all part of the OBD-II picture:

- Hardware to reduce emissions, and sensors to confirm that the hardware is functioning correctly
- The Check Engine Light
- The set of OBD-II self-tests (monitors) that run whenever the car is on
- The OBD-II Data Link Connector (DLC)
- The Diagnostic Trouble Codes (DTCs)
- The underlying bus and protocols
- The OBD-II code reader or scan tool
- How you use the codes to find and fix the problem

We will discuss all of these in detail.

OBD-II Cars Have Additional Hardware

In order for the OBD-II system to ensure emissions compliance, additional hardware is required, such as:

- Additional oxygen sensor on the downstream side of the catalytic converter. This is for monitoring catalytic converter efficiency.
- Electronically Erasable Programmable Read-Only Memory (EEPROM) chip that allows the ECM to be reflashed with new software.
- A modified evaporative system with self diagnostics to make sure there are no leaks in the fuel tank, fuel lines, or tank vents.
- EGR (Exhaust Gas Recirculation) systems with electric EGR valve and temperature sensor.
- Secondary air system that adds oxygen (air) to the exhaust gases to aid in oxidizing unburned hydrocarbons and carbon monoxide created by running rich at cold startup.
- Additional sensors for monitoring engine load and airflow.

Check Engine Light (CEL)

Although the SAE term for the OBD-II dashboard light is Malfunction Indicator Light (MIL), most people call it the Check Engine Light (CEL).

It is the light everyone loves to hate. When the CEL is illuminated, it means that some emission-related component is failing a self-diagnostic on-board test, and as a consequence, the car is in danger of possibly producing emissions that are 50% above OBD-II acceptable levels. Note the words "in danger of possibly." When the light goes on, it doesn't necessarily mean your car is killing frogs in the rain forest and laying waste to the planet, and it certainly doesn't mean that your car is in imminent danger of suffering violent meltdown.

Figure 1 The Check Engine Light indicates that your car may be polluting. When this light is on, it triggers and stores an OBD-II fault code. With a simple code reader, you can find out why the light is on.

The question of the degree of urgency with which you should regard the CEL ("Should I stop driving right now and call a tow truck?") causes much confusion. The causes are often minor in terms of both actual vehicle emissions and danger of damage to the car. The exception is very rich running conditions, which can harm the catalytic converter.

If there are no driveability issues, it is likely okay to drive the car until you have the codes read. But if the CEL remains lit, you need to find out why, which means getting an OBD-II code reader or scan tool, plugging it into the OBD-II socket, and reading the codes.

In SAE/EPA terms, the CEL illuminates under the following conditions:

- Bulb check (ignition key on before cranking starter).
- A fault has been detected in two consecutive drive cycles, indicating that a component malfunction might cause emissions to exceed 1.5 times OBD-II standards.
- A continuous misfire fault that might damage the catalytic converter has been detected. This is indicated by a blinking CEL.

Once illuminated, the CEL will remain on until the car completes three consecutive drive cycles without seeing the fault, at which time it will go out. The fault code, however, is still stored for 40 consecutive drive cycles. If the fault has not recurred at that point, it is cleared from memory. For a catalyst-damaging fault, this is increased to 80 drive cycles. (We'll discuss drive cycle in more detail later in this chapter.)

Tales From the Hack Mechanic: The Check Engine Light

OBD-II specifications do not standardize the shape of the CEL or the accompanying wording (see **Figure 1**). This has lead to much confusion. I will admit to having recently confused the CEL and the low coolant light in my 2001 E46 BMW 3 Series wagon. The two icons are very similar.

OBD-II Readiness

Monitors are self-tests (there are up to eleven of them) that are performed on a certain component or subsystem of the vehicle during each trip or drive cycle.

Readiness means that all the OBD-II monitors on your car have correctly run. Some monitors run quickly, and some will never run until you take a 55-mph drive on the freeway. It is also important to know that a few conditions must be satisfied before a monitor can run.

If your Check Engine Light is on, it is telling you that your car is not ready for its annual state inspection. In other words, one or more OBD-II readiness monitors are, well, not ready. You will need a code reader or a scan tool to know which monitors are not ready and to see the related fault codes. See **Figure 2**. Note that code readers use the term "I/M" to refer to inspection/maintenance readiness.

As an example, let's say you plug an OBD-II code reader into the OBD-II connector and you see that the Secondary Air Monitor is reporting INC (incomplete). You then check for codes and find you have a "P0491 Secondary Air Injection System Insufficient Flow Bank 1" fault code. Now you know which monitor has caused the fault, and where to start troubleshooting to get the car back to OBD-II readiness.

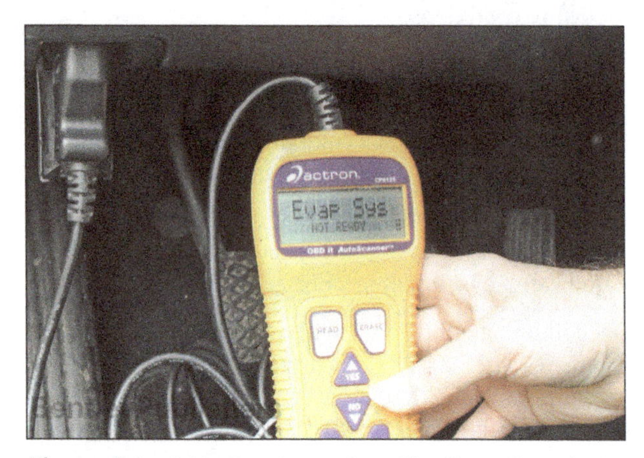

Figure 2 An OBD-II code reader will tell you if you have I/M readiness. Here the EVAP monitor is reporting not ready.

NOTE —

U.S. EPA guidelines allow for the car to pass inspection with up to two monitors to be in a "not ready" state for model year 1996 through 2000 vehicles and one monitor "not ready" for 2001 and newer model year vehicles.

Let's take a look at the OBD-II monitors in more detail. There are two different types of monitors: continuous and non-continuous.

Continuous Monitors

As the name implies, continuous monitors run all the time. Three continuous monitors are included in the OBD-II standard.

- **Misfire Monitor** – Misfire is defined as lack of combustion in a cylinder due to insufficient spark, poor fuel metering, poor compression, or other causes. Using the crankshaft sensor, this monitor looks for crankshaft acceleration changes to identify cylinder-specific misfires, which can generate unburned hydrocarbons and increase the load on the catalytic converter. A severe cat-damaging misfire condition will cause the Check Engine Light to flash on and off. The car should not be driven in this condition.
- **Fuel System Monitor** – This monitor continuously looks at fuel mixture, generically labeled short term and long term fuel trim. It uses oxygen sensor input and mapped fuel trim tables in the ECM for plausibility checks. If the specified range is exceeded (too rich or too lean), a fault will be set.
- **Comprehensive Component Monitor** – This covers engine management components and circuits and tests them in various ways depending on the hardware, software, and function. For example, analog inputs such as engine coolant temperature and throttle position are checked for open circuits, short circuits, and non-plausible values.

Non-Continuous Monitors

Non-continuous monitors run once during every trip or drive cycle. Each requires that certain conditions be met before a monitor test will execute. For example, the engine may need to be at operating temperature, with a road speed between above 55 and 65 mph, steady throttle and low load, for about five minutes before the Oxygen Sensor Monitor can run. This scenario is for illustrative purposes as a way of understanding the concept of enabling criteria. Each monitor has its own specific enabling criteria and conditions vary greatly between manufacturers.

Once enabling conditions are met AND the specified drive cycle condition is satisfied, the monitor will attempt to run. If it runs, that monitor is set to "ready." If the monitor aborts during the test, a pending code will be stored. If the fault is present during the next trip or drive cycle, the Check Engine Light will be lit.

Catalyst Monitor – The single most important piece of emissions-related hardware on your car is the catalytic converter. OBD-II puts computer eyeballs on the catalytic converter using the pre- and post-cat oxygen sensors for feedback. The presensor is used chiefly for fuel trim. The postsensor looks at what is coming out of the cat, and should have a steady flat response rate if the cat is working well. When the postcat sensor signal mirrors that of the precat signal, it is a sign that the efficiency of the catalytic converter has degraded over time, indicating that the cat may need to be replaced.

NOTE —

Code readers and scan tools will ID the precat sensor as sensor1bank 1 and the postcat as sensor1bank2. If your engine has two cats, the other sensors will be sensor2bank1 and sensor2bank2, respectively.

- **Heated Catalyst Monitor** – This one can be mostly ignored, unless you car has a high-performance V-12 engine (or something more exotic).
- **Evaporative (EVAP) System Monitor** – Many types of passive EVAP systems have been used over the

years. Consult your repair manual for details on the system on your car. Some monitors run when the engine is running and some run after the engine has been sitting for hours with the key out of the ignition. Some early cars used a fuel tank pressure sensor combined with engine vacuum to check the integrity of the fuel tank and lines. Then leak detection pump systems became popular. These systems pressurize the tanks and lines to check the decay rate (how long the pressure holds). Some of the latest systems have returned to engine vacuum (called Natural Vacuum Leak Detection or NVLD).

- **Secondary Air System Monitor** – Secondary air injection is a method of reducing emissions during the first few minutes of cold engine operation. The engine needs a rich mixture for cold running. This creates excess fuel, resulting in high quantities of carbon monoxide and unburned hydrocarbons. Since the oxygen sensors are off line during a cold start and the cat isn't operational yet, these emissions can escape out the tailpipe if not re-treated. Using an air pump, ambient air is injected into the exhaust manifold. This causes post-oxidation of the unburned hydrocarbons, creating harmless carbon dioxide and water. The heat generated as a result of this process also heats the catalytic converter to get it on line quickly. Most secondary air injection systems are comprised of an electric pump and a secondary air valve at the exhaust manifold. Consult your repair manual for specific system details. This monitor runs the pump and looks for a lean O2 sensor signal. If the sensor signal doesn't change, a fault is set.
- **Oxygen (O2) Sensor Monitor** – This monitor looks at the O2's sensor signal range, response, switching time, and sensor circuit integrity. Some common causes of O2 sensor fault are slow and sluggish response time owing to age, open or short circuited, or if the signal goes out of range (rich/lean limit exceeded). Note that a limit code does not necessarily mean the sensor is bad; it usually just means it can't deal with the whacked-out fuel mixture.
- **Oxygen Sensor Heater Monitor** – This monitor looks at the O2 heater current and how long it takes the sensor to warm up and start sending a good signal. A heater fault code can be triggered when the sensor takes too long to warm up. If you have a heater element circuit fault code, check the heater circuit fuse.

There are a few additional monitors for diesel cars only that we won't discuss in detail. They are as follows:

- EGR (Exhaust Gas Recirculation) and/or Variable Valve Timing (VVT) System monitor
- Non-Methane Hydrocarbon (NMHC) Catalyst monitor
- Nitric Oxide/Selective Catalytic Reduction (NOx/SCR) After Treatment monitor
- Boost Pressure monitor

When Monitors Won't Run

If after multiple drive cycles, your code reader indicates that you still have incomplete monitors, do some research on the enabling criteria required to allow the monitor to run. This information can be difficult to unearth, but enthusiasts' websites are a good place to start, and don't eliminate the domestic brands, since the OBD-II standard applies to all cars sold in the U.S. from the 1996 model year.

For example, depending on make and model year, the EVAP monitor may require a shutdown for up to eight hours before it will test. The catalyst monitor is also historically difficult to complete; the car will need the oxygen sensors to check in as "ready" before the catalyst monitor will run.

If the O2 sensor heater monitor is not ready, check to see how many other monitors are not ready. If the oxygen sensor and catalyst monitors are also not ready, the culprit is most likely a weak battery. If your battery is more than four years old, replace it and re-run the drive cycle. Even though your car may start just fine, the ECM is hypersensitive to battery state-of-health.

If only the oxygen sensor heater monitor is not ready, but all others are, the heater monitor should eventually set. Try loading the alternator (turn on A/C, rear window defroster, and headlights) with the transmission in D, parking brake set and the engine idling. When a heater circuit is old, it can be one of the last monitors to set.

If the EVAP monitor is not ready and you've got a large EVAP leak fault code, verify that your fuel cap is tight and the level of fuel is between 1/4 and 3/4 full when you try another drive cycle. A common EVAP issue is rotted or cracked vent hoses. Give the system a thorough visual inspection.

If the EVAP monitor just will not pass, have the system "smoked." The best tool for finding EVAP tough leaks is a smoke or fog machine. Smoke machines typically use compressed air to push a heated mineral oil or glycerin and water mixture into the sealed EVAP system. These tools were once exclusively the providence of repair shops, but DIY-quality smokers are now available for around $100. When you see the fog pouring out of a split in a hose, you'll have found the leak.

If you still can't get all the monitors to run and you do not have any fault codes, consider taking your vehicle to a shop that truly understands Mode 6 Diagnosis. Mode 6 is an advanced diagnostic mode within OBD-II that reports on how the sensors and other emission control components are functioning. It can only be accessed with a professional grade scan tool. The real beauty of Mode 6 is that it can verify a fix without having to wait days for OBD-II monitors to run.

NOTE —

Not all OBD-II vehicles have Mode 6, as it did not become commonplace until the 2000s.

Tales From the Hack Mechanic: Stubborn O2 Sensor Monitor

We had a BMW E60 5 Series that would not run the O2 monitor. Drive cycle after drive cycle, no pending codes, no Check Engine Light, no driveability issues, but NO READINESS.

We first tried erasing the engine management adaptations and did a battery reset (leave battery disconnected overnight, touch positive and negative terminals together away from battery, then reconnect the battery). Then we flashed (updated) the ECM software. We drove the car for a week, taking dozens of trips, and still no readiness. Since the engine had lots of miles with its original O2 sensors, we thought maybe the O2 sensors were aging, maybe not enough to throw a code, but enough to prevent readiness. Four new sensors later, this turned out to be incorrect thinking.

What else could it be? What are the manufacturer's enabling criteria for running the O2 sensor monitor? We knew coolant temp was in the fold, but we had no thermostat or operating temp plausibility codes. So we monitored coolant temp with a scan tool on the commute home. Sure enough, the temp dropped from 90°C (194°F) to below 60°C (140°F) as road speed exceeded 50 mph. A-HA! We needed at least 160°F in order to test to see if the sensor monitor executed. Next day, new thermostat. Day after, READINESS!

The moral of the story is that it pays to learn everything you can about what it takes to run a specific monitor, be it the drive cycle part or in the long list of enabling criteria. Check the boxes one by one.

Drive Cycle

Clearing diagnostic trouble codes and resetting the Check Engine Light using a scan tool or code reader will reset the I/M (Inspection/Maintenance) status to "not ready." Readiness status is also reset when the battery gets disconnected.

So how do you get the monitors set back to "Ready"?

Two words: Drive Cycle. A drive cycle is a standardized set of vehicle driving conditions during which the ECM is collecting enough data over a wide enough range of conditions for all of the monitors to be able to report "ready." You'll need to drive the car – a lot in some cases – from fully cold to fully warm, at low speed and high speed, braking, accelerating, and coasting down. You may even have to let the car sit for eight hours before it is ready to run the drive cycle routine again.

The concept of a drive cycle was originally introduced for two purposes. The first was to provide a uniform emissions testing standard for vehicle manufacturers. In other words, it isn't sufficient to say that emissions must not exceed certain levels. Specifying the conditions under which those levels are to be measured (idle? running? hot? cold? fast? slow?) is also necessary.

The second purpose was to provide standards for the length of time and the driving conditions that each of the OBD-II monitors need to run, before the monitors have collected enough data to allow them to report that they are ready. This creates something of a tamper-resistant function in OBD-II. That is, it prevents an owner or a technician from simply clearing a code to extinguish the CEL and immediately taking the vehicle in for inspection without repairing the underlying cause. If you do this and have not completed a drive cycle, the inspection computer will report "not ready."

A very specific industry-standard drive cycle, designed to simulate an urban drive around downtown LA (ground zero for smog), was originally part of the EPA's Federal Test Procedure (FTP) for vehicle emissions. It was performed on a dynometer (to the uninitiated, a kind of vehicle treadmill), and took 1,874 seconds (31.2 minutes). This test was eventually shortened to a 240-second dynomometer test referred to as IM240 (Inspection and Maintenance). EPA's and CARB's intent was for IM240 to be part of a federally mandated annual state inspection procedure, but due to public backlash and the high cost of dyno-based inspection testing equipment for garages, very few states ever adopted IM240 as an inspection standard.

Nonetheless, a manufacturer's drive cycle is loosely based on IM240. It typically combines a cold engine start, idling, acceleration, constant speed, braking, and engine turn-off. **Figure 3** gives one example of a manufacturer's drive cycle recommendation.

To professional technicians, the laborious pattern of "The Dreaded Drive Cycle" can be especially maddening. That is, a customer brings a car to the repair shop. The technician scans for codes (DTCs) and, based on the reported DTCs, he or she replaces a part, or two, and resets the Check Engine Light.

Then the fun begins. In order for the tech to verify that the new parts fixed the problem, a road test is required to run the monitors. Not so easy in some cases. The drive cycle can take a long as 40 minutes, and even then, there's no guarantee all the monitors will complete and report "ready."

Say it's rush hour in LA. You might have a hard time even finding a 10-mile stretch of open freeway to get the oxygen sensor and cat monitors to execute. Or if the car didn't cool down for long enough, another monitor may not run. The secondary air monitor, for example, needs a cold start.

To the owner and do-it-yourselfer, if you simply drive your car, you will eventually complete a drive cycle. But if you're trying to get the monitors to all report "ready" so you can take the car in to get it inspected, it is helpful to have some idea as to what the drive cycle parameters are.

You can find recipes online for drive cycles. They do vary by make, year, and model, but a generic drive cycle goes something like this:

- Fuel tank is between 1/4 and 3/4 full (if it's too full or too empty, the evaporative monitor won't run).
- Cold-start the vehicle (below 86 degrees F). Turn on the A/C and the rear window defroster to load the alternator. Let the engine warm up until coolant is at least 160 degrees F (typically done by idling for one to three minutes). Secondary air and oxygen sensor heater monitors should complete.
- Drive and accelerate to 40–55 mph at 1/4 throttle.
- Maintain speed for five minutes.
- Drive at speed of 55– 60 mph for five minutes, steady throttle. Oxygen sensors and catalyst monitors should complete.
- Decelerate without using brake (meaning coast down) to 20 mph or less.
- Stop the vehicle.
- Let the engine idle for ten seconds, then turn it off.
- Wait one minute before testing to see if all monitors report ready.

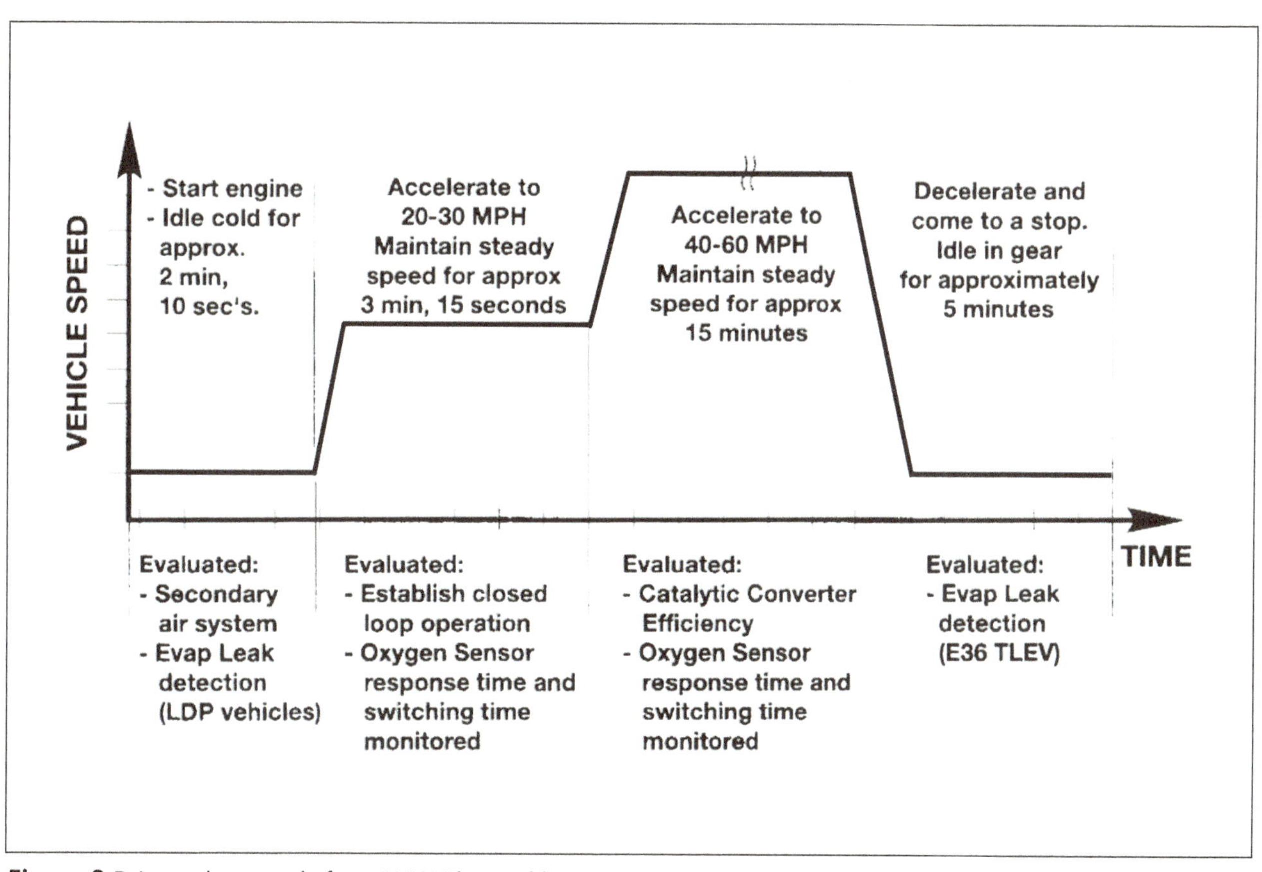

Figure 3 Drive cycle example from BMW. The enabling criteria need to be satisfied first before the test will run. If you have a stubborn monitor, do some research on the what conditions must be met before the self-test will run.

Tales From the Hack Mechanic: Drive Cycles

Many treatises on OBD-II don't make it clear that *OBD-II code readers do not have an option for "begin drive cycle."* It is not something that you have control over.

However, as we showed above, a decent OBD-II scanner will tell you the state of the monitors, and you can look at them one by one and see if they report "ready."

Note, though, that even with this, I've driven cars for an hour and had the EVAP monitor report "not ready." I've sometimes shrugged, taken the car in for inspection anyway, and had it pass, because as we said earlier, certain states allow one or two monitors to report not ready and still pass the vehicle.

OBD-II Connector

The OBD-II connector is the place you plug a code reader or scan tool in to get OBD-II I/M readiness and monitor status, and read DTCs, freeze frame data, and other emission-related live data depending on the functionality of your tool. See **Figure 4**.

The 16-pin OBD-II connector is more formally known as the Data Link Connector (DLC). It is always located within 16 inches of the steering wheel.

Depending on the make and year of the car, there may also be an additional proprietary diagnostic connector for a dealer-level scan tool. See **Figure 4**. By around the year 2000, most proprietary connectors were a thing of the past.

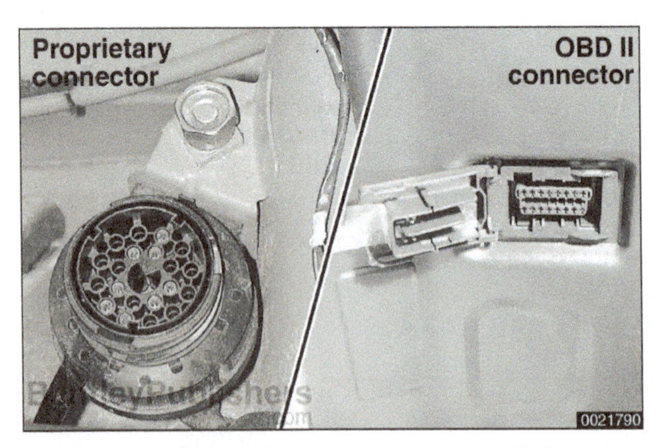

Figure 4 Proprietary BMW 20-pin connector (left) and standard OBD-II connector (right). *Photo by Xoneca courtesy of Creative Commons.*

Pinouts of the OBD-II connector are shown in **Figure 5**. Unless you're a hardware engineer looking to hack into the connector directly, the pin assignments and the signals on the pins aren't something you need to worry about, but it is useful to know that pin 16 is battery voltage, and pins 4 and 5 are ground. This is so a code reader can be powered directly from the connector.

If you plug in a code reader or scan tool and it doesn't power up, check that pin 16 has 12V, and that pins 4 and 5 are grounded. If there's no power, check the fuse powering the connector.

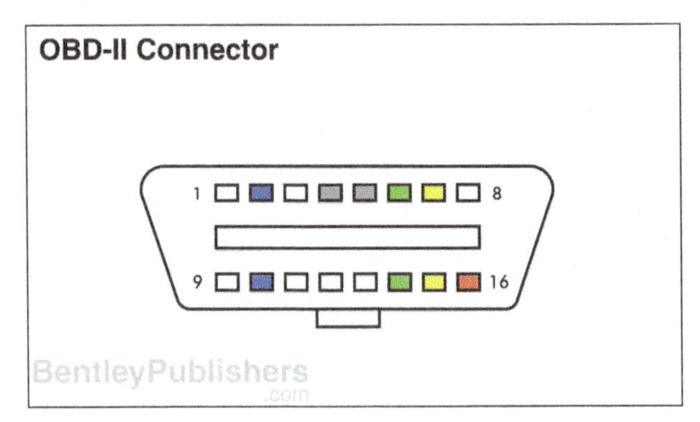

Figure 5 OBD-II connector pinout (color-coding for different bus protocols).

OBD-II Connector Pinout Chart

Pin	Description
1	Manufacturer discretion (VW/Audi: Switched +12 to tell a scan tool if ignition is on)
2	SAE J1850 Bus+ (PWM and VPW)
3	Manufacturer discretion
4	Chassis ground
5	Signal ground
6	CAN High (ISO 15765-4 and SAE J2284)
7	K-Line of ISO 9141-2 and ISO 14230-4
8	Manufacturer discretion (BMW: Second K-Line for non OBD-II (Body/Chassis/Infotainment) systems)
9	Manufacturer discretion
10	SAE J1850 Bus- PWM only
11-13	Manufacturer discretion
14	CAN Low (ISO 15765-4 and SAE J2284)
15	L-Line of ISO 9141-2 and ISO 14230-4
16	Battery voltage

OBD-II Connector Protocols

When the OBD-II standard was rolled out in 1996, it originally supported four different protocols. These were:

- SAE J1850 with pulse width modulation
- SAE J1850 with variable pulse width
- ISO 9142

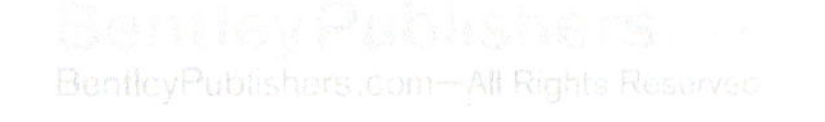

- ISO 14230

Conspicuously absent in that list is the CAN protocol (officially called ISO 15765), which, at the time, was too new. CAN is, by a substantial margin, the fastest of these protocols, and became an accepted OBD-II protocol in 2003. Moreover, in 2008, CAN became a *required* OBD-II protocol for all vehicles sold in the United States.

If you have a European car, it's highly likely that the OBD-II connector implements CAN, so if you are looking at buying a code reader or scan tool, you'd be wise to select one that supports CAN. If you look at the **OBD-II Protocols and Related Connector Pins Table**, CAN uses pins 6 and 14, so if these pins are populated in your OBD-II connector, you've got CAN.

OBD-II Protocols and Related Connector Pins				
Pin	J1850 PWM	J1850 VPW	ISO9141 /14230	ISO15765 (CAN)
2	Used	Used		
6				Used
7			Used	
10	Used			
14				Used
15			Optional	
16	Used	Used	Used	Used

Diagnostic Trouble Codes (DTCs)

OBD-II Diagnostic Trouble Codes (DTCs, which most people just call "codes") have a structure consisting of five alphanumeric characters. The first character is a letter. Most OBD-II codes are P-codes (powertrain). For example, P0300–Random or Multiple Misfire.

The second character is a number that tells you whether the code is a generic code or a manufacturer-specific code.

- 0 or 2 = Generic OBD-II
- 1 or 3 = OEM manufacturer defined

Some scan tools and code readers will not define the OEM or manufacturer codes, but an Internet search usually will. For example, P1300 is also a misfire-related code.

The third character tells you which system the code is for. This is not terribly important, as a) the code reader is probably going to give you an English language version of the code (e.g., "large EVAP system leak"), and b) you're likely going to look up the numerical code itself online to find likely causes and solutions.

The fourth and fifth fields are the actual DTC number that ranges from 00 through 99, a lengthy list.

Freeze Frame Data

Part of the OBD-II specification is that, when a code is thrown, the ECM stores *freeze frame data*, which, as the name implies, is a digital snapshot of a block of sensor data at the time of the code. The freeze frame data contains:

- DTC that was triggered
- Engine RPM
- Engine load as calculated by the ECM
- Short-term fuel trim(s)
- Long-term fuel trim(s)
- Intake manifold pressure
- Vehicle speed
- Coolant temperature
- Fuel system status

Live Data

Unlike freeze frame data, which is a historical snapshot of sensor readings when a code was thrown, *live data* are the real-time values from engine sensors – rpm, oxygen sensor, throttle opening, EGR, etc. Most inexpensive code readers don't have this capability, but some OBD-II scanners do. If you're buying one, you really want this capability, as, from a troubleshooting standpoint, it lets you verify that, for example, a voltage generated by the oxygen sensor is actually being received by the ECM and interpreted as a plausible, valid oxygen reading.

How to Scan for DTCs

- With the engine off and the key out of the ignition, locate the OBD connector. It should be located within 16 inches of the steering wheel.
- Plug the scan tool or code reader into the connector See **Figure 6**.

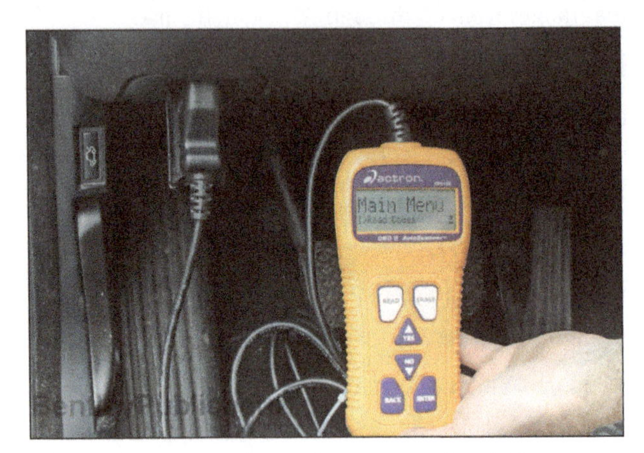

Figure 6 OBD-II code reader connected to OBD connector above the dead pedal on a BMW 5 Series.

- Turn the ignition on but do not start the engine. It is not necessary to start the engine on some tools, but it wouldn't hurt anything either.
- Follow the instructions on the tool to read the DTCs.

NOTE —

Erasing the OBD-II fault codes will turn the Check Engine Light out, but it will also reset all of the readiness monitors. Unless you fix the problem, the light will be back on after two trips (drive cycles).

Using DTCs to Tell You Where to Start Looking

In the **OBD-II Readiness** section above, we described a systematic "no shortcuts" approach of using codes to get to the root of a problem. Many people, however, do not have the patience to understand and untangle the web of monitors and enabling criteria. In this great big web-enabled world, the way that many people use the codes is to do a search for them in a enthusiast web forum. If you see many posts that say, "I had this code, and I replaced such and such a part, and cleared the code, and it never came back," and the part is inexpensive, there's nothing wrong with buying the part, installing it, and seeing if it works. We admit to having done this ourselves, and had it be successful more often than not.

But if the part is $300 instead of $30, you probably want to do more due diligence – independently test the component before you throw it away, and verify that it has power and ground and outputs or receives the signal it's supposed to. After all, the part could be fine and it could just be a loose connection. We cover testing the most common sensors in the chapters at the back of this book.

If replacing that part doesn't fix the problem, you may continue searching and try the next most popular successful solution. But by the time you reach the third iteration of this and have boxes of electrical parts you can't return, you may need to abandon the Edisonian technique in favor of the disciplined systematic approach.

Tales From the Hack Mechanic: The Gas Cap Myth

There's something of an urban myth that, when your CEL comes on, it's likely just because your gas cap is loose. The grain of truth in this is that the EVAP monitor is checking to make sure that the system of plumbing and valves intended to suck fumes from the gas tank and burn them in the engine is working, and if it senses a malfunction such as a vacuum leak, it lights the CEL. A missing or loose gas cap is certainly a major leak.

But in truth, I do not know a single person who has experienced the "my CEL came on, and I tightened the gas cap, and the light went out" moment that the Internet touts as a shared cultural experience. It's almost never that easy.

Boy, we've covered a lot of OBD-II-related ground. We'll leave you with an important piece of information.

As we said at the beginning of this chapter, OBD-II is an emissions-only diagnostic system. Although OBD-II has its hooks in many places in the engine management process, it has nothing to do with the ABS and airbag-related lights that often get tripped in a post-1996 car, or with diagnostics for other systems such as central locking, climate control, entertainment, etc. To find out the cause fault codes being set in those systems and to clear other dashboard warning lights, an OBD-II scanner won't work. You need a brand-specific scan tool. In the next chapter, we're going to talk more about OBD-II code readers, brand-specific scan tools, and the difference between them in more detail.

26

Scan Tools and Code Readers

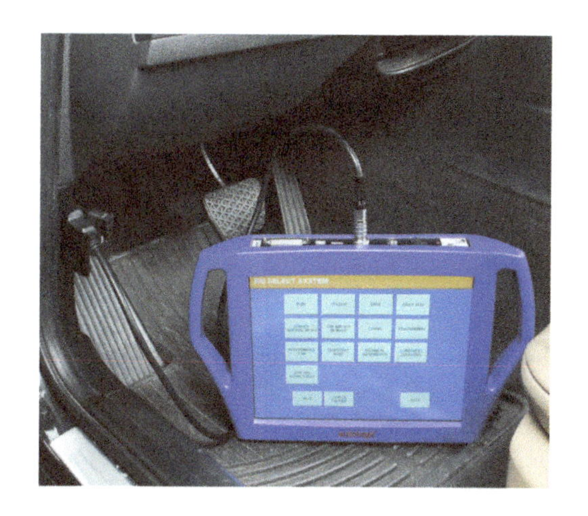

Diagnostic Tools

In this chapter we'll cover the various diagnostic tools ("scan tools") that are available to the shade tree mechanic. This list ranges from simple 20-dollar generic code readers to all-dancing-all-singing deep-pocket dealer options.

What's the Least I Need to Know?

The term "scan tool" is thrown around constantly. What does it mean? One of the most useful things to know is that "scan tool" is a very imprecise term. To understand its range of meanings, you need to understand the difference between OBD-II and the other proprietary diagnostic systems on your car. That way, once you've identified the problem you need to solve, you will be armed with the ability to select the correct tool.

- As discussed in the **OBD-II** chapter, on post-1996 cars, there is an OBD-II connector that allows you to connect an OBD-II code reader and obtain information about what emissions-related fault has caused the Check Engine Light (CEL) to be lit.
- In addition to the OBD-II connector, there is a proprietary diagnostic connector on many post-1990 cars that allows the dealer to connect a dealer-only scan tool and diagnose problems, communicate directly with each module, update software in the modules, change configurations, and more. The dealer-only scan tool is usually a desktop or laptop computer with a USB connection to an interface box that plugs into the car's diagnostic connector.
- By the early 2000s the proprietary diagnostic connector largely went away, and dealer scan tools began using the OBD-II connector. But even so, scan tools and OBD-II code readers represent two overlapping but distinct sets of functions.
- There are aftermarket scan tools that perform a subset of the functions of the dealer-only scan tool. Like the dealer-only tool, an aftermarket scan tool may be an actual computer with a USB connection to an interface box that plugs into the car's diagnostic connector, or it may be a handheld device like an OBD-II code reader that plugs directly into the diagnostic connector.
- To directly access the modules in your car and read and reset brand-specific fault codes, you need either the dealer-only scan tool or an aftermarket tool that mimics its capabilities. An OBD-II code reader is not going to allow you to directly access any of the modules in your car. It will only check the functionality of emissions-related sensors and read emissions-related codes that light the CEL.
- Whether you're using an OBD-II code reader or a scan tool, clearing a code won't make the problem go away; the CEL will simply re-light unless the underlying problem is fixed.
- Neither an OBD-II code reader nor a scan tool is going to fix your car. All they'll do is point you in the direction of something you probably need to test further, and, if necessary, replace.

So, can you use a scan tool to look at a sensor like an oxygen sensor or ABS sensor instead of troubleshooting it with a multimeter, an automotive multimeter, or a scope? It depends. If you have a brand-specific scan tool, almost certainly. An OBD-II code reader? No.

Four Classes of Diagnostic Tools

Looked at broadly, there are four classes of diagnostic tools. **The Diagnostic Scan Tool Classes Table** summarizes the distinctions. We're going to use the terms "OBD-II code reader," "OBD-II scanner," "brand-specific scan tool," and "dealer-level scan tool," but these designations are our own and do not represent industry-standard terminology. That is, you may find the words "scan tool" printed on a device that is only an OBD-II code reader. Some tools may not perform all the functions listed. Check the manufacturer's website for specific tool details before purchasing.

Diagnostic Scan Tool Classes	
Product	**Capability**
OBD-II Code Reader	Display Diagnostic Trouble Codes (DTCs), both generic and brand-specific. Erase DTCs (probably).
OBD-II Scanner	Same as OBD-II Code Reader plus enhanced functions depending on tool, such as: View freeze frame data Retrieve I/M readiness and I/M monitor status Reset I/M monitors Graphing View live data
Aftermarket Scan Tool (brand-specific)	Enhanced functions, access to some vehicle modules
Dealer-Level Scan Tool	Full function (coding, programing, test functions), access to all vehicle modules

OBD-II Tools

We dedicated the entire previous chapter to OBD-II. And that's appropriate, as regardless of how you feel about emission controls, if the CEL is illuminated, it will stop you cold on inspection day. Even though OBD-II tools will only access emissions-related devices and report and clear emissions-related codes that light the CEL, that is an important set of functionality. See **Figure 1**.

Code Readers Versus Scanners

Even within OBD-II tools, there is a wide range of capabilities. We enumerate them below. The least expensive, least capable OBD-II tools are generally called "OBD-II code readers," whereas those with more capabilities are often called "OBD-II scanners," though these terms are imprecise. Decide what's important to you, then verify that the OBD-II tool you're considering meets your needs.

Figure 1 An older OBD-II handheld scanner we have at the shop. This one has monitor status and live data display but lacks graphing capability.

- ***A simple OBD-II code reader.*** If the CEL is lit, plugging in a simple OBD-II code reader will tell you the code or codes that are stored. Even a basic code reader should tell you both the DTC number (e.g., "P0455") and the English meaning (e.g., "major EVAP leak detected").
- ***Clearing codes and resetting CEL.*** Don't just assume that a code reader can do this. Older inexpensive units may not. Verify it. Clearing codes will turn the CEL out. However, the code is almost certain to return and re-illuminate the CEL unless you repair the underlying cause. Also remember that, when codes are cleared, I/M readiness monitors are set to "not ready." The car must successfully complete a drive cycle (or two, or three) to get the monitors set to "ready" in order to pass an annual state inspection.
- ***Displaying I/M monitor status.*** If you want to know if the car is ready to pass inspection, having a code reader that displays whether the monitors have checked in as "ready" is very valuable.
- ***Live data display.*** While you're trouble-shooting a problem, live data display is extremely useful. For example, to check the throttle positioning sensor, you can move the throttle through its full range of opening and verify that the corresponding live data changes accordingly.
- ***CAN bus protocol.*** CAN is a high-speed data link that runs 50 times faster than the previous OBD-II communication protocols. It was implemented on most 2004 and later cars. As you get into more data-intensive functions like live data, it's all the more important to verify that the tool you purchase supports the fastest protocol, meaning CAN bus.
- ***Freeze frame data.*** As we've said in the **OBD-II** chapter, freeze frame data provide the overall digital context in which the code was reported.

NOTE —

While there is no officially recognized demarcation line between OBD-II code readers and OBD-II scanners, in our opinion, once a tool has monitors, live data, and freeze frame data capability, it has jumped across the threshold from code reader to scanner.

- ***Graphing.*** Some OBD-II scanners let you select a live data element – the oxygen sensor, for example – and graph its output over time. This can be a powerful tool in verifying the correct functionality of key sensors.
- ***Recording and playing back data.*** In addition to freeze frame data and live data, some tools allow you to log and play back data. This may help in trouble-shooting problems that only occur during certain circumstances such as cold starting or wide open throttle.
- ***A large easily readable display.*** The more heavily you use live data or graphing, the more you need a scanner with a large display. In fact, a clue whether an OBD-II scanner includes live data capability is the size of the device and its screen. Scanners with very small screens usually do little more than read and clear codes.
- ***Upgradability.*** If you are spending decent money on a tool, check that software is upgradable. Some of the more expensive handheld scanners have this capability, typically implemented through a USB port.
- ***Online help.*** The numerical code itself is rarely useful. "P0455," for example, tells you nothing. The English language translation often isn't helpful either, as "major EVAP leak detected" could mean anything. You need to know what this code means for your year, make, and model of car. What many people do is search for the code in an online model-specific enthusiast forum. There are many OBD-II scanners that provide online support and diagnosis by querying a database of codes and actions for your make and model car, and giving you the most likely underlying causes and solutions. If you actively read users forums and know how to search them, this capability is probably not of any value to you, but if you're not facile with forums, it might be helpful.

Handheld or Laptop?

In addition to the list of capabilities above, there is also the handheld versus Windows-based question. Most OBD-II scanners are handheld as shown in **Figure 1**. As is the case with PC-based oscilloscopes, however, you can buy a PC-based OBD-II scanner. There are three physical pieces: software that runs on a PC, the PC itself (a laptop or tablet is a good choice for the garage), and a USB plug-in interface to the OBD-II connector. See **Figure 2**.

Figure 2 The elements of a PC-based OBD-II scanner, with installation software, and the OBD-II plug-in module and its USB cable. The PC itself is not shown.

Keep in mind, though, that even a tablet PC is much bigger than a handheld scanner. It's not something you can easily hold with one hand. You have to set it down on something and use a pointing device and keyboard to navigate a menu. Plus, with a PC-based scanner, you need to stretch an extension cord into the car, or make sure the laptop's batteries are fully charged. Unless you identify some needed capability that is resident only in a PC-based OBD-II tool, handheld OBD-II tools are usually the better bet.

Brand-Specific Scan Tools

As we've said repeatedly, OBD-II is an emissions-only diagnostic system. If you need to reset the service or oil change light, or access the other systems in your car, or find out why an airbag or ABS light is on, or change the default setting of a module, or verify whether a module is going to sleep like it's supposed to, or upgrade the firmware in a module, or code a replacement module to work with the VIN of your car, you need a brand-specific scan tool that can access some or all the other modules in your car.

For example, a post-1996 European car has ABS-related onboard diagnostics that often cause the ABS and traction control lights to illuminate. It has airbag on-board diagnostics that cause the SRS light to illuminate. It probably also has transmission control, door locking, and lighting onboard diagnostics. These systems are not reached by a generic OBD-II tool because the modules in which they reside, and the commands to access those modules, vary by manufacturer.

The procedure for troubleshooting non-emissions-related problems is the same as it is with OBD-II-related issues: Pull the code and do a thorough web search to learn what other people with your car have experienced. If that leads you to a particular device, check the device as thoroughly as you can – power, ground, signal, connectors– to verify its status as good or bad. Replace or repair as necessary.

Figure 3 An oldie but a goodie–this BMW SRS (airbag only) scan tool plugs into the OBD-II port.

Tales From the Hack Mechanic: Success with an SRS Scan Tool

The airbag lights on both my 1999 BMW Z3 Roadster and my 1999 BMW Z3 M Coupe were on. Years back, I purchased a very inexpensive tool to read and reset codes for the supplemental restraint system (SRS) only (e.g., not a general-purpose scan tool; see **Figure 3**). It reported that the passenger side seat belt tensioner was bad on one car, and the driver's side tensioner was bad on the other car. A slate of posts on an enthusiast forum indicated that it's more likely for the connection beneath the seat to be bad than for the tensioner itself to go bad. On both cars, I lifted the seat, found the connector, reseated it, and cleared the code. And on both cars, the problem has not recurred. Sometimes it really is that easy.

Brand-specific scan tools may have some or all of the following capabilities:

- Basic access to modules (read and clear trouble codes).
- Clear adaptations (things like engine management fuel trim corrections, or automatic transmission pressure adaptations for shifting).

- Change default settings (for example, turning on or off daytime running lights, enabling or disabling the automatic locking of doors at 10 mph, etc.).
- Display live data like an OBD-II tool, plus a greater variety of data from additional modules.
- Component activations: Turn individual components on and off (e.g., cycle the ABS solenoids for brake, move adaptive headlights lights up and down for checking, turn injectors and ignition coils on an off for testing).
- Register or retrofit a new battery. Retrofit other options the car was fitted with when new.
- Code new modules to the car's VIN.
- Upload firmware, flash individual modules.

Note that there's the same "PC or not PC" issue with brand-specific scan tools as there is with OBD-II scanners. A PC-based scan tool may have the most flexible user interface offering the widest array of options, but a handheld tool is usually the quickest and easiest to use. It all depends what you need to do.

Remember that, with brand-specific scan tools, we're no longer in OBD-II land, so the physical connection to the diagnostic connector may use more lines than just the two CAN bus lines. For example, recall that, in the OBD-II connector pinout, pin 8, listed as "Manufacturer Discretion," also had the note "BMW: Second K-Line for non OBD-II (Body/Chassis/Infotainment) systems)." So as you start to walk down the path of learning about brand-specific scan tools, you may see a tool advertising "supports dual K-line."

Professional-Level Scan Tools

A brand-specific aftermarket scan tool is designed to function like the factory tool. They never fully pull it off, but some come very close. Most, however, are easier to use and more intuitive.

There are professional-level scan tools built specifically for independent repair shops. Prices range from hundreds of dollars to as high as many thousands. Here at Bentley Publishers, in addition to a host of other handheld units, we have a professional Autologic tool. It is a touchscreen tablet, connects wirelessly to the Internet, has live technician assist, a real-time camera, and covers most of the Euro brands. This tool is premium priced, yet provides premium dealer-level diagnostics. High-end independent repair shops that are serious about repairing European cars own them. See **Figure 4**.

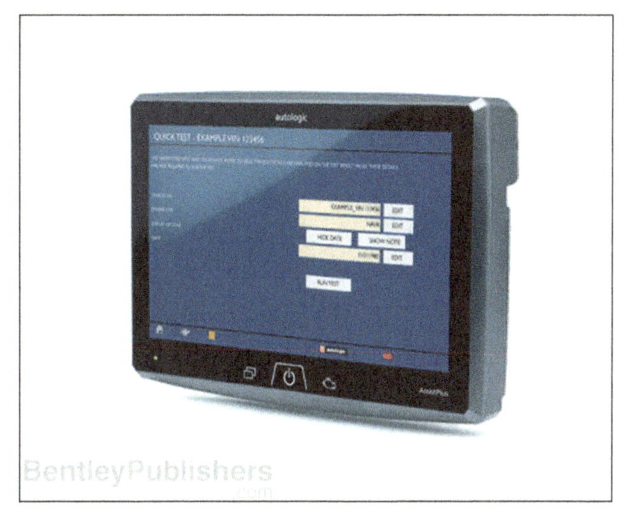

Figure 4 Autologic professional scan tool.

DIY-Level Scan Tools

A do-it-yourselfer won't spend thousands of dollars on a scan tool, but if the volume of work performed is great enough, he or she may spend hundreds. A popular tool in the semi-pro Volkswagen world is VCDS (formerly know as VAG-COM; note that "VAG" stands for "Volkswagen Audi Group") from Ross-Tech, who has been developing tools that straddle the line between professional-level and DIY-level since 2000. VCDS is installed on a laptop and includes a dedicated OBD-II cable. See **Figure 5**.

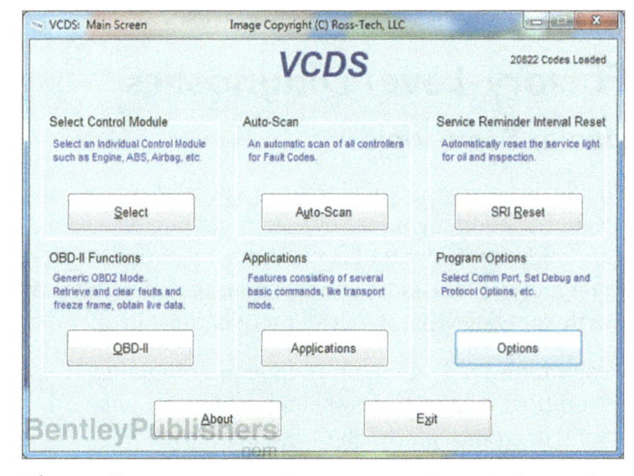

Figure 5 A screen shot of the Windows-based aftermarket VCDS diagnostic software for VW-Audi Group (VAG) cars. This semi-pro tool emulates the dealer tool, including measuring values, output tests, control module adaptations, and basic settings.

Since "VAG" became a term associated with Volkswagen diagnostics, there are now a variety of sub-$100 DIY-level handheld Volkswagen-Audi scan tools that use "VAG" in their name, available from a number of manufacturers. One such tool is shown in **Figure 6**.

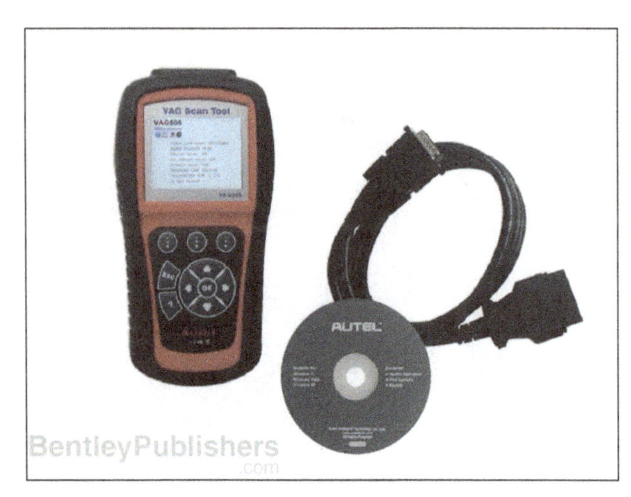

Figure 6 VW-Audi handheld scan tool.

A similar trend has occurred in the BMW diagnostics world, where the name of a well-reviewed aftermarket scanner became used by other companies. The line between name appropriation and knockoff can be fuzzy. A little web-searching to determine what you're actually buying is time well spent.

But, from a consumer standpoint, the fact that it is getting easier to find useful inexpensive brand-specific tools that perform many functions is a good thing. Dedicated tools can do a lot: reset oil service and service intervals, scan vehicle modules (such ABS, SRS, Automatic Transmission, HVAC), recalibrate brake pads after replacement, and much more. User forums are invaluable for finding which tools like-minded enthusiasts have had good experience with.

Factory-Level Diagnostics

Dealer Scan Tools

The dealer is always going to have the most up-to-date diagnostic equipment for your car, including scan tools. That said, the dealer tool is well out of the reach of the DIYer and even the independent repair facilities. But it is worth knowing what the dealer tool is capable of.

Dealer tools really do it all: coding, programming, retrofitting, vehicle warranty and recall info, onboard wiring diagrams, electrical component locations, technical service bulletins, repair manuals, OBD-II technical data, DTC lists, vehicle build data, and more.

Dealer tools have one other very special feature: Guided fault finding, also called test plans. This is a feature you can't get on an aftermarket tool, or not yet anyway. Based on a fault code, a test plan is generated that guides you, step-by-step, through all the tests you need to do to diagnose the fault. Given the price of new electronic modules and parts these days, it is a feature that is worth its weight in gold.

Dealer Knockoff Scan Tools

Copies of factory tools are illegal and are not recommend by us. Software can be copied and hardware designs can be reverse-engineered, so there are no shortages of what appear to be knockoff professional-level scan tools on eBay. Proceed at your own risk. Bootlegged, corrupt, or out-of-date software can quickly render a control module dead if you flash it with an outdated program. It will almost always be a better option to visit your local shop or dealer and pay the hourly rate than to burn up a module, or two or three.

PassThru Vehicle Programming (SAE Standard J2534)

Around the year 2000, the EPA and California ARB realized that many vehicles on the road were not meeting emissions standards because the engine management software was out of date. SAE J2534 PassThru was their solution to this problem. It's essentially OBD-II for Engine Control Module (ECM) reprogramming, a way for independent repair shops to reflash the existing software/firmware in emissions-related control modules with updated OEM software, something that was previously possible only for the dealer.

PassThru typically requires a fee-based subscription to a vehicle manufacturer's web site. Some manufacturers supply only the mandated ability to re-flash the ECM, but others supply software that works very much like the full-on dealer service tool, with extensive service functionality such as diagnostics, guided fault finding, coding, programming, and so on.

With a computer connected to the Internet, the factory software application installed, and a PassThru device connected with the appropriate OBD-II cable, up-to-date firmware can be installed. See **Figure 7**.

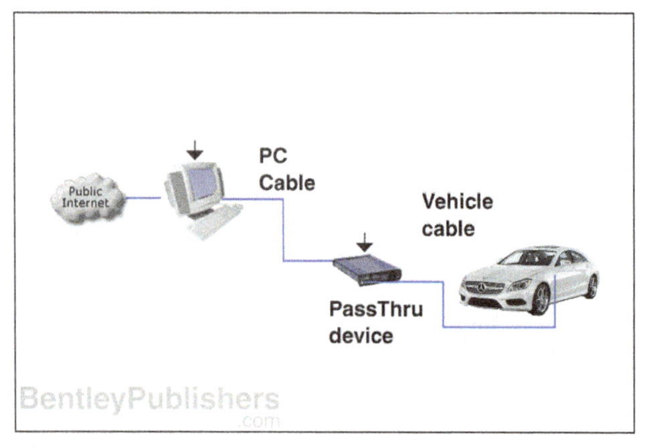

Figure 7 J2534 PassThru configuration. PassThru allows for OEM recalibration of engine management software and other emission-related systems. Some manufactures allow for full vehicle diagnostics and programming.

27

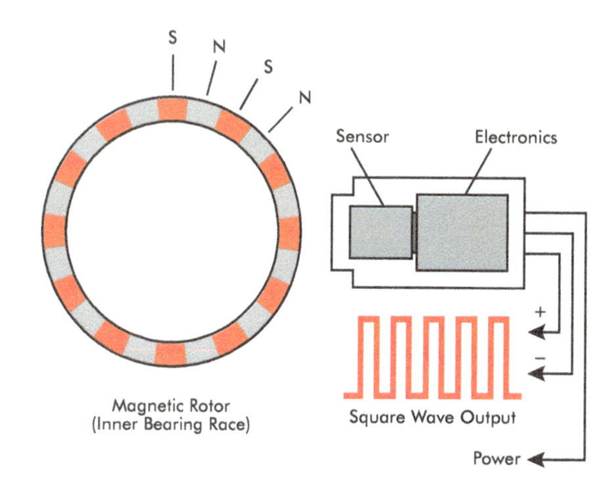

Introduction to Sensors

Introduction

In the chapters that follow this introductory chapter, we're going to discuss testing and troubleshooting common electrical engine management-related devices that are found, in one form or another, in most fuel-injected cars. Although we'll give each device its own chapter, it helps to understand that we're looking at two broad classes of devices – sensors and actuators.

Sensors and Actuators

Sensors are devices that *create a signal* that is sent to the ECM and other modules. As such, when we test a sensor, we need to verify that it is producing a signal. In the coming chapters, we will cover temperature sensors, throttle position sensors, oxygen sensors, wheel speed sensors, crankshaft and camshaft position sensors, and mass airflow sensors. These all output signals to the ECM.

In contrast, there are devices that *accept a signal* produced by the ECM. Fuel injectors, ignition coils, solenoids, fuel pumps, and other motors fall into this category. There isn't one overall universally accepted term that describes "things controlled by the ECM," but they are sometimes referred to as actuators.

So when we test a sensor, we verify that the sensor is producing a signal, but when we test an actuator, we verify that the actuator is receiving a signal. This will help you to see that many of the descriptions of how to troubleshoot sensors are similar to each other, and that, in fact, there's not much difference between troubleshooting sensors and troubleshooting actuators once you understand that you need to know what the signal is that the sensor outputs or the actuator is controlled by.

Test Tools

In the coming chapters, we're going to concentrate on troubleshooting a device by connecting the appropriate tool or tester – often an automotive multimeter – directly to the device to verify that, in the case of a sensor, it's producing the kind of signal it's supposed to, and in the case of an actuator, it's receiving the expected signal.

Looking at this in a general sense, with any of these sensors, there are two general fault mode possibilities:

- The sensor itself is bad. The multimeter, automotive meter, and oscilloscope are the tools needed to directly evaluate sensor functionality.
- The sensor is generating a valid signal and thus appears to be okay, but the ECM reports a problem with the signal. A scan tool is best to troubleshoot this.

Presumably there's some event that has led you to troubleshoot a sensor. Let's say, for example, that the ABS fault light is illuminated. The first thing you'd probably want to do is look at all four wheels and verify that none of the wires have ripped out of the ABS sensors or connectors. Next, you'd directly test the ABS sensors at the wheels (which we describe how to do). And, when you do, perhaps you find that the sensors themselves are fine. But you don't in fact know if the signals are getting back to the ECM.

The Role of the Scan Tool

If the car is post-1996 and has a diagnostic connector to which a manufacturer-specific scan tool can be connected, a bad sensor may produce an error code that can be read with a scan tool. Further, data produced by the sensor can often be directly examined with the scan tool. When you do this – look at sensor data on a scan tool – you're looking at the data received by the car's ECU and sent to the scan tool. So it's one level removed from the actual sensor.

In troubleshooting, this is important, because it is quite possible for the data to be properly generated by the sensor and not received by the ECU. When that's the case, you're into the realm of standard electrical troubleshooting – consulting a wiring diagram, verifying that the signal is going into a wire, and figuring out why it isn't coming out the other side.

There is, in fact, a third less likely possibility: The ECM is receiving the correct signal, but it is throwing a fault code that trips the CEL anyway. In this case, you likely have several evenings ahead of you acquiring freeze frame data with a scanner or code reader, looking at readings from additional sensors, understanding enabling criteria, and poring over user forums trying to understand what is going on.

There's another important thing to understand. If you have a pre-1996 car that doesn't have OBD-II and that's running badly, you usually have little choice but to roll up your sleeves, get in there with the multimeter, and start troubleshooting. Is the crankshaft positioning sensor receiving voltage and outputting a sine wave? Are the coils all firing? You need to start somewhere to figure it out for yourself.

But if you have a post-1996 car with OBD-II, you can take advantage of the fact that OBD-II is tasked with making sure that the car's emissions-related components all work. This includes damned near everything associated with digital engine management. As much as you might whine about the CEL getting tripped by minor vacuum leaks, the fact is that when you hook up a code reader and read the DTC that tripped the CEL, it often tells you *exactly* what's wrong. "Misfire in Cylinder #3." Boy, that's awfully clear, and often indicative of a bad stick coil (see the **Testing Stick Coils** chapter).

In addition, OBD-II's monitoring produces DTCs that tell you if a particular sensor has a short circuit or an open circuit, as well as "plausibility" codes that warn if a sensor is producing readings that, when cross-checked against other sensors, are implausible. Take advantage of this. It is doing some of your work for you.

That having been said, you shouldn't blindly assume all of OBD-II's dire conclusions are correct. Sometimes the code is more of a clue than an edict. The classic example is when you get an OBD-II code telling you the reading from your secondary oxygen sensor is out of range. This *could* indicate a bad sensor, but it also could be that the sensor is working perfectly and OBD-II is telling you that the catalytic converter is no longer functioning to spec. Testing the sensor independently as we describe in the **Oxygen Sensors** chapter should reveal what's what.

The point is that the OBD-II's monitoring and self-diagnostics can be real assets in sensor troubleshooting. If your car has it, use it.

Frequently Used Automotive Sensors

In the coming sensor-specific chapters, two sensor types will come up repeatedly – Variable Reluctance Transducers (VRTs) and Hall Effect Sensors. We're going to introduce them and talk about how they work here.

VRT Sensor

The Variable Reluctance Transducer, or VRT sensor, is a simple rugged device that was the main automotive rotation measurement sensor. VRTs were used for decades as electronic ignition triggers, ABS sensors, and crankshaft and camshaft positioning sensors.

The VRT circuit has two parts – a stationary magnet wrapped in a coil of wire (the sensor), and a rotating toothed metal wheel (the reluctor wheel). The reluctor wheel can have square teeth and look like a gear, or it can be star-shaped.

As the toothed wheel rotates through the magnetic field created by the magnet, when a tooth is at its closest approach to the magnet, the magnetic field is bent most strongly in the direction of the tooth.

A changing magnetic field creates an electric field. The bending of the magnetic field as the tooth makes its closest approach constitutes a change, which creates an electric field, which induces a voltage in the coil of wire. As the tooth passes, the magnetic field is relaxed – another change, but in the other direction – which induces a voltage change opposite to the first. The result is an up-and-down signal which is shaped approximately like a sine wave. When the tooth-gap spacing on a reluctor wheel is regular, as is generally the case on ABS sensors and crankshaft positioning sensors, the output signal fairly closely matches a smoothly varying sine wave, as shown in **Figure 1**. However, when a VRT is used in an ignition module, the shape of the reluctor wheel is star-shaped, with small teeth and wide gaps, and although the resulting signal oscillates with a clear frequency, the signal is more jagged.

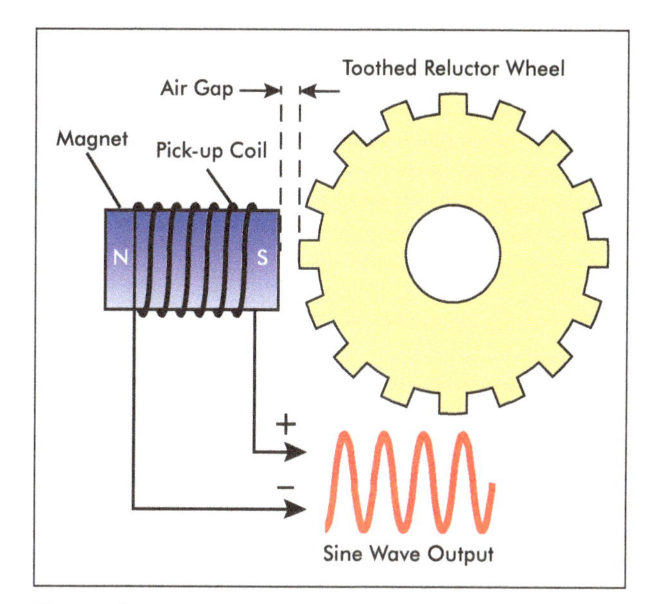

Figure 1 Variable Reluctance Transducer (VRT) sensor outputting a sine wave signal.

You can intuitively see how, as the toothed wheel spins faster, the sine wave will speed up, meaning that its frequency will increase. But less intuitive is the fact that the sine wave's *amplitude* also increases. This is because, in a VRT, the strength of the induced electric field – and thus the voltage – is proportional to how fast the magnetic field is changing. If the toothed wheel is spun very slowly, the magnetic field changes very slowly, which doesn't create a big electric field, which in turn doesn't create a big voltage. For this reason, VRTs don't work well at low speeds.

When a VRT is used as a crankshaft position sensor, one particular tooth on the reluctor wheel is missing, thus providing a way for the ECM to determine where Top Dead Center (TDC) is for cylinder #1. When graphed on an oscilloscope, this gap shows up as a missing peak.

VRT Sensor Circuit Testing

The VRT is a *passive sensor.* It does not require power. In fact, the sine wave that it generates *is* power – it's a small AC (alternating current) signal.

A VRT has two wires – signal and ground. VRTs have no power wire. To test a VRT, resistance across the two wires can be checked to verify that the pick-up coil is neither broken nor shorted, but this is not a definitive test of the sensor.

The presence of the VRT's sine wave signal can be verified with a multimeter set to measure AC voltage. As the engine or wheel is spun faster, the sine wave's amplitude will increase. If the multimeter has the ability to measure AC frequency, the frequency of the VRT's sine wave can be observed to increase as the engine or wheel is spun faster.

Hall Effect Sensors

The Hall Effect sensor is another device to measure rotation. It is more immune to electromagnetic interference than the VRT, and has the advantage that it outputs a signal regardless of how slowly things are rotating. For the most part, manufacturers phased out VRTs and replaced them with Hall Effect sensors in about the year 2000, sometimes even midway through a particular model's product cycle.

The underlying physics of the Hall Effect sensor is more complex than for the VRT. When an electric current is applied to a semiconductor that's in the presence of a magnetic field, it will create a potential difference (meaning it will induce a voltage) in the semiconductor at a 90° angle to the applied current.

Dr. Edwin Hall discovered the odd effect in 1879, but for 90 years it was a phenomenon in search of an application. When the use of integrated circuits became widespread and the cost of silicon-based semiconductors dropped, a silicon chip began being used as the semiconductor in devices that implement the Hall Effect. Turning on and off the magnetic field that is in close proximity to the semiconductor will start and stop the Hall Effect (e.g., it will create, then stop, the voltage from the semiconductor).

One way to enable and disable the magnetic field is to shield the semiconductor from the field, so one kind of Hall Effect sensor uses a rotating ring with windows in it. When a window lines up between the semiconductor and the magnet, the magnetic field is let through, and the semiconductor experiences the Hall Effect, rapidly ramping up the voltage. This creates the off-on-off square wave. These sensors with rotating interrupter rings were commonly used in electronic ignition mechanical distributors.

Another, more modern application is a rotating toothed ring that looks very similar to a VRT's reluctor ring. In this case, the teeth interrupt the magnetic field, shutting the Hall Effect off, but leaving it on the rest of the time. Sensors with toothed rings like this may output an inverted square wave. The inverted signal may be passed to the ECM, or it may be reversed by the sensor's internal circuitry before it reaches the ECM.

However, most Hall Effect sensors employed in European cars use a rotating ring of magnets to disrupt the magnetic field from the stationary magnet, stopping the Hall Effect. These magnets are often pressed into an inner bearing race or sleeve as shown in **Figure 2**.

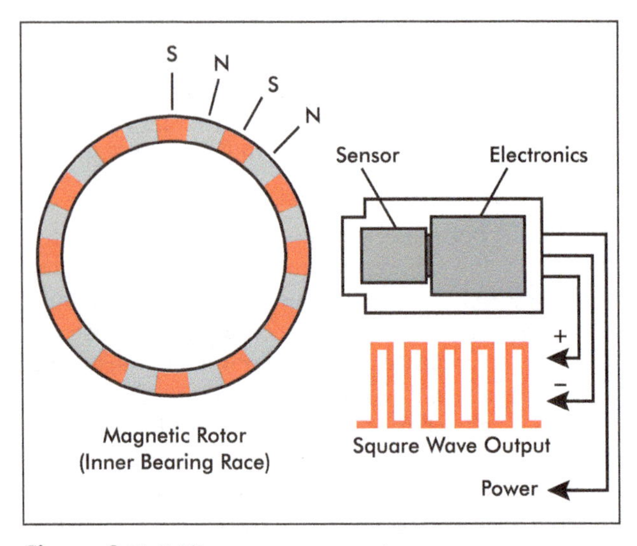

Figure 2 Hall Effect sensor outputting a square wave signal.

Note that the Hall Effect is not a "digital effect." The effect itself is analog. Every automotive Hall Effect sensor has integrated electronics that takes the analog signal from the effect, amplifies it, cleans it up, and uses it to create the output square wave. Part of the "digital" confusion comes from the fact that, because there is a semiconductor used for the Hall Effect itself, they're often referred to as "solid state sensors."

Hall Effect Sensor Circuit Testing

Hall Effect sensors are *active* sensors. They require power. For this reason, there is a third wire, usually supplying a voltage of 5, 10.2, or 12 volts. However, some newer Hall Effect sensors on post mid-2000s cars use only two wires by combining signal and ground on the same wire.

A resistance test is not valid on a Hall Effect sensor. To test a Hall effect sensor, check that voltage is present on the voltage supply wire, and that the sensor outputs a frequency-varying square wave.

The presence of the Hall Effect sensor's square wave signal can be verified with an automotive multimeter such as a Fluke 88V that's capable of detecting a square wave and measuring its frequency. As the engine or wheel is spun faster, the square wave's frequency will increase.

For additional VRT and Hall Effect sensor testing and troubleshooting, see the following chapters:

- **Testing Wheel Speed Sensors**
- **Testing Crankshaft Position Sensors**
- **Testing Camshaft Position Sensors**

AMR Sensors

Beginning in approximately the mid-2000s, a third kind of rotation measurement sensor began to be used for wheel speed and steering angle sensing applications. Anisotropic Magnetoresistive (AMR) sensors are active sensors similar to Hall Effect sensors, but they can detect the direction of rotation. This ability has become increasingly important as wheel speed sensors are used not only for ABS but also for features such as vehicle stability control and parking assist.

Other sensors we've discussed thus far has a signal that is a dynamically changing voltage. The signal from an AMR sensor is different–it is a dynamically changing *current,* a PWM square wave that toggles between 7 mA and 14 mA. Without a high sensitivity current clamp connected to an oscilloscope, the only way for a do-it-yourselfer to evaluate the health of these sensors is to use a scan tool.

Sensors That Are Modules

Also beginning in approximately the mid-2000s, certain sensors became available that connect directly to the CAN or other vehicle bus and output a digital message. Note that this is fundamentally different from a sensor outputting an analog signal that can be seen on an oscilloscope or detected using an automotive multimeter. As with the AMR sensor, the only way for a do-it-yourselfer to evaluate the health of an "on the bus" sensor is to use a scan tool.

Variable Resistors and Reference Voltage

There are a number of sensors – temperature sensors, throttle position sensors, vane-style airflow meters – that create a variable resistance, which is then converted to a voltage-based signal by the use of a reference voltage. Let's take a moment and explain why that's necessary and how it works.

As we cover in the **Temperature Sensors** chapter, a temperature sensor is usually a Negative Temperature Coefficient (NTC) thermistor whose resistance decreases as temperature goes up. In the **Throttle Position Sensors** chapter, we describe how a variable resistor (potentiometer) works, moving an arm along a resistance track to mechanically vary resistance. What both of these sensors have in common is that they create a changing resistance, one in response to changing temperature, the other in response to mechanical motion.

In both cases, you might think the variable resistance is a "signal." It's not. It needs an extra step to be turned into a signal the ECM can use. Why?

Because resistance isn't something that the ECM understands. Resistance, by itself, is passive. It isn't until a voltage is applied that current actually flows. Without a flow of current, there's no "signal" for the ECM to measure.

Recall from our **Common Multimeter Tests** chapter that resistance is measured with a circuit unpowered, and that the way a multimeter internally measures resistance is to apply a small current and measure the resulting drop in voltage across the resistance. In your car, a variable resistance is usually converted into a voltage in a very similar way, as follows.

With a resistance track device like a throttle position sensor, the TPS is part of a circuit where a reference voltage, typically 5 volts, is supplied by the ECM. This voltage flows through the resistance track, which drops the voltage by an amount proportional to the resistance. It is this output voltage, typically running from close to 0 volts to nearly the 5-volt reference voltage, that is reported to the ECM.

With a temperature sensor, it's similar, but there's an extra step because the thermistor is never intended to drop all the voltage by itself. With the 5 volt reference voltage applied, current flows. The current typically first passes through a separate resistor of a fixed value, whose resistance causes a voltage drop. From there, the current then passes through the thermistor which drops the rest of the voltage, and then to ground. The ECM typically monitors the voltage at the inbound side of the thermistor and uses this, along with an internal calibration, to determine the temperature.

In both cases, the reference voltage has allowed a variable resistance to be turned into a voltage that the ECM can understand as a signal.

The Difference Between Reference Voltage and Supply Voltage

Note that "supply voltage" is not the same thing as reference voltage. As we've described above, reference voltage is present specifically to turn a resistance into a voltage. In contrast, "supply voltage" simply means that the sensor or other device is an active piece of electronics (i.e., that it requires voltage in order to work). For example, a Hall Effect sensor is an active sensor. It has electronics inside. It needs a supply voltage. But the sensor is not using that voltage to turn a resistance reading into a voltage signal.

Sensor Signals

The tests we're going to use in the upcoming chapters to verify correct operation of a variety of sensors have a certain rhythm to them. Typically we'll verify the presence of a supply voltage, a reference voltage, or the resistance across a sensor, then we'll look for its signal, which may be static, slowly changing, quickly changing, or a periodic square wave or sine wave. Here we are going to introduce a set of graphic icons that will be used the upcoming chapters to make it easier to identify what kind of quantity you're measuring in each test.

- We will use the icon below if we are measuring a ***static value***. The static value could be a supply voltage, a reference voltage, the resistance across a coil, or a current, but in all cases, it shouldn't be changing while we're measuring it. As such, a conventional multimeter will be sufficient.

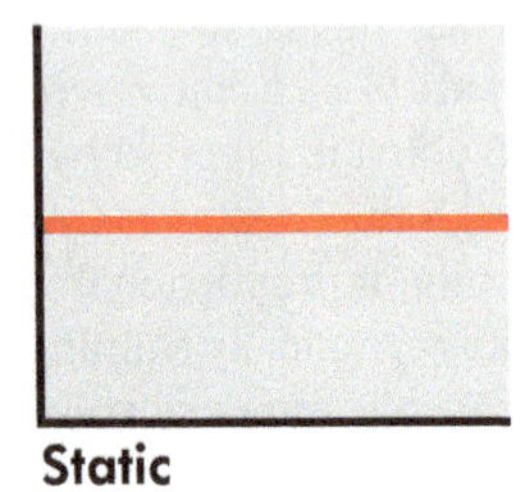

Static

- We will use the icon below if we are measuring a ***slowly changing value***. This will be the case for temperature sensors and throttle position sensors. Because the value will be changing slowly in response to some controllable stimulus (e.g., the engine warming up, depressing the accelerator pedal), a conventional multimeter will be sufficient. We will then apply that stimulus (e.g., warm up the engine, depress the accelerator pedal), and verify that the sensor responds appropriately and that the reading on the multimeter changes in the way it should.

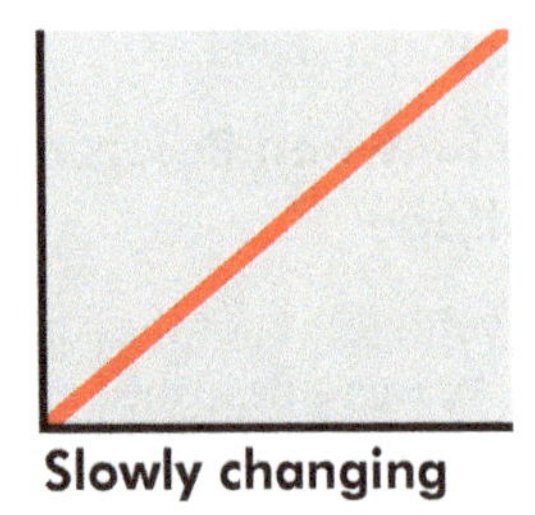

Slowly changing

- We will use the icon below if we are measuring a ***quickly changing non-periodic value***. This will be the case for oxygen sensors that oscillate about fives times per second between 0.1 and 0.9 volts. In this case, a conventional multimeter is sufficient to catch the maximum and minimum values of the change, but if you have another tool such as a graphing scan tool or an oscilloscope, you can use it to your advantage.

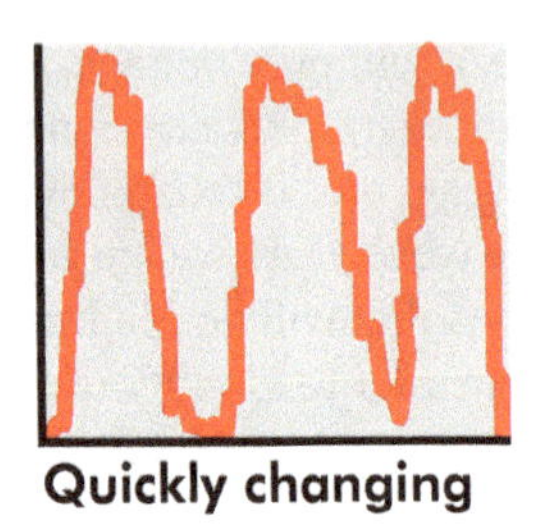

Quickly changing

- We will use the icon below if the signal we are measuring is a ***sine wave***. This is the case for VRT sensors used as wheel speed sensors and camshaft and crankshaft sensors, For these, the necessary tool is the automotive multimeter or oscilloscope. The parameter of interest is usually the frequency. Most of the time we simply need to verify that the signal is there and that it changes in response to stimulus (e.g., spinning a wheel, revving the engine).

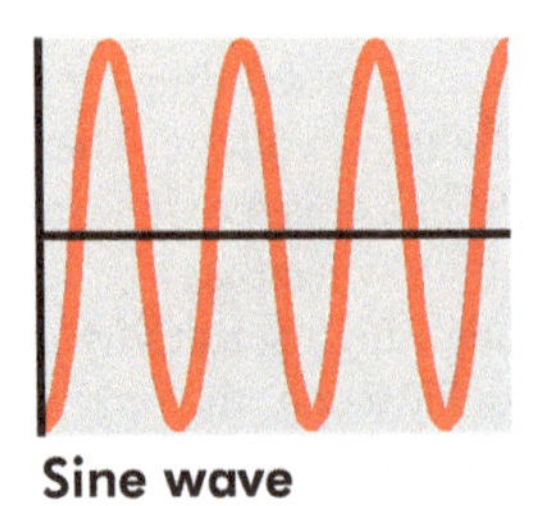

Sine wave

- We will use the icon below if the signal we are measuring is a ***square wave***. This is the case for Hall Effect sensors used as wheel speed, camshaft, and crankshaft sensors, for digital mass airflow sensors, for stick coils and fuel injectors, and for pulse width modulated (PWM) fuel pumps and solenoids. Like the sine wave, the necessary tool is the automotive multimeter or oscilloscope. The parameters of interest are usually the frequency and perhaps the duty cycle or pulse width. Most of the time we simply need to verify that the signal is there and that it changes in response to stimulus (e.g., spinning a wheel, revving the engine).

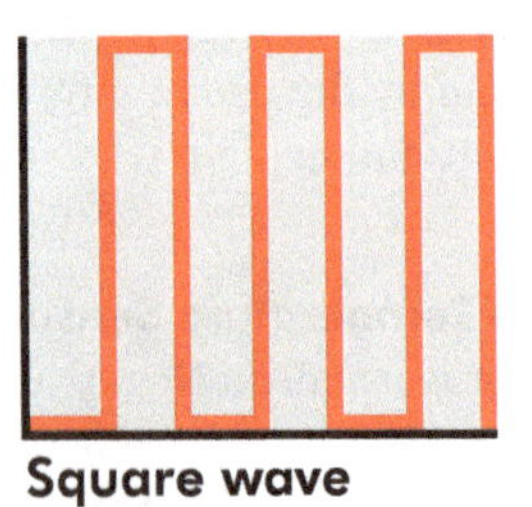

Square wave

Let's go test some sensors.

28

Testing Temperature Sensors

Engine Coolant Temperature (ECT) Sensor

Multimeter Test Set-Ups

⚠ WARNING —

Make personal safety your top priority. Watch what you are doing, stay alert. Do not work on your car if you are tired, not feeling well, or in a bad mood. Common sense, alertness, and good judgment are crucial to safe work on your vehicle.

Protect yourself from shock, trauma, and hazardous chemicals: Wear safety glasses, work boots, and protective gloves when working on your vehicle. Have hearing protection available for work that generates noise.

Be sure the work environment is well lit, properly ventilated, and that the vehicle is on a level surface. The floor, vehicle, and components inside the vehicle should be dry to minimize the possibility of shocks.

This book is designed as background technical training and as a general overview of how to do electrical diagnosis and service. It is designed to be used in conjunction with the appropriate service information for your vehicle, not to replace factory service information.

If you lack the skills, tools and equipment, or a suitable workshop for any procedure described in this book, have the job done by a qualified shop or an authorized dealer.

For additional warnings and cautions, see the **Safety** *chapter and the warnings, cautions, and notices in your vehicle manufacturer's service information.*

ECT Reference Voltage

Test Set-Up **28-A**

See page 28-5 for step-by-step procedure

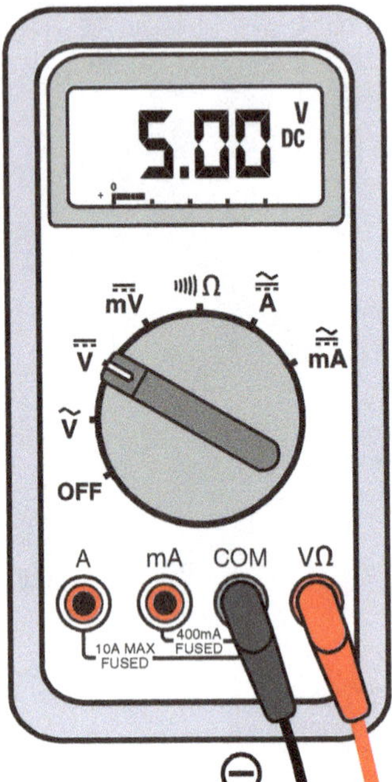

Testing for Sensor Supply Voltage

Test Conditions

Dial set to:	V DC Volts
Red input terminal:	VΩ
Black input terminal:	COM
Red lead:	power wire (+)
Black lead:	chassis ground (–)

Connector disconnected, probed at power wire

Engine off, ignition ON

Typical Normal Values

Voltage: Approximately 4.5 to 5 volts

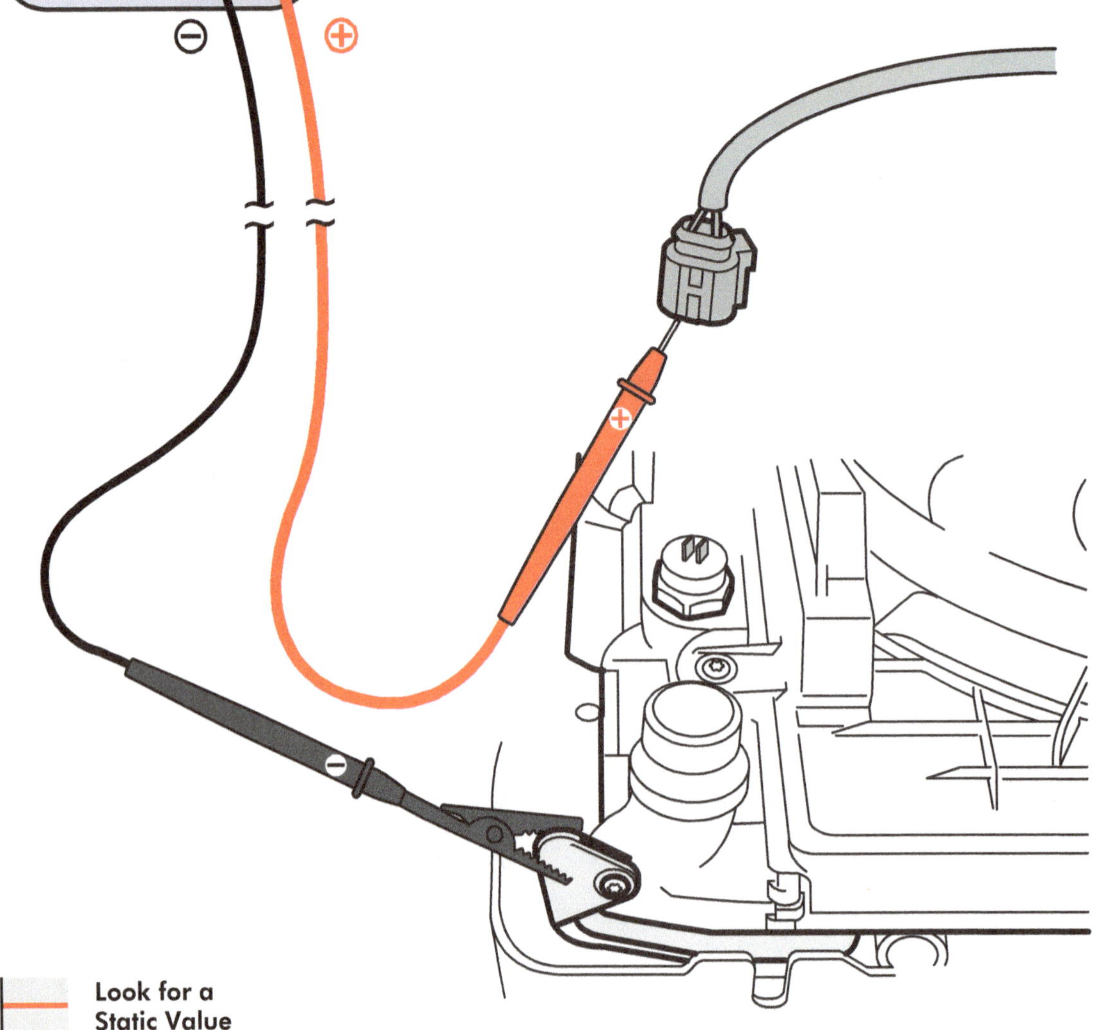

Look for a Static Value

ECT Resistance

See page 28-6 for step-by-step procedure

Test Set-Up **28-B**

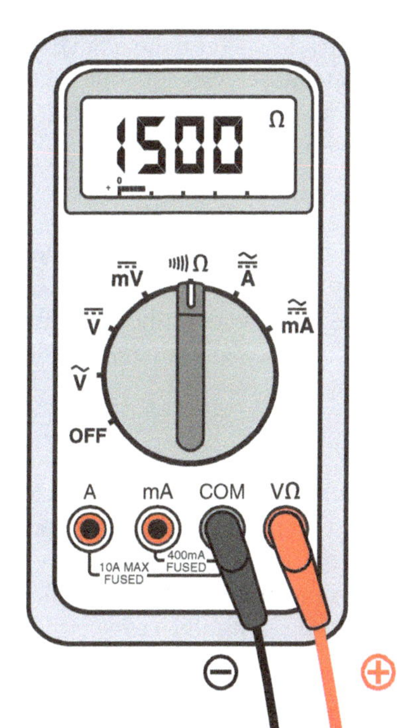

Testing Temperature Sensor Resistance

Test Conditions

Dial set to:	Ω resistance
Red input terminal:	VΩ
Black input terminal:	COM
Red lead:	sensor signal pin (+)
Black lead:	sensor ground pin (–)

Connector disconnected, testing at sensor

Engine off

Typical Normal Values

7000 - 12000 ohms @ 14°F (–10 °C)
2000 - 3000 ohms@ 68°F (20 °C)
700 - 1000 ohms @ 122°F (50 °C)
250 - 400 ohms @ 176°F (80°C)

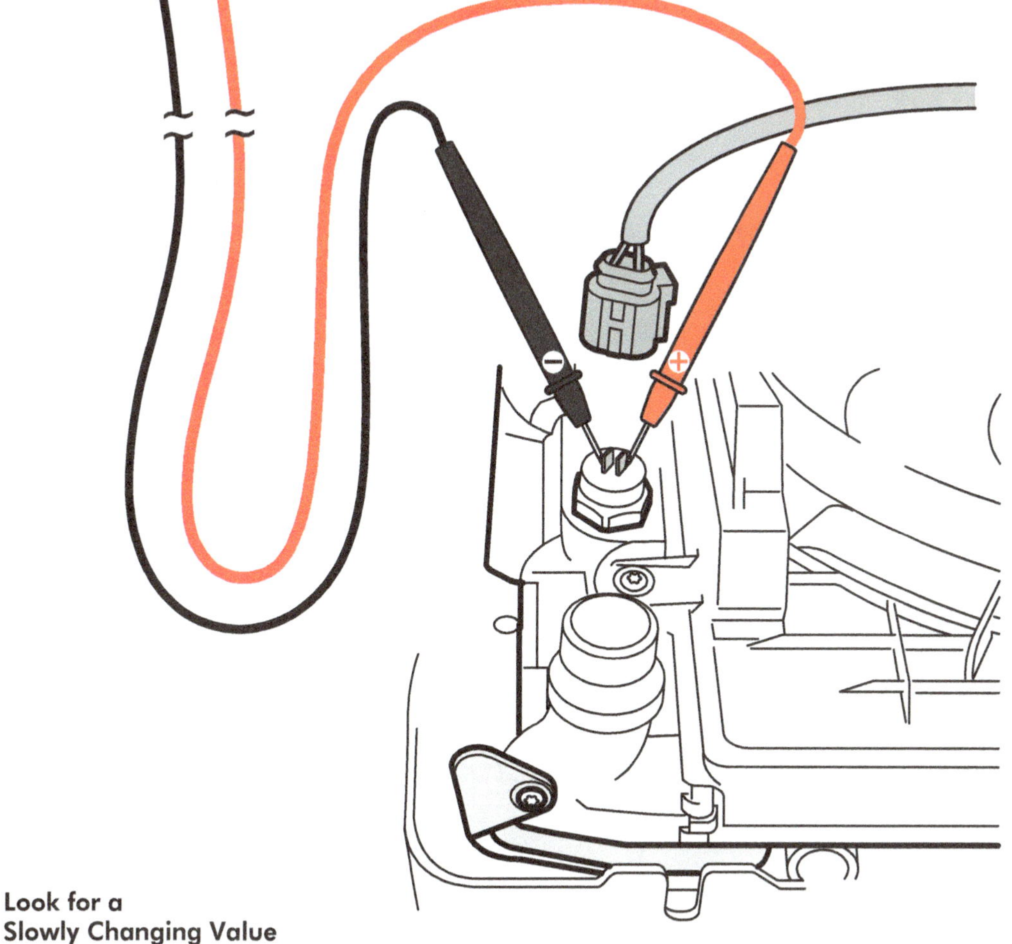

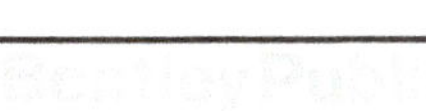

Look for a Slowly Changing Value

ECT Sensor

In pre-1975 cars, there was a single temperature sensor that supplied input to the temperature gauge. But ever since the advent of electronic fuel injection, temperature has been a crucial input to the digital engine management process and is used in fuel trim, ignition timing, idle speed control, emission controls, and automatic transmission adaptation.

In this chapter, we concentrate on the Engine Coolant Temperature (ECT) sensor, which monitors the internal temperature of the engine. Note that several temperature sensors may be employed in addition to the ECT. For example, there is often an additional temperature sensor located at the radiator. The information in this chapter should help you to troubleshoot not only the ECT but also any temperature sensor. See **Figure 1**.

Owners may be led to suspect a bad temperature sensor due to the temperature gauge not working, a cooling fan not turning on, or by a fault code.

Figure 1 Cylinder-head-mounted ECT sensor (**arrow**) on a 2013 BMW 328i.

Operation

Most automotive temperature sensors are Negative Temperature Coefficient (NTC) thermal resistors (thermistors) whose resistance decreases as the temperature increases. Thus, resistance is high when the engine is cold and drops slowly until the engine reaches normal operating temperature. Note that this behavior is opposite to that of standard copper wire, whose resistance increases with rising temperature.

Signal

The temperature sensor's resistance decreases slowly as temperature increases. This resistance is converted into a voltage. Thus, in the framework of the **Dynamic Analog Signals** chapter, the signal to the ECM is a slowly changing voltage.

Circuit

A 5-volt reference voltage is sent from the ECM to the ECT sensor. As coolant temperature changes, the internal resistance of the sensor changes and correspondingly alters the output voltage signal to the ECM based on a -volt scale (typically 3 to 5V range). For more on how reference voltage works, see the **Introduction to Sensors** chapter.

Most cars use two-wire ECT sensors. On pre-1975 cars, the temperature gauge sensor may use only a single wire, with the thermistor grounding directly to the chassis through contact with the engine. Some temperature sensors may have four wires. In this case it is usually because there are two sensors mounted in a single housing, one for engine coolant and one for the temperature gauge. If in doubt, consult the appropriate wiring diagram to determine the location and number of wires.

Figure 2 shows a sample sensor wiring diagram showing both an ECT and a temperature sensor at the radiator. Note that in this example, the ground side of the circuit is switched by the ECM.

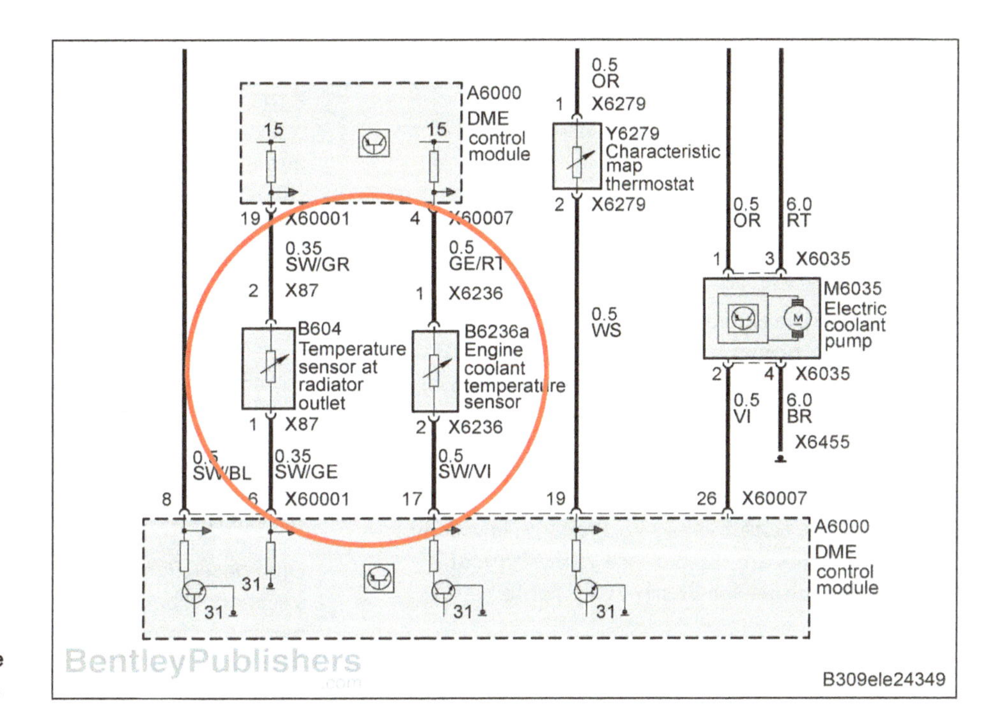

Figure 2 ECT and radiator outlet temperature NTC sensors (**red circle**) for a 2008 BMW. A reference voltage (terminal 15) is fed through the sensor and back to the ECM (ground, terminal 31). Note that the ground side may be switched by the ECM.

Testing

Because the signal of the ECT sensor is changing slowly, it can be easily checked using a multimeter.

Before testing, make sure the temperature sensor itself is not cracked. Disconnect the connector from the sensor. Physically inspect both sides of the connector. Make sure no wires are broken and no pins are pushed out or bent.

Determining Pinouts

Pinouts are best determined using factory service information (a wiring diagram). However, we have found that the following procedure is useful for confirming that the factory service information is correct, or determining pinouts if a wiring diagram is not available.

For a single-wire temperature sensor, the wire carries the signal, and the sensor is grounded through contact with the engine.

For a two-wire temperature sensor:

- Locate the ECT connector and disconnect it.
- Turn the key to ignition ON, but do not start engine.
- Set the meter to measure DC voltage.
- Connect the black multimeter probe to any chassis ground.
- On the wiring harness side of the connector, touch the red multimeter probe to each of the pins.
- If one of them reads approximately 5 volts, that pin is the supply voltage. The other pin is ground. In addition, as with nearly any European car, the brown wire is usually the ground wire.
- Turn the ignition off.

If the temperature sensor has four wires, the voltage and ground pins should be paired on the connector (e.g., two voltage pins side by side, two ground pins side by side).

ECT Reference Voltage Test

See Test Set-Up on page 28-2

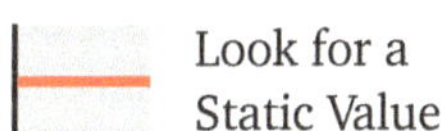

Look for a Static Value

- Locate the ECT sensor connector and disconnect it.
- Set the meter to measure DC voltage.
- Connect the black multimeter probe to any chassis ground.
- Turn the key to ignition ON, but do not start engine.
- Working at the wiring harness side of the connector, touch the red multimeter probe to each of the pins.
- The multimeter should read approximately 4.5 to 5V on the wire carrying the reference voltage. If there is no voltage on the reference voltage wire, the problem must be troubleshot by consulting a wiring diagram to trace the wire back to the ECM to determine where the voltage is being interrupted.
- Turn the ignition off.

If the temperature sensor has four wires, the voltage and ground pins should be paired on the connector.

ECT Resistance Test

See Test Set-Up on page 28-3

Look for a
Slowly Changing Value

- Disconnect the connector from the sensor.
- Configure the multimeter to measure resistance.
- If the sensor has two terminals, put the two multimeter probes across the two terminals. If the sensor has only a single terminal, put one probe on the terminal and the other probe on the body of the sensor.
- Examine the resistance reading. Typical normal temperature sensor values for Bosch ECTs used in European cars are summarized in the **Typical Normal Values for Bosch ECTs Table**.

Typical Normal Values for Bosch ECTs

Temperature	Resistance (ohms)
14°F (-10 °C)	7000 - 12000
68°F (20 °C)	2000 - 3000
122°F (50 °C)	700 - 1000
176°F (80°C)	250 - 400

- If the resistance is zero, the sensor is likely internally shorted. If the reading is infinite ("OL"), the sensor is broken. In either case, the sensor should be replaced.
- Additional testing can be conducted by cooling the sensor with ice and heating it with a hair dryer and verifying that the reading changes. This can be done either with the sensor in the car or with the sensor removed.

Supplemental Testing

If the sensor resistance readings are not correct, the sensor is faulty and should be replaced. If the sensor is outputting a signal when measured directly, but there's still a problem (for example, a scan tool is still reporting there's no signal from the sensor), there may be a broken wire or bad connection between the sensor and the ECM.

- Use a multimeter set to measure continuity to verify continuity between each wire on the connector and the ECM. This will require a wiring diagram for your car, knowledge of the location of the ECM, ECM pin designations, and no power in the circuit when measuring resistance.

Is a Scan Tool Useful?

On a post-1996 car, yes. If the ECT is working correctly, a scan tool should directly report the actual temperature reading as interpreted by the ECM. In addition, if the ECT or its wiring is faulty, the scan tool should report a fault code.

Is an Oscilloscope Useful?

Generally not. Because the output of the temperature sensor changes slowly, a multimeter is perfectly adequate and an oscilloscope is not necessary.

However, it is possible for the temperature sensor's internal thermistor to be cracked and for the sensor's output to be erratic. It may be possible to see this on the meter or on a scan tool, but it may be more obvious on an oscilloscope's graphic display.

Additional Information

Related Chapters

- **Common Multimeter Tests** for instructions on using a multimeter to measure voltage and resistance.
- **Introduction to Sensors** for a discussion on reference voltage.

29

Testing Throttle Position Sensors (TPS)

Multimeter Test Set-Ups

⚠ WARNING —

Make personal safety your top priority. Watch what you are doing, stay alert. Do not work on your car if you are tired, not feeling well, or in a bad mood. Common sense, alertness, and good judgment are crucial to safe work on your vehicle.

Protect yourself from shock, trauma, and hazardous chemicals: Wear safety glasses, work boots, and protective gloves when working on your vehicle. Have hearing protection available for work that generates noise.

Be sure the work environment is well lit, properly ventilated, and that the vehicle is on a level surface. The floor, vehicle, and components inside the vehicle should be dry to minimize the possibility of shocks.

This book is designed as background technical training and as a general overview of how to do electrical diagnosis and service. It is designed to be used in conjunction with the appropriate service information for your vehicle, not to replace factory service information.

If you lack the skills, tools and equipment, or a suitable workshop for any procedure described in this book, have the job done by a qualified shop or an authorized dealer.

For additional warnings and cautions, see the **Safety**, *chapter and the warnings, cautions, and notices in your vehicle manufacturer's service information.*

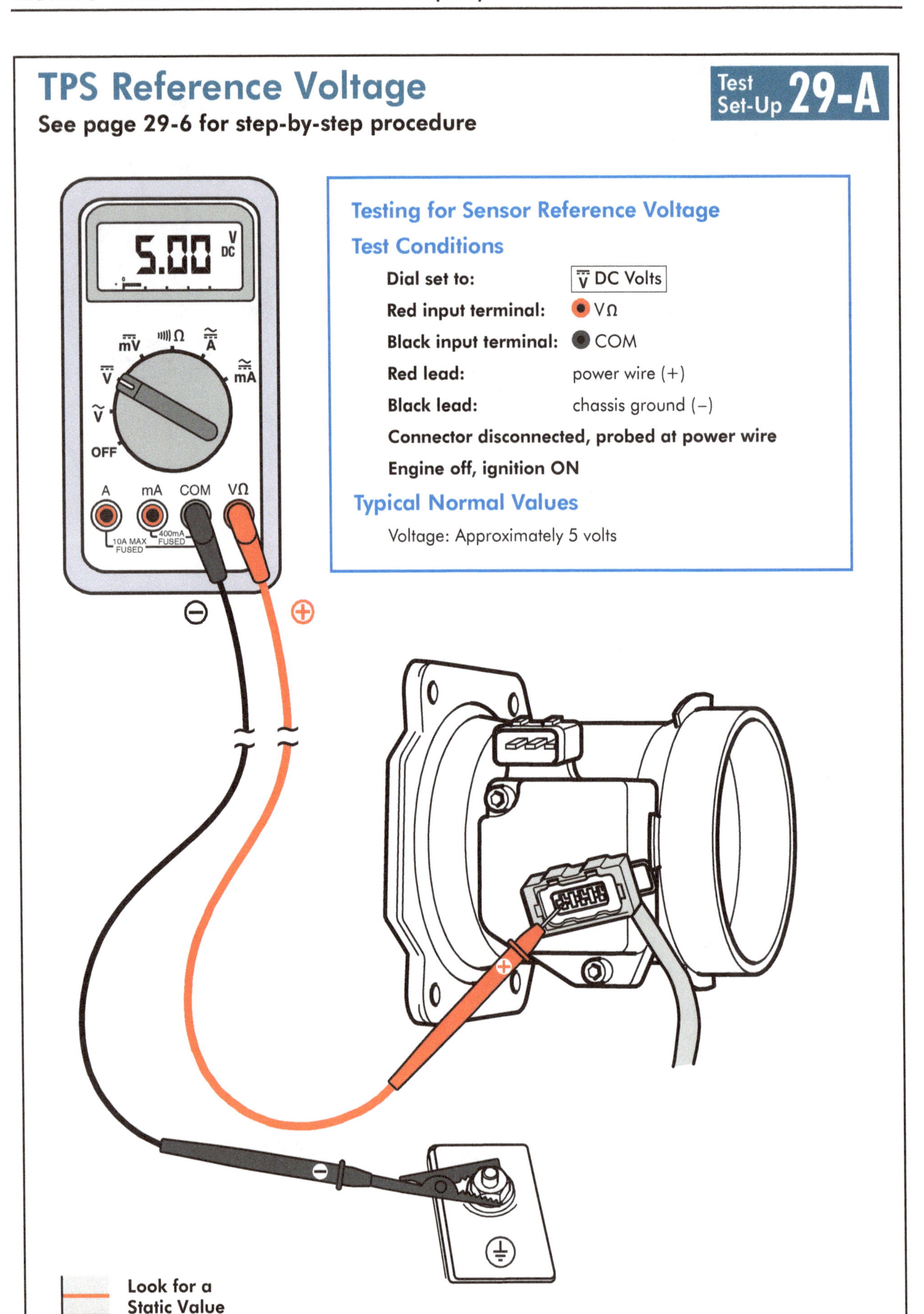
TPS Reference Voltage
See page 29-6 for step-by-step procedure
Test Set-Up 29-A
Testing for Sensor Reference Voltage
Test Conditions
Dial set to: DC Volts
Red input terminal: VΩ
Black input terminal: COM
Red lead: power wire (+)
Black lead: chassis ground (–)
Connector disconnected, probed at power wire
Engine off, ignition ON
Typical Normal Values
Voltage: Approximately 5 volts
5.00
V DC
mV
Ω
A
V
mA
V
OFF
A
mA
COM
VΩ
10A MAX FUSED
400mA FUSED
Look for a Static Value

TPS Resistance

See page 29-6 for step-by-step procedure

Test Set-Up 29-B

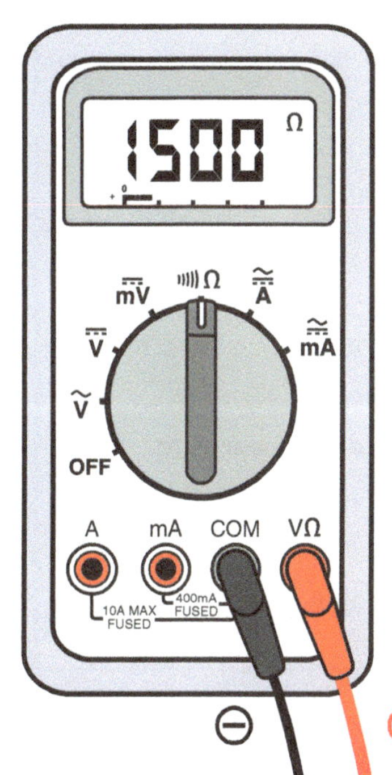

Testing Resistance Across Throttle Position Sensor

Test Conditions

Dial set to:	Ω resistance
Red input terminal:	VΩ
Black input terminal:	COM
Red lead:	sensor signal pin (+)
Black lead:	sensor ground pin (–)

Connector disconnected, testing at sensor

Engine off

Typical Normal Values

Not zero, not "OL," decreasing as throttle rotated open

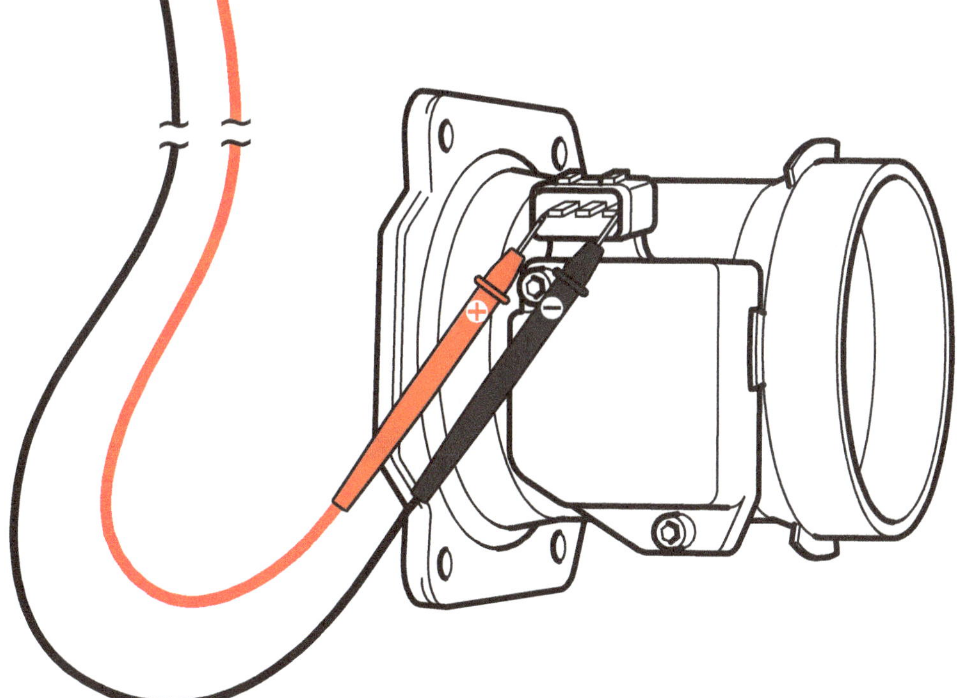

Look for a Slowly Changing Value

TPS Signal Voltage

Test Set-Up 29-C

See page 29-6 for step-by-step procedure

Testing for Voltage Signal

Test Conditions

Dial set to:	V DC Volts
Red input terminal:	VΩ
Black input terminal:	COM
Red lead:	power wire (+)
Black lead:	chassis ground (–)

Connector disconnected, probed at power wire

Engine off, ignition ON

Typical Normal Values

Voltage: 0.5 to 4.5 volts, varies smoothly as throttle opens

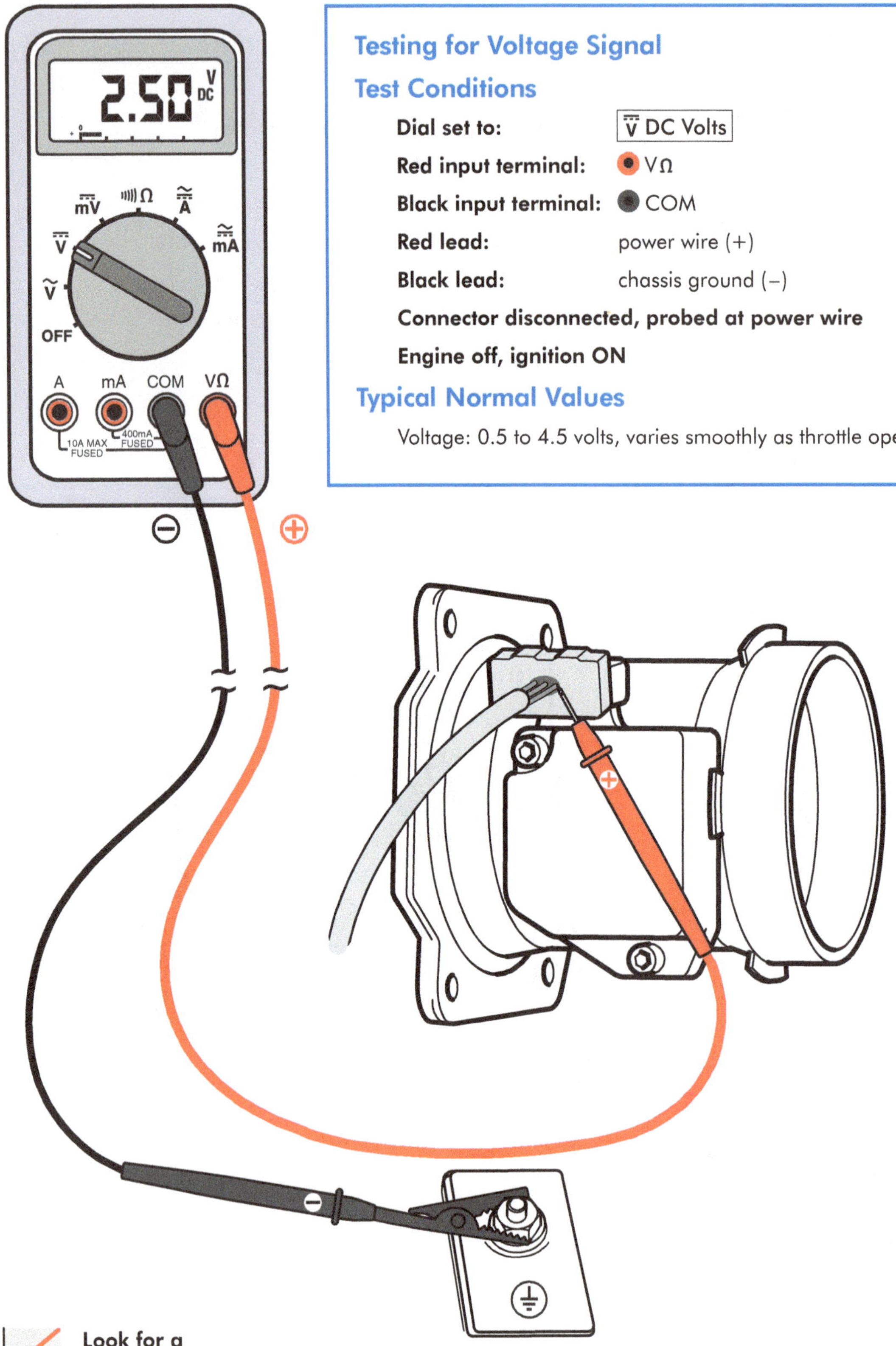

Look for a Slowly Changing Value

Throttle Position Sensor

On an electronically fuel-injected car, a Throttle Position Sensor (TPS), also sometimes called a Throttle Valve Sensor (TVS), is employed to measure the position of the throttle (the butterfly valve that controls the amount of air being allowed into the engine) and provide that information to the ECM.

Owners may be alerted to a bad TPS by flat spots in acceleration, stumbling, or a code on a scan tool that implicates the TPS.

Operation

A TPS is a potentiometer – a variable resistor like a volume control on a piece of electronics – with a wiper arm that moves across a resistance track. Current flows through the wiper arm and onto the resistance track wherever the arm happens to be positioned. The current then flows from there to the end of the track. Thus, when the wiper arm is at the very beginning of the track, current must flow through the entire resistance track, so the resistance is high. As the wiper arm moves farther along the resistance track, it encounters less and less of the high-resistance track, thereby decreasing the total resistance seen by the current.

When a potentiometer is used as a throttle position sensor, the throttle shaft is literally attached to the potentiometer's knob, as if hitting the gas is raising the volume. With the throttle closed, the arm is at the beginning of the resistance track, so the resistance is high. As the throttle is opened farther, the arm moves farther along the resistance track, decreasing the resistance.

Signal

Because a resistance value is not a signal that an ECM understands, a reference voltage is applied to the resistance track to convert the resistance into a voltage. The voltage coming through the end of the track is then sent to the ECM. For more information on reference voltage, see the **Introduction to Sensors** chapter.

As such, the response from a TPS is a static voltage value until the throttle position changes. Then it is a dynamic but slowly changing non-periodic voltage value. It should be constant when the throttle position is constant, and should vary when the throttle position varies.

Circuit

Most TPSs are three-wire sensors, with wires for reference voltage, signal, and ground.

A specific wiring example for the throttle valve sensor on a 1996 Porsche 911 (993) is shown in **Figure 1**.

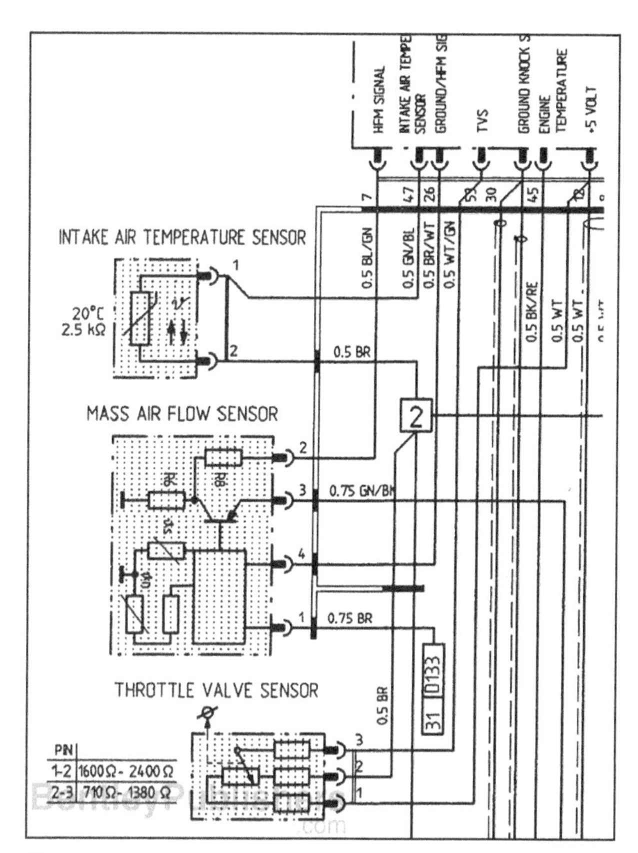

Figure 1 Wiring for the throttle valve sensor on a 1996 Porsche 911 (993). The small icon in the center of the TVS depicts the wiper on the resistance track, to which pins 2 and 3 are connected. Pin 3 is the signal wire connecting to the ECM. Pin 2 is a common ground. Pin 1 is the 5V reference signal.

Testing

Because the signal from the TPS is a slowly changing resistance converted to a slowly changing voltage, the TPS can be checked with any conventional multimeter.

Before testing, make sure the sensor itself is not cracked. Disconnect the connector from the sensor. Physically inspect both sides of the connector. Make sure no wires are broken and no pins are pushed out or bent.

Determining Pinouts

Pinouts are best determined using factory service information (a wiring diagram). However, we have found that the following procedure is useful for confirming that the factory service information is correct, or determining pinouts if a wiring diagram is not available.

- Locate the connector to the TPS and disconnect it.
- Set the multimeter to measure DC voltage.
- Connect the black lead to any chassis ground.
- Turn the key to ignition ON, but do not start engine.
- On the wiring harness side of the connector, touch the red lead to each of the pins on the connector on the wiring harness side. The pin that reads 5 volts should be the reference voltage.
- Turn the key off.
- Set the meter to measure resistance.
- On the wiring harness side of the connector, touch the red lead to each of the remaining pins. The pin at which the multimeter registers continuity (near zero ohms) should be ground.
- The remaining pin should be signal.

TPS Reference Voltage Test

See Test Set-Up on page 29-2

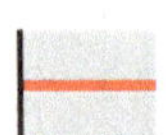

Look for a
Static Value

- Disconnect the connector from the sensor.
- Configure the multimeter to measure voltage.
- Connect the black multimeter lead to chassis ground.
- Turn the key to ignition on, but do not start engine.
- Working at the wiring harness side of the connector, touch the red multimeter probe to each of the pins.
- The multimeter should read 5V on the wire carrying the reference voltage. If there is no voltage on the reference voltage wire, the problem must be troubleshot by consulting a wiring diagram to trace the wire back to the ECM to determine where the voltage is being interrupted.

TPS Resistance Test

See Test Set-Up on page 29-3

Look for a
Slowly Changing Value

- With ignition off, disconnect the connector from the sensor.
- Configure the multimeter to measure resistance.
- Put the two multimeter probes across the ground and signal terminals of the TPS.
- Examine the resistance reading. If the resistance is zero, the TPS is internally shorted. If the reading is infinite, the TPS is likely broken. In either case, the sensor should be replaced.
- Rotate the throttle open. The exact reading is unimportant as we will measure voltage in the next test, but the resistance reading should decrease as the throttle opens.

TPS Signal Voltage Test

See Test Set-Up on page 29-4

Look for a
Slowly Changing Value

The voltage test is more accurate than the resistance test. The voltage values with the reference voltage applied should be approximately as shown in **Typical Normal Values for TPS Table**, and should vary smoothly as the throttle is opened and closed.

Typical Normal Values for TPS

Condition	Value
Throttle fully closed	0.5V – 0.9V
Throttle fully open	3.5V – 4.5V

- Connect the connector to the TPS.
- Configure the multimeter to measure voltage.
- Connect the black multimeter lead to chassis ground.
- Back-probe the signal wire on the connector and connect the red multimeter lead to the back-probed connector.
- Turn the key to ignition ON, but do not start engine.
- Open and close the throttle slowly and steadily.
- The voltage should smoothly vary between about 0.5V and about 4.5V. If the voltage does not change at all, or if it drops or jumps erratically, the TPS is likely defective and should be replaced.

Supplemental Testing

If you have direct confirmation via a multimeter that the sensor is outputting a signal, but there's still a problem (for example, a scan tool is still reporting there's no signal from the sensor), there may be a broken wire or bad connection between the sensor and the ECM.

- Use a multimeter set to measure continuity to verify continuity between the signal wire on the connector and the ECM. This will require a wiring diagram for your car, knowledge of the location of the ECM, and knowledge of which pin on the ECM connector carries the signal from the TPS sensor.
- Use a multimeter set to measure continuity to verify continuity between the negative wire on the connector and ground. If the meter does not show continuity, troubleshoot the problem by consulting a wiring diagram to trace the wire to determine where the ground is being interrupted.

Is a Scan Tool Useful?

Yes. A scan tool may be useful in alerting you to a TPS-related trouble code. In addition, a scan tool that displays live data may be useful in allowing you to open and close the throttle and verify that its measured position is changing appropriately.

Is an Oscilloscope Useful?

Under most circumstances an oscilloscope is not necessary. Graphing the signal on an oscilloscope, however, may enable you to see erratic readings that are difficult to see on a multimeter.

Additional Information

Related Chapters

- **Common Multimeter Tests** for a discussion on how to use a multimeter to measure voltage and resistance.
- **Introduction to Sensors** for a discussion on how reference voltage works.

Technical Notes

Up through the late 1990s, cars had their accelerator pedal connected to the throttle body via a mechanical linkage such as a cable or a series of rods and levers. Then, in approximately the late 1990s, cars began to have the mechanical throttle linkage replaced by a so-called electronic throttle control or "throttle by wire" system. In most cases, the accelerator pedal also has a position sensor, and the pedal's position is then relayed to the ECM, which sends a signal to a motor on the throttle body which opens the throttle. Although electronic throttle control systems are outside the scope of this chapter, many of them use a potentiometer to measure the position of the accelerator pedal, and function very similarly to the TPS described in this chapter.

Note that, because vane-style (barn door) airflow meters utilize a resistance track, they work in the same way as throttle position sensors.

30

Testing Oxygen Sensors

Multimeter Test Set-Ups

⚠ WARNING —

Make personal safety your top priority. Watch what you are doing, stay alert. Do not work on your car if you are tired, not feeling well, or in a bad mood. Common sense, alertness, and good judgment are crucial to safe work on your vehicle.

Protect yourself from shock, trauma, and hazardous chemicals: Wear safety glasses, work boots, and protective gloves when working on your vehicle. Have hearing protection available for work that generates noise.

Be sure the work environment is well lit, properly ventilated, and that the vehicle is on a level surface. The floor, vehicle, and components inside the vehicle should be dry to minimize the possibility of shocks.

This book is designed as background technical training and as a general overview of how to do electrical diagnosis and service. It is designed to be used in conjunction with the appropriate service information for your vehicle, not to replace factory service information.

If you lack the skills, tools and equipment, or a suitable workshop for any procedure described in this book, have the job done by a qualified shop or an authorized dealer.

For additional warnings and cautions, see the **Safety**, *chapter and the warnings, cautions, and notices in your vehicle manufacturer's service information.*

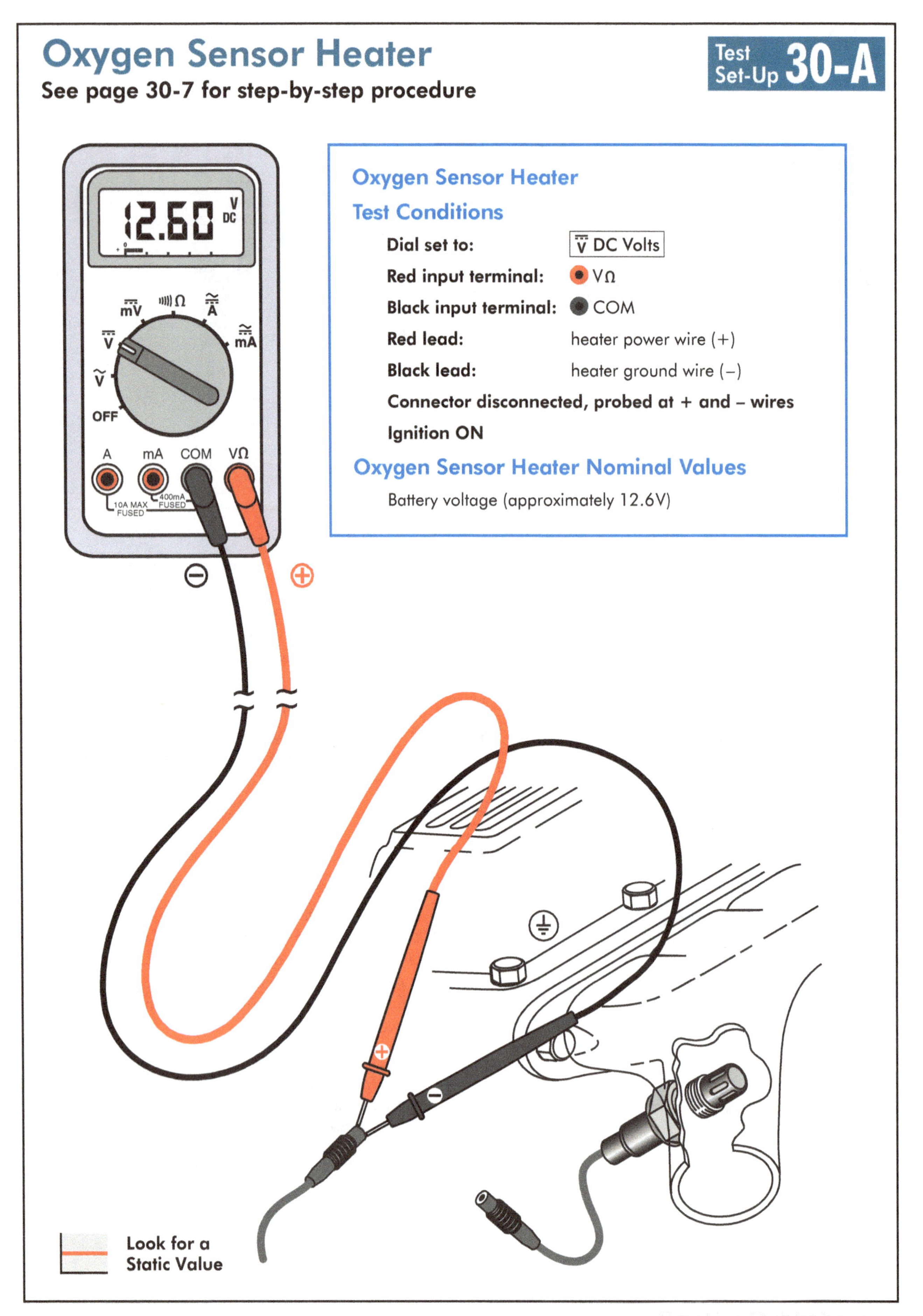
Oxygen Sensor Heater
See page 30-7 for step-by-step procedure
Test Set-Up 30-A
Oxygen Sensor Heater
Test Conditions
Dial set to: V DC Volts
Red input terminal: VΩ
Black input terminal: COM
Red lead: heater power wire (+)
Black lead: heater ground wire (–)
Connector disconnected, probed at + and – wires
Ignition ON
Oxygen Sensor Heater Nominal Values
Battery voltage (approximately 12.6V)
12.60
V
DC
mV
A
V
mA
OFF
A
mA
COM
VΩ
10A MAX FUSED
400mA FUSED
Look for a Static Value

Oxygen Sensor Signal

See page 30-8 for step-by-step procedure

Test Set-Up 30-B

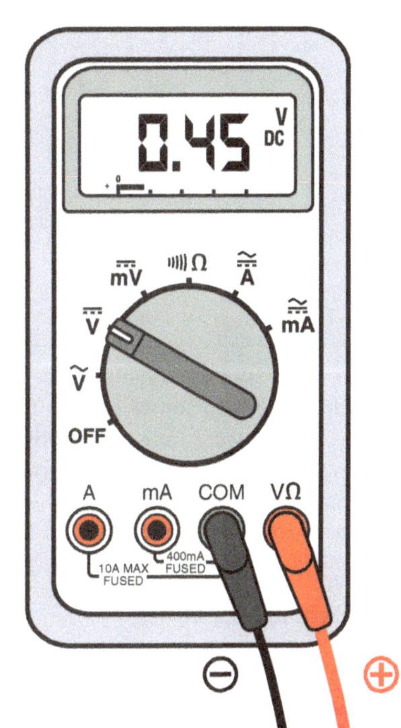

Oxygen Sensor Signal

Test Conditions

Dial set to:	V DC Volts
Red input terminal:	VΩ
Black input terminal:	COM
Red lead:	sensor signal wire (+)
Black lead:	chassis ground (–)

Connector remains connected, back probed at signal wire

Engine running

Oxygen Sensor Signal Nominal Values

Precat (Primary) Oxygen Sensor
Min and Max Voltage: Approximately 0.1V to 0.9V
Average (Center) Voltage: Approximately 0.45V

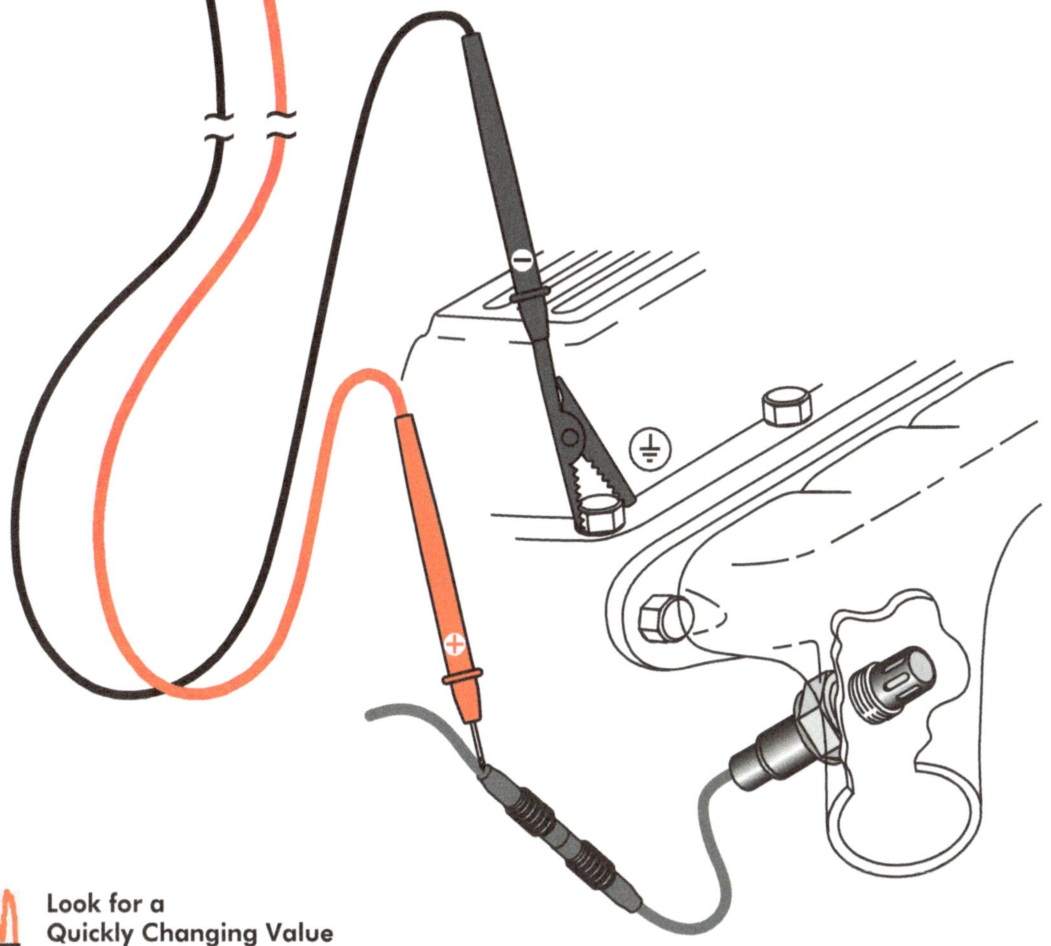

Look for a Quickly Changing Value

Oxygen Sensor

The engine management system on nearly every electronically fuel-injected car built since the mid to late 1970s uses an oxygen(O2) sensor in the exhaust stream as part of a feedback control loop. When in closed loop operation, the car's ECM uses the signal from the oxygen sensor to make fuel trim corrections in order to balance power, emissions, and fuel economy.

Most owners are alerted to an oxygen sensor problem because the Check Engine Light is on and a scan tool has reported an oxygen sensor-related code.

There are several different types of oxygen sensors. In this chapter, we will concentrate on a commonly used type, the narrow-band Zirconia sensor. Information on the other types of sensors is contained in the **Technical Notes** section at the end of this chapter.

Operation

As the name implies, oxygen sensors measure the amount of oxygen in the exhaust. They do this by employing a ceramic element in the exhaust stream that produces a signal whose voltage is proportional to the amount of oxygen. See **Figure 1**. Detail on the mechanism by which they do this is contained in the **Technical Notes** section at the end of this chapter.

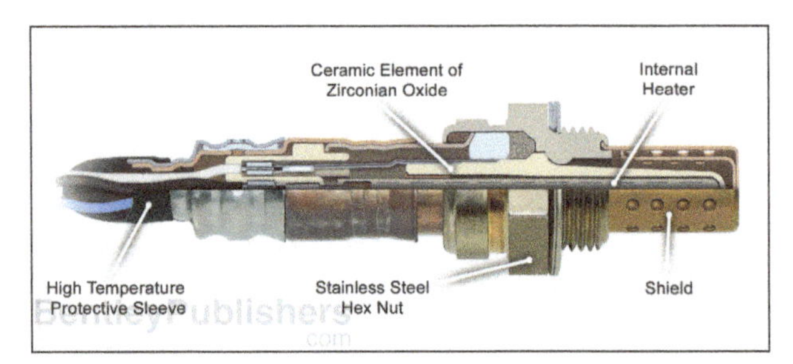

Figure 1 A typical 4-wire narrow-band-Zirconia oxygen sensor is threaded into the exhaust system and generates a small DC voltage (0–1 volts) based on the residual oxygen content in the exhaust. The sensor's ceramic material is electrically heated to about 600°F to produce a reliable signal. *Courtesy of Walker Productions, Inc.*

Oxygen sensors employed since the early 1980s incorporate an internal heater to bring them up to operating temperature quickly. OBD-II cars built after 1996 employ a secondary oxygen sensor after the catalytic converter to verify that the converter did its job in catalyzing the exhaust gas (see **Figure 2**).

Types

Oxygen sensors are divided by their ceramic sensing elements into Zirconia sensors and Titania sensors. Zirconia sensors are much more common.

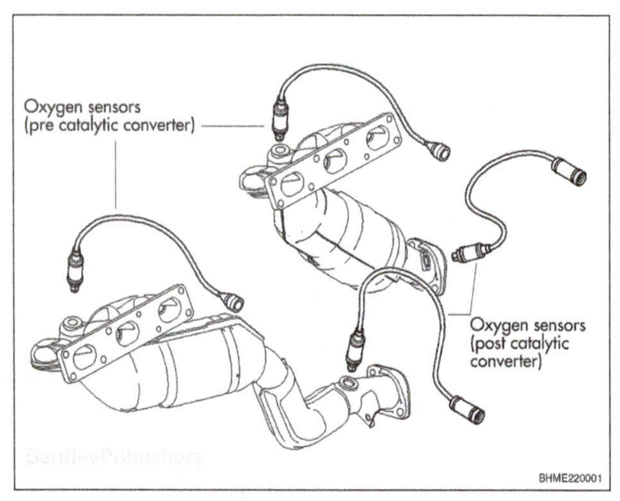

Figure 2 1996 and later cars use two oxygen sensors per catalytic converter. The front (precat) sensor is the main sensor for fuel trim. The rear (postcat) sensor is used primarily as a monitoring sensor for catalytic converter efficiency. The signal from the rear sensor should sense very little oxygen and have a steady (high) voltage signal output.

Within the group of Zirconia sensors are:

- Single-wire unheated narrow-band sensors used by the earliest electronic fuel-injected cars.
- Three-wire and four-wire heated narrow band sensors used from the early 1980s through about the mid-2000s.
- Heated wide-band sensors used in most cars since the mid-2000s.

You may need to consult a service manual to determine which type of oxygen sensor you have, but the information can usually be determined by examining the number of wires and the signal voltage as follows:

- Only a handful of early 1990s cars used Titania sensors. But if a voltage near 5V is present on any of the wires, the sensor is likely a Titania sensor.
- If the sensor has five or six wires, and if a voltage near 2.6V is present on any of the wires, the sensor is likely a heated wide-band (air/fuel) Zirconia sensor.
- Otherwise, the sensor is likely a narrow-band Zirconia sensor. Its signal should range from 0.1 to 0.9 volts.
- If the sensor has only a single wire, it is, obviously, a single-wire unheated Zirconia sensor.
- If the sensor has three or four wires, it is likely a heated Zirconia narrow-band sensor.

Many cars use Zirconia wide-band sensors as the primary (precat) sensors, but employ Zirconia narrow-band sensors as the secondary (postcat) sensors.

Signal

The signal from the narrow-band oxygen sensor is unique as compared with other automotive sensors in that it is a dynamic but non-periodic signal. That is, the signal changes several times per second, but it doesn't follow an exact repeating pattern.

You may read that the signal from the oxygen sensor has a sinusoidal (sine wave-like) shape and that the readings "oscillate." The readings *do* oscillate, but the signal is not, strictly speaking, a sine wave. If the signal were truly sinusoidal, it would be centered on zero, the frequency would change in response to some stimulus, and there would be information in the frequency that is used by the car's ECU. None of these are the case.

If you measure the oxygen sensor's voltage with a multimeter, it should be centered at about 0.45 volts, and change from full lean (0.1V) to full rich (0.9V) approximately one to five times per second. We could say that the changes occur at a frequency of 1 Hz to 5 Hz. But, to be careful, the frequency itself is not a parameter that carries information in the same way as it does with a sine wave or a square wave.

The transitions between the lean and rich conditions should occur quickly, within about 0.1 seconds. As the sensor ages, the response time can slow due to contamination.

Figure 3 shows the signal from a narrow-band oxygen sensor plotted on an oscilloscope.

If the car is a post-1996 OBD-II car and has a secondary oxygen sensor, when car's emission control system is working correctly, the second oxygen sensor should read a relatively consistent value – it should not be oscillating like the primary sensor.

Circuit

Early (pre-early-1980) narrow-band Zirconia oxygen sensors may be unheated and have only one wire, but most O2 sensors on post-OBD-II European cars have four wires – signal and ground for the sensor itself, and signal and ground for the heater. Some sensors may use the body of the sensor as the heater ground to produce a three-wire sensor.

In some vehicles, the oxygen sensor's heater is always powered whenever the key is turned, but others use a Pulse Width Modulated (PWM) signal on the negative leg of the heater circuit to modulate the heater on and off. You may need to consult the manufacturer's wiring diagram for the specifics on your vehicle to learn whether your O2 sensor's heater is run this way.

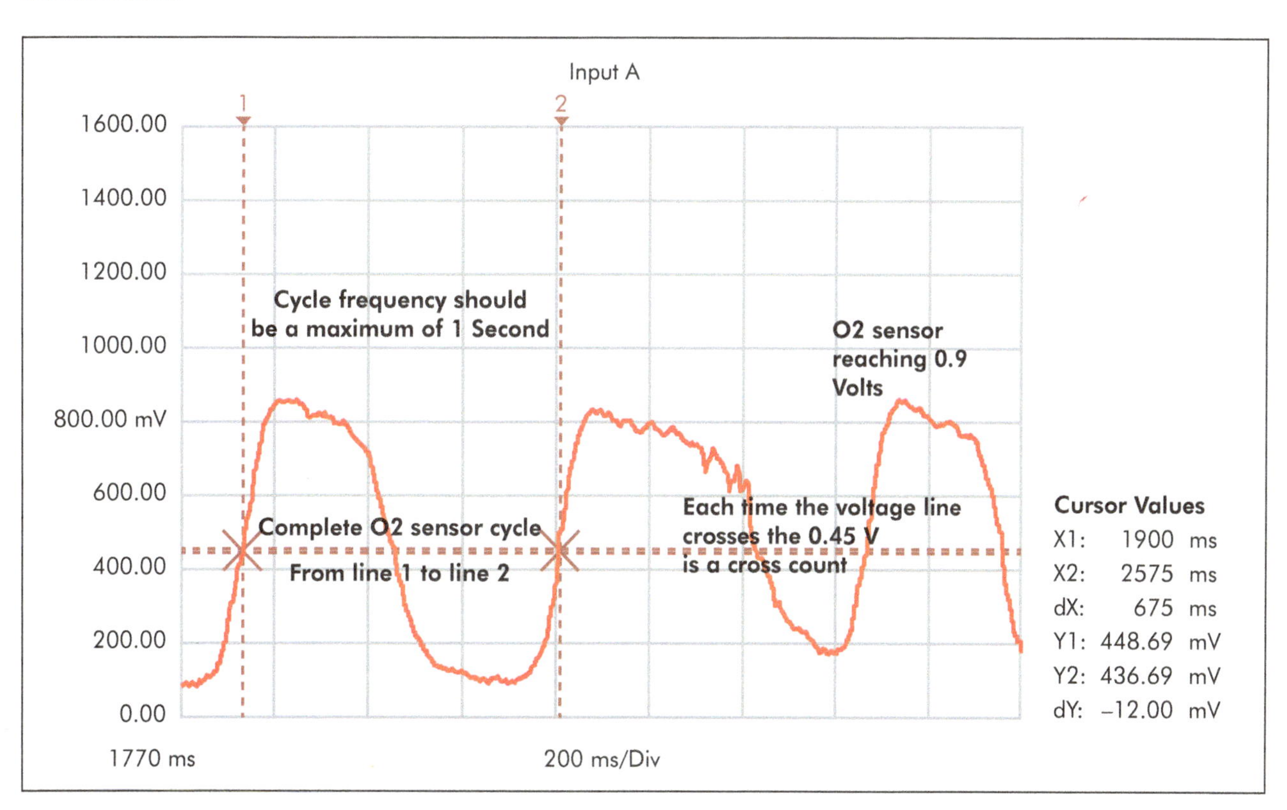

Figure 3 This scope trace shows the precat oscillating narrow-band O2 sensor signal switching rich-lean-rich-lean. The ECM is constantly adjusting fuel trim (injector pulse width) using feedback from the O2 sensor.

You may need to consult a repair manual for the connector location. The connector is sometimes hidden inside a tray with a snap-on cover that protects other connectors from the elements. Following the cable that comes out of the sensor will usually reveal the location of the connector.

Note that cars may have had their OEM oxygen sensors replaced with generic equivalents, so the wire colors may not match what is specified on the wiring diagram.

Figure 4 shows a wiring diagram for the 4-wire oxygen sensor in a 2005 BMW M3. A few things worth noting while reviewing the diagram:

- A 30-amp fuse protects the sensor heater circuit: red/white (+) wire and brown (–) wire.
- The 4-pin connector at the sensor is identified a X62101.

Oxygen Sensor Connector

Connector X62101	Description
Pin 1 (red/white)	Heater positive (+)
Pin 2 (brown)	Heater ground (–)
Pin 3 (black)	Sensor ground (–)
Pin 4 (yellow)	Sensor signal

- In this diagram, the ECM supplies both the ground for the heater circuit and the ground for the sensor itself. Note the heater ground may be a Pulse Width Modulated (PWM) signal. In other words, it may be turned on and off by the ECM based on engine operating conditions.

Testing

We will concentrate on testing four-wire narrow-band Zirconia sensors. Notes on testing the other types of sensors are contained in the **Technical Notes** section at the end of this chapter.

Any multimeter can be used to test the oxygen sensor, but because the signal changes 1 to 5 times a second between about 0.1V and about 0.9V, it is difficult to see on a meter's display. Thus, it is enormously helpful if the multimeter has a "bar graph" function that emulates a swing needle, and a MIN-MAX function so the minimum and maximum can be easily recorded and examined.

The primary oxygen sensor may be located in the exhaust manifold and accessible from under the hood. The secondary oxygen sensor may only be accessible from beneath the car. The location of the connectors can vary. Consult a repair manual for your car for specific information.

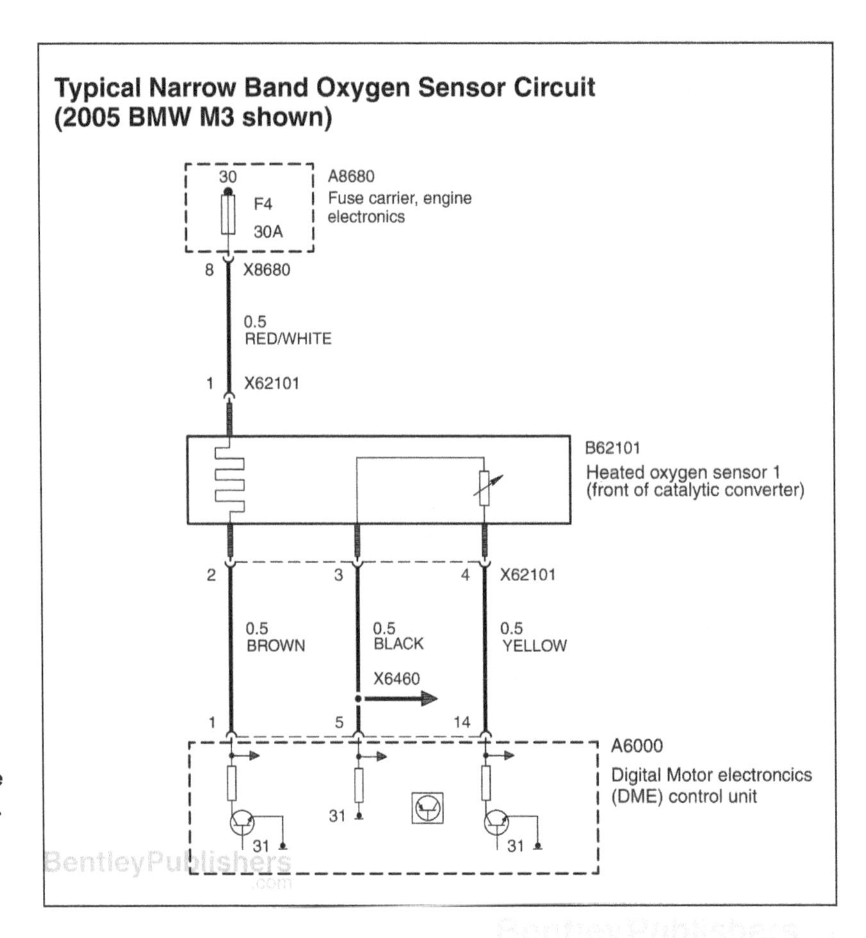

Figure 4 Wiring schematic for BMW M3 showing a 4-wire sensor. A 30-amp fuse protects the heater circuit (Pins 1 and 2). Pin 4 in connector X62101 is the signal wire to the ECM. Pin 3 is the sensor ground wire.

The following warnings and cautions apply to all tests.

WARNING —

It may be necessary to jack up the car in order to access the oxygen sensors or their connectors from below. ALWAYS jack up the car with a high-quality floor jack, and then support the car on jack stands using the support points as described in the car's factory manual. NEVER crawl underneath a car that is supported only by a jack, as death may result.

CAUTION —

Oxygen sensors are easily damaged. Do not drop or bang on oxygen sensors. To avoid possible engine damage, do not mix up oxygen sensor connectors. Mark connectors before disassembling.

Use extreme care when working at or near hot exhaust system components or hot engine parts. Wear heavy work gloves, a long-sleeve work shirt, and eye protection.

Disconnect the connector from the oxygen sensor. Physically inspect both sides of the connector. Make sure no wires are broken and no pins are pushed out.

Determining Pinouts

Pinouts are best determined using factory service information (a wiring diagram). However, we have found that the following procedure is useful for confirming that the factory service information is correct, or determining pinouts if a wiring diagram is not available.

- Some but not all narrow-band sensors use two wires of the same color for the heater circuit (e.g., two black or two white wires). If you see two wires of the same color on the sensor pigtail, note which pins they go to, then unplug the connector, set the multimeter to measure voltage attach, the negative probe to the chassis of the car, turn the ignition key on, and probe the wiring harness connector with the positive multimeter lead.
- When you detect battery voltage, that should be the pin for the voltage to the heater, and the other wire of the same color should be heater ground.
- If you don't see two wires of the same color, go through the same steps, probing all four pins to find the pin with battery voltage on it. That will be the power to the heater.

- Turn the ignition off, set the multimeter to measure continuity, connect the black lead to chassis ground, and probe the remaining three pins on the wiring harness to find two ground wires by checking for continuity with ground. One ground will be for the heater, the other for the sensor.
- The third wire will be signal for the sensor.
- Use trial and error to determine which ground wire is for the heater and which is for the sensor.

Oxygen Sensor Heater Test

See Test Set-Up on page 30-2

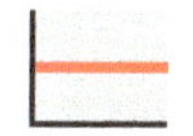

Look for a
Static Value

- Locate and disconnect the connector from the oxygen sensor.
- Set the multimeter to measure DC volts.
- Connect the red lead to the heater(+) pin on the wiring harness side of the connector.
- Connect the black lead to the heater (–) pin on the wiring harness side of the connector.
- Turn the key to ignition ON, but do not start engine.
- Check that battery voltage (12.6V) is present at the connector.

NOTE —

Some ECMs will cycle the heater voltage on and off based on engine operating conditions. If voltage is not present, turn the key on and off or run the engine and recheck for voltage.

- If battery voltage is not present on the heater wires, review the appropriate wiring diagram to troubleshoot the faulty circuit. If a scan tool shows that heater malfunction DTCs are present, and if power is present on the connector, replace the sensor.

Oxygen Sensor Signal Test

See Test Set-Up on page 30-3

Look for a
Quickly Changing Value

As discussed above, the primary narrow-band oxygen sensor should oscillate between 0.1 and 0.9 volts about 1 to 5 times per second with a response time of no longer than 0.1 seconds, but the secondary oxygen sensor should have a much flatter response centered around 0.45 volts. These values are depicted in the **Typical Normal Values for Oxygen Sensor Signal Table**.

Typical Normal Values for Oxygen Sensor Signal (Narrow-Band Zirconia Sensor)	
Data	**Value**
Precat (Primary)	
Min and Max Voltage	0.1V to 0.9V
Average (Center) Voltage	0.45V
Cycle time	1 to 5 times per second
Transition Time	0.1 seconds
Postcat (Secondary)	
Min and Max Voltage	0.1V to 0.9V, but slower oscillations
Average (Center) Voltage	0.45V or higher

To check for this signal:

- Locate the connector from the oxygen sensor. The sensor must remain connected for the feedback loop to function.
- Set the multimeter to measure DC voltage. If the meter has a "bar graph" function and a MIN/MAX function, activate these functions.
- Back-probe the meter's red (+) lead into the wire corresponding to the sensor signal wire.
- If the sensor is a single-wire sensor, connect the meter's black (–) lead to chassis ground.
- If the sensor is a 3-wire or 4-wire sensor, it may be necessary to back-probe the signal ground wire instead of using chassis ground.
- Start the engine and allow it to warm up. Raise the idle speed a few times to help get the sensor hot and into closed loop operation.
- Check that the voltage is fluctuating between 0.1 volts and 0.9 volts, with an average of 0.45. volts. This is easiest to verify with the meter's bar graph and in the MIN/MAX function.
- An unlit propane torch can be used to introduce a small amount (two seconds) of propane into the engine's intake to create a rich mixture. This should drive the sensor toward 0.9 volts. If it does not, the sensor is likely defective.
- A vacuum line can be disconnected from the engine to create a lean mixture. This should drive the sensor toward 0.1 volts. If it does not, the sensor is likely defective.
- Turn the ignition off and remove the test leads.

NOTE —

Voltage from the secondary oxygen sensor (sensor 2, postcat) should be, on average, higher than sensor 1, and the signal should be more flatlined. The presence of the catalytic converter should take most of the oxygen fluctuations out of the signal. If the signal at the secondary sensor is the same as the signal at the primary sensor, it does not represent a sensor failure – it likely represents a catalytic converter failure and a cat efficiency fault code will likely be stored.

Supplemental Testing

In addition to the two tests described above, the following tests can be performed.

- The resistance across the heater wires can be checked to verify the heater is neither open nor shorted.
- The sensor can be removed from the car, clamped in a vice, and heated with a propane torch which drives the sensor rich, and its reading can be measured with a multimeter to confirm it approaches 0.9 volts and drops rapidly when the flame is pulled away.

NOTE —

The "heat the sensor with a torch" approach isn't as crazy as it sounds. It's actually a fairly effective way of performing a direct measurement. Plus, the act of heating does have some cleaning effect, burning impurities off the sensor.

If the sensor passes the above tests, and you have an Oxygen Sensor Slow Response" diagnostic trouble code, and you don't have a graphing scan tool or a scope, then you can't measure the response time, and thus you may have no choice but to replace the sensor. If the sensor has more than 120k miles on it, then it's a good idea to replace it. Performance diminishes with age as contaminants accumulate on the sensor tip and gradually reduce its ability to produce voltage.

Is a Scan Tool Useful?

On a post-1996 car, yes. In fact, a code reader or scan tool is usually the *only* way an owner is alerted to a bad oxygen sensor.

In addition to reporting fault codes, if you have a scan tool that directly reports live data from the oxygen sensors, you should be able to see the actual numerical readings from the oxygen sensor. If you have a scan tool that can graph the changing signal, so much the better. Whether the graph from the scan tool is good enough to verify that the response time of the sensor is less than 0.1 seconds depends on the data bus and the update rate of the scan tool. In addition, if the O2 or its wiring is faulty, the scan tool should report a fault code.

Is an Oscilloscope Useful?

Yes. If you have a scope, you can plot the data from the primary and secondary oxygen sensors and verify that the primary one is varying like it's supposed to and the graph of the secondary one is much flatter. If you absolutely need to verify that the response time of the sensor is less than 0.1 seconds, this is the best way to do it.

Additional Information

Related Chapters

- **Common Multimeter Tests** for a discussion on how to use a multimeter to measure voltage and resistance.

Technical Notes

Below is additional information on the different types of oxygen sensors, and on the mechanism by which oxygen sensors produce their signal.

Zirconia Sensors (Narrow-Band)

The most common type of oxygen sensor is the narrow-band Zirconia oxygen sensor. This type of sensor creates a voltage in much the way a battery does. It uses a zirconium oxide ceramic material which becomes conductive for oxygen once the material is heated above 300 degrees C. The ceramic and oxygen act together a bit like the acid and lead plates in a battery. A portion of the ceramic protrudes into the exhaust stream, while the other end hangs out into ambient air.

If the two ceramic ends are exposed to the same amount of oxygen, then oxygen ions are evenly distributed, there's no flow of ions, and no voltage is generated. But if the engine is running lean (too much oxygen in the exhaust) or too rich (too little oxygen in the exhaust), the imbalance of oxygen between the in-exhaust part and the ambient air part causes a flow of charge, creating a voltage.

For optimum fuel efficiency, the ideal ratio of air and fuel is 14.7:1. This 14.7:1 air/fuel ratio is known as a stoichiometric ratio or Lambda $(\lambda) = 1.0$. The Zirconia oxygen sensor is designed so that, when the air/fuel mixture is at its optimum 14.7:1 stoichiometric ratio, the sensor outputs 0.45 volts, and deviations from this ideal point generate a voltage that ranges from about 0.1 volts (lean) to about 0.9 volts (rich). It is referred to as a *narrow-band sensor* because the voltage signal goes up or down quickly when the mixture wanders even a small amount away from the ideal 14.7:1 ratio, making it so that a rich or lean condition results in a sharp voltage change. This sharp change is shown in **Figure 5**.

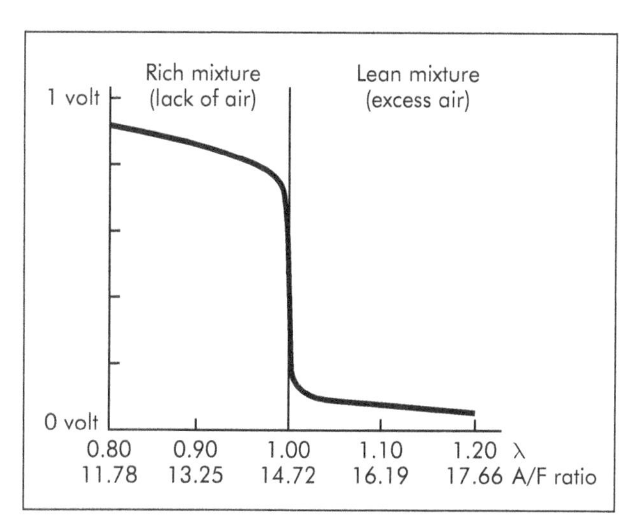

Figure 5 The narrow-band oxygen sensor is basically a rich/lean sensor. This graph shows how the output voltage quickly jumps to 0.9 volts when the air/fuel mixture goes rich, and quickly drops to 0.1 volts when lean.

Single-Wire Unheated Narrow-Band Sensors

The single-wire narrow-band O2 sensor, widely known as the "Lambda sensor," was introduced in 1976 on Volvo and Saab cars. As its name implies, it had just one wire and no heater. All vehicles sold in California got the Lambda sensor in 1980. Federal emission laws made O2 sensors required on all cars and light trucks built since 1981.

The single-wire O2 sensor also produces a 0.1 to 0.9 volt signal once warmed up. The only difference between it and the heated O2 sensors that followed is that the single-wire sensor needs to be heated by the exhaust gas. This requires raising the idle speed for a few minutes before checking the sensor's output voltage.

3- and 4-Wire Heated Narrow-Band Sensors

By the early 1980s, manufacturers were replacing the single-wire sensors with heated O2 sensors. Rather than wait for exhaust gas to heat them up to operating temperature, these sensors employ an internal heater to warm them up and thus be able to manage emissions as quickly as possible. Most heated narrow-band sensors use four wires (signal and ground, and heater power and ground), but some employ three wires, using the car's chassis as heater ground.

Wide-Band Sensors

The wide-band Zirconia-type sensor, also know as Air/Fuel (A/F) sensors, was developed to meet the demands for increased fuel economy and lower emissions. As the name implies, it is capable of measuring a much wider range of air/fuel ratios. It was introduced in 1992 and has become widespread in its use as the primary upstream (precat) sensor. Note that postcat sensors are still often narrow-band sensors.

The wide-band sensor incorporates an internal electrochemical gas pump called a Nernst Cell. An electronic circuit controls the gas pump current to keep the sensor element voltage constant. So in effect, the pump current is what indicates the oxygen content in the exhaust. The current flow is small, about 0.02 amps or less. This sensor does not have the inherent lean-rich limits of the narrow-band sensor, thus allowing for more responsive fuel trim. The wide band Zirconia sensor is also used in diesel engines to comply with ULEV (Ultra Low Emission Vehicle) emission limits.

Wide-band sensors typically present a steady voltage (usually around 2.6V). A rich signal results in a negative current with a voltage signal slightly below 2.6V. A lean signal results in a positive current with a voltage slightly above 2.6V. If there is no change in current and voltage is steady at 2.6V, the air/fuel mixture is at stoichiometry.

In contrast with 4-wire narrow-band sensors, wide-band sensors typically employ five or six wires. A sample wiring diagram for a wide-band O2 sensor on a 2010 BMW X3 is shown in **Figure 6**.

Because the "signal" is actually the current, and owing to the small operating currents and the electronic circuitry, the best way to accurately check a wide-band sensor is by using a scan tool. The OBD-II software will detect most problems that affect the sensors and set a code.

Resistive-Jump (Titania) Sensors

A less common type of narrow-band Lambda heated sensor was the *resistive-jump* or *Titania* sensor. Titania sensors used a different ceramic element that responded faster but was more expensive than Zirconia sensors. They were used for a small number of years beginning around the mid-1990s.

This resistive-jump sensor does not generate its own voltage like the Zirconia sensor. Instead, it requires a 5-volt reference supply from the ECM. The signal voltage is then altered according to the engine's air/fuel ratio, with a lean mixture having a voltage as low as 0.4 volt to a rich mixture, producing a voltage in the region of 4.0 volts.

Unlike Zirconia sensors, Titania sensors do not need a pocket of atmosphere around the outside of the sensor. For this reason, they could be employed in off-road vehicles whose oxygen sensors might otherwise have suffered if water or mud accumulated around the exterior of the sensor.

To test this sensor, consult the appropriate wiring diagram for the heater and sensor pin designation. Be sure to back-probe the connected connector when checking the sensor signal to the ECM.

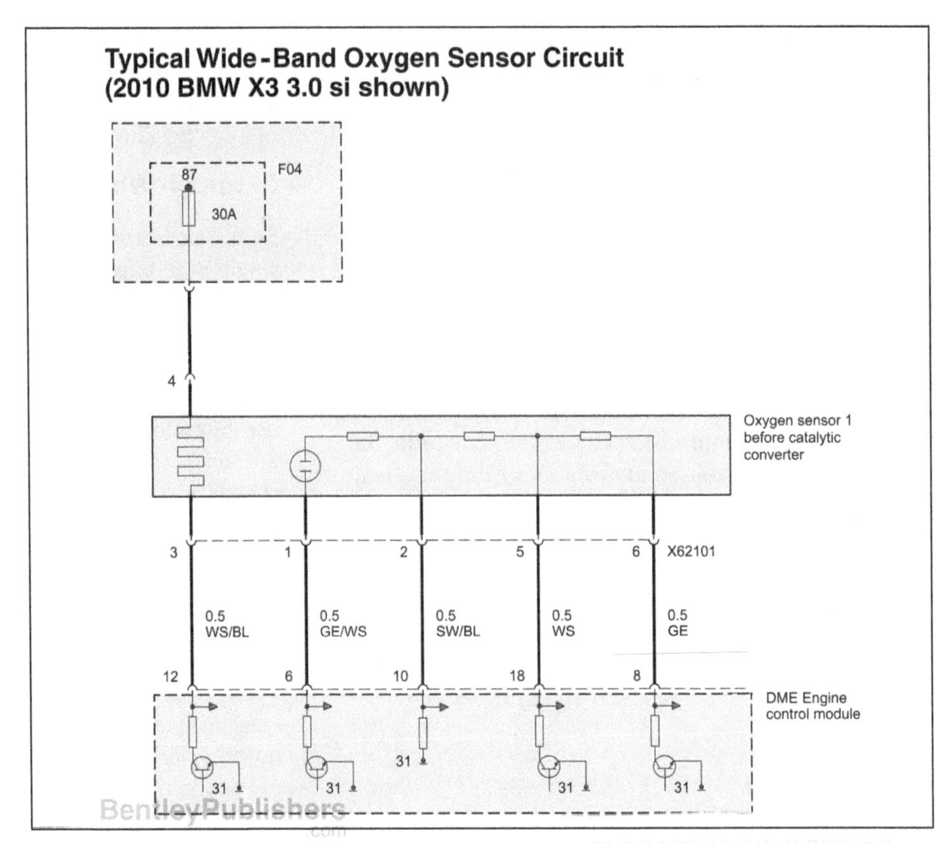

Figure 6 Wide-band oxygen sensor (aka A/F sensor) wiring schematic. The wide-band sensor typically has five or six wires. These advanced sensors should be checked using a scan tool. If faulty, the Check Engine Light (CEL) will illuminate and a Diagnostic Trouble Code (DTC) will be stored.

31

Testing Wheel Speed Sensors (WSS)

Multimeter Test Set-Ups

WARNING —

Make personal safety your top priority. Watch what you are doing, stay alert. Do not work on your car if you are tired, not feeling well, or in a bad mood. Common sense, alertness, and good judgment are crucial to safe work on your vehicle.

Protect yourself from shock, trauma, and hazardous chemicals: Wear safety glasses, work boots, and protective gloves when working on your vehicle. Have hearing protection available for work that generates noise.

Be sure the work environment is well lit, properly ventilated, and that the vehicle is on a level surface. The floor, vehicle, and components inside the vehicle should be dry to minimize the possibility of shocks.

This book is designed as background technical training and as a general overview of how to do electrical diagnosis and service. It is designed to be used in conjunction with the appropriate service information for your vehicle, not to replace factory service information.

If you lack the skills, tools and equipment, or a suitable workshop for any procedure described in this book, have the job done by a qualified shop or an authorized dealer.

For additional warnings and cautions, see the **Safety** *chapter and the warnings, cautions, and notices in your vehicle manufacturer's service information.*

31

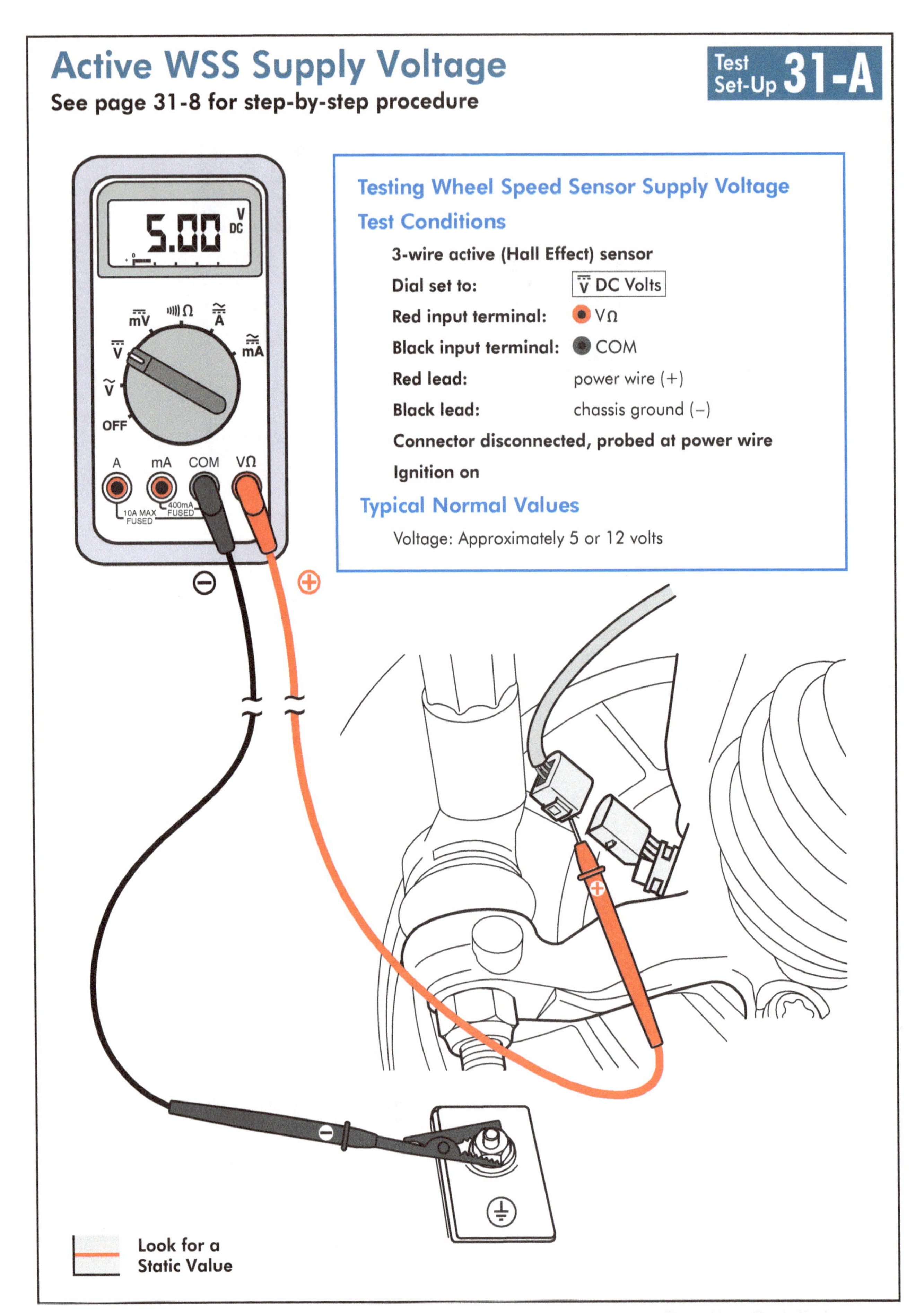

Active WSS Supply Voltage
See page 31-8 for step-by-step procedure
Test Set-Up 31-A
Testing Wheel Speed Sensor Supply Voltage
Test Conditions
3-wire active (Hall Effect) sensor
Dial set to: V DC Volts
Red input terminal: VΩ
Black input terminal: COM
Red lead: power wire (+)
Black lead: chassis ground (–)
Connector disconnected, probed at power wire
Ignition on
Typical Normal Values
Voltage: Approximately 5 or 12 volts
5.00
V DC
mV
Ω
A
V
mA
OFF
A
mA
COM
VΩ
10A MAX FUSED
400mA FUSED
Look for a Static Value

Active WSS Signal

See page 31-8 for step-by-step procedure

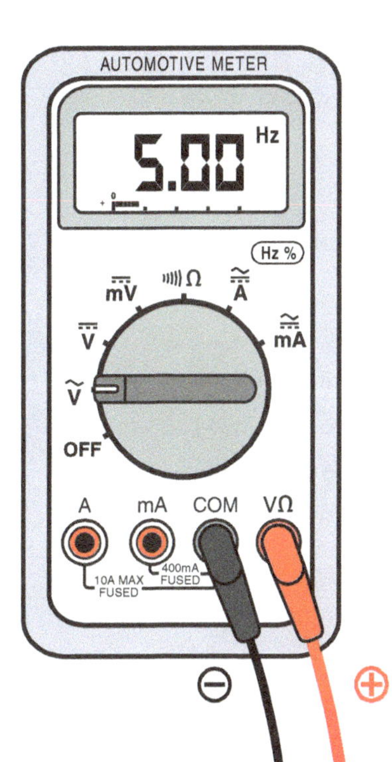

Testing Wheel Speed Sensor Signal

Test Conditions

3-wire active (Hall Effect) sensor

Automotive meter that reports frequency of a square wave

Dial set to:	Ṽ AC Volts | Hz %
Red input terminal:	VΩ
Black input terminal:	COM
Red lead:	sensor signal wire (+)
Black lead:	chassis ground (–)

Connector remains connected, back-probed at signal wire

Ignition on, wheel spinning

Typical Normal Values

Signal Frequency: Varies with wheel speed

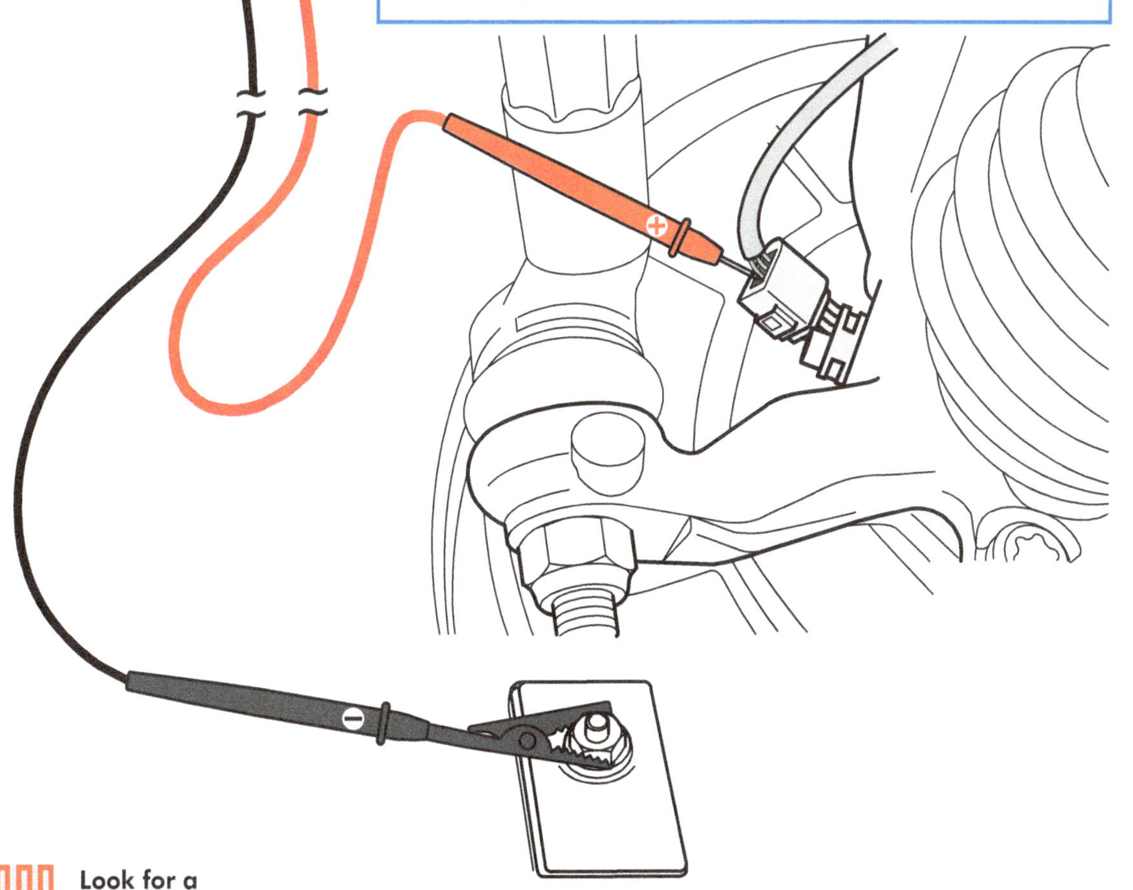

Look for a Square Wave

Passive WSS Resistance

See page 31-9 for step-by-step procedure

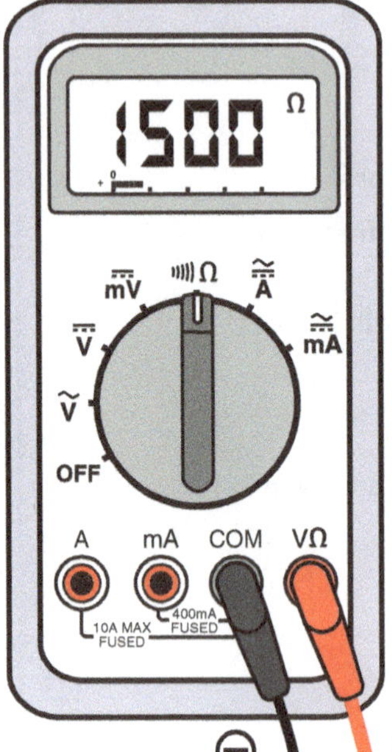

Testing Wheel Speed Sensor Resistance

Test Conditions

2-wire passive (VRT) sensor

Dial set to:	Ω resistance
Red input terminal:	VΩ
Black input terminal:	COM
Red lead:	sensor signal pin (+)
Black lead:	sensor ground pin (–)

Connector disconnected, testing at sensor

Engine off

Typical Normal Values

Resistance approximately 1500 Ω

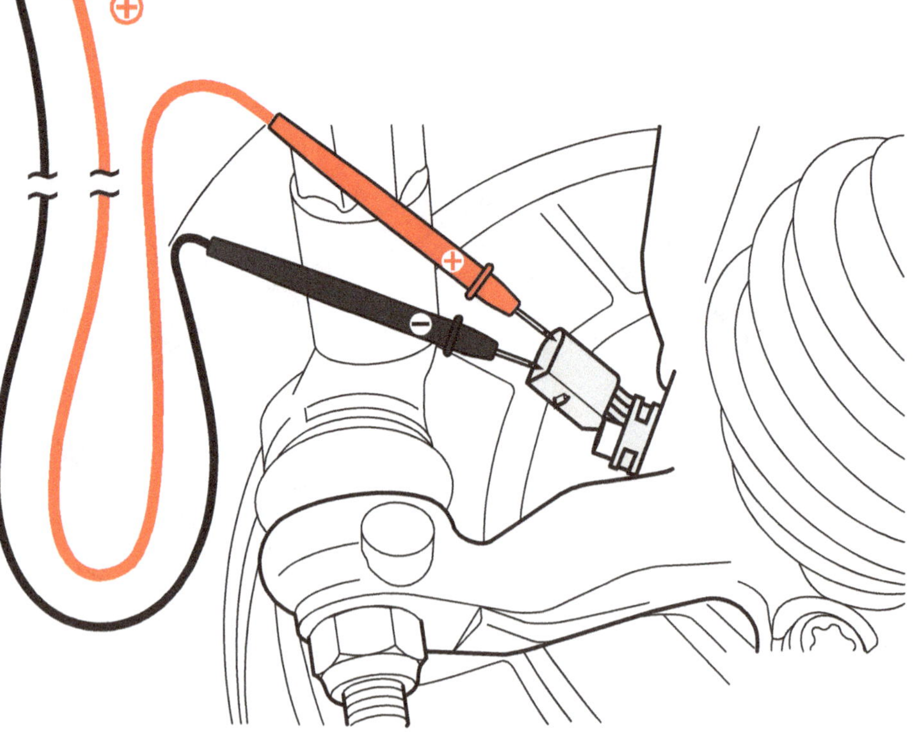

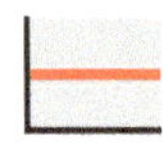

Look for a Static Value

Passive WSS Signal

See page 31-9 for step-by-step procedure

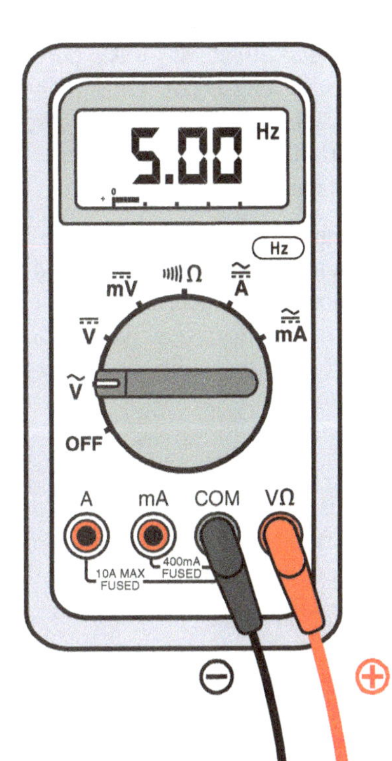

Testing Wheel Speed Sensor Signal

Test Conditions

2-wire passive (VRT) sensor

Meter that measures AC frequency

Dial set to: Ṽ AC Volts | Hz

Red input terminal: VΩ

Black input terminal: COM

Red lead: sensor signal wire (+)

Black lead: chassis ground (–)

Connector remains connected, back-probed at signal wire

Wheel spinning

Typical Normal Values

Signal Amplitude: 1 to 5 volts

Signal Frequency: Varies with wheel speed

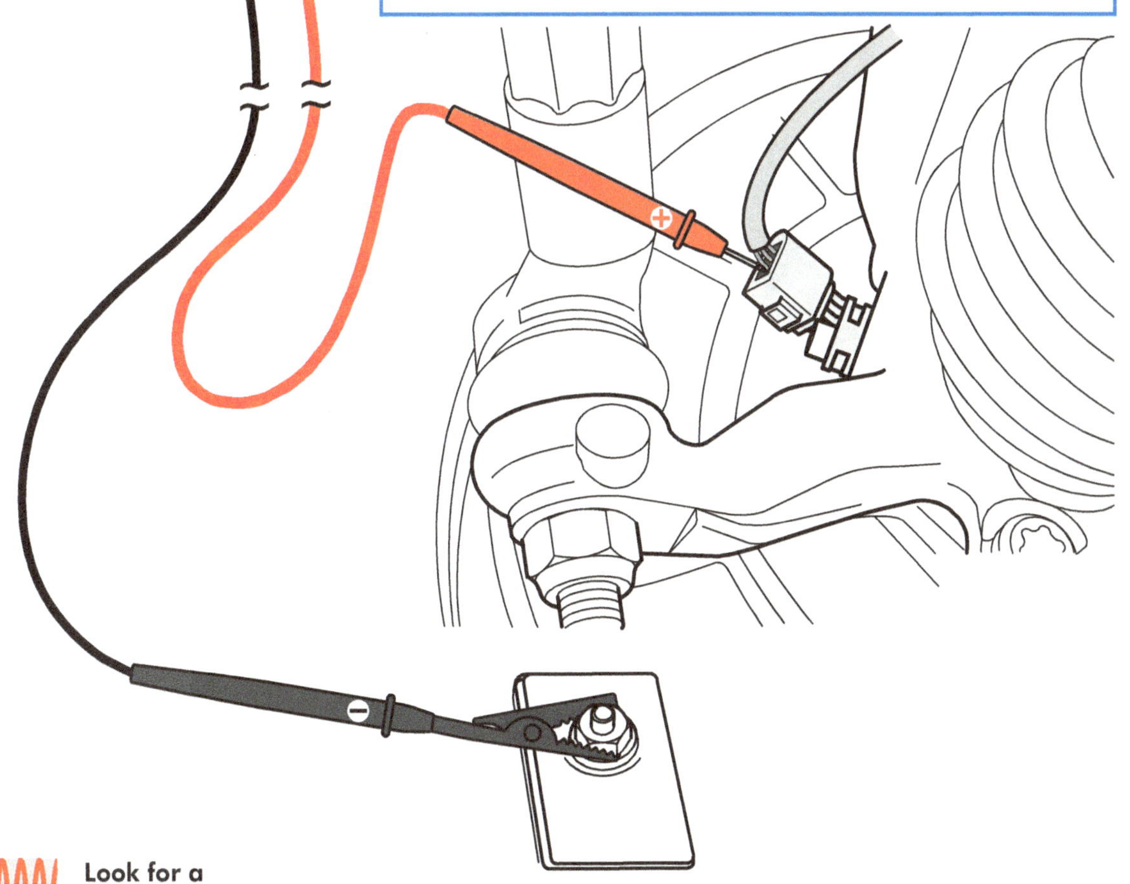

Look for a Sine Wave

Wheel Speed Sensor

Antilock brake systems (ABS) became standard equipment on most European cars in the mid-1980s. Sensors are employed to determine when the wheels are about to lock up. The ABS hydraulics are then triggered to automatically pulse the brakes rapidly, allowing the vehicle to be steered during panic braking.

An ABS sensor, also called a Wheel Speed Sensor (WSS), is present on all four wheels. The wheel speed sensors were originally used only by the ABS, but now are commonly used by a variety of other systems on the car (traction control, for example). See **Figure 1**.

Most owners are alerted to ABS problems by an ABS or a traction control dashboard light being lit. Unfortunately, these lights do not tell you which wheel speed sensor is causing a problem, or even whether the problem is due to a wheel speed sensor as opposed to some other ABS-related component. Depending on the age of the car, connecting a scanner and reading codes may provide more information, but it may be necessary to test the sensors directly to verify whether any of them are the cause of the cause of the malfunction.

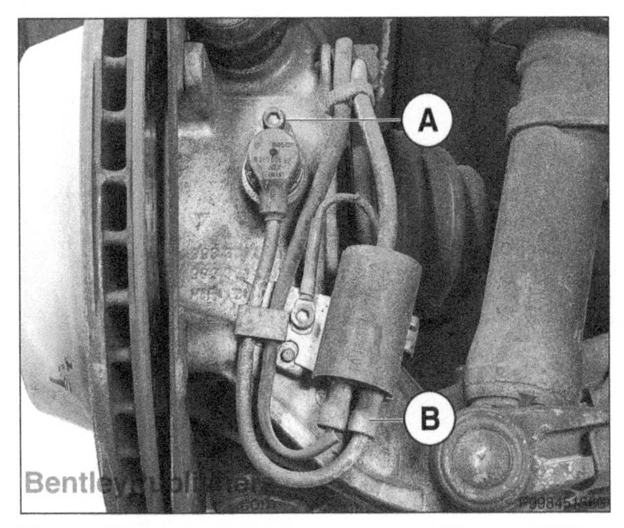

Figure 1 Wheel speed sensor connector (**B**) on a 1997 Porsche 911 (993). Wheel speed sensor mounting bolt shown at (**A**).

Operation

There are two kinds of wheel speed sensors: Active sensors that use the Hall Effect principle, and passive sensors that rely on Variable Reluctance Transducers (VRT). The workings of both types of sensors are discussed in the **Introduction to Sensors** chapter.

Types

To test a wheel speed sensor, you need to know which type of sensor you have. Most manufacturers changed their wheel speed sensors in approximately the year 2000 from passive VRT sensors to active Hall Effect sensors. The passive VRT sensor has a large external clearly visible toothed reluctor wheel on the outside of the wheel's hub, whereas the active Hall Effect sensor generally has small magnets embedded someplace inside the hub that is not plainly visible.

In addition, the active Hall Effect sensor is a powered sensor whereas the passive VRT is not. Thus, if any wire to the sensor carries power, it is a Hall Effect sensor. For more specific information, see the **Determining Pinouts** section in this chapter.

Beginning in approximately the mid-2000s, a third kind of rotation measurement sensor began to be used for wheel speed sensing applications Anisotropic Magnetoresistive (AMR) sensors are active sensors similar to Hall Effect sensors, but with the ability to detect direction of rotation.

Signal

Like any other sensor using a VRT or a Hall Effect device, wheel speed sensors produce dynamic analog signals that are periodic. In both cases, the frequency is the parameter used by the ECM.

The active AMR sensor produces a signal that is a dynamically-changing current in the form of a PWM square wave that toggles between 7mA and 14mA. Without a high-sensitivity current clamp connected to an oscilloscope, the only way for to evaluate the health of an AMR sensor is to connect a brand-specific scan tool to the car. Thus, independently testing the signal generated by an AMR sensor is outside the scope of this chapter.

The active Hall Effect sensor produces a square wave signal whose frequency increases as wheel speed increases.

The passive VRT sensor produces a sine wave signal whose amplitude and frequency both increase as wheel speed increases. Note that when rotation speed is very slow, there may be no visible amplitude to the sine wave. In contrast, the active Hall Effect sensor produces a square wave at any rotation speed.

Circuit

Since the Hall Effect is an active sensor, it has power supplied, usually on a third wire. However, some Hall Effect sensors combine signal and ground on the same wire, thus requiring only two wires.

Since the VRT is a passive sensor, it has only two wires – signal and ground.

Non-model-specific wiring and signal information are summarized in the **WSS Signal and Wires Table**.

WSS Signal and Wires

	Active (Hall Effect)	Passive (VRT)
Wires	Usually 3 (signal, ground, power), but may be two (signal/ground, power)	2 (signal, ground)
Signal	Square wave	Sine wave
Important Parameter	Frequency	Frequency

A specific wiring example for the wheel speed sensor on a 1997 Porsche 911 (993) is shown in **Figure 2**. Each sensor has two wires, which means the sensor type could be a passive VRT sensor or a two-wire active Hall Effect sensor. However, on this wiring diagram, the two small rectangular icons shown inside each sensor represent a magnet and a coil, indicating that these are passive VRT sensors.

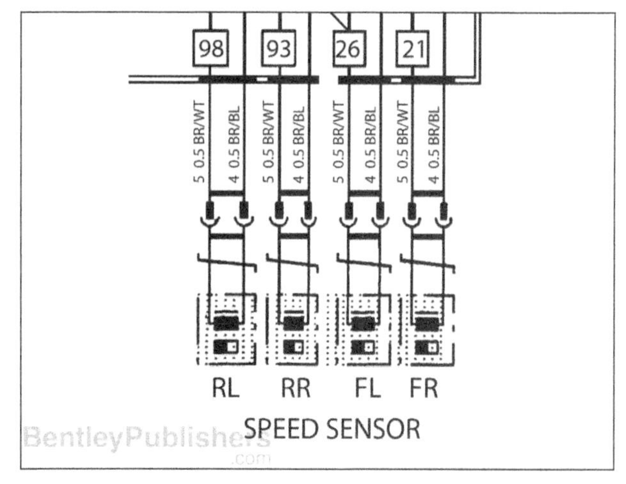

Figure 2 Wiring for the wheel speed sensors on a 1997 Porsche 911 (993). This shows the sensors as having two wires. The small icons for each sensor represent a magnet and a coil, identifying the sensor type as passive VRT.

Testing

For each sensor type (active or passive), there is a static test (wheel not spinning) and a signal test (wheel spinning). The tests have different equipment requirements, as shown in the **Equipment Needed for WSS Testing Table.**

Equipment Needed for WSS Testing

	Active (Hall Effect)	Passive (VRT)
Static test	Multimeter	Multimeter
Signal test	Automotive multimeter that detects a square wave and reports its frequency, or oscilloscope	Multimeter that measures AC frequency, or oscilloscope

If you have an active (Hall Effect) sensor, conduct the two tests below labeled "Active."

If you have a passive (VRT) sensor, conduct the two tests below labeled "passive."

Note that, in testing, we don't care what the actual frequency value is; we just need to verify that a square wave or sine wave is present and that its frequency increases with increasing wheel speed.

You may need to consult a repair manual to learn the location of the connector, as it may be in the wheel housing, in the engine compartment, in the luggage compartment, or even under the rear seat. Note that there is substantial variation in the form factor of the connector. In the illustrations in this chapter, we show a fixed socket mounted proud of the wheel's hub with a plug that mates to it, but on many European cars, the two connector halves are like a pen and its cap that snap together, and the mated pair then snaps into a weather-protecting plastic housing.

Before testing, make sure the sensor itself is not cracked. Disconnect the connector from the sensor. Physically inspect both sides of the connector. Make sure no wires are broken and no pins are pushed out or bent.

In the case of a VRT sensor, inspect the toothed wheel to make sure no teeth are broken or missing.

Determining Pinouts

Pinouts are best determined using factory service information (a wiring diagram). However, we have found that the following procedure is useful for confirming that the factory service information is correct, or determining pinouts if a wiring diagram is not available.

As discussed above, if the sensor has three wires, it is almost certainly a Hall Effect, but if it has two wires, it could be either a VRT or a Hall Effect.

To determine the pinouts:

- Safely raise the wheel so that the wheel can be spun by hand.

WARNING —

Make sure the vehicle is stable and well supported at all times. Use a professional automotive lift or jack stands. A floor jack is not adequate support.

- Locate the connector and disconnect it.
- Turn the key to ignition ON, but do not start engine.
- Set the meter to measure DC voltage.
- Connect the black multimeter probe to any chassis ground.
- On the wiring harness side of the connector, touch the red multimeter probe to each of the pins.
- If one of them reads 5 or 12 volts, that pin is the supply voltage to a Hall Effect sensor, even if there are only two wires. Note that this supply voltage is sometimes referred to as a "reference voltage," even though that term is really meant for resistance-based sensors, and this isn't one.
- Turn the ignition off.

To determine the other pinouts:

- Set the meter to measure resistance.
- Connect the black multimeter probe to any chassis ground.
- On the wiring harness side of the connector, touch the red multimeter probe to each of the other pins. The pin at which the multimeter registers continuity (near zero ohms) should be the ground pin.
- If the sensor has only two wires and you've found voltage on one of the wires, the ground pin is also the signal pin of a Hall Effect sensor.
- But if the sensor has three wires and you've identified one as power and one as ground, the remaining pin should be the signal pin.

Active WSS Supply Voltage Test

See Test Set-Up on page 31-2

Look for a Static Value

In this test, we check for supply voltage to the active (Hall Effect) sensor.

Typical Normal Values for Active WSS Supply Voltage
5V or 12V, depending on engine management application

- Set the multimeter to measure DC voltage.
- Safely raise the wheel so that the wheel can be spun by hand.

WARNING —

Make sure the vehicle is stable and well supported at all times. Use a professional automotive lift or jack stands. A floor jack is not adequate support.

- Disconnect the connector to the wheel speed sensor.
- Turn the key to ignition ON, but do not start engine.
- Working at the wiring harness side of the connector, touch the red multimeter probe to each of the pins.
- The multimeter should read 5V or 12V on the wire carrying the supply voltage. If there is no voltage on the supply voltage wire, the problem must be troubleshot by consulting a wiring diagram to trace the wire back to the ECM to determine where the voltage is being interrupted. The Hall Effect sensor will not work unless there is a supplied voltage.

Active WSS Signal Test

See Test Set-Up on page 31-3

Look for a Square Wave

In this test, we spin the wheel and check that the active (Hall Effect) sensor outputs a square wave whose frequency varies with wheel speed.

Typical Normal Values for Active WSS Signal (Square Wave)	
Amplitude	About half the supply voltage (e.g., about 2.5V or 6V)
Frequency	Increases with wheel speed

- Plug the connector back in.
- If the Hall Effect sensor has three wires, back-probe the signal wire and connect the red probe to it. Connect the black probe to chassis ground.
- If the Hall Effect sensor has two wires, whichever wire is not power is combined signal and ground. Back-probe that wire and connect the red probe to it, and connect the other probe to chassis ground.
- Set the automotive meter to detect the frequency of a square wave. The exact method is highly meter-dependent. It may involve using the pulse width or % duty cycle setting. We have generically indicated it via the "Hz %" button on the illustration below. See the **Tools for Dynamic Analog Signals** chapter for more information.
- Turn the key to ignition ON, but do not start engine.
- Spin the wheel and vary its speed. You should see the automotive meter display a frequency whose value increases and decreases with wheel speed.

Passive WSS Resistance Test

See Test Set-Up on page 31-4

Look for a Static Value

In this test, we check to see if the passive (VRT) coil is broken or short-circuited by measuring its resistance.

Typical Normal Values for Passive WSS Resistance
Approx. 1,500 ohms (will vary by manufacturer)

- Set the multimeter to measure resistance.
- Safely raise the wheel so that the wheel can be spun by hand.

WARNING —

Make sure the vehicle is stable and well supported at all times. Use a professional automotive lift or jack stands. A floor jack is not adequate support.

- Disconnect the connector to the wheel speed sensor.
- Touch the multimeter probes to the two pins of the wheel speed sensor (at the sensor, not at the wiring harness) to measure the resistance of the VRT coil. It doesn't matter which probe goes to which of the two pins.
- The resistance should be about 1,500 ohms. If the resistance is infinite (e.g., the meter reads OL), the coil in the VRT is broken. If the resistance is close to zero (e.g., the meter reads than one ohm), the coil is shorted out. In either case, the sensor is likely bad and should be replaced.

Passive WSS Signal Test

See Test Set-Up on page 31-5

Look for a Sine Wave

In this test, we spin the wheel and check that the passive (VRT) sensor produces a sine wave whose frequency varies with wheel speed.

Typical Normal Values for Passive WSS Signal (Sine Wave)	
Amplitude	About one volt
Frequency	Increases with wheel speed

- Reconnect the connector so the ECM receives the signal from the wheel speed sensor.
- Back-probe the connector, connecting the signal and ground wires to the multimeter.
- If your meter can measure the frequency of an AC voltage signal, set it to measure AC frequency. The setting for this varies meter to meter. In the illustration, we show it as an "Hz" button.
- Spin the wheel. You may need to get it spinning quickly before it outputs a voltage high enough that the multimeter can detect it. Then vary its speed. You should see the meter display a frequency whose value increases and decreases with changing wheel speed.
- If your meter can't measure AC frequency, it still should be able to measure AC voltage. You may see the AC voltage increase slightly as you raise wheel speed. The voltage will be small, less than one volt.

31

- If there's not a voltage or a frequency reading from the VRT sensor, check the air gap between it and the toothed reluctor wheel. On some cars, this gap is adjustable. The VRT coil should be very close to the teeth on the reluctor wheel but not touch them.
- If the VRT coil is close to the reluctor wheel, you're certain you have the multimeter configured and connected correctly, and you receive no signal, then the sensor is likely defective and should be replaced.

Supplemental Testing

If you have direct confirmation via an automotive multimeter or a scope that the sensor is outputting a signal, but there's still a problem (for example, a scan tool is still reporting there's no signal from the sensor), there may be a broken wire or a bad connection between the sensor and the ECM.

- Use a multimeter set to measure resistance to verify continuity between the ground wire on the connector and chassis ground.
- Use a multimeter set to measure resistance to verify continuity between the signal wire on the connector and the ECM. This will require a wiring diagram for your car, knowledge of the location of the ECM, and knowledge of which pin on the ECM connector carries the signal from the wheel speed sensor.

Is a Scan Tool Useful?

Possibly. Many people will be directed to the wheel speed sensor because the ABS warning light is on and a scan tool reports a code implicating the ABS. Note that, because the wheel speed sensor is not an emissions-related component, an OBD-II code reader may not provide any fault code information about it, particularly on an older car where the wheel speed sensor was used only for ABS. However, a brand-specific scan tool should give you information about the wheel speed sensors, and may tell you which sensor is responsible for the fault. Note that the wheel speed sensor reported as bad by a scan tool may be fine, and the problem may be rooted in a broken wire or unseated connector.

Is an Oscilloscope Useful?

Yes. If you have access to a oscilloscope, using it is better than using an automotive meter. Nothing is as good as spinning the wheel and actually seeing the sine wave or square wave from the wheel speed sensor. Plus, visual examination of the signal on an oscilloscope may reveal noise or an intermittent characteristic that is not detectable on an automotive multimeter.

Additional Information

Related Chapters

See the **Common Multimeter Tests** chapter for a discussion on how to use a multimeter to measure voltage and resistance.

- **Dynamic Analog Signals** for a discussion on square waves and sine waves.
- **Tools for Dynamic Analog Signals** for a discussion on how to use an automotive multimeter to detect sine waves and square waves.
- **Introduction to Sensors** for a detailed explanation on how active (Hall Effect) sensors and passive (VRT) sensors work.

Technical Notes

Whether the sensor is an active (Hall Effect) outputting a square wave or a passive (VRT) outputting a sine wave, the amplitude is not a relevant parameter. The car is not going to stop faster if the signal is half a volt higher. The amplitude needs to be high enough that the ECM detects the signal, but there's no information in the amplitude. The information is in the frequency.

When testing either the active (Hall Effect) sensor or the passive (VRT) sensor, note that we don't care about the actual frequency; we just need to verify that a signal is present and that its frequency changes with wheel speed. For this reason, if you're testing a Hall Effect sensor and your meter does not have a "square wave frequency" setting, you may be able to use a setting such as pulse width, verify that the meter is displaying a pulse width reading, and verify that the pulse width gets shorter (that the reading gets smaller) as the wheel is spun faster.

In addition to outputting a frequency-varying square wave, the Hall Effect sensors on some post-mid-2000-era cars may also output an "I am alive but not moving" pulse at about a 1Hz rate when the wheels are stationary. This frequency is so low that this pulse may not be picked up by an automotive meter and may only be visible on an oscilloscope.

32

Testing Crankshaft Position (CKP) Sensors

Multimeter Test Set-Ups

⚠ WARNING —

Make personal safety your top priority. Watch what you are doing, stay alert. Do not work on your car if you are tired, not feeling well, or in a bad mood. Common sense, alertness, and good judgment are crucial to safe work on your vehicle.

Protect yourself from shock, trauma, and hazardous chemicals: Wear safety glasses, work boots, and protective gloves when working on your vehicle. Have hearing protection available for work that generates noise. Be sure the work environment is well lit, properly ventilated, and that the vehicle is on a level surface. The floor, vehicle, and components inside the vehicle should be dry to minimize the possibility of shocks.

This book is designed as background technical training and as a general overview of how to do electrical diagnosis and service. It is designed to be used in conjunction with the appropriate service information for your vehicle, not to replace factory service information.

If you lack the skills, tools and equipment, or a suitable workshop for any procedure described in this book, have the job done by a qualified shop or an authorized dealer.

For additional warnings and cautions, see the **Safety**, *chapter and the warnings, cautions, and notices in your vehicle manufacturer's service information.*

Active CKP Sensor Supply Voltage

Test Set-Up **32-A**

See page 32-10 for step-by-step procedure

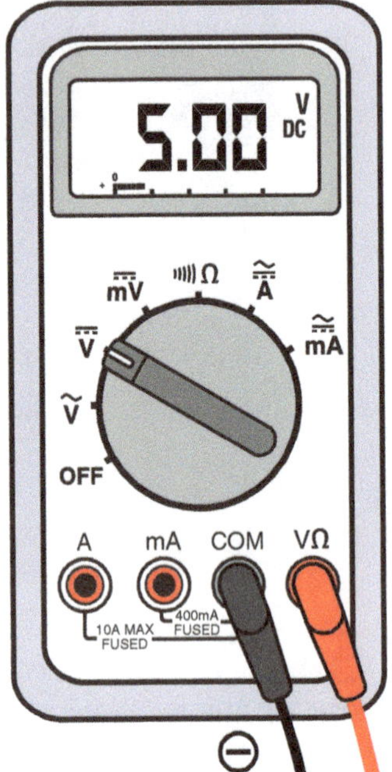

Testing for Sensor Supply Voltage

Test Conditions

3-wire active (Hall Effect) sensor

Dial set to: V DC Volts

Red input terminal: VΩ

Black input terminal: COM

Red lead: power wire (+)

Black lead: chassis ground (–)

Connector disconnected, probed at power wire

Ignition ON

Typical Normal Values

Voltage: Approximately 5 or 12 volts

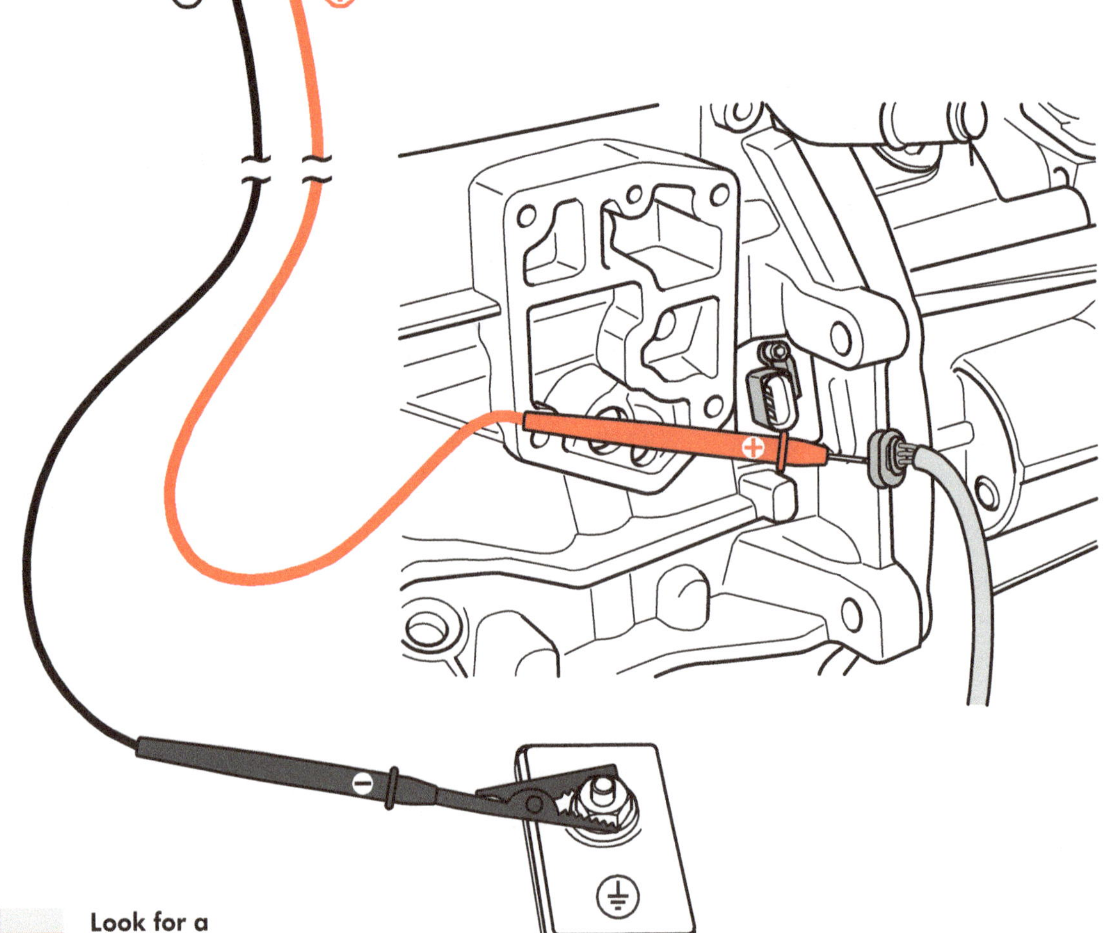

Look for a Static Value

Active CKP Sensor Cranking Signal

Test Set-Up **32-B**

See page 32-10 for step-by-step procedure

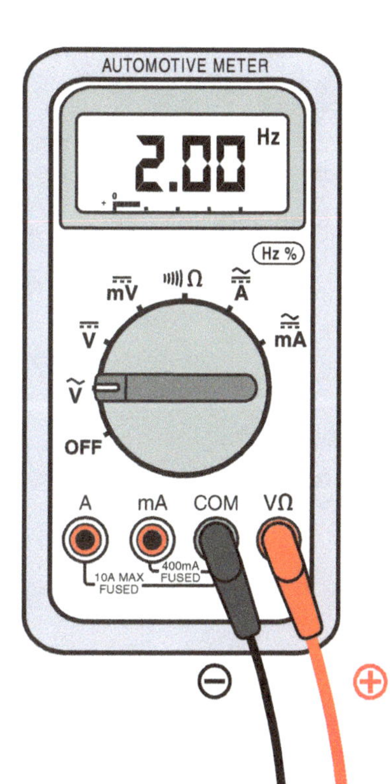

Testing for Presence of a Square Wave

Test Conditions

3-wire active (Hall Effect) sensor

Automotive meter that reports frequency of a square wave

Dial set to:	V AC Volts, Hz %
Red input terminal:	VΩ
Black input terminal:	COM
Red lead:	sensor signal wire (+)
Black lead:	chassis ground (–)

Connector remains connected, back-probed at signal wire

Engine cranking

Typical Normal Values

Square wave detected (non-zero frequency shown)

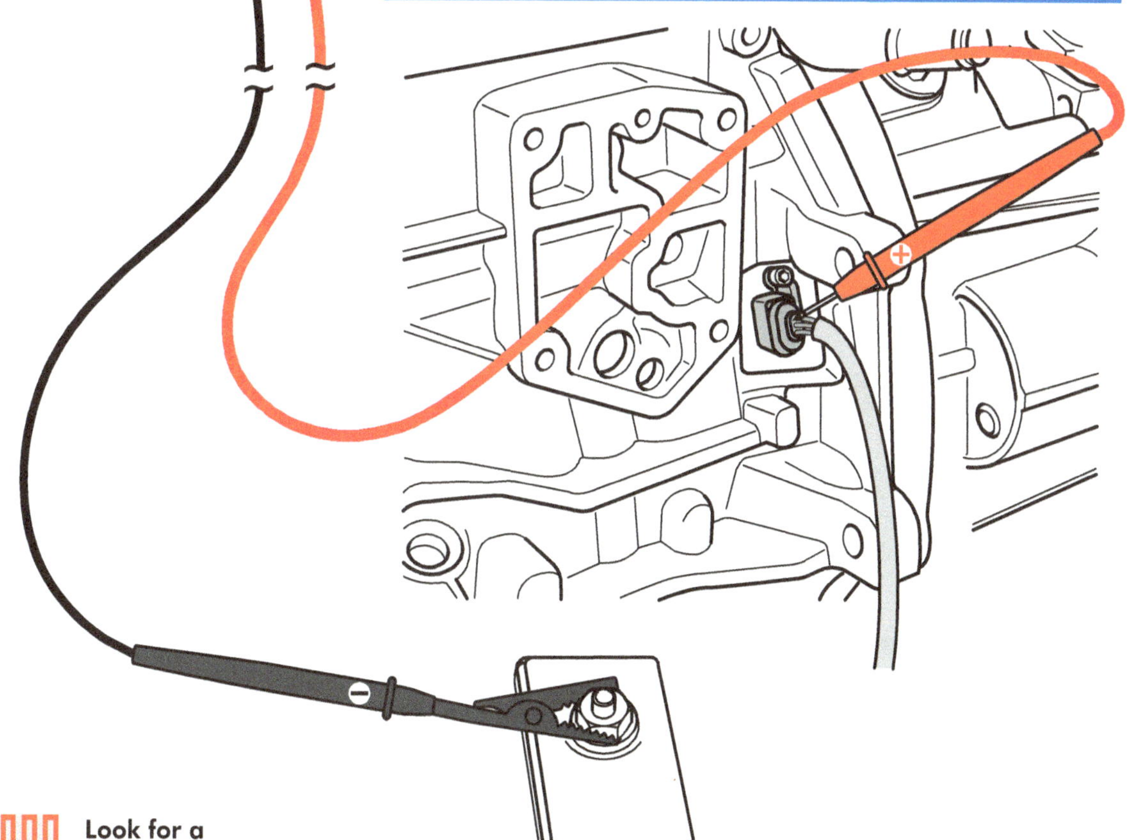

Look for a Square Wave

Active CKP Sensor Running Signal

Test Set-Up 32-C

See page 32-10 for step-by-step procedure

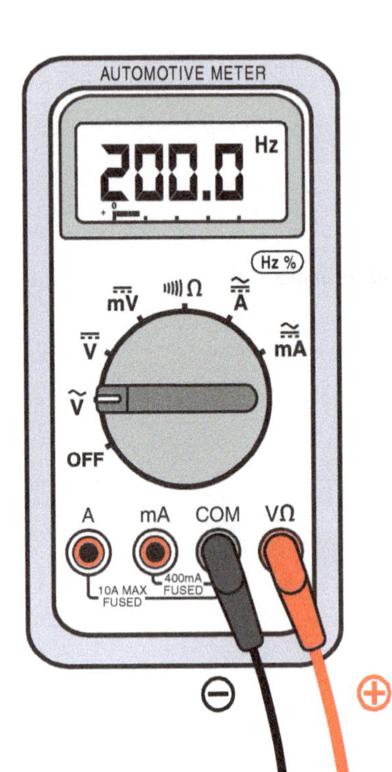

Testing for Varying Frequency of Square Wave

Test Conditions

3-wire active (Hall Effect) sensor

Automotive meter that reports frequency of a square wave

Dial set to:	Ṽ AC Volts / Hz %
Red input terminal:	VΩ
Black input terminal:	COM
Red lead:	sensor signal wire (+)
Black lead:	chassis ground (–)

Connector remains connected, back-probed at signal wire

Engine running

Typical Normal Values

Signal frequency: Varies with engine rpm

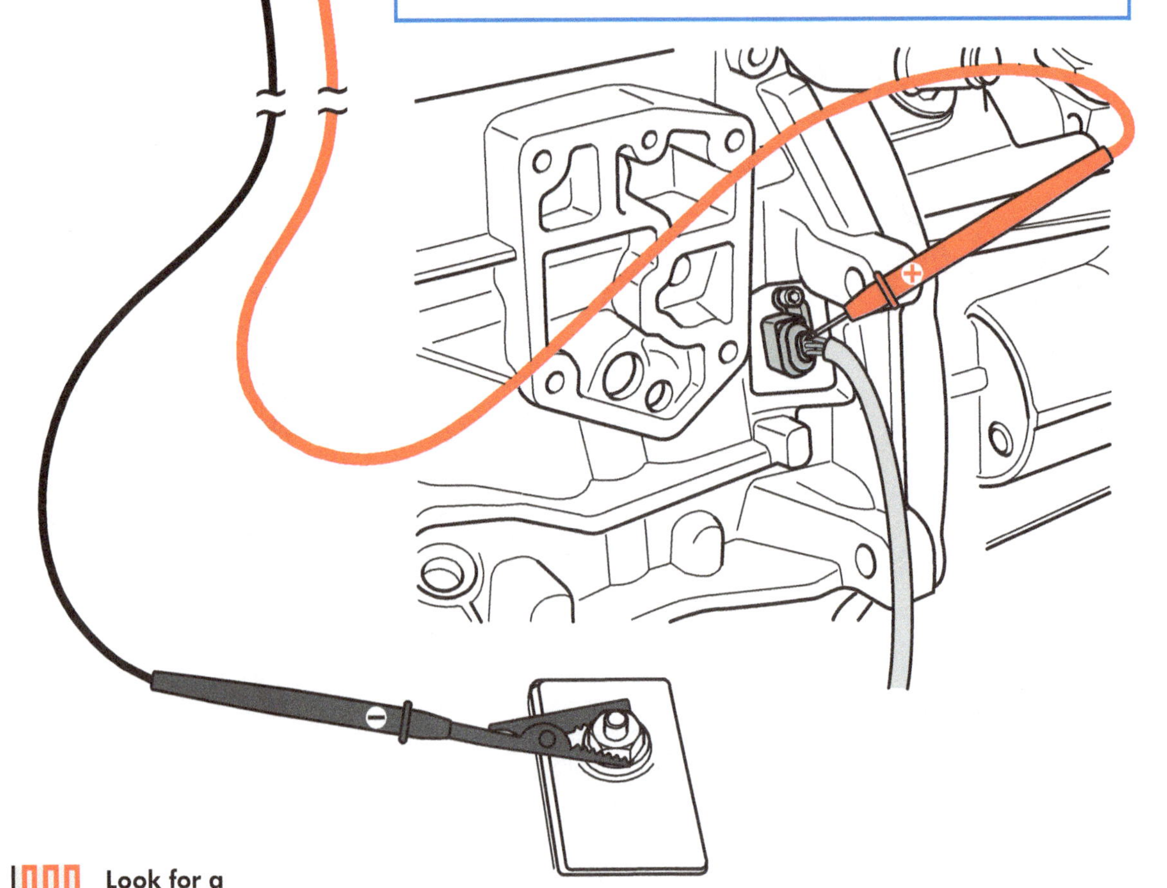

Look for a Square Wave

Passive CKP Sensor Resistance

See page 32-11 for step-by-step procedure

Test Set-Up 32-D

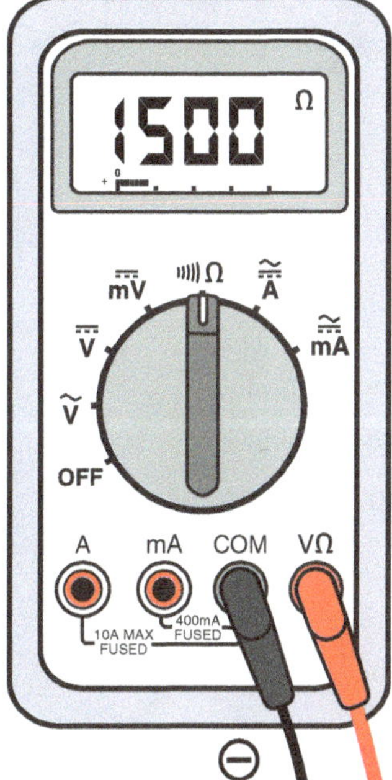

Testing Resistance Across VRT Coil

Test Conditions

2-wire passive (VRT) sensor

Dial set to: Ω resistance

Red input terminal: VΩ

Black input terminal: COM

Red lead: sensor signal pin (+)

Black lead: sensor ground pin (–)

Connector disconnected, testing at sensor

Engine off

Typical Normal Values

Resistance approximately 1500 Ω

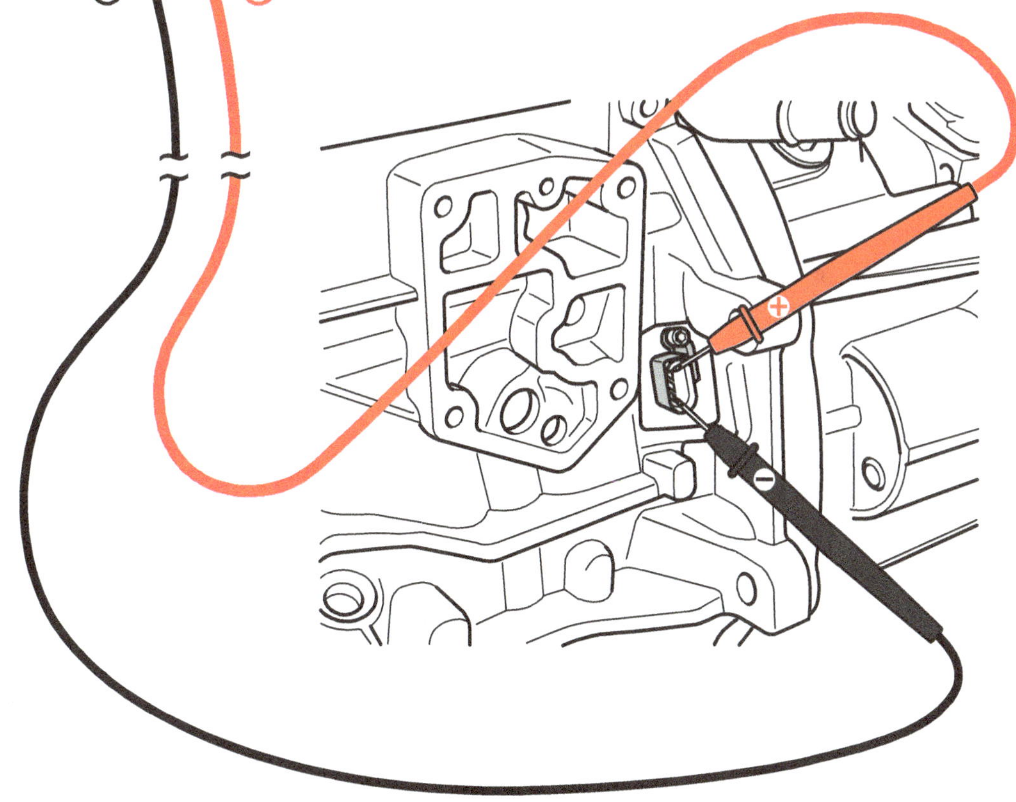

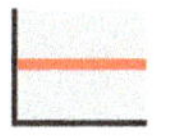

Look for a Static Value

Passive CKP Sensor Cranking Signal

Test Set-Up **32-E**

See page 32-11 for step-by-step procedure

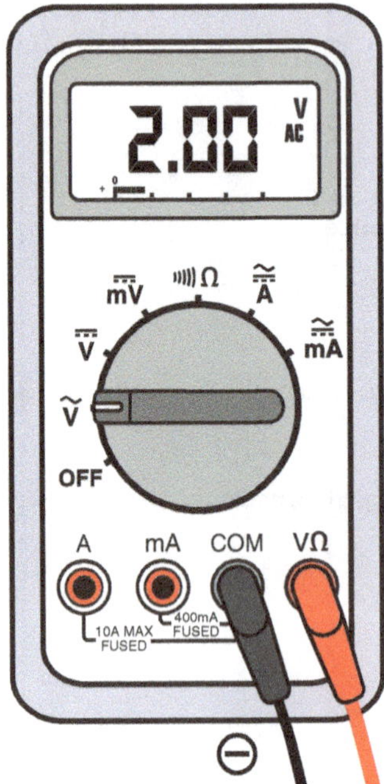

Testing for Presence of AC Signal

Test Conditions

2-wire passive (VRT) sensor	
Dial set to:	Ṽ AC Volts
Red input terminal:	VΩ
Black input terminal:	COM
Red lead:	sensor signal pin (+)
Black lead:	sensor ground pin (–)
Connector disconnected, testing at sensor	
Engine cranking	

Typical Normal Values

Voltage: Approximately 1 to 5 volts

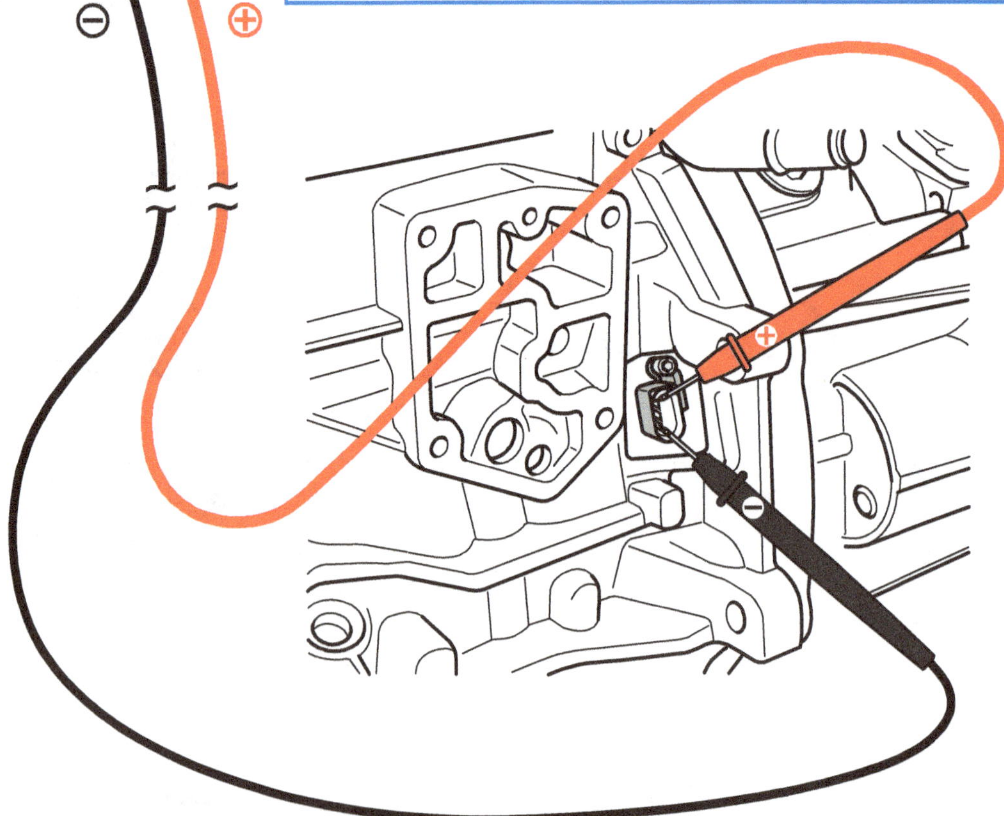

Look for a Sine Wave

Passive CKP Sensor Running Signal

Test Set-Up 32-F

See page 32-11 for step-by-step procedure

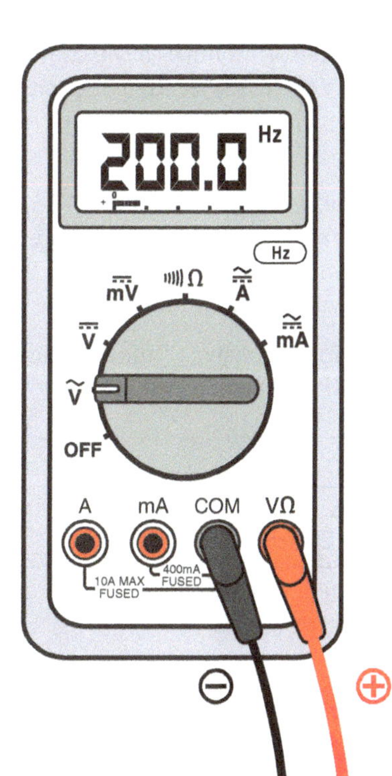

Testing for Varying Frequency of AC Signal

Test Conditions

2-wire passive (VRT) sensor

Meter that measures AC frequency

Dial set to:	Ṽ AC Volts | Hz
Red input terminal:	VΩ
Black input terminal:	COM
Red lead:	sensor signal wire (+)
Black lead:	chassis ground (–)

Connector remains connected, back-probed at signal wire

Engine running

Typical Normal Values

Signal Amplitude: About 1 to 5 volts

Signal frequency: Varies with engine rpm

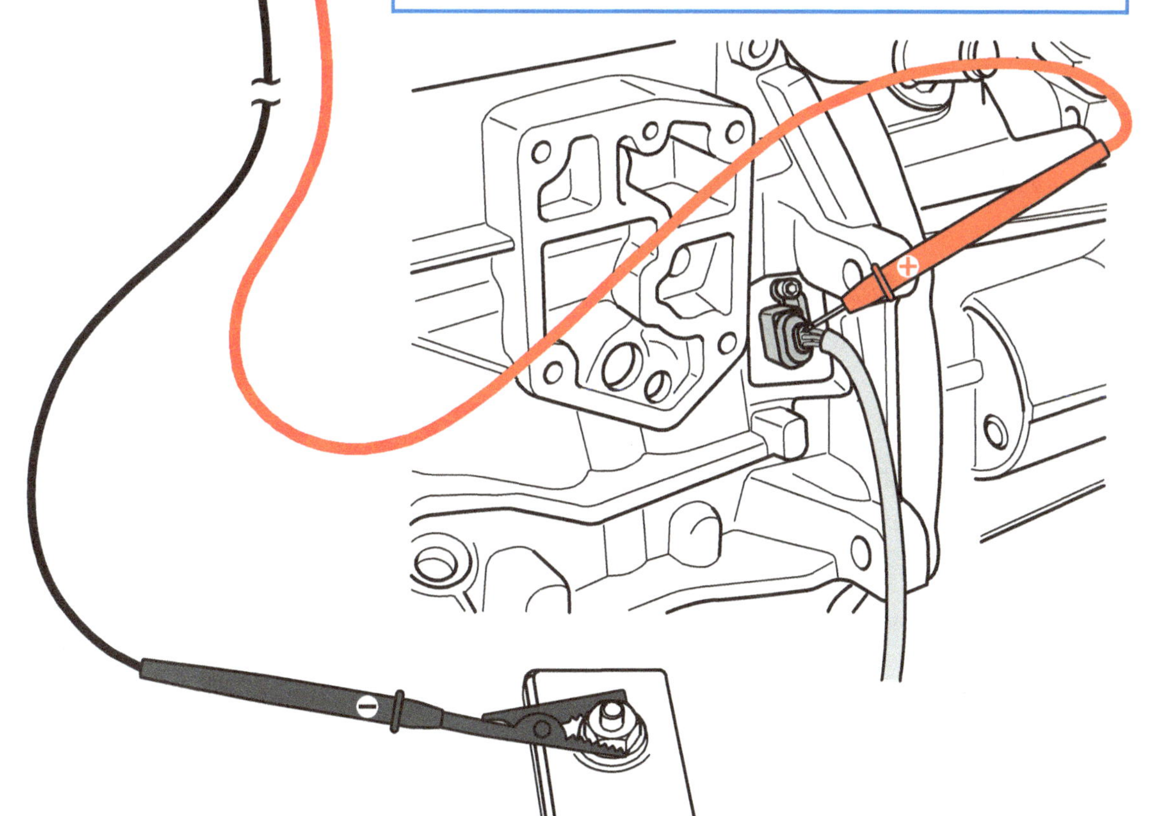

Look for a Sine Wave

Crankshaft Position Sensor

The Crankshaft Position (CKP) sensor, also often called the engine speed sensor, reports the position and speed of the crankshaft to the ECM. Engine speed and crank top dead center (TDC) position are inputs to the ECM for control of spark and fuel delivery. The CKP sensor is also used by the ECM for misfire detection on 1996 and later cars. In both operation and troubleshooting, the CKP sensor differs little from the Camshaft Position (CMP) sensor. Unlike the CMP, however, the CKP has a reference mark telling the ECM when the engine is at TDC.

Owners are alerted to a crankshaft position sensor problem because the car won't start, or runs horribly, or the Check Engine Light is on and a scan tool has reported a code that has directed them to the CKP.

CKP Sensor Location

The crankshaft position sensor is most often located on the rear of the cylinder block. See **Figure 1**. On early 1990s cars, it may be located in at the front of the engine. Consult a repair manual for your car for specific location information.

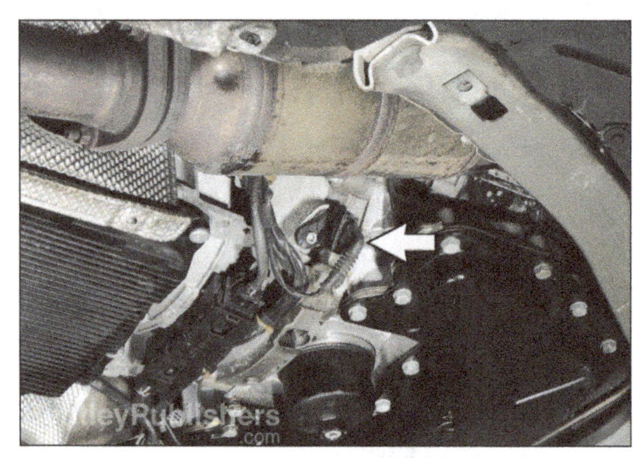

Figure 1 Crankshaft Position (CKP) Sensor (**arrow**) on a 2004 BMW E60 545i.

CKP Sensor Types

There are two types of crankshaft position sensors:

- Active sensors that use the Hall Effect principle
- Passive sensors called Variable Reluctance Transducers (VRT)

As a general rule, cars from model year 1996 used Hall effect sensors, while 1995 and earlier cars used the VRT type sensor.

Hall effect sensors have three wires (signal, ground, power). Passive VRT sensors have two wires (signal and ground).

The workings of both types of sensors are discussed in the **Introduction to Sensors** chapter.

CKP Sensor Signal

Like any other sensor using a VRT or a Hall Effect device, crankshaft position sensors produce dynamic analog signals that are periodic. In both cases, the frequency is the parameter used by the ECM.

The active Hall Effect sensor produces a square wave signal whose frequency increases as engine speed increases. The Hall Effect sensor produces a square wave at any rotation speed.

The passive VRT sensor produces a sine wave signal whose amplitude and frequency both increase as engine speed increases. Note that when engine speed is very slow, there may be no visible amplitude to the sine wave.

Non model-specific wiring and signal information are summarized in the **CKP Sensor Signal Table**.

CKP Sensor Signal

	Active (Hall Effect)	Passive (VRT)
Wires	3 (signal, ground, power)	2 (signal, ground)
Signal	Square wave	Sine wave
Important Parameter	Frequency	Frequency

A wiring example for the CKP on a 2005 Jetta is shown in **Figure 2**. The presence of three wires in the diagram indicates that the sensor is an active (Hall Effect) sensor.

Testing

For each sensor type (active or passive), there is a static test (engine stopped) and two signal tests (engine cranking and engine running). Different test equipment is required depending on the sensor type.

CKP Sensor Testing Equipment

Test	Active (Hall Effect)	Passive (VRT)
Static test	Multimeter	Multimeter
Signal tests	Automotive multimeter that detects a square wave and reports its frequency, or oscilloscope	Multimeter that measures AC frequency, or oscilloscope

Note that, in testing, we don't care what the actual frequency value is; we just need to verify that a square wave or sine wave is present and that its frequency increases with increasing engine speed.

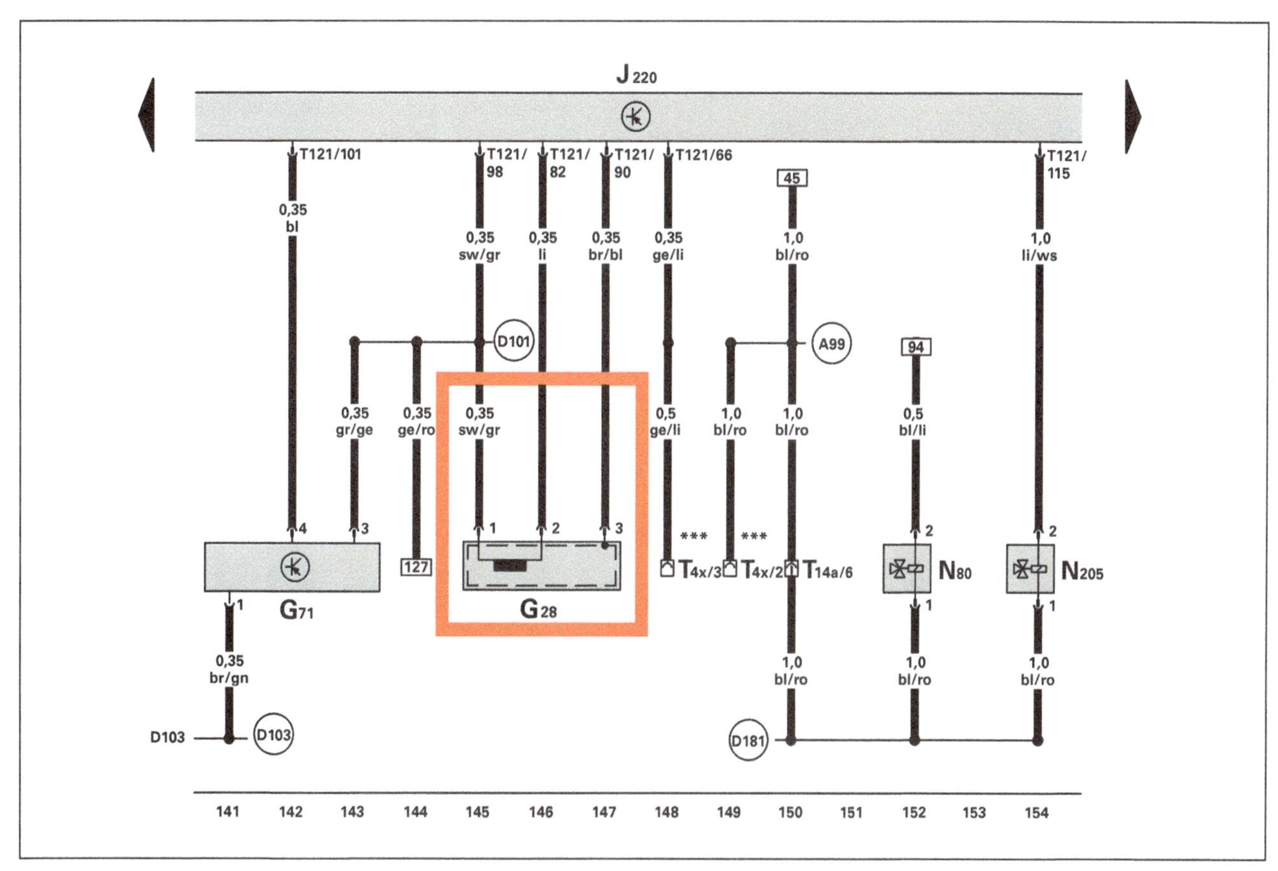

Figure 2 Wiring for a 3-wire (Hall effect) crankshaft position sensor on a 2005 VW Jetta. The wiring diagram does not show which pins are power, signal, or ground, but the presence of three wires indicates that it is a Hall Effect sensor.

Before testing, make sure the sensor itself is not cracked. Disconnect the connector from the sensor. Physically inspect both sides of the connector. Make sure no wires are broken and no pins are pushed out or bent.

Inspect the toothed wheel to make sure no teeth are broken or missing. Note that there is an intentional gap in the teeth so the ECM can tell, from the missing signal peak, where Top Dead Center (TDC) is, but no other teeth should be missing.

On a post-OBD-II car, one testing approach is to use a scan tool and simply replace the CKP if the CEL is on, the car has posted a CKP-related code, and the scan tool shows problems with data from the CKP. Note, however, that the sensor itself may be fine, and the problem may be in the wiring. The procedures below are to test the sensor itself.

Determining Pinouts

Pinouts are best determined using factory wiring diagrams. However, we have found that the following procedure is useful for confirming that the factory wiring is correct, or determining pinouts if a wiring diagram is not available.

As discussed above, if the sensor has three wires, it is almost certainly an active (Hall Effect) sensor; if it has two wires and is on a 1995 and earlier car, it is a passive (VRT) sensor.

To determine pinouts on a 3-wire (active) sensor:

- Locate the connector and disconnect it.
- Turn the key to ignition ON, but do not start engine.
- Set the meter to measure DC voltage.
- Connect the black multimeter probe to any chassis ground.
- On the wiring harness side of the connector, touch the red multimeter probe to each of the pins.
- The one that reads voltage (5 or 12 volts) is the supply voltage to the active sensor.
- Turn the ignition off.

To determine the other pinouts:

- Set the meter to measure resistance.
- Connect the black multimeter probe to any chassis ground.

- On the wiring harness side of the connector, touch the red multimeter probe to each of the other pins. The pin at which the multimeter registers continuity (near zero ohms) should be the ground pin.
- At this point, you've identified power and ground. Therefore the remaining pin is the signal wire.

Active CKP Sensor Supply Voltage Test

See Test Set-Up on page 32-2

Look for a Static Value

In this test, we check for supply voltage to the active (Hall Effect) sensor.

Typical Normal Values for Active CKP Sensor Supply Voltage
5V or 12V, depending on sensor/vehicle

- Set the multimeter to measure DC voltage.
- Disconnect the connector at the CKP.
- Turn the key to ignition ON, but do not start engine.
- Working at the wiring harness side of the connector, touch the red multimeter probe to each of the pins.
- The multimeter should read 5V or 12V on the wire carrying the supply voltage. If there is no voltage on the supply voltage wire, the problem must be troubleshot by consulting a wiring diagram to trace the wire back to the ECM to determine where the voltage is being interrupted. The Hall Effect sensor will not work unless there is voltage at the connector.

Active CKP Sensor Cranking Signal Test

See Test Set-Up on page 32-3

Look for a Square Wave

In this test, we use the starter to spin the crankshaft and check that the active (Hall Effect) sensor produces a square wave. The sensor remains connected for this test, and you'll need to back-probe the connector.

Typical Normal Values for Active CKP Sensor Square Wave Signal	
Amplitude	About half the supply voltage (e.g., about 2.5V or 6V)
Frequency	Not important; meter just needs to detect the presence of a square wave

- Re-connect the connector at the CKP.
- Back-probe the connector, connecting the signal and ground wires to an automotive multimeter.
- Set the automotive multimeter to detect a square wave. The exact method is meter dependent. It may involve using the pulse width or % duty cycle setting. We have generically indicated it via the Hz % button / AC volts in the Test Set-Up 32-B illustration. See the **Tools for Dynamic Analog Signals** chapter for more information.
- Crank the engine. The Hall Effect sensor generates a signal even at very low speeds. Operate the starter motor. You should see the frequency of the square wave displayed as a numerical reading on the meter.
- If there is no signal while the engine is being spun, and the sensor is receiving its supply voltage, and the connector is properly seated, the sensor is likely defective and should be replaced.

Active CKP Sensor Running Signal Test

See Test Set-Up on page 32-4

Look for a Square Wave

In this test, we check to see that the frequency of the square wave varies with engine rpm.

Typical Normal Values for Active CKP Square Wave Signal
Frequency that increases with engine rpm

- If the engine is capable of running, leave the automotive meter connected to the back-probed connector.
- Set the automotive meter to detect the frequency of a square wave. The exact method is meter dependent. It may involve using the pulse width or % duty cycle setting. We have generically indicated it via the Hz % button / AC volts in the Test Set-Up 32-C illustration.
- Start the engine, and vary its speed. You should see the automotive meter display a frequency whose value increases and decreases with engine rpm.

Passive CKP Sensor Resistance Test

See Test Set-Up on page 32-5

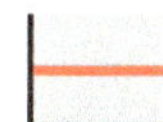

Look for a
Static Value

In this test, we check to see if the VRT coil is shorted or open by measuring its resistance.

Typical Normal Values for Passive CKP Sensor Resistance
Varies by manufacturer, but about 1,500 ohms

- Set the multimeter to measure resistance.
- Disconnect the connector to the CKP sensor.
- Touch the multimeter probes to the two pins of the CKP sensor (at sensor, not at wiring harness) to measure the resistance of the VRT coil. It doesn't matter which probe goes to which of the two pins.
- The resistance should be about 1,500 ohms. If the resistance is infinite (e.g., the meter reads OL), the coil in the VRT is broken. If the resistance is close to zero (e.g., the meter reads less than one ohm), the coil is shorted out. In either case, the sensor is likely bad and should be replaced.

Passive CKP Sensor Cranking Signal Test

See Test Set-Up on page 32-6

Look for a
Sine Wave

In this test, we use the starter to spin the crankshaft and check that the passive (VRT) sensor is producing an AC signal.

Typical Normal Values for Passive CKP Sensor Signal (Sine Wave)	
Amplitude	About 1 to 5 volts
Frequency	Not important; meter just needs to detect the presence of an AC voltage

- Leave the multimeter probes connected to the two pins of the CKP.
- Set the meter to measure AC volts.
- Crank the engine to rotate the crankshaft using the starter motor.
- The sine wave signal output by the VRT should register on the multimeter as an AC voltage of between 1 and 5 volts.
- If there's not a voltage reading from the VRT sensor, check the air gap between it and the toothed reluctor wheel. On some cars, this gap is adjustable. The VRT coil should be very close to the teeth on the reluctor wheel but not touch them.
- If the VRT coil is close to the reluctor wheel, you're certain you have the multimeter configured and connected correctly, and you receive no signal, then the sensor is likely defective and should be replaced.

Passive CKP Sensor Running Signal Test

See Test Set-Up on page 32-7

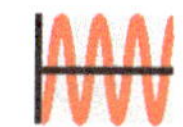

Look for a
Sine Wave

In this test, we run the engine and verify that the frequency of the AC signal varies with engine RPM.

Typical Normal Values for Passive CKP Sensor Signal (Sine Wave)
Frequency that increases with engine RPM

- If the engine is capable of running, reconnect the connector so the ECM receives the signal from the CKP sensor.
- Back-probe the connector, connecting the signal and ground wires to the multimeter.
- If your meter can measure the frequency of an AC voltage signal, set it to measure AC frequency. The setting for this varies meter to meter. In the illustration, we show it as an "Hz" button.
- Start the engine and vary its speed. You should see the meter display a frequency whose value increases and decreases with changing engine rpm.
- If your meter can't measure AC frequency, it still should be able to measure AC voltage. You may see the AC voltage increase slightly as you raise engine rpm.

Supplemental Testing

If you have direct confirmation via an automotive multimeter or a scope that the sensor is outputting a signal, but there's still a problem (for example, a scan tool is still reporting there's no signal from the sensor), there may be a broken wire or bad connection between the sensor and the ECM.

- Use a multimeter set to measure resistance to verify continuity between the ground wire on the connector and chassis ground.
- Use a multimeter set to measure resistance to verify continuity between the signal wire on the connector and the ECM. This will require a wiring diagram for your car, knowledge of the location of the ECM, and knowledge of which pin on the ECM connector carries the signal from the CKP.

Is a Scan Tool Useful?

Yes. Many people will be directed to the CKP sensor because the Check Engine Light is on and a scan tool reports a code implicating the CKP sensor. Both an OBD-II scanner that displays live data and a brand-specific scan tool should give you information about the crankshaft position sensor. However, a CKP sensor reported as bad by a scan tool may be fine, and the problem may be rooted in a broken wire or unseated connector.

Is an Oscilloscope Useful?

Yes. If you have access to a oscilloscope, using it is better than using an automotive meter. Nothing is as good as spinning the engine and actually seeing the sine wave or square wave from the crankshaft sensor. Plus, visual examination of the signal on an oscilloscope may reveal noise or an intermittent characteristic that is not detectable on an automotive multimeter.

Additional Information

Related Chapters

- **Common Multimeter Tests** for a discussion on how to use a multimeter to measure voltage and resistance.
- **Dynamic Analog Signals** for a discussion on square waves and sine waves.
- **Tools for Dynamic Analog Signals** for a discussion on how to use an automotive multimeter to detect sine waves and square waves.
- **Introduction to Sensors** for a detailed explanation on how active (Hall Effect) sensors and passive (VRT) sensors work.

Technical Notes

Whether the sensor is an active (Hall Effect) outputting a square wave or a passive (VRT) outputting a sine wave, the amplitude is not a relevant parameter. The amplitude needs to be high enough that the ECM detects the signal, but there's no information in the amplitude. The information is in the frequency.

When testing either the active (Hall Effect) sensor or the passive (VRT) sensor, note that we don't care about the actual frequency; we just need to verify that a signal is present and that its frequency changes with engine speed. For this reason, if you're testing a Hall Effect sensor and your meter does not have a "square wave frequency" setting, you may be able to use a setting such as pulse width, verify that the meter is displaying a pulse width reading, and verify that the pulse width gets shorter (that the reading gets smaller) as the engine is spun faster.

33

Testing Camshaft Position (CMP) Sensors

Multimeter Test Set-Ups

⚠ WARNING —

Make personal safety your top priority. Watch what you are doing, stay alert. Do not work on your car if you are tired, not feeling well, or in a bad mood. Common sense, alertness, and good judgment are crucial to safe work on your vehicle.

Protect yourself from shock, trauma, and hazardous chemicals: Wear safety glasses, work boots, and protective gloves when working on your vehicle. Have hearing protection available for work that generates noise.

Be sure the work environment is well lit, properly ventilated, and that the vehicle is on a level surface. The floor, vehicle, and components inside the vehicle should be dry to minimize the possibility of shocks.

This book is designed as background technical training and as a general overview of how to do electrical diagnosis and service. It is designed to be used in conjunction with the appropriate service information for your vehicle, not to replace factory service information.

If you lack the skills, tools and equipment, or a suitable workshop for any procedure described in this book, have the job done by a qualified shop or an authorized dealer.

For additional warnings and cautions, see the **Safety,** *chapter and the warnings, cautions, and notices in your vehicle manufacturer's service information.*

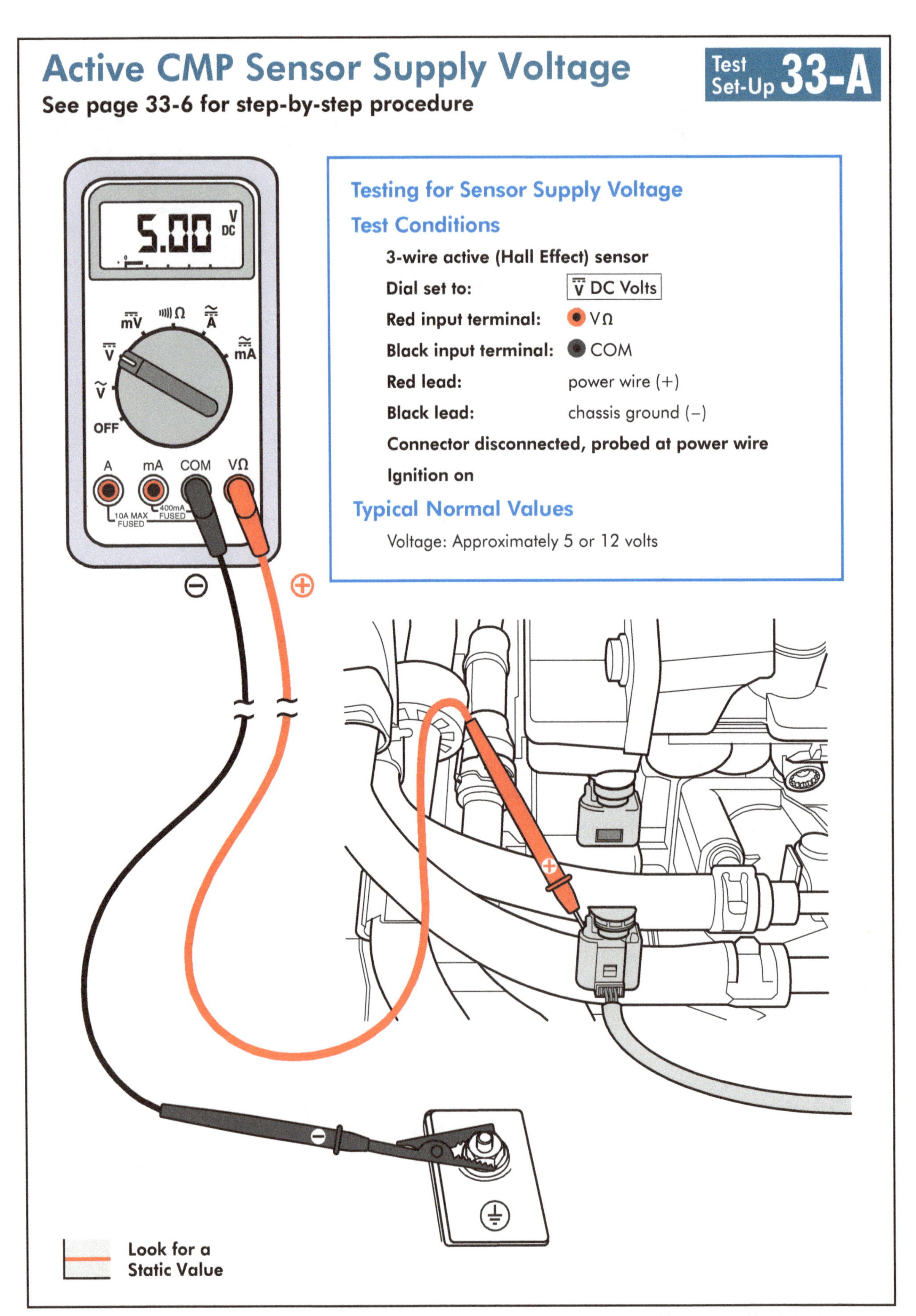

Active CMP Sensor Supply Voltage
Test Set-Up 33-A
See page 33-6 for step-by-step procedure
Testing for Sensor Supply Voltage
Test Conditions
3-wire active (Hall Effect) sensor
Dial set to: V DC Volts
Red input terminal: VΩ
Black input terminal: COM
Red lead: power wire (+)
Black lead: chassis ground (–)
Connector disconnected, probed at power wire
Ignition on
Typical Normal Values
Voltage: Approximately 5 or 12 volts
5.00
V DC
mV
Ω
A
V
mA
OFF
A
mA
COM
VΩ
10A MAX FUSED
400mA FUSED
Look for a Static Value

Active CMP Sensor Cranking Signal

Test Set-Up **33-B**

See page 33-7 for step-by-step procedure

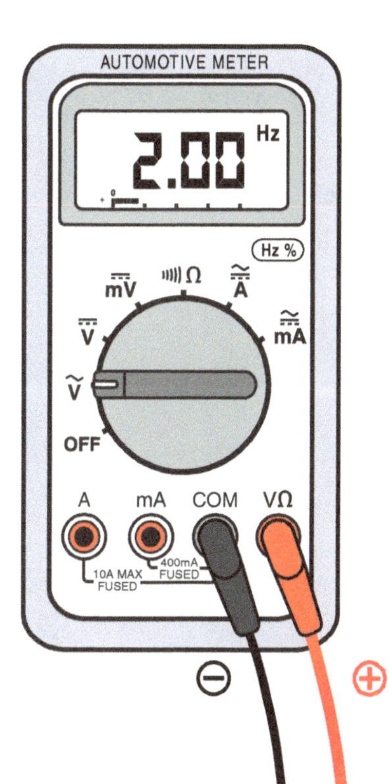

Testing for Presesence of Square Wave

Test Conditions

3-wire active (Hall Effect) sensor

Automotive meter that reports frequency of a square wave

Dial set to:	Ṽ AC Volts / Hz %
Red input terminal:	VΩ
Black input terminal:	COM
Red lead:	sensor signal wire (+)
Black lead:	chassis ground (–)

Connector remains connected, back-probed at signal wire

Engine cranking

Typical Normal Values

Signal frequency: Varies with engine rpm

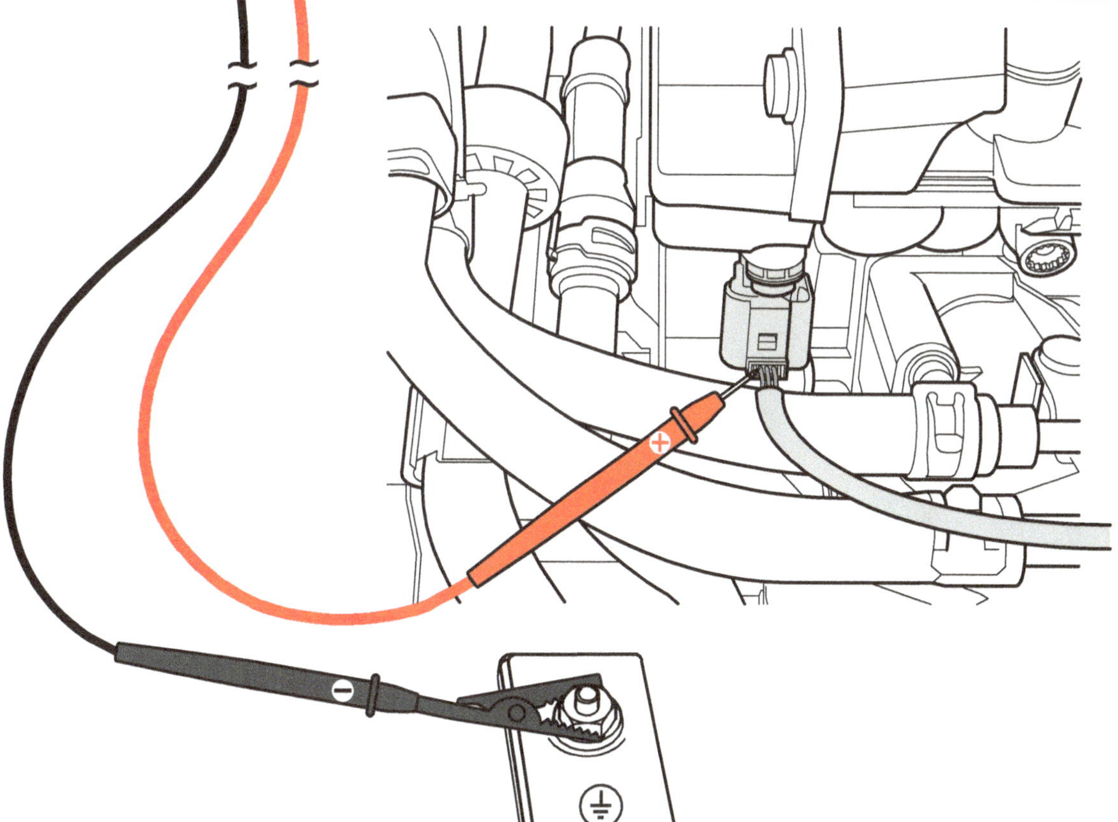

Look for a Square Wave

Active CMP Sensor Running Signal

Test Set-Up **33-C**

See page 33-7 for step-by-step procedure

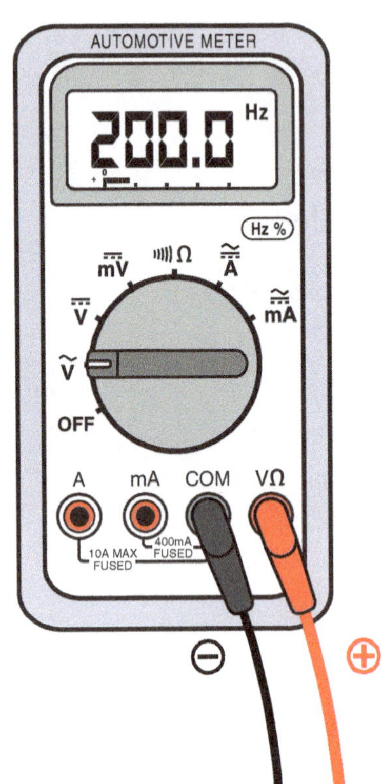

Testing for Varying Frequency of Square Wave

Test Conditions

3-wire active (Hall Effect) sensor

Automotive meter that reports frequency of a square wave

Dial set to:	Ṽ AC Volts / Hz %
Red input terminal:	VΩ
Black input terminal:	COM
Red lead:	sensor signal wire (+)
Black lead:	chassis ground (–)

Connector remains connected, back-probed at signal wire

Engine running

Typical Normal Values

Signal frequency: Varies with engine rpm

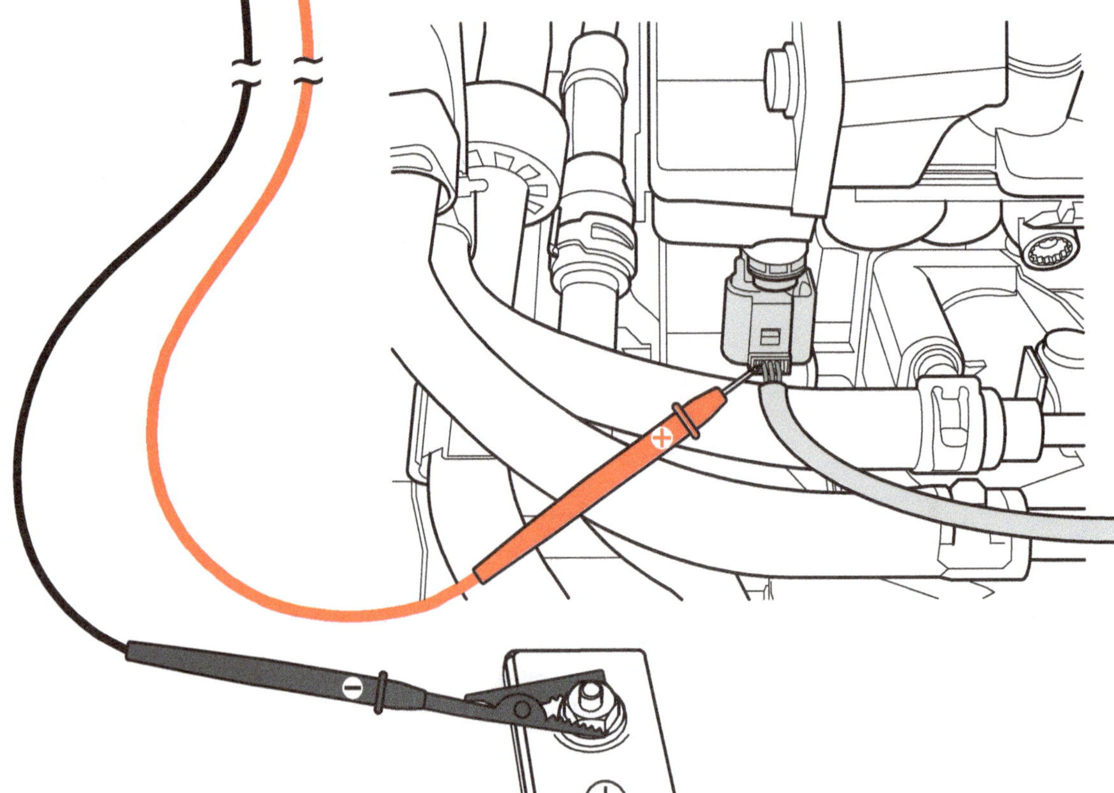

Look for a Square Wave

Camshaft Position Sensor

The Camshaft Position (CMP) sensor reports the position of the camshafts to the ECM, allowing the ECM to synchronize digital engine management functions including spark and fuel delivery.

Most owners are alerted to a camshaft position sensor problem because the car won't start, or runs horribly, or the check engine light is on and a scan tool has reported a code that has directed them to the CMP sensor.

CMP Sensor Location

The camshaft position sensors are most commonly located at either the front or rear of the camshaft. The CMP sensor connector may be located underneath the intake manifold or engine covers, making it difficult to access. See **Figure 1**. Consult a repair manual for your car for specific information.

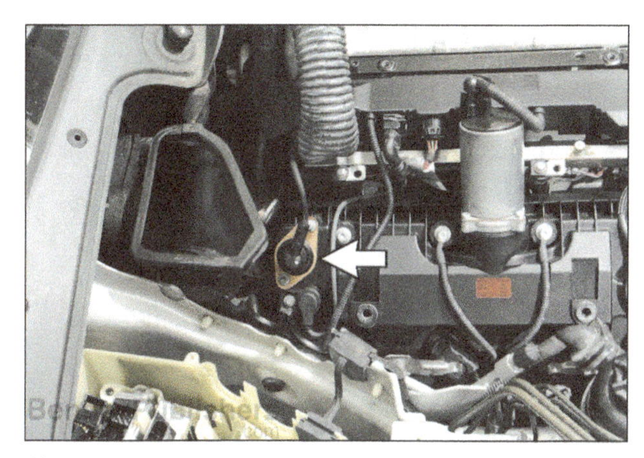

Figure 1 Camshaft Position (CMP) Sensor **(arrow)** on a 2004 BMW E60 545i.

CMP Sensor Types

The most common type of camshaft position sensor is the Hall Effect type sensor. However, a small number of early 1990s (pre-OBD-II) cars use a passive (2-wire) VRT sensor. Testing a VRT-based camshaft position sensor is the same as testing a VRT-based crankshaft position sensor; see the **Testing Crankshaft Position (CKP) Sensors** chapter.

 **NOTE —**

Hall Effect sensors have three wires (signal, ground, power). Passive VRT sensors have two wires (signal and ground). The workings of both types of sensors are discussed in the **Introduction to Sensors** *chapter.*

CMP Sensor Signal

Like other Hall Effect devices, camshaft position sensors produce a dynamic analog signal that is periodic. The frequency is the parameter used by the ECM.

The active Hall Effect sensor produces a square wave signal whose frequency increases as engine speed increases. Non-model-specific wiring and signal information are summarized in the table below.

CMP Sensor Signal

Type	Active Hall Effect
Wires	3 (signal, ground, power)
Signal	Square wave
Important Parameter	Frequency

Figure 2 shows a wiring example for a 3-wire Hall Effect CMP sensor on a 2005 Jetta. The presence of three wires indicates that the sensor is an active (Hall Effect) sensor.

Testing

CMP sensor testing includes a static test (engine stopped) and two signal tests (engine cranking and engine running). A multimeter is the only tool required for the tests.

Equipment Needed for CMP Sensor Testing

Test	Equipment
Static test	Multimeter
Signal tests	Automotive multimeter that detects a square wave and reports its frequency, or oscilloscope

Note that, in testing, we don't care what the actual frequency value is; we just need to verify that a square wave is present and that its frequency increases with increasing engine speed.

Before testing, make sure the sensor itself is not cracked. Disconnect the connector from the sensor. Physically inspect both sides of the connector. Make sure no wires are broken and no pins are pushed out or bent.

On a post-OBD-II car, one testing approach is to use a scan tool and simply replace the CMP if the CEL is on, the car has posted a CMP-related code, and the scan tool shows problems with data from the CMP. Note, however, that the sensor itself may be fine, and the problem may be in the wiring. The procedures below are to test the sensor itself.

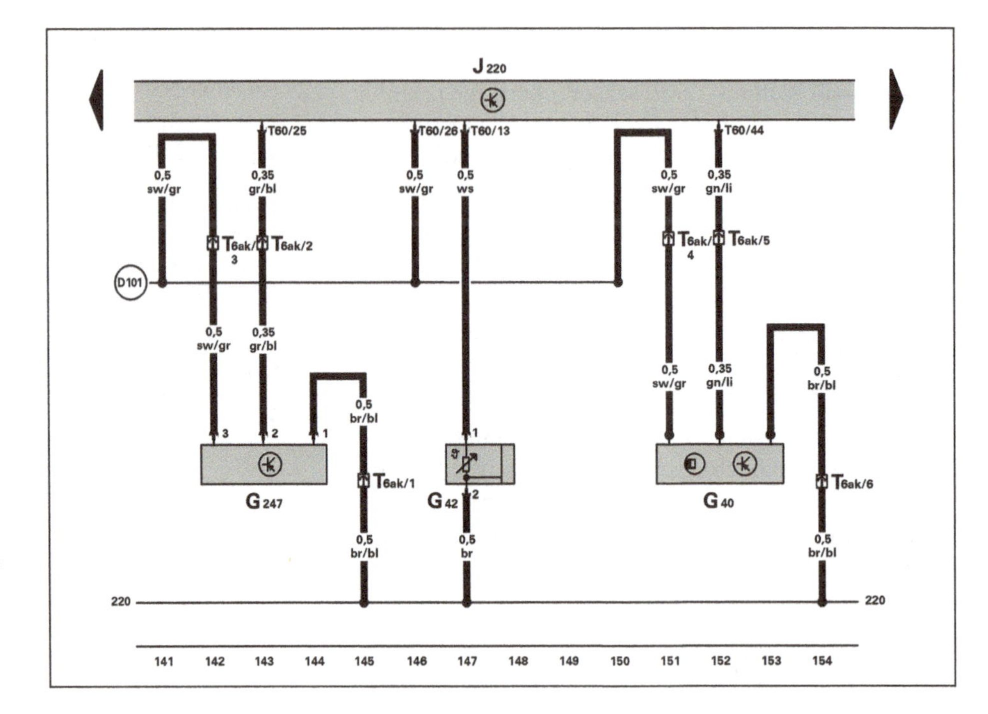

Figure 2 Wiring for a 3-wire Hall Effect camshaft position sensor (G40) on a 2005 Volkswagen Jetta. This shows the CMP using pins 4-6 of the six pin connector T6ak, with pin 4 being power, pin 5 being signal to the ECM, and pin 6 being ground.

Determining Pinouts

Pinouts are best determined using factory wiring diagrams. However, we have found that the following procedure is useful for confirming that the factory wiring is correct, or determining pinouts if a wiring diagram is not available.

To determine pinouts on a 3-wire (active) sensor:

- Locate the connector and disconnect it.
- Turn the key to ignition ON, but do not start engine.
- Set the meter to measure DC voltage.
- Connect the black multimeter probe to any chassis ground.
- On the wiring harness side of the connector, touch the red multimeter probe to each of the pins.
- The one that reads voltage (5 or 12 volts) is the supply voltage to the active sensor.
- Turn the ignition off.

To determine the other pinouts:

- Set the meter to measure resistance.
- Connect the black multimeter probe to any chassis ground.
- On the wiring harness side of the connector, touch the red multimeter probe to each of the other pins. The pin at which the multimeter registers continuity (near zero ohms) should be the ground pin. The remaining pin is the signal wire.

Active CMP Sensor Supply Voltage Test

See Test Set-Up on page 33-2

Look for a Static Value

In this test, we check for supply voltage to the active (Hall Effect) sensor.

Typical Normal Values for Active CMP Sensor Supply Voltage
5V or 12V (dependent on model)

- Set the multimeter to measure DC voltage.
- Disconnect the connector to the CMP.
- Connect the black multimeter probe to chassis ground.
- Turn the key to ignition ON, but do not start engine.
- Working at the wiring harness side of the connector, touch the red multimeter probe to each of the pins.
- The multimeter should read 5V or 12V on the wire carrying the supply voltage. If there is no voltage on the supply voltage wire, the problem must be troubleshot by consulting a wiring diagram to trace the wire back to the ECM to determine where the voltage is being interrupted. The Hall Effect sensor will not work unless there is a supplied voltage.

Active CMP Sensor Cranking Signal Test

See Test Set-Up on page 33-3

Look for a Square Wave

In this test, we use the starter to spin the camshaft and verify that the active (Hall Effect) sensor produces a square wave. Because a Hall Effect sensor is an active sensor, it will need to stay connected to the supply voltage while you look at the sensor's output. Thus, you'll need to back-probe the connector.

Typical Normal Values for Active CMP Sensor Signal (Square Wave)	
Amplitude	About half the supply voltage (e.g., about 2.5V or 6V)
Frequency	Not important; meter just needs to detect the presence of a square wave

- Re-connect the connector to the CMP.
- Connect the black multimeter probe to chassis ground.
- Back-probe the connector, connecting the signal wire to an automotive multimeter.
- Set the automotive multimeter to detect a square wave. The exact method is highly meter-dependent. It may involve using the pulse width or % duty cycle setting. We have generically indicated it via the "Hz %" button in the Test Set-Up 33-B illustration. See the **Tools for Dynamic Analog Signals** chapter for more information.
- Crank the engine. Because Hall Effect sensors generate a signal even at very low speeds, you could do this by either hand with a wrench on the engine, but it's much easier to spin the starter motor. You should see the frequency of the square wave displayed as a numerical reading on the meter.
- If there is no square wave generated while the engine is being spun, and the sensor is receiving its supply voltage, and the connector is properly seated, the sensor is likely defective and should be replaced.

Active CMP Sensor Running Signal Test

See Test Set-Up on page 31-4

Look for a Square Wave

In this test, we look to see that the frequency of the square wave varies with engine rpm.

Typical Normal Values for Active CMP Sensor Signal (Square Wave)
Frequency that increases with engine rpm

- Connect the black multimeter probe to chassis ground.
- If the engine is capable of running, leave the automotive meter connected to the signal pin of the back-probed connector.
- Set the automotive meter to detect the frequency of a square wave. The exact method is highly meter-dependent. It may involve using the pulse width or % duty cycle setting. We have generically indicated it via the "Hz %" button in the Test Set-Up 33-C illustration. See the **Tools for Dynamic Analog Signals** chapter for more information.
- Start the engine, and vary its speed. You should see the automotive meter display a frequency whose value increases and decreases with engine rpm.

Supplemental Testing

If you have direct confirmation via an automotive multimeter or a scope that the sensor is outputting a signal, but there's still a problem (for example, a scan tool is still reporting there's no signal from the sensor), there may be a broken wire or bad connection between the sensor and the ECM.

- Use a multimeter set to measure resistance to verify continuity between the ground wire on the connector and chassis ground.
- Use a multimeter set to measure resistance to verify continuity between the signal wire on the connector and the ECM. This will require a wiring diagram for your car, knowledge of the location of the ECM, and knowledge of which pin on the ECM connector carries the signal from the CMP.

Is a Scan Tool Useful?

Yes. Many people will be directed to the CMP because the Check Engine Light is on and a scan tool reports a code implicating the CMP. Both an OBD-II scanner that displays live data and a brand-specific scan tool should give you information about the camshaft position sensor. However, a CMP sensor reported as bad by a scan tool may be fine, and the problem may be rooted in a broken wire or unseated connector.

Is an Oscilloscope Useful?

Yes. If you have access to a oscilloscope, using it is better than using an automotive meter. Nothing is as good as spinning the engine and actually seeing the sine wave or square wave from the camshaft sensor. Plus, visual examination of the signal on an oscilloscope may reveal noise or an intermittent characteristic that is not detectable on an automotive multimeter.

Additional Information

Related Chapters

- **Common Multimeter Tests** for a discussion on how to use a multimeter to measure voltage and resistance.
- **Dynamic Analog Signals** for a discussion on square waves and sine waves.
- **Tools for Dynamic Analog Signals** for a discussion on how to use an automotive multimeter to detect sine waves and square waves.
- **Introduction to Sensors** for a detailed explanation on how active (Hall Effect) sensors and passive (VRT) sensors work.

Technical Notes

Whether the sensor is an active (Hall Effect) outputting a square wave or a passive (VRT) outputting a sine wave, the amplitude is not a relevant parameter. The amplitude needs to be high enough that the ECM detects the signal, but there's no information in the amplitude. The information is in the frequency.

When testing either the active (Hall Effect) sensor or the passive (VRT) sensor, note that we don't care about the actual frequency; we just need to verify that a signal is present and that its frequency changes with engine speed. For this reason, if you're testing a Hall Effect sensor and your meter does not have a "square wave frequency" setting, you may be able to use a setting such as pulse width, verify that the meter is displaying a pulse width reading, and verify that the pulse width gets shorter (that the reading gets smaller) as the engine is spun faster.

34

Testing Mass Airflow (MAF) Sensors

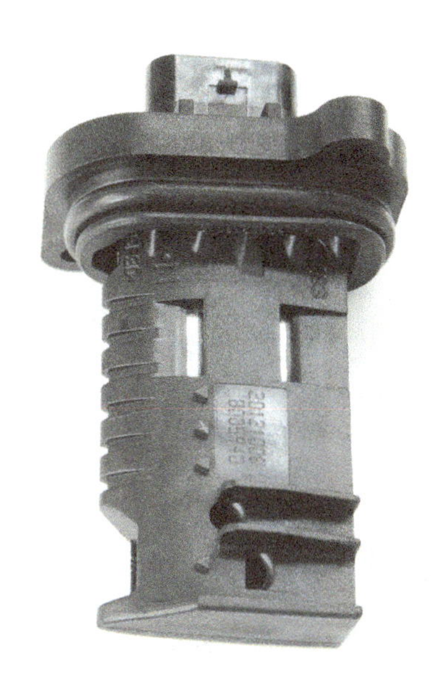

Multimeter Test Set-Ups

⚠ *WARNING —*

Make personal safety your top priority. Watch what you are doing, stay alert. Do not work on your car if you are tired, not feeling well, or in a bad mood. Common sense, alertness, and good judgment are crucial to safe work on your vehicle.

Protect yourself from shock, trauma, and hazardous chemicals: Wear safety glasses, work boots, and protective gloves when working on your vehicle. Have hearing protection available for work that generates noise.

Be sure the work environment is well lit, properly ventilated, and that the vehicle is on a level surface. The floor, vehicle, and components inside the vehicle should be dry to minimize the possibility of shocks.

This book is designed as background technical training and as a general overview of how to do electrical diagnosis and service. It is designed to be used in conjunction with the appropriate service information for your vehicle, not to replace factory service information.

If you lack the skills, tools and equipment, or a suitable workshop for any procedure described in this book, have the job done by a qualified shop or an authorized dealer.

For additional warnings and cautions, see the **Safety**, *chapter and the warnings, cautions, and notices in your vehicle manufacturer's service information.*

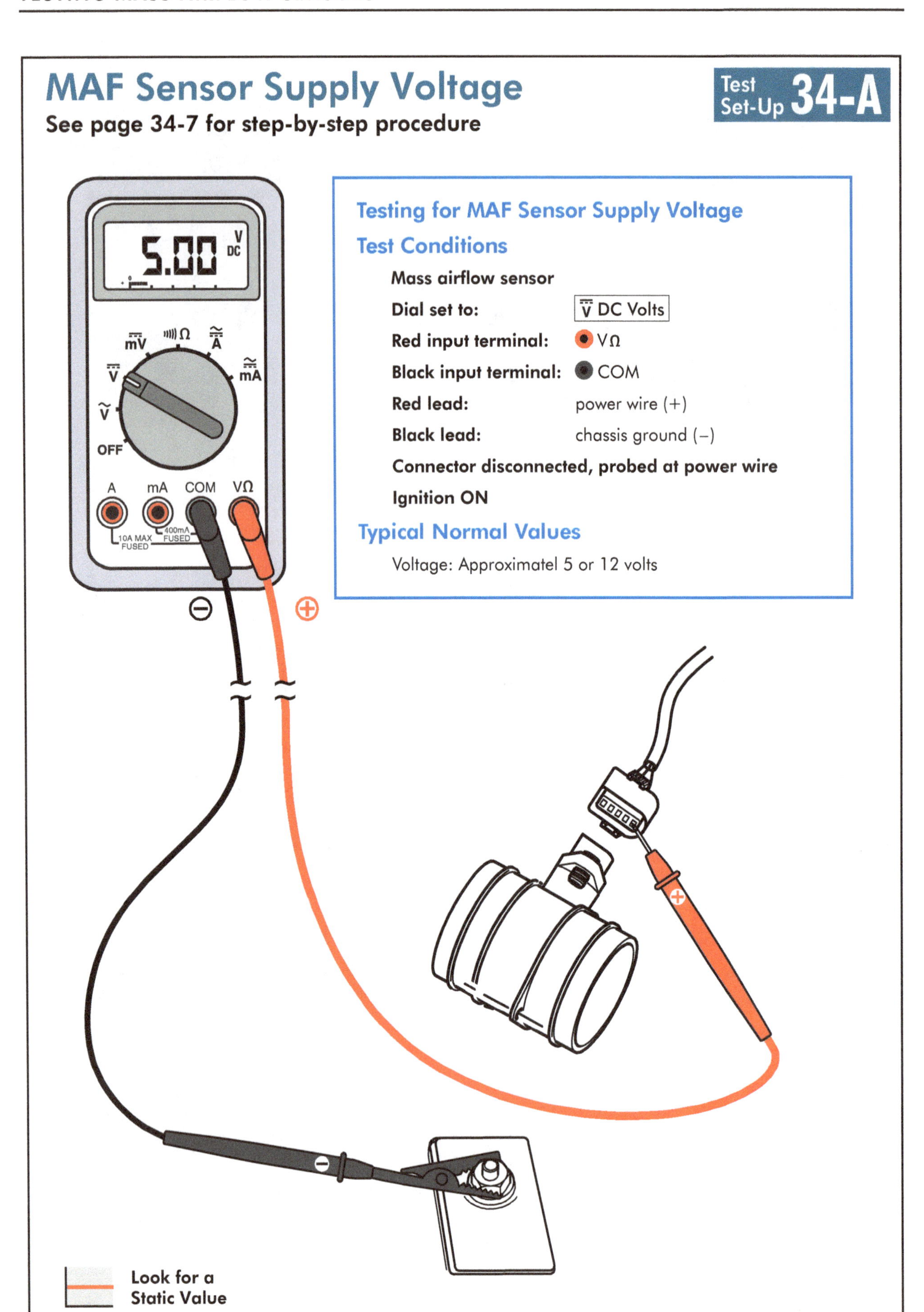
MAF Sensor Supply Voltage
See page 34-7 for step-by-step procedure
Test Set-Up 34-A
Testing for MAF Sensor Supply Voltage
Test Conditions
Mass airflow sensor
Dial set to: V DC Volts
Red input terminal: VΩ
Black input terminal: COM
Red lead: power wire (+)
Black lead: chassis ground (–)
Connector disconnected, probed at power wire
Ignition ON
Typical Normal Values
Voltage: Approximatel 5 or 12 volts
5.00
V
DC
mV
Ω
A
mA
V
OFF
A
mA
COM
VΩ
10A MAX FUSED
400mA FUSED
Look for a Static Value

Analog MAF Sensor Signal

See page 34-7 for step-by-step procedure

Test Set-Up 34-B

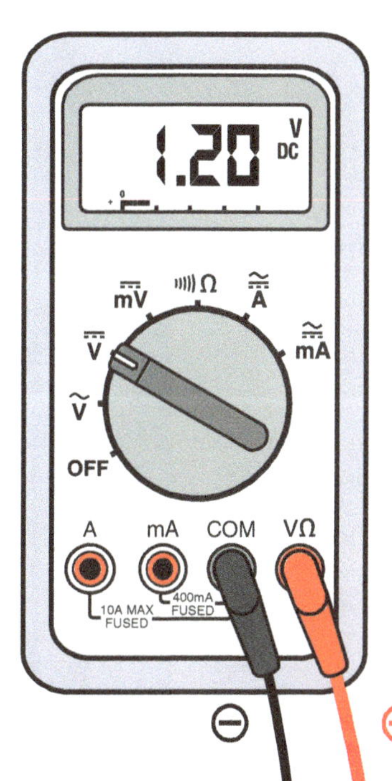

Testing for Anaolog MAF Sensor Signal

Test Conditions

Mass airflow sensor

Dial set to:	V DC Volts
Red input terminal:	VΩ
Black input terminal:	COM
Red lead:	power wire (+)
Black lead:	chassis ground (–)

Connector remains connected, back-probed at signal wire

Engine running

Typical Normal Values

Voltage: Between 0.5 and 4.5 volts

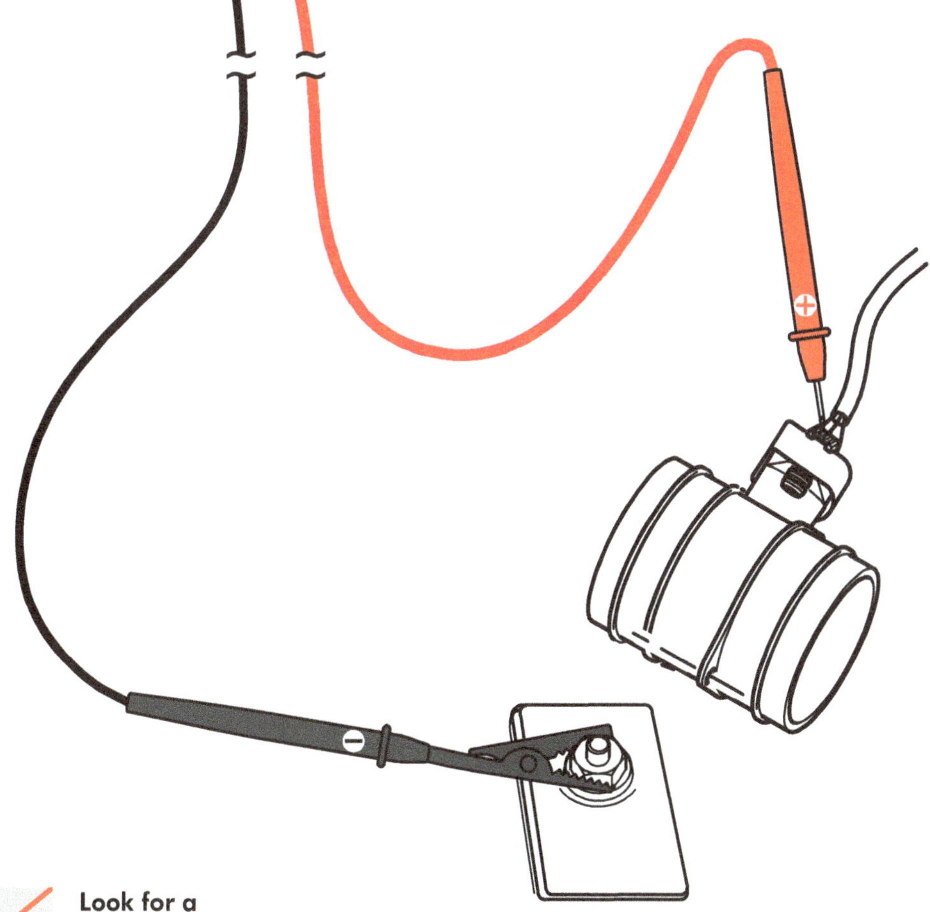

Look for a Slowly Changing Value

Digital MAF Sensor Signal

Test Set-Up 34-C

See page 34-8 for step-by-step procedure

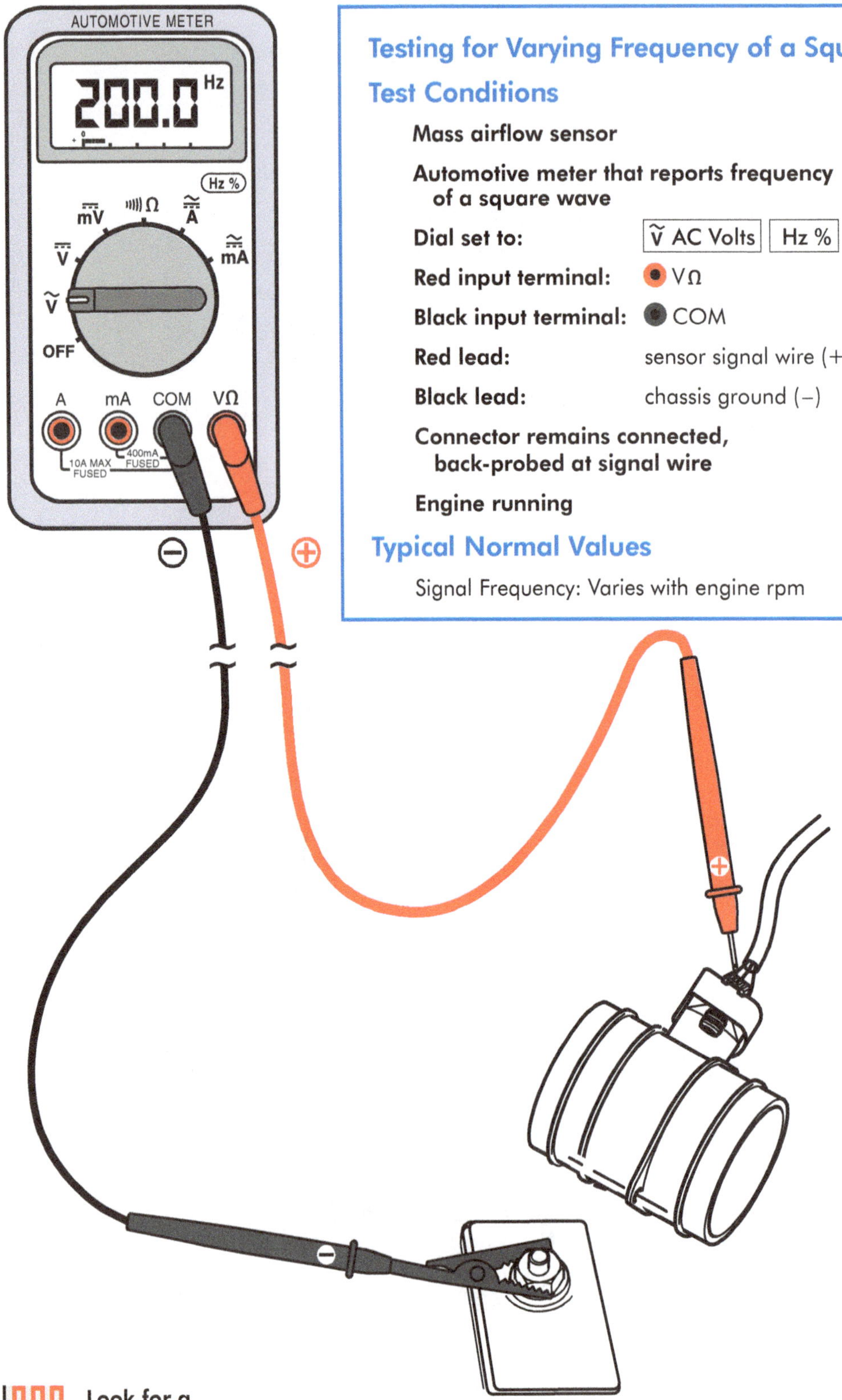

Testing for Varying Frequency of a Square Wave

Test Conditions

Mass airflow sensor

Automotive meter that reports frequency of a square wave

Dial set to:	Ṽ AC Volts / Hz %
Red input terminal:	VΩ
Black input terminal:	COM
Red lead:	sensor signal wire (+)
Black lead:	chassis ground (–)

Connector remains connected, back-probed at signal wire

Engine running

Typical Normal Values

Signal Frequency: Varies with engine rpm

Look for a Square Wave

Mass Airflow Sensor

Sensors that measure the amount of engine intake air are used on every car with electronic fuel injection. Those installed from the mid-1970s through the 1980s employed a door that swung open as the in-rushing air pushed past it. The amount of door swing was measured on a resistance track. These so-called vane-style devices were effective, but the door itself affected the flow of air, and the resistance track was subject to wear. These problems were addressed with the development of the mass airflow (MAF) sensor, which began appearing on European cars in the late 1980s. See **Figure 1**.

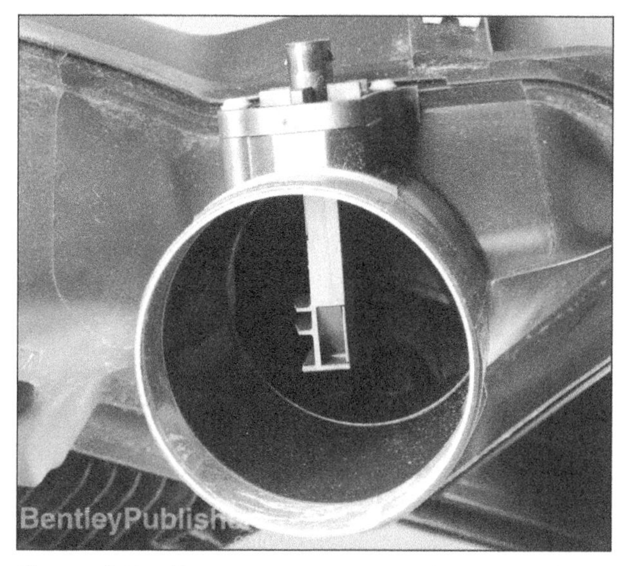

Figure 1 Hot-film mass MAF sensor in the air box of a 2013 BMW 328i xDrive.

Owners are often alerted to a bad MAF by hard starting, stalling, or stumbling under load, or by the CEL being lit and scan tool reading a code that implicates the MAF.

MAF Sensor Operation

Early mass airflow sensors are also known as hot wire sensors. A platinum wire is suspended in the path of the air that is flowing into the engine. In addition, a thermistor (temperature sensor) is used to measure the temperature of the incoming air. With the air temperature known, the wire is then heated to a temperature that is a fixed amount above the temperature of the incoming air.

The air flowing past the wire has a natural cooling effect on the wire, causing it to lose heat. The wider the throttle is open, the greater the amount of air being pulled into the engine, and, in turn, the greater the cooling effect on the heated wire. In order to keep the wire at the specified temperature, current must constantly be applied to the wire. For this reason, the sensor uses the current to measure the air coming into the engine.

Because the humidity of the air also cools the wire, the current becomes a metric of both airflow and air mass, which is a more accurate value for the ECM to use to calculate the corresponding amount of fuel needed.

The MAF internally converts the current into a signal, either a voltage or a square wave with varying frequency (see the **MAF Sensor Signal** section below), and sends the signal to the ECM.

Hot wire sensors incorporate a burn-off circuit to pass a high current through the wire for a short period of time after driving to burn off contamination.

An improved technology, hot film sensors, gradually replaced hot wire sensors in approximately the mid-2000s. A thin film, usually on a plastic backing to provide rigidity, protrudes into the airstream. The mechanics of the thin film measurement is very similar to that of the hot wire, except that hot film sensors do use burn-off.

MAF Sensor Types

If you remove the mass airflow sensor from the car, hold it up to the light, and look through it, you should either see a thin wire, in which case you have a hot wire sensor, or a small tab about the size of a fingernail, in which case you have a hot film sensor. If the MAF isn't part of the open tube, it is most likely a hot film sensor.

However, as we'll explain in the next paragraph, the type of sensor (hot wire versus film) isn't as important as the type of signal it outputs.

MAF Sensor Signal

A MAF outputs either a slowly changing voltage or a frequency-varying square wave. You'll sometimes see this distinction referred to as "analog MAFs" and "digital MAFs," and the frequency-varying square wave referred to as a "digital signal." In this book, we use "digital signal" only to refer to signals of messages whose data can't be decoded without use of a computer. However, we do need a way of differentiating between the sensors that output a voltage signal and those that produce a square wave signal, and, unfortunately, that distinction does not fall along hot wire versus hot film lines. Therefore, for the sake of clarity, we will refer to "analog MAFs" as those that output a voltage, and "digital MAFs" as those that output a square wave.

For an analog MAF that outputs a voltage, the signal is a value between 0.5 and 4.5 volts. When the throttle position changes and more air enters the engine, the signal is a slowly changing non-periodic voltage value. It should be constant or very lightly fluctuating when the throttle position is constant, and should increase as the throttle position and resulting airflow (air mass) increases.

For a digital MAF that outputs a square wave, the signal is a dynamically varying square wave whose frequency is the important parameter. The frequency should be constant or very lightly fluctuating for a fixed amount of airflow (air mass), and should increase as more air enters the engine with increasing engine rpm.

These distinctions in sensor type and signal are summarized in the **Signal from Different MAF Sensor Types Table**.

Signal from Different MAF Sensor Types

	Analog Hot Wire	Analog Hot Film	Digital Hot Film
Signal	Slowly changing voltage	Slowly changing voltage	Frequency-varying square wave

MAF Sensor Circuit

Because of the variety of MAF sensors, you may need to consult a wiring diagram for circuit and wiring information for your car. There can be as few as three wires (supply voltage, signal, and ground), or as many as six (adding power and ground for the burn-off circuit, and the output of the internal temperature sensor).

The pinouts for some sensors include both a battery voltage and a reference voltage. A wiring diagram for the digital MAF sensor in a 2006 Volkswagen A5 Jetta is shown in **Figure 2**.

Testing

Because the analog and digital MAFs output different signals, the equipment required to test them is slightly different, as per the **Equipment Needed for MAF Sensor Testing Table**.

Equipment Needed for MAF Sensor Testing

Test	Analog MAF	Digital MAF
Voltage test	Multimeter	Multimeter
Signal test	Automotive multimeter that detects a square wave and reports its frequency, or oscilloscope	Multimeter that measures AC frequency, or oscilloscope

Before testing, make sure the sensor itself is not cracked. Disconnect the connector from the sensor. Physically inspect both sides of the connector. Make sure no wires are broken and no pins are pushed out or bent.

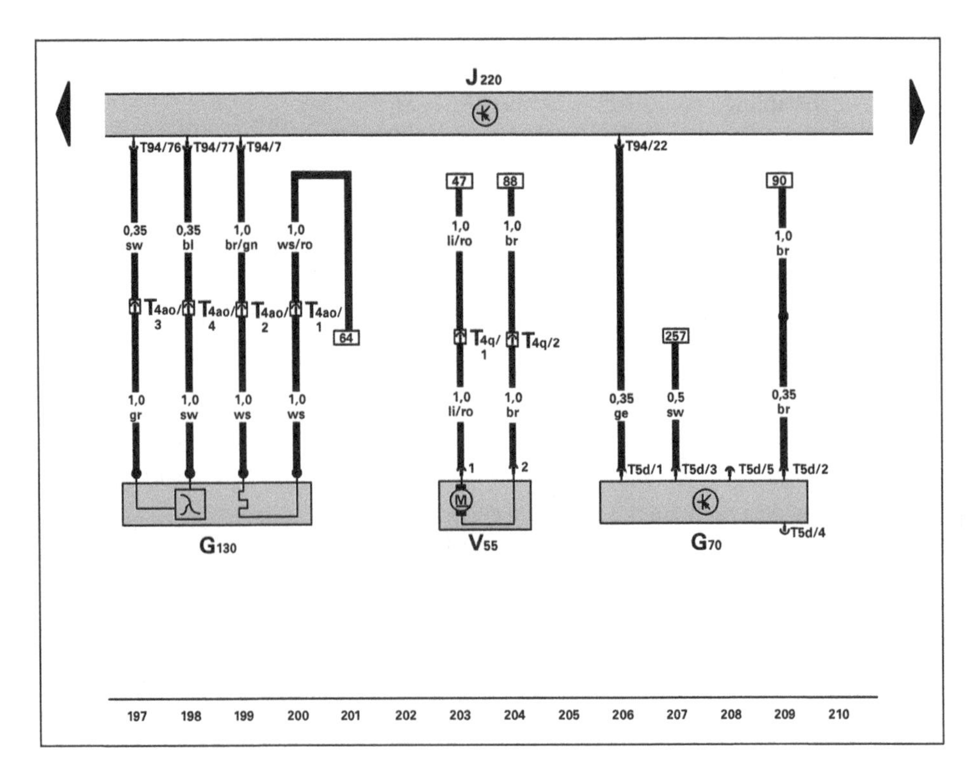

Figure 2 Wiring for mass airflow (MAF) sensor (G70) on a 2006 Volkswagen A5 Jetta with a 2.0L engine. Diagram shows five connections at the MAF. The transistor symbol indicates this is a digital MAF.

Determining Pinouts

Pinouts are best determined using a wiring diagram. However, we have found that the following procedure is useful for confirming that the factory service information is correct, or determining pinouts if a wiring diagram is not available.

- Locate the connector to that MAF and disconnect it.
- Turn the key to ignition ON, but do not start engine.
- Set the multimeter to measure DC voltage.
- Connect the black lead to any chassis ground.
- On the wiring harness side of the connector, touch the red lead to each of the pins. If any of them read 12 volts, that pin is the supply voltage. If any of them read 5 volts, that pin is a reference voltage.
- Turn the ignition off.
- Set the meter to measure resistance.
- Connect the black lead to any chassis ground.
- On the wiring harness side of the connector, touch the red lead to each of the other pins. The pin(s) at which the multimeter registers continuity (near zero ohms) should be ground.

The following steps only apply to hot wire sensors.

- Set the meter to measure voltage.
- Position the red multimeter probe on the remaining pins on the harness side of the connector one at a time.
- Turn the key to ignition ON, but do not start engine.
- If a pin has voltage that persists for several seconds but then drops to zero, that pin is the burn-off voltage.

The remaining pin(s) should be signal, or temperature (if it's there). You should be able to pair the remaining wires with the ground wire and see if one of them has a signal that's either a voltage that changes with engine speed, or a square wave whose frequency changes with engine speed.

MAF Sensor Supply Voltage Test

See Test Set-Up on page 34-2

Look for a Static Value

In this test, we check for supply voltage to the MAF sensor.

Typical Normal Values for MAF Sensor Supply Voltage
5V or 12V, depending on model

- Set the multimeter to measure DC voltage.
- Disconnect the connector to the MAF.
- Turn the key to ignition ON, but do not start engine.
- Connect the black multimeter probe to chassis ground.
- Working at the wiring harness side of the connector, touch the red multimeter probe to each of the pins.
- The multimeter should read 5V or 12V on the wire carrying the supply voltage. If it is not present, consult a wiring diagram and troubleshoot the wiring.

Analog MAF Sensor Signal Test

See Test Set-Up on page 34-3

Look for a Slowly Changing Value

In this test, we look for a changing voltage signal from an analog MAF while the engine is running. The MAF sensor will need to stay connected during the test. Thus, you'll need to back-probe the connector.

Typical Normal Values for Analog MAF Sensor Signal
About 0.5V (engine idling) to 4.5V (high rpm)

- Set the multimeter to measure voltage.
- Connect the black multimeter probe to chassis ground.
- Use the red multimeter probe to back-probe the signal pin on the connector.
- Start the car and let it idle.
- Verify that the voltage at idle is near 0.5V, and as you rev the engine higher and draw more air into the engine, it smoothly increases, approaching 4.5 volts. Note that the transition from 0.5 to 4.5V may have a peak in the middle, as the sensor is outputting values that are later calibrated (mapped) by the ECM. The readings should not abruptly fall or jump.

- Tap the MAF lightly with the handle of a screwdriver or similar tool. Observe the reading on the meter. If the idle stumbles and the reading abruptly jumps or drops when the MAF is tapped, the MAF is likely defective and should be replaced.
- If the readings are erratic, try cleaning the MAF with a MAF-specific cleaner from an auto parts store. If, after cleaning, the readings are still jumpy, the MAF sensor is likely defective and should be replaced.

Digital MAF Sensor Signal Test

See Test Set-Up on page 33-3

Look for a Square Wave

In this test, we look for the frequency-varying square wave signal while the engine is running. The MAF sensor will need to stay connected during the test. Thus, you'll need to back-probe the connector. The digital sensor outputs a square wave whose frequencies are either in the low hundreds of hertz or in the thousands of hertz, depending on the generation of the sensors. These values are summarized below.

Typical Normal Signal Values for Digital MAF Sensor	
1st generation	About 100Hz (engine idling) to 300 Hz (high rpm)
2nd generation	About 1000Hz (engine idling) to 9000 Hz (high rpm)

- Configure an automotive multimeter to detect a square wave. The exact method is meter dependent. It may involve using the pulse width or % duty cycle setting. We have generically indicated it via the "Hz %" button in the Test Set-Up 34-C illustration.
- Connect the black multimeter probe to chassis ground.
- Use the red multimeter probe to back-probe the signal pin on the connector.
- Start the car.
- Tap the MAF lightly with the handle of a screwdriver or similar tool. Observe the reading on the meter. If the idle stumbles and the reading abruptly jumps or drops when the MAF is tapped, the MAF is likely defective and should be replaced.
- Verify that the frequency changes smoothly as you rev the engine and draw more air into it. If it does not, the MAF sensor is likely defective and should be replaced.
- If the readings are erratic, you can try cleaning the MAF with a MAF-specific cleaner from an auto parts store. If, after cleaning, the readings are still jumpy, the MAF sensor is likely defective and should be replaced.

Supplemental Testing

- On most post-1996 OBD-II cars, if the MAF sensor is disconnected, the ECM will use data from other sensors instead. Therefore, if the car has a rough idle, disconnect the MAF sensor. If the idle smooths out when the MAF sensor is unplugged, it is very likely the MAF is dirty or defective.
- A model-specific repair manual can be consulted to learn when the burn-off cycle comes on, and its presence can be visually verified. If the burn-on cycle does not come on, the pins on the connectors can be checked with a multimeter set to register voltage to verify that burn-off voltage is present. If the voltage is present and burn-off is not occurring, the sensor is probably bad.

If you have direct confirmation via an automotive multimeter or a scope that the sensor is outputting a signal, but there's still a problem (for example, a scan tool is still reporting there's no signal from the sensor), there may be a broken wire or a bad connection between the sensor and the ECM.

- Use a multimeter set to measure resistance to verify continuity between the ground wire on the connector and chassis ground.
- Use a multimeter set to measure resistance to verify continuity between the signal wire on the connector and the ECM. This will require a wiring diagram for your car, knowledge of the location of the ECM, and knowledge of which pin on the ECM connector carries the signal from the MAF.

Is a Scan Tool Useful?

Yes. If your car is post-1996, OBD-II fault codes should be present for MAF sensor malfunction. In addition, if you have an OBD-II code reader that displays live data, or a brand-specific scan tool, it should be able to display the data from the MAF for you to verify correct operation.

Is an Oscilloscope Useful?

For a voltage-based MAF sensor, an oscilloscope is overkill, though you can use it to display the ramp-up and ramp-down of the voltage reading as you change the amount of airflow.

For a frequency-based sensor, you can directly look at the square wave with a scope.

Additional Information

Related Chapters

- **Common Multimeter Tests** for a discussion on how to use a multimeter to measure voltage and resistance.
- **Dynamic Analog Signals** for a discussion on square waves and sine waves.
- **Tools for Dynamic Analog Signals** for a discussion on how to use an automotive multimeter to detect sine waves and square waves.

Technical Notes

European cars use predominantly Bosch electrical components, and for many years, Bosch MAF sensors were analog voltage-based hot wire sensors. The change from voltage to frequency did not appear to occur coincident with the wire/film switch.

Generally speaking, for a given make and model, the trend was that voltage-based hot wire sensors were replaced by voltage-based hot film sensors, which in turn were replaced by digital frequency-based hot film sensors.

For example, the BMW E36 3 Series had a voltage-based hot wire sensor, updated to a voltage-based hot film sensor. Use of voltage-based hot film sensors continued through the E46 3 Series into the early E90 3 Series cars, but appears to have changed to frequency-based ("digital") hot film sensors early in the E90 product run.

35

Testing Stick Coils

Multimeter Test Set-Ups

⚠ WARNING —

Make personal safety your top priority. Watch what you are doing, stay alert. Do not work on your car if you are tired, not feeling well, or in a bad mood. Common sense, alertness, and good judgment are crucial to safe work on your vehicle.

Protect yourself from shock, trauma, and hazardous chemicals: Wear safety glasses, work boots, and protective gloves when working on your vehicle. Have hearing protection available for work that generates noise.

Be sure the work environment is well lit, properly ventilated, and that the vehicle is on a level surface. The floor, vehicle, and components inside the vehicle should be dry to minimize the possibility of shocks.

This book is designed as background technical training and as a general overview of how to do electrical diagnosis and service. It is designed to be used in conjunction with the appropriate service information for your vehicle, not to replace factory service information.

If you lack the skills, tools and equipment, or a suitable workshop for any procedure described in this book, have the job done by a qualified shop or an authorized dealer.

For additional warnings and cautions, see the **Safety,** *chapter and the warnings, cautions, and notices in your vehicle manufacturer's service information.*

Stick Coil Resistance

Test Set-Up **35-A**

See page 35-8 for step-by-step procedure

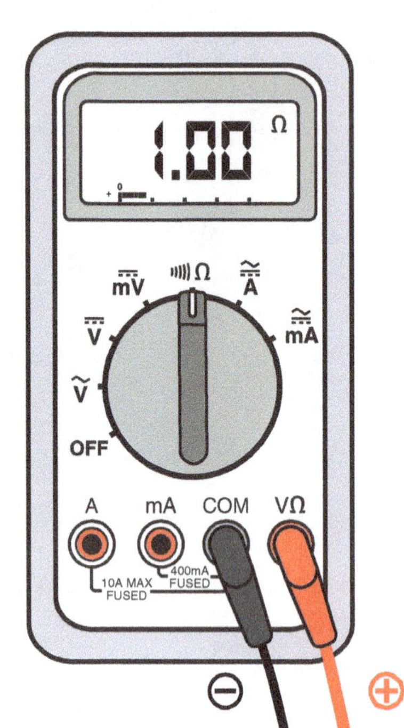

Testing Resistance Across Stick Coil

Test Conditions

Stick coil

Dial set to:	Ω resistance
Red input terminal:	VΩ
Black input terminal:	COM
Red lead:	sensor power pin (+)
Black lead:	sensor ground pin (–)

Connector disconnected, testing at sensor

Engine off

Typical Normal Values

Resistance: Approximately 1.0Ω

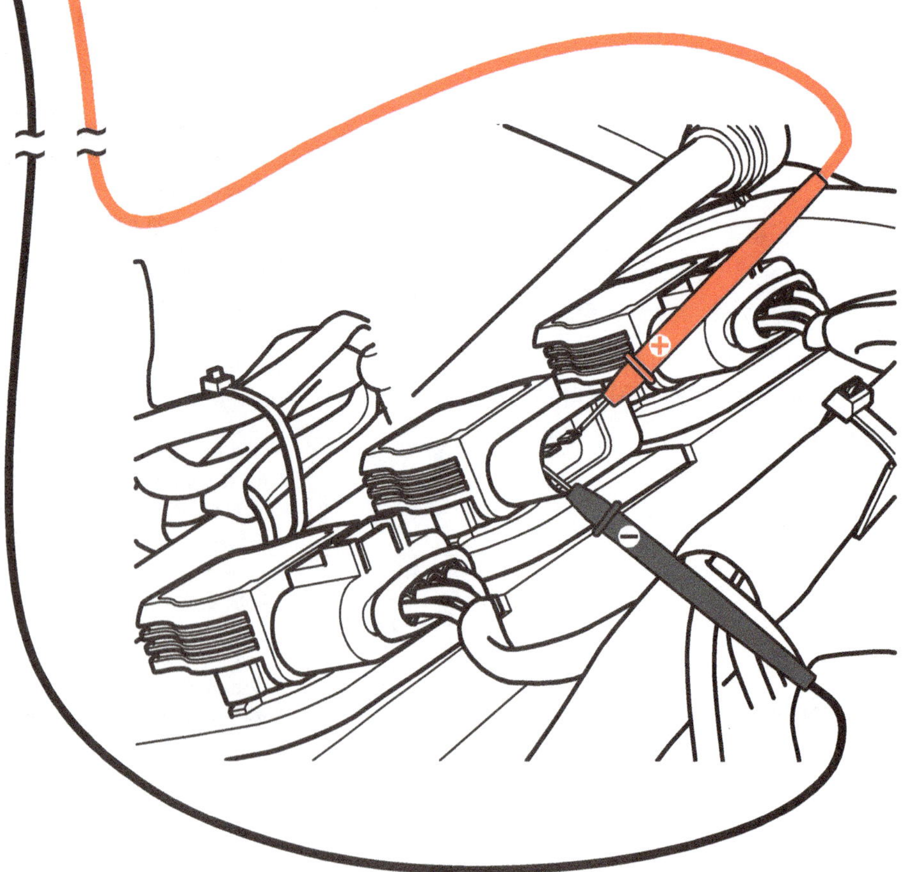

Look for a Static Value

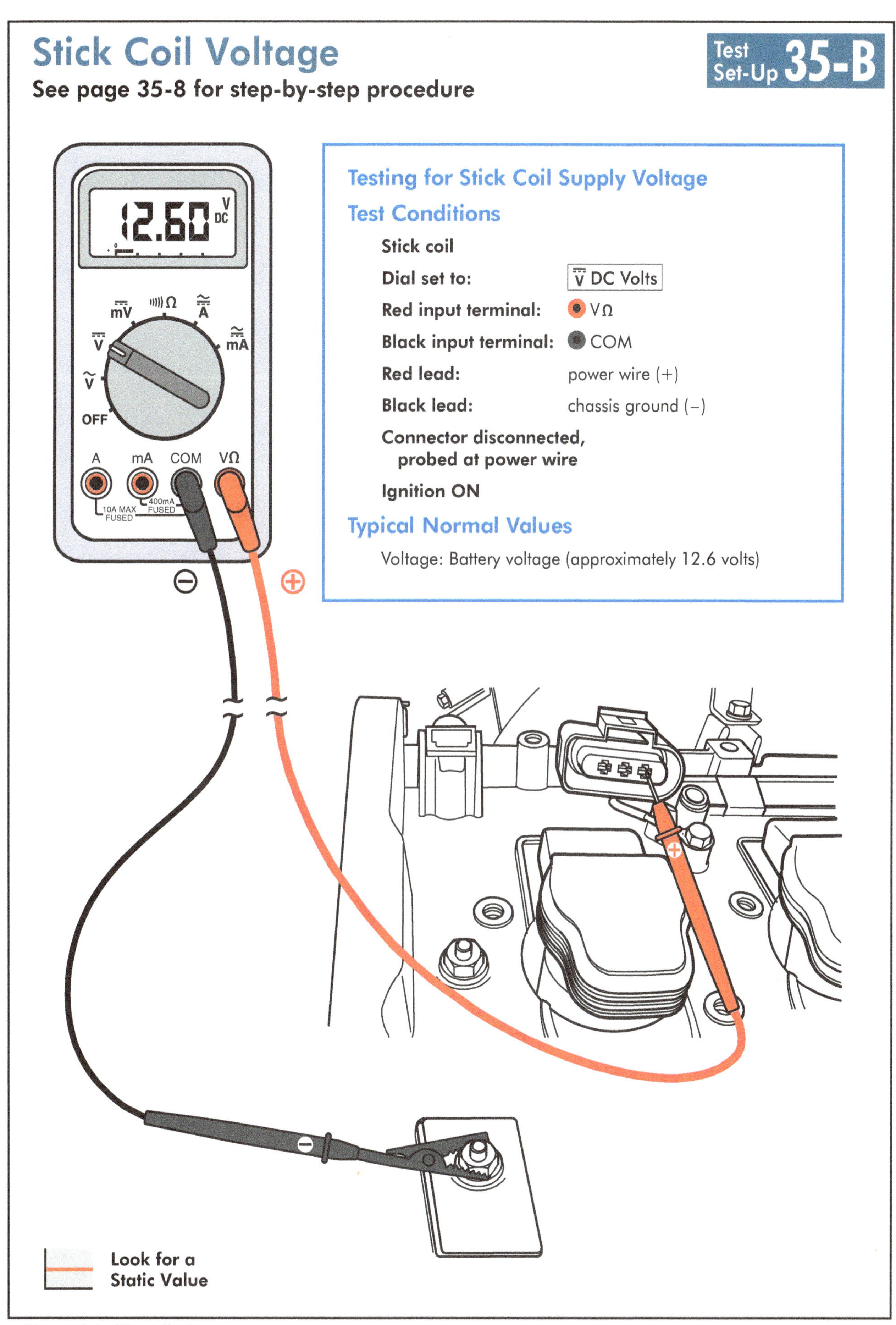
Stick Coil Voltage
See page 35-8 for step-by-step procedure
Test Set-Up 35-B
Testing for Stick Coil Supply Voltage
Test Conditions
Stick coil
Dial set to: V DC Volts
Red input terminal: VΩ
Black input terminal: COM
Red lead: power wire (+)
Black lead: chassis ground (–)
Connector disconnected, probed at power wire
Ignition ON
Typical Normal Values
Voltage: Battery voltage (approximately 12.6 volts)
12.60
V DC
mV
Ω
A
mA
V
OFF
A
mA
COM
VΩ
10A MAX FUSED
400mA FUSED
Look for a Static Value

Stick Coil Trigger Signal

See page 35-9 for step-by-step procedure

Test Set-Up 35-C

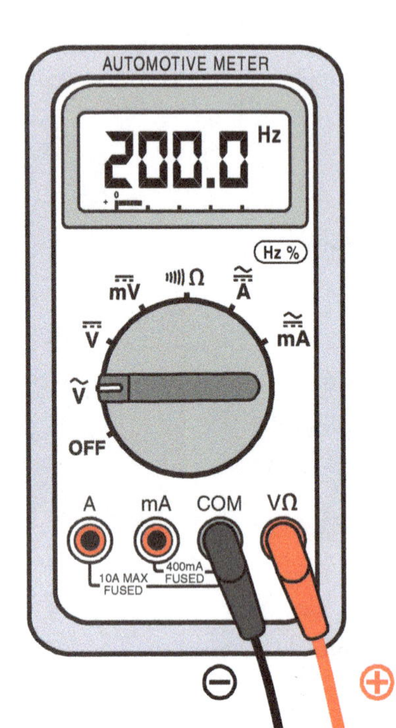

Testing for Varying Frequency of Square Wave

Test Conditions

Stick coil

Automotive meter that reports frequency of a square wave

Dial set to: Ṽ AC Volts | Hz %

Red input terminal: VΩ

Black input terminal: COM

Red lead: sensor signal wire (+)

Black lead: chassis ground (–)

Connector remains connected, back-probed at signal wire

Engine running

Typical Normal Values

Signal Frequency: Varies with engine rpm

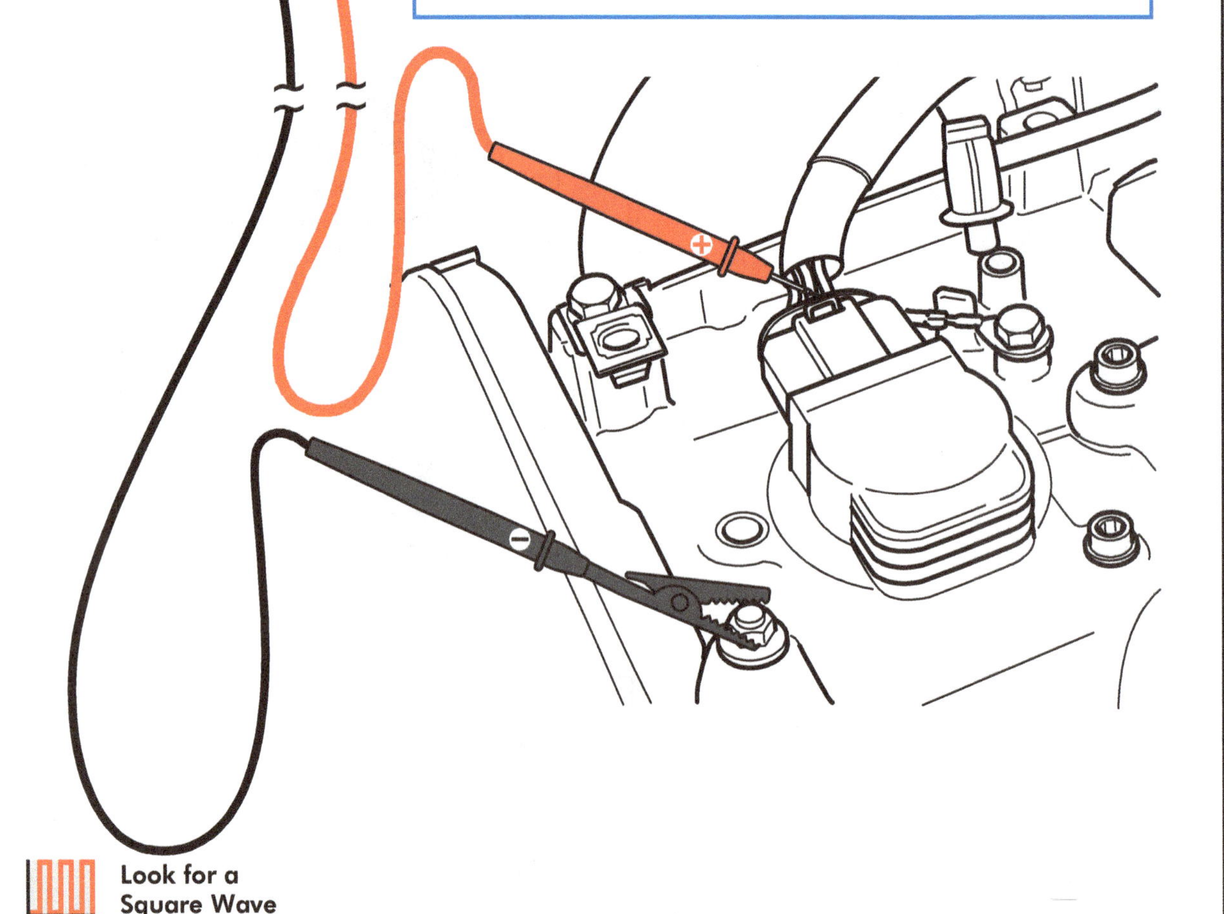

Look for a Square Wave

Stick Coil

Every gasoline engine needs a method to transform battery voltage into the tens of thousands of volts needed to jump a gap in a spark plug. Since the early 1990s, most cars have used Direct Ignition (DI) systems with one coil per cylinder. The most common DI configuration uses stick coils where the coil and the spark plug connector form a single rigid unit, as shown in **Figure 1**.

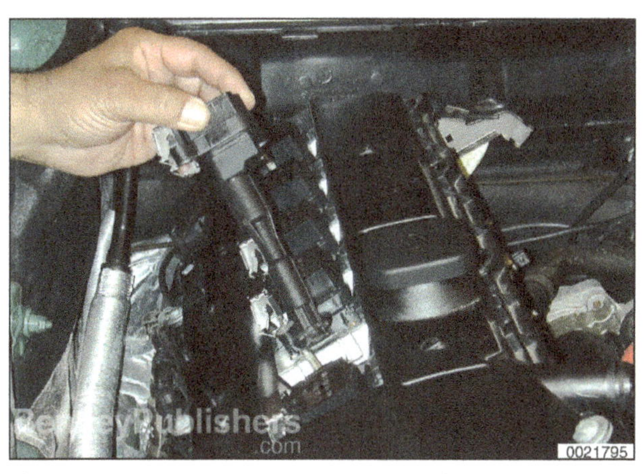

Figure 1 Stick coils are installed atop each spark plug. Each coil is individually controlled by the Engine Control Module (ECM). Common OBD-II Diagnostic Trouble Codes (DTCs) for faulty coils are usually misfire codes, such as P0301 Cylinder #1 Misfire Detected.

Most owners are alerted to one bad stick coil by a rough idle and a slight loss of power. In a post-1996 OBD-II car, the CEL will almost certainly be lit, and a scan tool will likely report a misfire code on a particular cylinder.

Operation

Each coil has a primary and a secondary winding. The primary winding is supplied battery voltage and a path to ground. The flowing current sets up a magnetic field in the core. When the ground path is interrupted, the magnetic field collapses (shuts off). Faraday's law (a changing magnetic field creates an electric field) then does its thing, which induces hundreds of volts in the primary winding, which in turn induces tens of thousands of volts in the secondary winding. The resulting current flows out the secondary winding, through the spark plug connector, to the top of the spark plug, and finally to ground by creating a spark when it jumps across the gap in the spark plug, which is indeed the whole point.

Types

There are several direct ignition configurations.

"Stick coil" refers to the configuration shown in **Figure 1** where the coil and the plug connector form a single rigid unit, typically inserted into a recess in the middle of the camshafts in the head and hidden beneath a cover.

"Coil on plug" (COP) is an imprecise term that can mean a stick coil, but can also mean an earlier DI configuration where each coil is attached to its plug via a short flexible plug wire. They are usually used on single camshaft engines where the plugs are accessible on the exhaust side rather than in the center of the head.

"Coil pack" is also an imprecise term. It may refer to a stick coil, or i to a configuration where multiple coils are mechanically packaged together and connect to the individual spark plugs with flexible plug wires.

"Cassette" refers to a configuration where individual coils and plug connectors are mechanically packaged together in a structure and mount together on top of multiple spark plugs (for example, two cassettes with one on each side of a V8 or V12).

Visual inspection will easy reveal which type your car has. Note that only stick coils and COPs can be swapped cylinder-to-cylinder to aid in troubleshooting of isolating a bad coil (see the **Testing** section later in this chapter).

Signal

The signal controlling each stick coil is a frequency-varying square wave. The frequency increases with increasing engine rpm (that is, the plugs fire faster). The rising edge of the square wave signals the coil's triggering electronics to start current flowing into the primary winding. The falling edge of the square wave shuts off the primary current and causes the magnetic field to collapse, creating the current in the secondary winding and thus the spark.

Circuit

Stick coils can have two, three, or four wires. Two-wire coils (positive and ground) are found primarily on Chrysler and Ford products. With these, the ECM is directly switching the coil's ground connection with square wave timing. That is, the ground wire is also the signal wire. This is typically the same method of triggering used for fuel injectors.

The stick coils found on most European cars have three wires. The terminals are power, ground, and a trigger signal from the ECM. Depending on the car, these may be labeled on the wiring diagram using the standard DIN numbers 15, 1, and 4a. 15 is battery voltage and 1 is ground, and correspond to the input and output ends of the primary winding just like on a conventional ignition coil. 4a is the trigger signal. This is a frequency-varying square wave described above, typically 5 volts, received by a driving circuit in the head of the stick coil, which then switches the primary winding in the coil on and off. These are summarized in the **Pinout for 3-Wire Stick Coil Table**.

Pinout for 3-Wire Stick Coil

Terminal	Signal
15	Power
1	Ground
4a	Square wave trigger signal

The terminals on the three-wire connector are usually numbered as shown in **Figure 2**. Note that this shows the connector on the wiring harness side.

If a fourth wire is present, it often contains a confirmation square wave that is present if the coil has fired and is used by the ECM to set and clear codes.

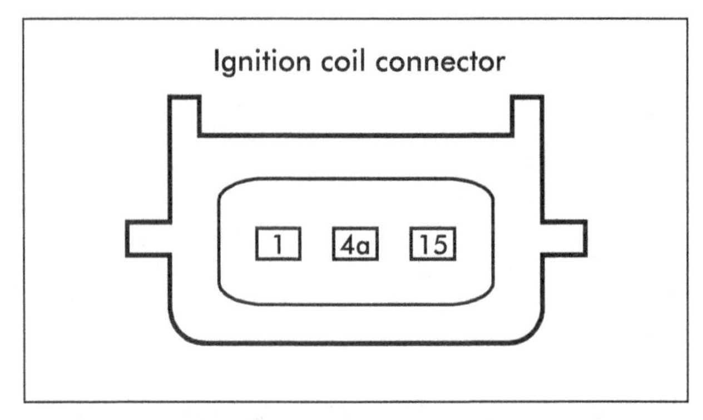

Figure 2 The ground, trigger signal, and power terminal on the wiring harness side of a 3-pin stick coil connector used on many European cars.

A specific wiring example for the stick coil on a 2003 BMW 3 Series is shown in **Figure 3**.

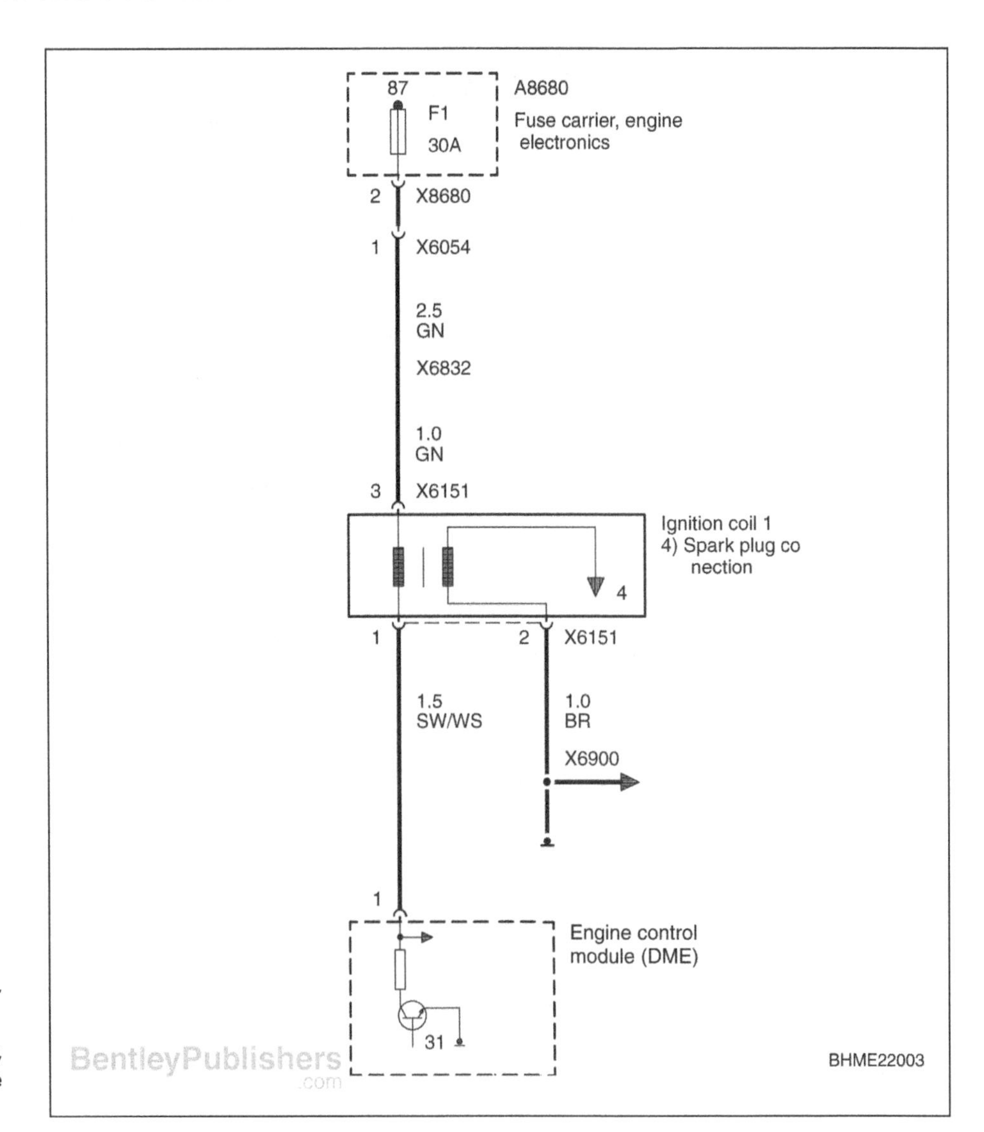

Figure 3 Stick coil wiring for a 2003 BMW E46 3 Series. Three wires are used – power, ground, and the square wave signal from the ECM.

Testing

The voltage to and the resistance across a stick coil can be verified with any multimeter, but in order to verify that the coil is receiving the square wave pulse train from the ECM, an automotive multimeter that can detect a square wave and measure its frequency, such as the Fluke 88V, can be employed.

However, on post-1996 OBD-II cars, most people are alerted to a bad coil by the CEL being lit and the scan tool reporting a misfire code. So, a highly effective technique is to take the coil that the misfire code reports, swap it with the adjacent stick coil, and see if the problem stays with the cylinder or moves with the coil. If the problem moves with the coil, the coil is probably bad. Its resistance can be checked, but that is not a definitive test. If the problem stays with the cylinder, the power and signal to that cylinder must be checked.

Equipment Needed for Stick Coil Testing

Test	Tool
Voltage and resistance	Multimeter
Reading / clearing codes	For a post-OBD-II car, OBD-II scanner
Running test	Automotive multimeter that detects a square wave and reports its frequency, or oscilloscope

Disconnect the connector from the coil. Physically inspect both sides of the connector. Make sure no wires are broken and no pins are pushed out. Inspect the coil itself to make sure it isn't cracked. Verify that the coil isn't covered with, or sitting in, oil or antifreeze. Not only does this indicate a leak which should be addressed, but also, the fluid can adversely affect both the triggering electronics and the high-voltage spark.

Determining Pinouts

Pinouts are best determined using factory service information (a wiring diagram). If the stick coil's wiring does not match the connector shown in Figure 2, we have found that the following procedure is useful for confirming that the factory service information is correct, or determining pinouts if a wiring diagram is not available.

- Locate the connector to a stick coil and disconnect it. Typically this requires removing a false valve cover protecting the coils and their wiring harness.
- Turn the key to ignition ON, but do not start engine.
- Set the multimeter to measure DC voltage.
- Connect the black lead to any chassis ground.
- On the wiring harness side of the connector, touch the red lead to each of the pins. If any of them read 12 volts, that pin is the supply voltage.
- Turn the ignition off.

To determine the other pinouts:

- Set the meter to measure resistance.
- Connect the black lead to any chassis ground.
- On the wiring harness side of the connector, touch the red lead to each of the other pins. The pin at which the multimeter registers continuity (near zero ohms) should be ground.

If the stick coil has three terminals, the remaining pin should be the signal pin. If it has four terminals, you may need to consult a wiring diagram.

If the car has a true multi-coil pack instead of individual stick coils, there is likely a common power wire, a common ground wire, and individual signal wires for each cylinder. The power and ground pins could be ascertained as described above, but a wiring diagram would need to be consulted to be certain.

Determining Which Coil Is Misfiring

The first thing you need to do is to determine which cylinder is causing the problem. If the CEL is lit, you can plug in an OBD-II code reader, read the code, and it will likely tell you which cylinder is causing the problem. But if the car is pre-OBD-II, or if you don't have a code reader, you may need to determine which coil is malfunctioning yourself.

There is a technique of pulling the plug wires and connectors off the spark plugs on a running car one plug at a time. If the engine runs worse, you've disconnected a plug that was firing, but if the engine runs about the same with a plug wire off, that cylinder probably wasn't firing.

However, even using rubber gloves for insulated protection, we do not recommend doing this with stick coils. The battery voltage feeding the coil is unlikely to hurt you, but the coil's secondary winding can generate tens of thousands of volts, and that can kill you. Even though high voltage shouldn't be present at the stick coil's connection to the wiring harness, and even though it makes the troubleshooting a little murkier, for safety reasons we strongly recommend shutting off the car before touching any stick coil, including disconnecting or reconnecting its connector!

WARNING —

IGNITION COILS GENERATE HIGH VOLTAGE THAT CAN KILL YOU! DO NOT TOUCH THE COILS WHILE THE CAR IS RUNNING!

Swapping Coils to Isolate the Problem

Stick coils are usually located beneath the plastic engine covers. Consult a repair manual to learn what needs to be removed in order to remove the cover(s) and access the stick coils.

NOTE —

You can only swap stick coils if you actually have stick coils! If you instead have a multi-coil pack or a cassette, this technique is not applicable.

CAUTION —

Use care when working at or near hot exhaust system components or hot engine parts. Wear heavy work gloves, a long sleeve-work shirt, and eye protection. Work in a well-ventilated area.

- Determine which coil is suspect as described above.
- On an OBD-II car, use an OBD-II scanner to clear the codes.
- With the engine switched off, remove the engine cover and disconnect the suspect coil connector. Physically inspect the connector and the coil itself. Make sure no wires are broken, no pins are pushed out, and the connector is corrosion free.
- Swap the suspect coil with the one next to it.
  - Label the coils with tape (e.g., "coil A was in cyl 1").
  - Disconnect the connector to each of the two coils.
  - Remove whatever mechanical attachment is affixing each coil to the head (typically one or two small bolts).
  - Exchange the positions of the two stick coils.
  - Reattach and re-connect both coils. Be sure to bolt down any ground wires removed.
- Repeat the test to see which cylinder is misfiring, either using an OBD-II code reader / scan tool or disconnecting one coil connector at a time.
- If the problem moves to a different cylinder (e.g., if you originally had the code "P0301 cylinder 1 misfire detected" and now have the code "P0302 cylinder 2 misfire detected"), then the coil ("coil A") is very likely defective and should be replaced.
- But if the problem stays at the same cylinder (e.g., "P0301 cylinder 1 misfire detected"), then the wiring (circuit) for that cylinder must be checked.

Stick Coil Resistance Test

See Test Set-Up on page 35-2

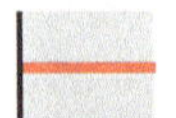

Look for a Static Value

The resistance across the stick coil's primary winding (DIN terminals 1 and 15) should be similar to that of a conventional coil. You should see approximately 1 ohm. If the resistance between 1 and 15 shows no continuity, the primary winding is likely broken. However, coil failures are usually not in the primary winding – they're usually in the secondary winding. And, unfortunately, because the integrated driver circuitry sits between the coil itself and the terminals on the connector, you can't make a direct measurement of the resistance of the secondary winding on a stick coil with a three-wire connector. Note that if the resistance is correct, it doesn't guarantee the coil is good; it simply means the primary winding isn't broken or shorted.

Typical Normal Stick Coil Resistance
Primary resistance (Terminals 1 and 15) Approximately 1 ohm

- Use a multimeter set to measure resistance.
- Touch one multimeter lead to the stick coil's voltage terminal (15).
- Touch the other multimeter lead to the stick coil's ground terminal (1).
- Verify that the meter shows approximately 1 ohm.
- If it instead shows 0 ohms (short circuit) or "OL" (open circuit), the coil is likely defective and should be replaced.

Stick Coil Voltage Test

See Test Set-Up on page 35-3

Look for a Static Value

With the ignition on, the coil's connector should be receiving battery voltage on pin 15.

Typical Normal Stick Coil Voltage
(Terminal 15 and ground) Approximately 12.6V

- Disconnect the connector from the stick coil.
- Use a multimeter set to measure DC voltage.
- Connect the black lead to any chassis ground.
- Turn the key to ignition ON, but do not start engine.
- On the wiring harness side of the connector, touch the red lead to the voltage pin (15). Verify that battery voltage (about 12.6V) is present.
- Set the multimeter to measure resistance.
- On the wiring harness side of the connector, touch the red lead to the ground pin (1). Verify that there is continuity to ground.

NOTE —

If ground is not present, it may be because it is a switched ground from the ECM. Confirm circuit logic using a wiring digram for your specific car. If the ground wire is coming from the ECM, there's a good chance it is a switched ground.

- If either voltage or ground are missing, consult a wiring diagram for the car and troubleshoot the problem.

Stick Coil Trigger Signal Test

See Test Set-Up on page 35-4

Look for a Square Wave

As we discussed, with the engine running, there should be a square wave signal on the coil's pin 15.

Typical Normal Stick Coil Signal
Frequency–varying square wave

- Use an automotive meter set to detect a square wave and measure frequency, or an oscilloscope.
- Back-probe the stick coil's connector and connect the red lead to terminal 4a.
- Connect the black lead to chassis ground.
- Start the car.
- With the car running, the meter should detect the presence of the square wave.
- As you increase engine rpm, you should see the frequency increase.

Supplemental Testing

If you've verified there's power, ground, and signal at a coil, and you've swapped the coil with another coil that isn't throwing a misfire code, then it's possible that there may be a non-coil-related cause, such as low compression.

If you are a professional, you may have an oscilloscope and a current clamp connection to the scope so you can look at the amperage drawn by the coil as it fires and judge if the ramp-up in amperage looks normal. But absent that, the best way to tell if the coil is firing is to see if it creates a spark on a spark plug.

Because it is not safe to have exposed high-voltage spark, we strongly recommend against the old-school method of putting an exposed spark plug in a stick coil and grounding the tip, and instead strongly recommend the use of a spark tester. This is an inexpensive ($15) device that plugs into the coil in place of the plug and presents a flashing light if the coil is firing.

Unless there is massive physical damage to a spark plug (like a melted electrode), it would be unusual for a coil to test out as good with a spark tester but not fire the actual plug in a cylinder. But if the power, ground, signal, and coil all check out, the plug should be examined. Physical examination, swapping plugs with another cylinder, or replacement with new plugs should address any question.

If all of the above check out but the misfire persists, the problem is likely not an ignition problem at all. The fuel injector, cylinder compression, and carbon buildup on the valves would need to be checked.

Is a Scan Tool Useful?

It is more than useful – in a post-1996 OBD-II car, a scan tool is indispensable for misfire troubleshooting. Even the least expensive OBD-II code reader can be used to pull a misfire code, lead you to the offending cylinder, clear the code, and help you perform the swap test.

Is an Oscilloscope Useful?

Yes. If you have a scope, ignition problems are one of the best places to use it. An inexpensive pocket oscilloscope can be used to verify the presence of the square wave. If you have a professional level scope with a current clamp attachment, you can look at the current drawn by the coil, the square wave input signal, and the actual high-voltage secondary pulse itself, but this is beyond the reach of most do-it-yourselfers.

Additional Information

Related Chapters

- **Electronic Ignition** for a discussion of direct ignition.
- **Dynamic Analog Signals** for a discussion on square waves and sine waves.
- **Tools for Dynamic Analog Signals** for a discussion on how to use an automotive multimeter to detect sine waves and square waves.

Technical Notes

The frequency controlling the stick coil is a frequency-varying square wave. That means the pulse width is fixed. It *is* fixed, but its width is important. The pulse width of the square wave controlling a stick coil is none other than our old friend ignition dwell. That's the engine rotation angle over which the ignition points "dwell together," which governs how long the primary winding has to charge up the coil.

36

Testing Fuel Injectors

Multimeter Test Set-Ups

WARNING —

Make personal safety your top priority. Watch what you are doing, stay alert. Do not work on your car if you are tired, not feeling well, or in a bad mood. Common sense, alertness, and good judgment are crucial to safe work on your vehicle.

Protect yourself from shock, trauma, and hazardous chemicals: Wear safety glasses, work boots, and protective gloves when working on your vehicle. Have hearing protection available for work that generates noise.

Be sure the work environment is well lit, properly ventilated, and that the vehicle is on a level surface. The floor, vehicle, and components inside the vehicle should be dry to minimize the possibility of shocks.

This book is designed as background technical training and as a general overview of how to do electrical diagnosis and service. It is designed to be used in conjunction with the appropriate service information for your vehicle, not to replace factory service information.

If you lack the skills, tools and equipment, or a suitable workshop for any procedure described in this book, have the job done by a qualified shop or an authorized dealer.

For additional warnings and cautions, see the **Safety**, *chapter and the warnings, cautions, and notices in your vehicle manufacturer's service information.*

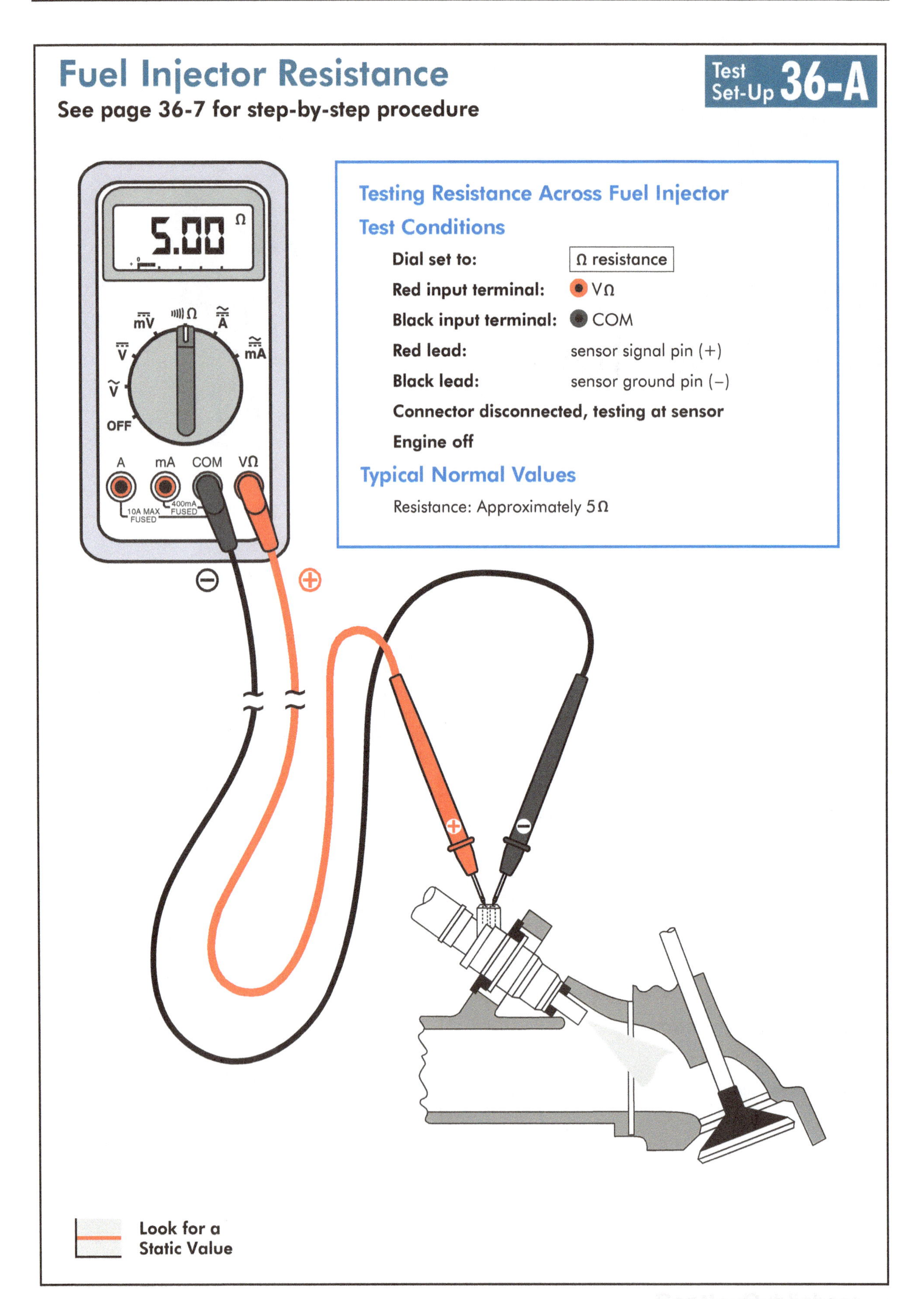
Fuel Injector Resistance
See page 36-7 for step-by-step procedure
Test Set-Up 36-A
5.00 Ω
mV
Ω
A
V
mA
V
OFF
A
mA
COM
VΩ
10A MAX FUSED
400mA FUSED
Testing Resistance Across Fuel Injector
Test Conditions
Dial set to: Ω resistance
Red input terminal: VΩ
Black input terminal: COM
Red lead: sensor signal pin (+)
Black lead: sensor ground pin (–)
Connector disconnected, testing at sensor
Engine off
Typical Normal Values
Resistance: Approximately 5 Ω
Look for a Static Value

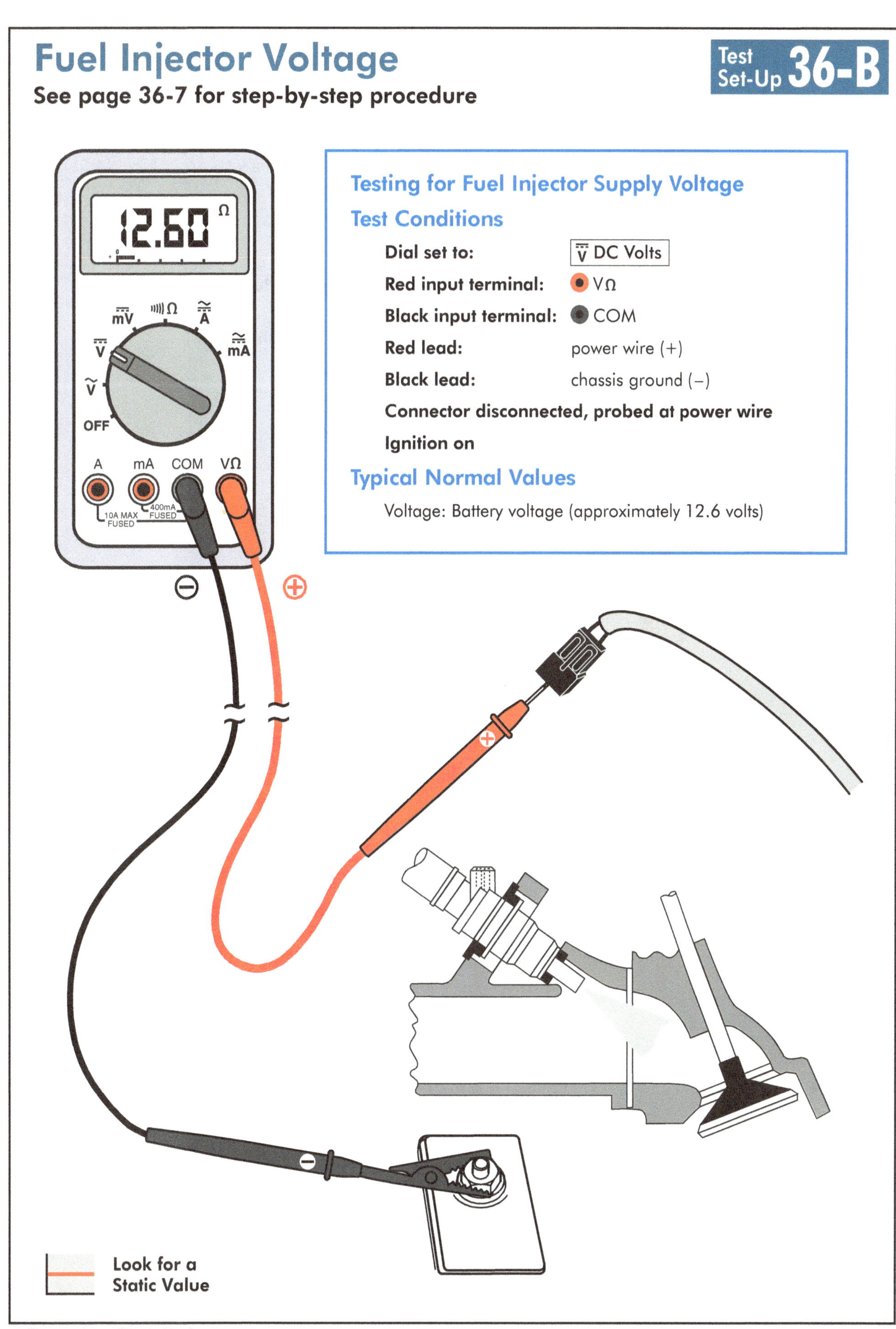
Fuel Injector Voltage
See page 36-7 for step-by-step procedure
Test Set-Up 36-B
Testing for Fuel Injector Supply Voltage
Test Conditions
Dial set to: V DC Volts
Red input terminal: VΩ
Black input terminal: COM
Red lead: power wire (+)
Black lead: chassis ground (–)
Connector disconnected, probed at power wire
Ignition on
Typical Normal Values
Voltage: Battery voltage (approximately 12.6 volts)
12.60
Ω
mV
A
mA
V
OFF
A
mA
COM
VΩ
10A MAX FUSED
400mA FUSED
Look for a Static Value

Fuel Injector Trigger Signal

See page 36-7 for step-by-step procedure

Test Set-Up 36-C

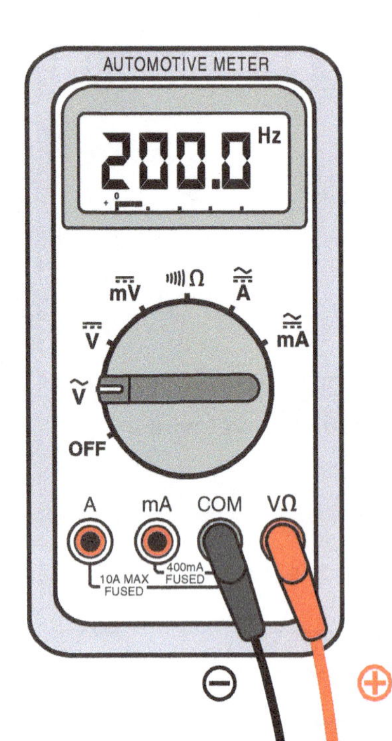

Testing for Varying Frequency of Square Wave

Test Conditions

Fuel injector

Automotive meter that reports duty cycle of a square wave

Dial set to:	Ṽ AC Volts / Hz %
Red input terminal:	VΩ
Black input terminal:	COM
Red lead:	sensor signal wire (+)
Black lead:	chassis ground (−)

Connector remains connected, back-probed at signal wire

Engine running

Typical Normal Values

Signal Frequency: Varies with engine rpm

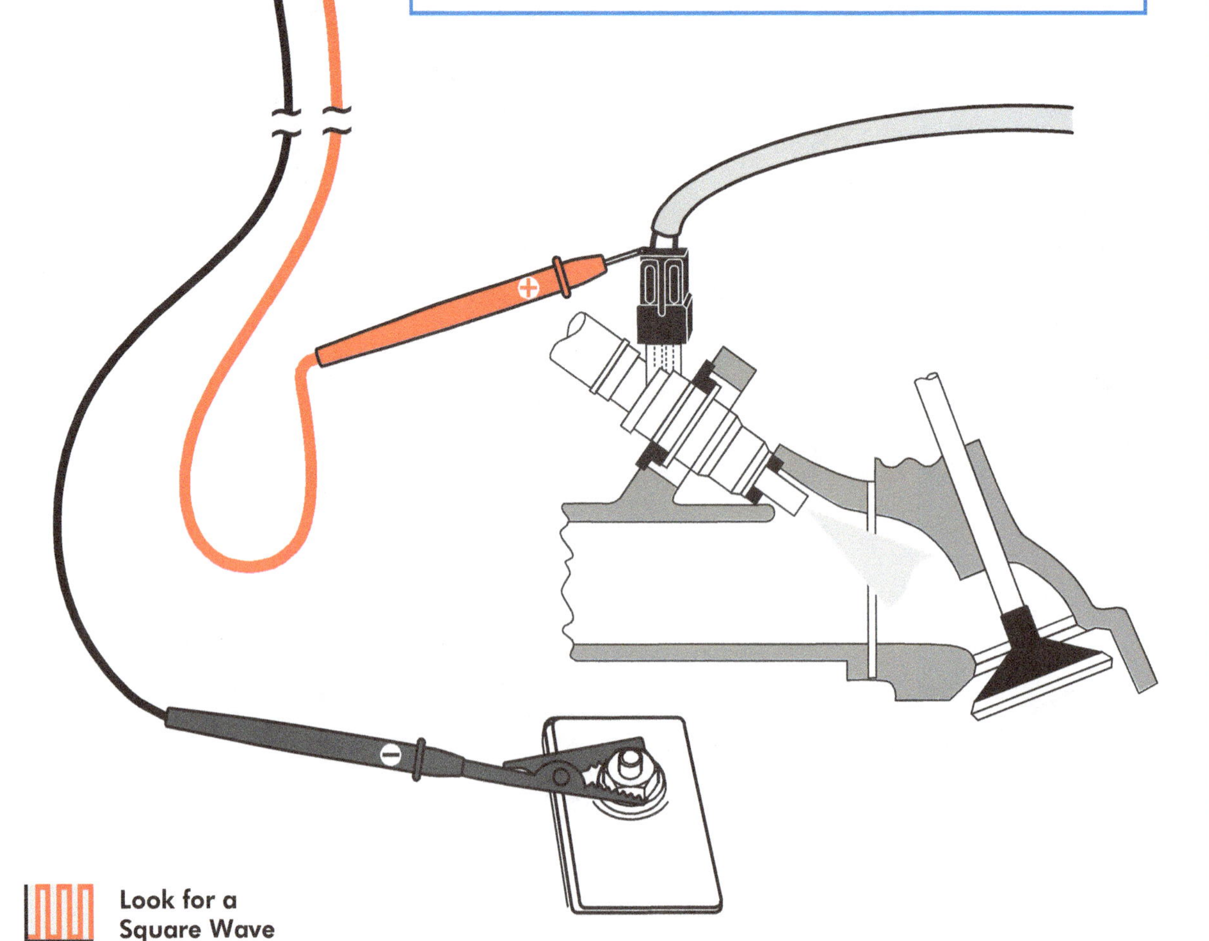

Look for a Square Wave

Fuel Injector

Fuel injectors spray pressurized fuel into the engine's cylinders, where it mixes with the proper amount of air that has been let into the engine and is ignited by the spark plugs. See **Figure 1**.

Rough running isolated to one cylinder that is not due to a bad stick coil. may be a sign of a bad fuel injector. On post-1996 cars with OBD-II, users may be led to a particular fuel injector by the Check Engine Light and an injector-related trouble code.

Figure 1 Injectors (**arrows**) and fuel rail on a 1998 Porsche Boxster.

Operation

On certain rare pre-1975 cars, injectors are purely mechanical, but most cars manufactured since the adoption of emission controls have electronic fuel injection which uses electronically actuated injectors.

Each injector has a solenoid inside that acts as a valve to stop and start the spraying of fuel into the intake manifold. Fuel is supplied to the injectors by the fuel pump and is present at an essentially constant pressure, waiting to squirt in when the injector opens. There's an electrical connection on the outside of the injector. Voltage is typically always present on one connector terminal. When the ground path to the other terminal is completed, the solenoid is energized, which opens the valve, allowing the injector to spray fuel into the cylinder. The faster the engine spins, the more frequently the injectors fire. The greater the load is on the engine (for example, going uphill), the more fuel needs to be delivered per injection, so the solenoid is left open longer.

NOTE —

This chapter covers 12V solenoid-based fuel injectors. If your injectors are short, like the one pictured on the first page of this chapter, and are mounted in the intake manifold, they are almost certainly solenoid-based injectors. In approximately the mid- to late-2000s, direct injection began to be used which often utilized long piezoelectric fuel injectors to spray fuel directly into the combustion chamber. Piezo injectors typically operate at a much higher voltage (approximately 140V) and fuel pressure. Piezoelectric injectors are not covered in this chapter.

Signal

The signal that controls an injector is often described as a Pulse Width Modulated (PWM) square wave, but that's a bit of a misnomer. If you look back at the **Dynamic Analog Signals** chapter, you'll see that PWM signals are square waves whose duty cycle (pulse width) changes but whose frequency remains constant. In the case of injectors, however, *both the frequency and the duty cycle (pulse width) are changing*. Here's why.

For each cylinder, the injector is generally firing once for every other crank revolution. So if injector firing is controlled by a square wave, as engine rpm changes, the square wave's frequency must change. The duty cycle (pulse width), on the other hand, is varied with engine load. If you're in a period of high load (going up a hill, for example), more fuel is injected per cycle by lengthening the injector "on time," which is another way of saying increasing the duty cycle (pulse width). The duty cycle is also changed by the ECM as part of the closed-loop reporting of the oxygen sensor.

Circuit

Injectors have two wires – power and ground. The power side is usually supplied with constant voltage, and the ground side is usually the side that is switched. That is, the car's ECM is connecting and disconnecting each injector's circuit to ground with timing in the shape of a square wave.

A wiring example for the injector on a 2008 MINI is shown in **Figure 2**.

Testing

The voltage to and the resistance across a fuel injector can be verified with any multimeter, but in order to verify that the injector is receiving the square wave pulse train from the ECM, an automotive multimeter that can detect a square wave and measure its frequency, such as the Fluke 88V, can be employed.

Disconnect the connector from the injector. Physically inspect both sides of the connector. Make sure no wires

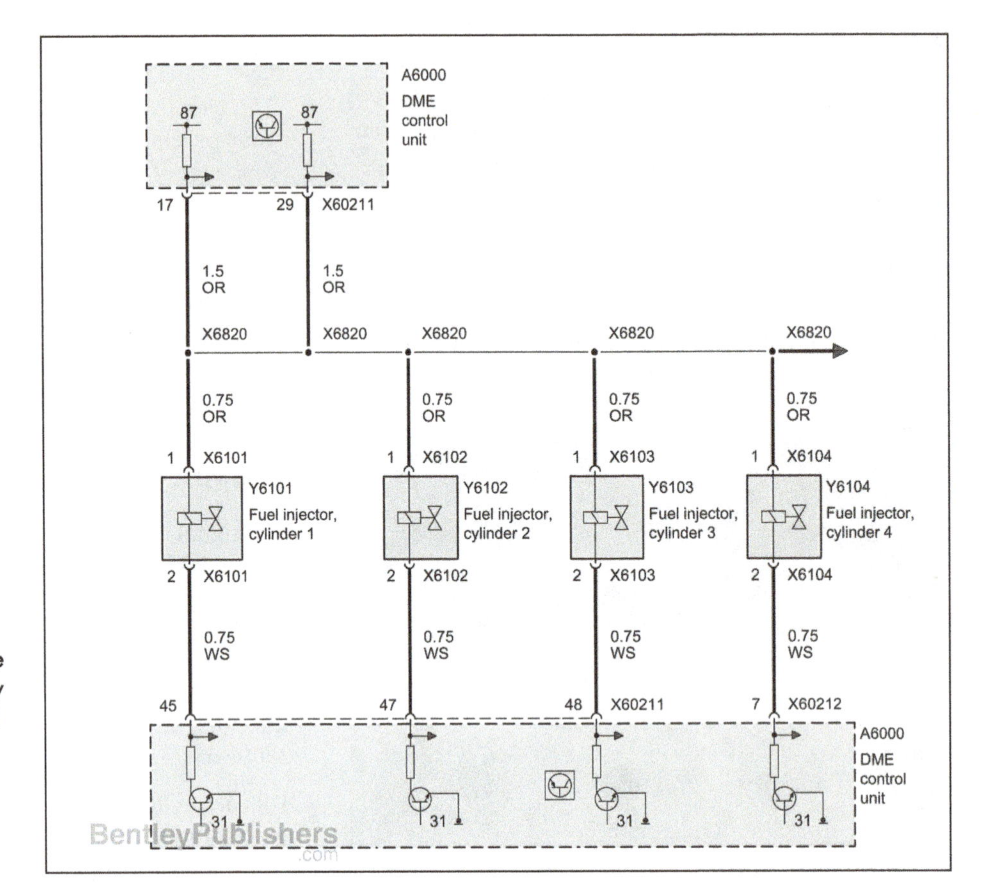

Figure 2 Fuel injector wiring for a 2008 MINI. Injectors have two wires. Power is supplied by the orange (OR) wire. Ground is on the white (WS) wire. The toggling on and off of the ground signal is indicated by the transistor symbol in the DME control unit.

Equipment Needed for Injector Testing

Test	Tool
Voltage and resistance	Multimeter
Running test	Automotive multimeter that detects a square wave and reports its pulse width or frequency, or oscilloscope

are broken and no pins are pushed out. Inspect the injector itself to make sure it isn't cracked.

Determining Pinouts

Pinouts are best determined using factory service information (a wiring diagram). However, we have found that the following procedure is useful for confirming that the factory service information is correct, or determining pinouts if a wiring diagram is not available.

- Locate the connector to an injector and disconnect it.
- Set the multimeter to measure DC voltage.
- Turn the key to ignition ON, but do not start engine.
- Connect the black lead to any chassis ground.
- On the wiring harness side of the connector, touch the red lead to each of the pins. The one that reads battery voltage (about 12.6V) is the power pin. The other pin is the switched ground pin.
- Turn the ignition off.

Determining Which Injector Is Misfiring

The first thing you need to do is to determine which cylinder is causing the problem. If the CEL is lit on a post-1996 OBD-II car, you can plug in an OBD-II code reader, read the code, and it will likely tell you which cylinder is causing the problem. But if the car is pre-OBD-II, or if you don't have a code reader, you may need to determine which injector is malfunctioning yourself.

The old-school technique of pulling the connectors off the injectors on a running car one plug at a time is quite effective. If the engine runs obviously worse, you've disconnected an injector that was firing, but if the engine runs about the same with an injector unplugged, that cylinder probably wasn't firing.

 **CAUTION —**

Use care when working at or near hot exhaust system components or hot engine parts. Wear heavy work gloves, a long-sleeve work shirt, and eye protection. Work in a well-ventilated area.

Fuel Injector Solenoid Test

As with any solenoid or relay, the presence or absence of an audible *click* when the solenoid is energized is a valuable troubleshooting tool. You can connect an injector's two terminals directly to the car's battery using a long pair of wires with small alligator clips on the ends, but care must be taken to be certain the clips don't touch each other when you're attaching them to the injector terminals. For this reason, using a spare injector connector and splicing in long wires that will reach the battery is preferable. These are readily available on eBay and other sources.

- Remove the injector connector and replace it with the test connector.
- Briefly touch the positive and negative leads to the battery's positive and negative terminals. Do not leave it on for long; the solenoids in injectors are designed to cycle on and off.
- You should hear and feel the solenoid inside the injector click.
- If there is battery voltage and ground present at the solenoid but the injector does not click, then the solenoid inside it is likely defective and the injector should be replaced.

Fuel Injector Resistance Test

See Test Set-Up on page 36-2

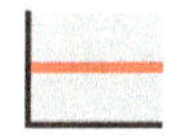

Look for a Static Value

The resistance across the terminals of an injector (across the coil of the solenoid inside the injector) can be in the single number of ohms for a low impedance injector, and in the 10–20 ohm range for a high impedance injector. If the resistance between the two terminals on the connector shows no continuity, the winding is likely broken. Note that if the resistance is correct, it doesn't guarantee the injector is good; it simply means the coil in the solenoid isn't broken or shorted.

Typical Normal Injector Resistance
1–5 or 10–20 ohms

- Use a multimeter set to measure resistance.
- Unplug the connector from the injector.
- Touch one multimeter lead to one of the injector's terminals.
- Touch the other multimeter lead to the injector's other terminal.
- Verify that the meter shows a reading in the correct range. If it instead shows 0 ohms (short circuit) or "OL" (open circuit), the injector is likely defective and should be replaced.

Fuel Injector Voltage Test

See Test Set-Up on page 36-3

Look for a Static Value

With the ignition on, the injector's connector should be receiving battery voltage on pin 15.

The voltage input to a simple solenoid should be the car's charging voltage (about 14.2 volts) with the engine running. A square wave should be present on the ground side, toggling the voltage path to ground.

Typical Normal Injector Voltage
Charging voltage (about 14.2V)

- Disconnect the connector from the injector.
- Use a multimeter set to measure DC voltage.
- Connect the black lead to any chassis ground.
- Turn the key to ignition ON, but do not start engine.
- On the wiring harness side of the connector, touch the red lead to each pin. The one showing battery voltage (about 12.6V) is the voltage pin.
- If voltage is missing, consult a wiring diagram for the car and troubleshoot the problem.

Fuel Injector Trigger Signal Test

See Test Set-Up on page 36-4

Look for a Square Wave

As we discussed, with the engine running, there should be a square wave signal on the injector's ground wire.

Typical Normal Injector Signal
Frequency and duty cycle–varying square wave on the ground wire

- Use an automotive meter set to detect a square wave and measure frequency. For more information, see the **Tools for Dynamic Analog Signals** chapter.
- Use the positive meter lead to back-probe the negative terminal of the injector connector.
- Connect the meter's negative lead to ground.
- With the car running, the meter should detect the presence of the square wave.
- As you increase engine rpm, you should see the frequency increase, and the duty cycle (pulse width) slightly increase.
- If there is no square wave detected at the meter, consult a wiring diagram for the car and troubleshoot the problem.

36

Supplemental Testing

If you've verified there's power and signal at an injector, and that the solenoid functions, then it's possible that there may be a non-injector-related cause.

If the solenoid *does* click but poor performance persists, the injector may need to be cleaned. It can be removed and taken to a shop that cleans injectors.

If all of the above check out but the misfire persists, the problem is likely not a injector problem at all. The ignition, cylinder compression, and carbon buildup on the valves would need to be checked.

Is a Scan Tool Useful?

On a post-1996 car, yes; injector malfunction should generate a code specific to that cylinder, and the code should be displayed on an OBD-II scan tool. A model-specific scan tool may provide additional information, such as whether the fault is on the power or ground side.

Is a Noid Light Useful?

Probably. A "Noid light" ("noid" for "soleNOID") is a small incandescent light, usually mounted in a connector with pins that are designed to be plugged directly into a solenoid's connector. Testing fuel injector operation is a primary application for Noid lights. If the Noid light flashes, the injector is receiving a signal. If the Noid light stays on solid, the signal is not being toggled. If the Noid light does not come on, the injector is not receiving a signal. Noid lights are usually available in a set with different lights for different applications. You would need to verify that a set has a Noid light for your specific injector.

Is an Oscilloscope Useful?

A scope can be used to verify the presence of the square wave at the injector. Remember that this is *not* a classic PWM square wave. As engine rpm increases, the *frequency* of the square wave increases, *not the duty cycle (pulse width)*. The duty cycle (pulse width) will increase as engine load increases.

Additional Information

Related Chapters

- **Common Multimeter Tests** for a discussion on how to use a multimeter to measure voltage and resistance.
- **Dynamic Analog Signals** for a discussion on square waves and sine waves.
- **Tools for Dynamic Analog Signals** for a discussion on how to use an automotive multimeter to detect sine waves and square waves.
- **Testing Solenoids** for more information on testing solenoids.

Technical Notes

As is the case with stick coils, on post-1996 OBD-II cars, most people are alerted to a bad injector by the CEL being lit and the scan tool reporting a misfire code for a specific cylinder. Therefore, in theory, you could swap the injector with the adjacent one, and see if the problem stays with the cylinder or moves with the injector. Unfortunately, because the access to the injectors is often difficult, and because removing them from the fuel rail is often non-trivial, they are often considerably more difficult to swap than stick coils. As such, swapping injectors is not the go-to test it is with stick coils.

Despite the fact that injectors are solenoids controlled by a PWM signal, note that injectors are not technically PWM solenoids. That is, unlike a true PWM solenoid, the injector does NOT use the duty cycle of the square wave to open the solenoid a proportional amount. Instead, injectors are either fully open or fully closed, and toggle between the two states in response to the timing of the square wave.

37

Testing Fuel Pumps

Multimeter Test Set-Ups

⚠ WARNING —

Make personal safety your top priority. Watch what you are doing, stay alert. Do not work on your car if you are tired, not feeling well, or in a bad mood. Common sense, alertness, and good judgment are crucial to safe work on your vehicle.

Protect yourself from shock, trauma, and hazardous chemicals: Wear safety glasses, work boots, and protective gloves when working on your vehicle. Have hearing protection available for work that generates noise.

Be sure the work environment is well lit, properly ventilated, and that the vehicle is on a level surface. The floor, vehicle, and components inside the vehicle should be dry to minimize the possibility of shocks.

This book is designed as background technical training and as a general overview of how to do electrical diagnosis and service. It is designed to be used in conjunction with the appropriate service information for your vehicle, not to replace factory service information.

If you lack the skills, tools and equipment, or a suitable workshop for any procedure described in this book, have the job done by a qualified shop or an authorized dealer.

For additional warnings and cautions, see the **Safety**, *chapter and the warnings, cautions, and notices in your vehicle manufacturer's service information.*

Fuel Pump Supply Voltage

Test Set-Up 37-A

See page 37-7 for step-by-step procedure

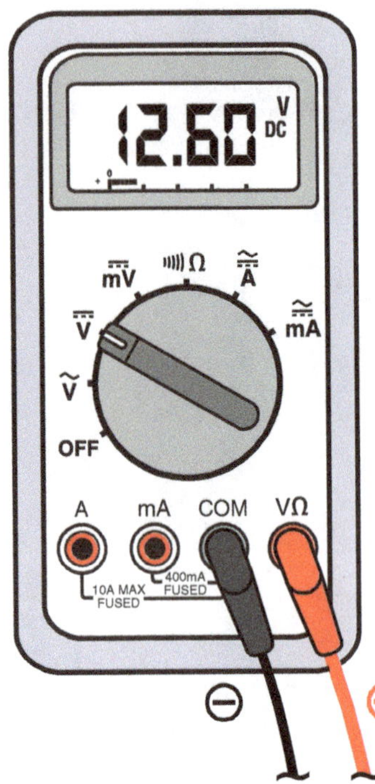

Testing Fuel Pump Supply Voltage

Test Conditions

Fuel pump	
Dial set to:	V DC Volts
Red input terminal:	VΩ
Black input terminal:	COM
Red lead:	power wire (+)
Black lead:	chassis ground (–)

Connector connected, or disconnected and probed at power wire if necessary

Ignition ON, relay jumpered if necessary

Typical Normal Values

Voltage: Battery voltage (approximately 12.6 volts)

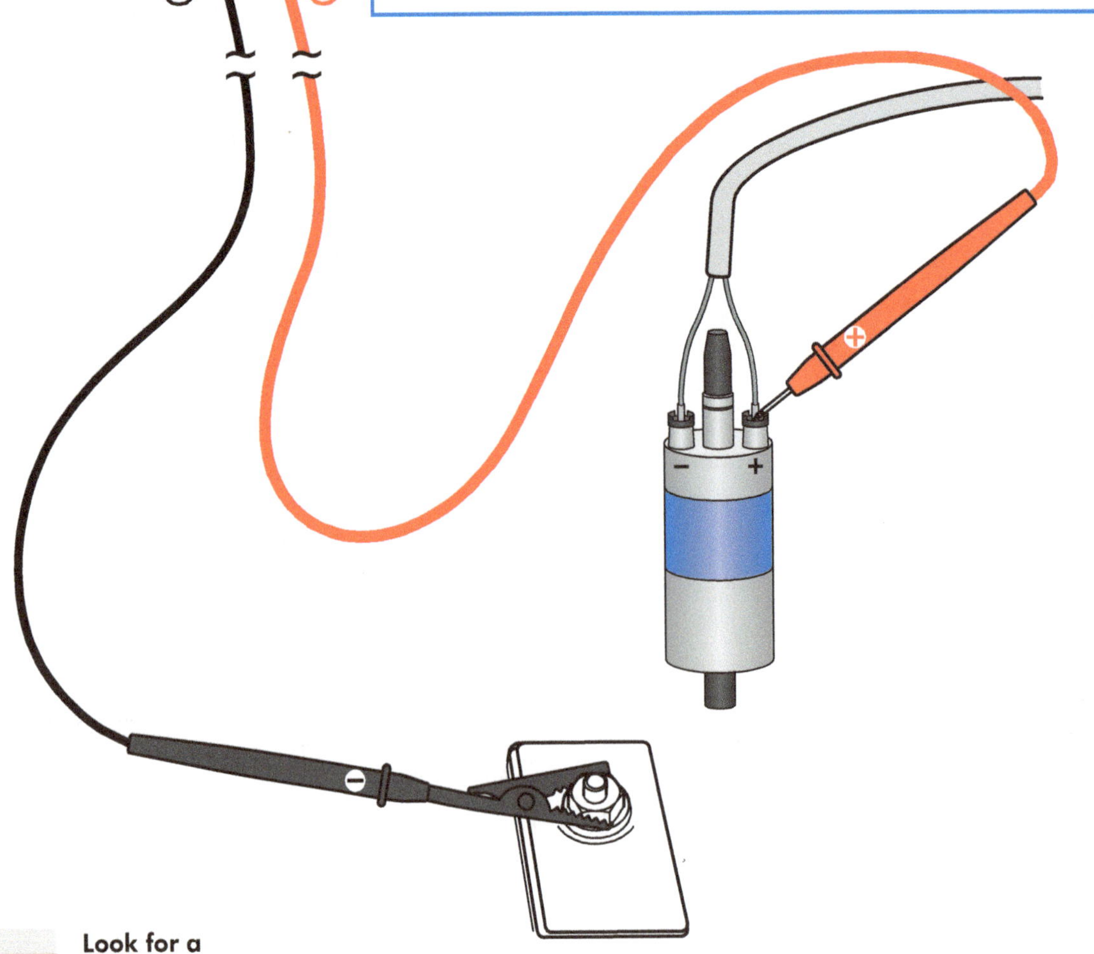

Look for a Static Value

Fuel Pump Resistance

See page 37-8 for step-by-step procedure

Test Set-Up **37-B**

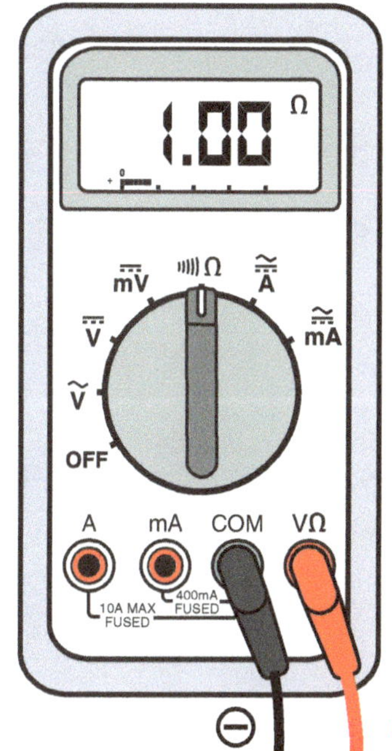

Testing Resistance Across Fuel Pump

Test Conditions

Fuel pump	
Dial set to:	Ω resistance
Red input terminal:	VΩ
Black input terminal:	COM
Red lead:	fuel pump power pin (+)
Black lead:	fuel pump ground pin (–)
Connector disconnected, testing at fuel pump	
Engine off	

Typical Normal Values

Resistance: Approximately 1Ω

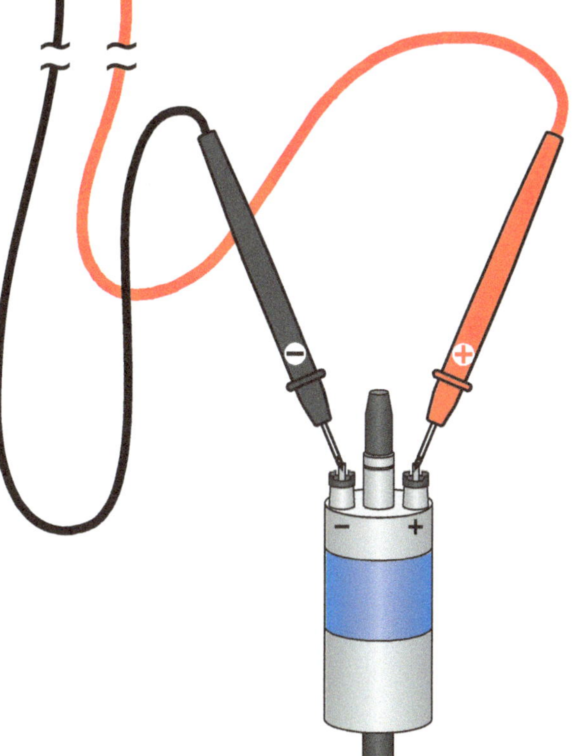

Look for a Static Value

Fuel Pump PWM Signal

See page 37-8 for step-by-step procedure

Test Set-Up 37-C

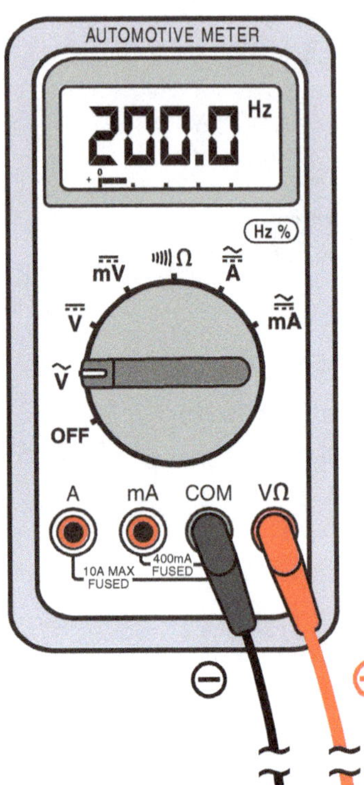

Testing Fuel Pump PWM Signal

Test Conditions

Fuel pump

Automotive meter that reports duty cycle of a square wave

Dial set to:	Ṽ AC Volts / Hz %
Red input terminal:	VΩ
Black input terminal:	COM
Red lead:	sensor signal wire (+)
Black lead:	chassis ground (–)

Connector remains connected, back-probed at signal wire

Engine running

Typical Normal Values

Signal Duty Cycle: Varies with engine load

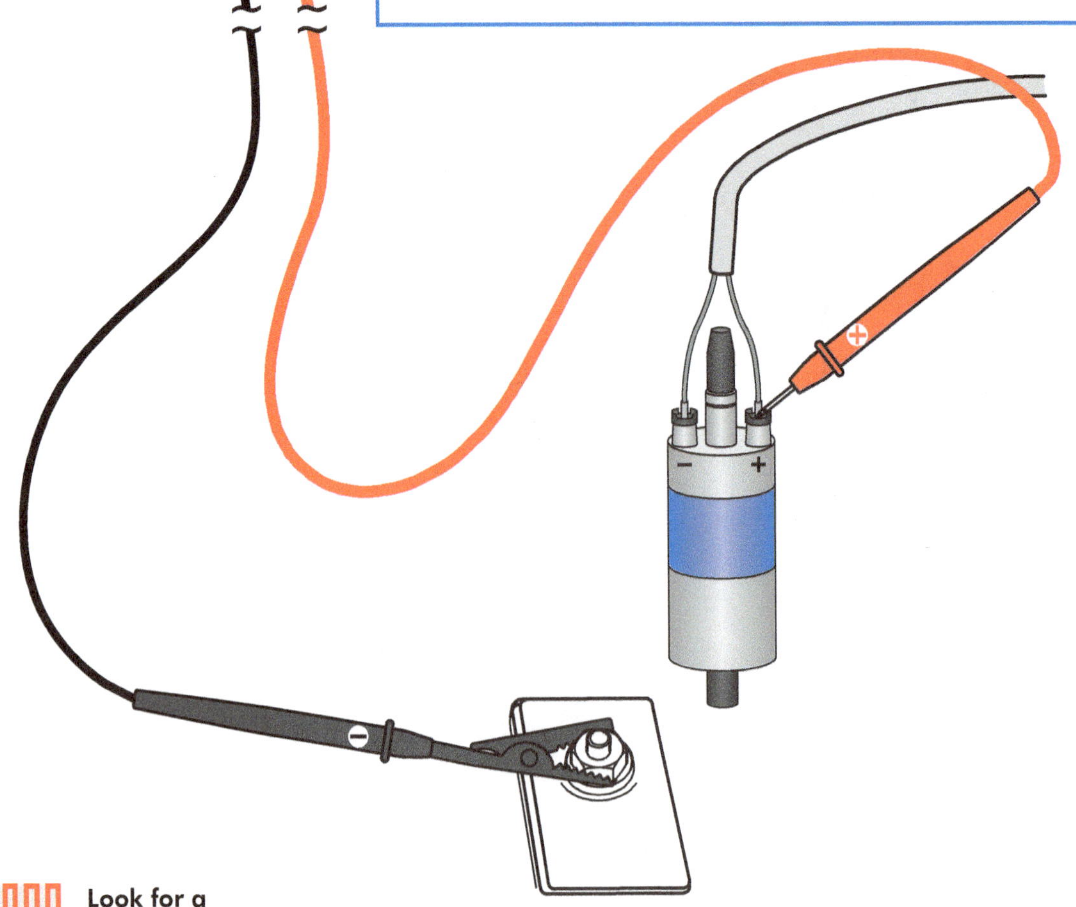

Look for a Square Wave

Fuel Pump Current Draw

See page 37-9 for step-by-step procedure

Test Set-Up 37-D

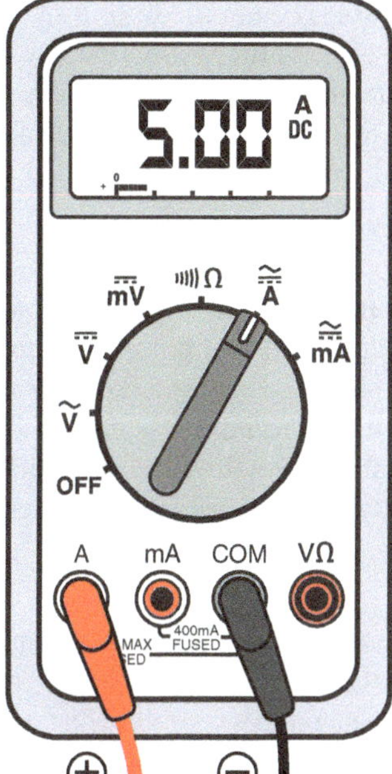

Testing for Fuel Pump Current Draw

Test Conditions

Fuel pump

Dial set to:	A Current
Red input terminal:	A
Black input terminal:	COM
Red lead:	fuel pump ground connection (+)
Black lead:	chassis ground (–)

Ignition ON

Typical Normal Values

About 1 amp per 10 psi of fuel pressure

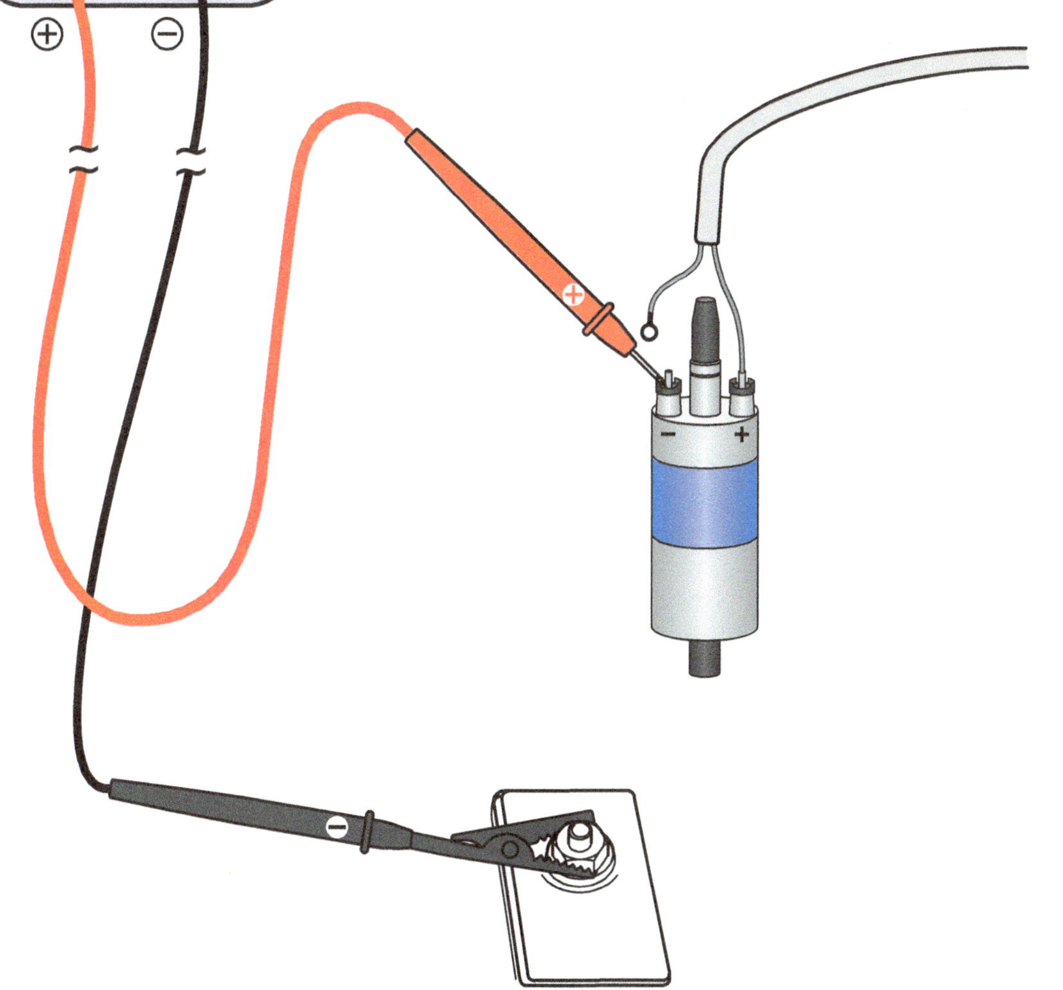

Fuel Pump

Fuel pumps, obviously, pump fuel. Those on old carbureted cars are usually mechanical in nature, using a diaphragm driven by a pushrod from the camshaft. These fall outside the scope of an auto electrics book. Because fuel injection systems require fuel to be delivered to the injectors at higher pressure, electric fuel pumps largely replaced mechanical ones as fuel injection overtook carburetion in the mid to late 1970s. It is these we will cover.

Depending on the make and model of your car, the electric fuel pump may be inside the fuel tank, or externally mounted beneath the car, or mounted on the engine. Or the car may have more than one fuel pump. See **Figure 1**. You'll need to consult a repair manual or an enthusiast forum to learn your car's fuel pump configuration.

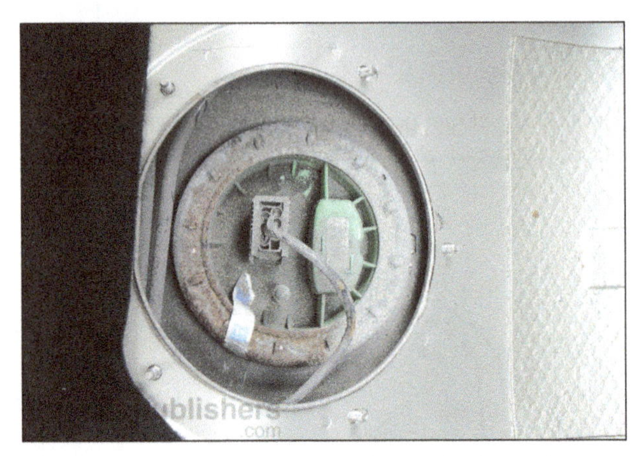

Figure 1 Fuel pump located under the back seat bottom of a 2007 MINI Cooper S.

Most owners are alerted to a bad fuel pump in the most dramatic fashion possible–the car cranks but won't start, or dies while being driven in a way that feels like it's running out of gas. This latter symptom in particular makes the fuel pump or a related fuse or relay immediately suspect.

Operation

Most electric fuel pumps are simply electric motors with a wound armature spinning inside a set of magnets like any other electric motor. The windings in the electric motor of a fuel pump are actually cooled by the fuel passing through them, which is a bit counterintuitive until you remember that gas itself isn't explosive–it's the mixture of gas and air that's explosive.

As the pump spins, it sucks gas out of the tank and pumps it to the engine. The actual pressure of the fuel supplied to the injectors is controlled by a fuel pressure regulator in the engine compartment. Any fuel not used by the engine is sent back to the tank via a return fuel line.

Types

Up through the 1990s, electric fuel pumps were supplied a constant voltage. Once powered, these run at a fixed speed and deliver a fixed fuel pressure which is higher than that required by the engine. Beginning in about 2000, some manufacturers began incorporating fuel pumps that utilize a Pulse Width Modulated (PWM) signal from the ECM to control the pump's delivery pressure. This allows the fuel pressure to more closely match engine demand.

Signal

Conventional electric fuel pumps are powered by a simple constant voltage. PWM fuel pumps are instead controlled by a PWM signal, which is a fixed frequency square wave signal with a varying pulse width. This is the same thing as a varying duty cycle. For additional information, see the **Dynamic Analog Signals** chapter.

Circuit

Both constant voltage fuel pumps and PWM pumps use two wires carrying voltage and ground. However, because some in-tank fuel pumps combine the pump and a fuel level sensor, there may be more than two wires on the fuel pump connector.

On most cars built since the late 1970s, for safety reasons, the fuel pump is controlled by a relay. This enables the pump to run only when the engine is running, preventing fuel from continuing to pump if the engine is off. 1970s and 1970s-era fuel-injected cars typically had a micro-switch incorporated into the airflow meter that signaled the fuel pump relay to turn on the pump when air was being drawn into the engine.

Modern cars typically have the ECM directly control the relay. This is the case with the fuel pump wiring example shown in **Figure 2**.

Testing

Locate the fuel pump or pumps. Disconnect the connector and physically inspect both sides. Make sure no wires are broken and no pins are pushed out. Inspect the fuel pump itself to make sure it isn't cracked.

The voltage to, the current drawn by, and the resistance across a fuel pump can be verified with any multimeter, but if the fuel pump is a PWM pump and is receiving the square wave pulse train from the ECM, an automotive multimeter that can detect a square wave and measure its duty cycle, such as the Fluke 88V, can be employed. See the **Equipment Needed for Fuel Pump Testing Table**.

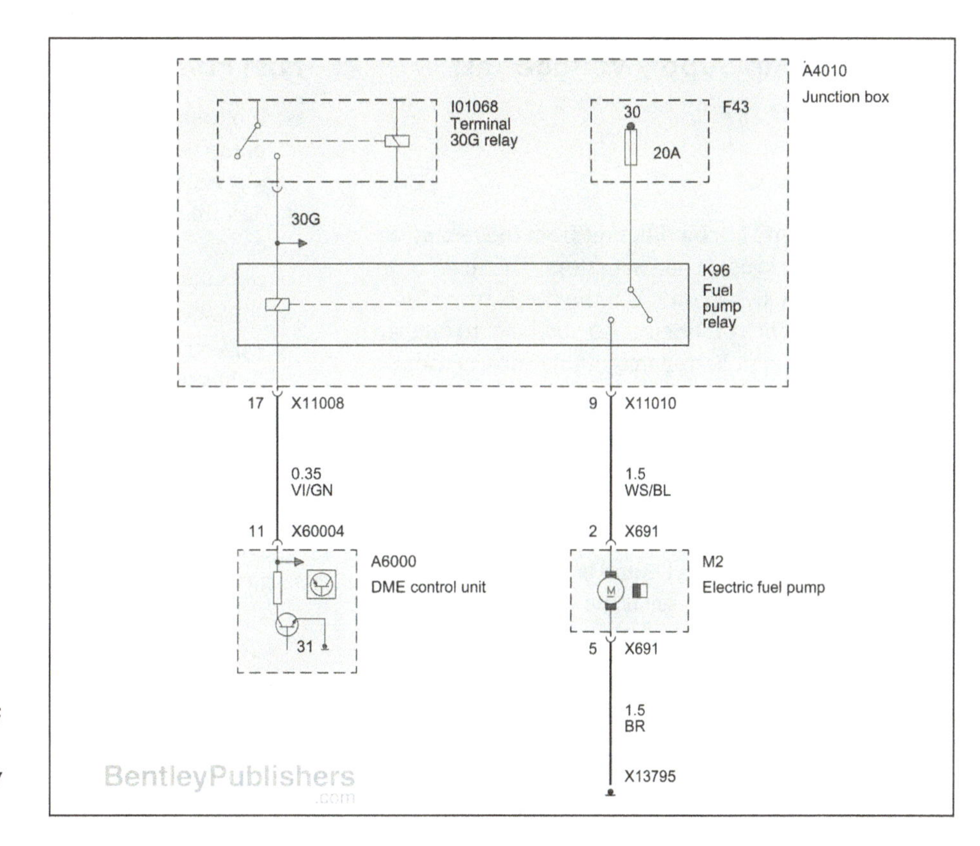

Figure 2 Fuel pump and relay wiring on a 2007 MINI Cooper. The fuel pump has two wires, the voltage side of which is actuated by a relay, which is, in turn, controlled by the DME.

WARNING —

When testing the fuel pump, be absolutely certain that no spilled or leaking fuel is present. When you disconnect and reconnect the fuel pump connector, it can create a spark which could ignite any spilled or leaking fuel, potentially causing death. Wear rubber gloves and have a fire extinguisher handy at all times when testing the fuel pump.

If the fuel pump is leaking fuel, do not proceed with the electrical test! Wear rubber gloves and isolate the source of the leak. If it is at a hose, tighten the hose until the leak stops, dry all gas, and wash the area before proceeding. If the leak is at the body of the fuel pump or the electrical connector, do not test the fuel pump. Replace it instead.

Equipment Needed for Fuel Pump Testing

Test	Tool
Voltage, amperage, and resistance tests	Multimeter
Square wave signal test	Automotive multimeter that detects a square wave and reports its duty cycle, or oscilloscope

Determining Pinouts

Pinouts are best determined using factory service information (a wiring diagram). However, we have found that the following procedure is useful for confirming that the factory service information is correct, or determining pinouts if a wiring diagram is not available.

- Locate the connector to the fuel pump and disconnect it.
- Set the meter to measure DC voltage.
- Connect the black lead to any chassis ground.
- Turn the key to ignition. Depending on the age and model of the car, it may be necessary to enable the fuel pump via additional means, such as propping open the door to the airflow meter, or jumpering across the fuel pump relay.
- On the wiring harness side of the connector, touch the red lead to each of the pins. The pin that reads battery voltage (about 12.6 volts) should be the supply voltage. The other pin should be ground.
- Turn the key off.

Fuel Pump Supply Voltage Test

See Test Set-Up on page 37-2

Look for a Static Value

Testing a fuel pump is a bit different from the testing we describe of other sensors and actuators. The fuel pump generally doesn't switch on when the key is turned to ignition (or switches on briefly), so you have to take an extra step to power it. Fortunately, unlike most sensors, a non-PWM fuel pump can be easily tested by wiring it directly to the battery.

- Look up the location of the fuel pump fuse in your owner's manual. Locate the fuse and verify it is not blown. If it's blown, replace it and see if it blows again. If it does, the fuel pump is likely seized. If the new fuse doesn't blow, see if the car starts. But if it's not blown, continue.
- Look up the location of the fuel pump relay in a repair manual or an online forum. If the relay is an ISO standard relay, remove the relay and jumper across terminals 30 and 87 in the relay socket as described in the **Switches and Relays** chapter. If it is not a standard ISO relay, you'll need to consult a wiring diagram to see which terminals you need to jumper across to supply voltage to the fuel pump.
- Try to start the car. If it now starts but did not before, you have a bad relay. You can test the relay itself following the steps in the **Switches and Relays** chapter. But if it doesn't start, leave the fuel pump relay jumpered and continue.
- Turn the key to ignition to energize relay terminal 30, which should (because the relay socket is jumpered) send voltage out terminal 87 to the fuel pump.
- Put your ear near the fuel pump (or, if it's external to the gas tank, lay your hand on it). You should be able to hear and feel the fuel pump run.
- If you don't hear the fuel pump run, use a multimeter set to measure voltage to verify the presence of battery voltage (about 12.6 V) across the power and ground terminals of the connector to the pump.
  - If the pump has threaded post terminals, the voltage and ground can be checked at those terminals without disconnecting anything.
  - If the pump has a socketed connector, turn off the ignition, pull off the connector, turn on the ignition, and check for voltage and ground at the connector.
- If there is not voltage across the terminals, use the techniques described in the **Troubleshooting** chapter to determine whether the fault is on the power side or on the ground side. Find the cause of the fault and fix it.

Fuel Pump Direct Battery Connection Test

- If battery voltage and ground are present at the fuel pump but the fuel pump still does not run, the pump, if it is not a PWM pump, can be tested directly. As described in the **Troubleshooting** chapter, you can connect the fuel pump's two terminals directly to the car's battery using a long pair of wires with alligator clips on the ends.
  - If you do this, care must be taken not to short the clips against the body of the car or each other.

WARNING —

Do not do this test if there is any spilled or leaking gas, as touching a battery wire to the fuel pump connector may generate a small spark that could ignite any spilled fuel. Wear rubber gloves and have a fire extinguisher handy at all times.

  - Slide a piece of heat shrink tubing over each alligator clip to insulate them from touching each other.
  - Connect the two clips to the fuel pump. Verify that they are secure and do not touch.
  - Connect the clips to the battery, positive to positive and negative to negative.
  - With the fuel pump directly connected to the battery, you should hear and feel the fuel pump run. If it does not run when wired directly to the battery in this fashion, the fuel pump is likely defective and should be replaced.
- If battery voltage and ground are present on the connector but the fuel pump only runs when it is wired directly to the battery, it is possible there is a voltage drop problem. Consult the **Common Multimeter Tests** chapter and perform a voltage drop test on both the voltage and ground sides of the fuel pump.
- Old fuel pumps can sometimes be bound up by internal particulate matter, and can sometimes be persuaded to run by gently tapping on them with the handle of a screwdriver. If tapping the fuel pump successfully revives it, be aware that the pump is living on borrowed time. Drive the car home or to a repair shop, do not shut it off on the way, and replace the pump immediately afterward.

Fuel Pump Resistance Test

See Test Set-Up on page 37-3

Look for a Static Value

The resistance across the fuel pump's terminals should be about an ohm. If the resistance between the two terminals on the connector is infinite (shows no continuity), a brush inside the fuel pump is likely bad or the armature winding is likely broken somewhere. Note that if the resistance is correct, it doesn't guarantee the fuel pump is good; it simply means nothing inside is broken or shorted.

Typical Normal Values for Fuel Pump Resistance
About 1 ohm

- Use a multimeter set to measure resistance.
- Unplug the connector from the fuel pump.
- Touch one multimeter lead to one of the fuel pump's terminals.
- Touch the other multimeter lead to the fuel pump's other terminal.
- Verify that the meter shows approximately 1 ohm.

If it instead shows "OL" (open circuit) or 0 ohms (short circuit), the fuel pump is likely defective and should be replaced.

Fuel Pump PWM Signal Test

See Test Set-Up on page 35-4

Look for a Square Wave

If the pump signal is PWM, there should be a square wave signal on the fuel pump's signal wire.

Typical Normal Values for PWM Fuel Pump Signal
Fixed-frequency square wave. Duty cycle (pulse width) changes from about 50% at idle to near 100% under full engine load.

- Use an automotive meter set to detect a square wave and measure duty cycle, or an oscilloscope.
- Use the red lead to back-probe the signal pin on the fuel pump's connector.
- Connect the black lead to chassis ground.
- Start the car.
- With the car running, the meter should detect the presence of the square wave. The duty cycle at idle should be about 50%
- As you increase engine load, you should see the pulse width and duty cycle increase.

Fuel Pump Current Draw Test

See Test Set-Up on page 37-4

If there is a mechanical restriction in the fuel pump from, for example, worn bearings or the pump having ingested rust from the gas tank, the amount of current drawn by the pump may be higher than normal.

Typical Normal Values for Fuel Pump Current
The current drawn by the pump, as a rule of thumb, should be about 1 amp for each 10 psi in pressure. So, if the fuel pump pressure is 40 psi, it should draw about 4 amps of current.

CAUTION —

You can use the ammeter setting on your multimeter to measure pump current, provided the current doesn't exceed the rating of the meter (usually 10 amps).

- Shut the car off.
- Configure the multimeter to measure current on the high amperage setting (see the **Common Multimeter Tests** chapter).

Method #1: Measuring Between the Fuel Pump and Ground

This requires wiring the multimeter in series with the fuel pump, which can be a little tricky.

- Disconnect the ground connection to the fuel pump.
- If the fuel pump has two wires simply held on by ring or spade terminals, undo the ground wire, clip the red lead to the fuel pump's ground terminal, and clip the black lead to the ground wire (or vice versa). This will cause all current to flow through the meter before it flows to ground.
- If, however, the fuel pump has a push-on connector where both positive and negative wires must be removed at the same time, it will be necessary to fabricate cables with the necessary pins on the end to extend both wires so the positive one can go directly to the fuel pump, and the negative one can go to the ammeter and then to ground.

Method #2: Measuring Across the Relay Socket Terminals

- Because the first method is tricky, it is usually easier to insert the multimeter in place of the jumper across the fuel pump relay.
- Locate and remove the fuel pump relay.
- Insert spade connectors into terminals 30 and 87 of the relay socket (the purpose of the spade terminals is so the multimeter probes don't deform the contacts in the socket).
- With the multimeter configured to measure current on the high amperage setting, connect the two multimeter probes to the two spade connectors with alligator clips.

Method #3: Using a Current Clamp

- If you have a current clamp (either a clamp attachment to a multimeter, or a standalone "clamp meter") that measures DC amperage to an accuracy of 1 amp, that can be used instead. This has the advantage of being a non-invasive measurement (no cables or relays need to be detached).
- Position the clamp around the wires leading to the fuel pump.

Whichever of the three methods you use, start the car and observe the current drawn by the fuel pump. The rule of thumb of 1 amp per 10 psi is exactly that – a rule of thumb. If the current exceeds that but the fuel pump has no other symptoms, you'd be wise to see if the nominal fuel pump current is published in a repair manual or listed on an enthusiasts website before condemning and replacing the pump. If, however, the fuel pump is noisy or has suffered a won't-turn-until-smacked episode, the totality of evidence would point toward replacement being a wise path.

Supplemental Testing

If you've verified there's power at the fuel pump, and the fuel pump is spinning, but a problem persists, it may be that the fuel pump pressure is too low.

 **WARNING —**

When pressure-testing a fuel pump, the risk of fire exists from spilled fuel. Wear rubber gloves and have a fire extinguisher handy at all times.

Most mid-2000s and later European manufacturers use High-Pressure Direct Injected (HPDI) engines. Whereas previous electric fuel pumps operate at tens of psi, HPDI pumps may operate at thousands of psi. Under no circumstances should you attempt a fuel pressure or volume measurement on a high-pressure fuel pump.

To pressure test a fuel pump, you need a fuel pressure gauge. On a 1970s or 1980s era fuel-injected car whose rubber fuel lines are held on with hose clamps, you simply undo one of the clamps, put a Tee fitting with a short hose in line, and put the pressure gauge at the base of the Tee. Beginning approximately in the 1980s, however, hose clamp fittings began to be replaced with dedicated snap-lock pressure fittings, making connection of a fuel pressure gauge require special fuel line adapters. Consult a repair manual for your car to see what the fuel pressure is supposed to be and what adapters may be necessary. Be aware that a blocked line out of the fuel pump may cause low pressure, whereas a blocked return line may cause high pressure.

In addition to reporting the proper fuel pressure, some repair manuals also list the volume of fuel that the pump is supposed to produce in a given time frame (e.g., one gallon in 20 seconds). If this information is provided by your repair manual, it is worth verifying the pump's performance against it before replacing the pump.

Is a Scan Tool Useful?

On a post-1996 car, yes; incorrect fuel pump operation may generate a code, and the code should be displayed on an OBD-II scan tool. A model-specific scan tool may provide additional information.

Is an Oscilloscope Useful?

Only for a PWM pump. A scope can be used to verify the presence of the square wave at the pump. As engine load increases, the duty cycle (pulse width) should increase.

Additional Information

Related Chapters

- **Common Multimeter Tests** for a discussion on how to use a multimeter to measure voltage and resistance.
- **Dynamic Analog Signals** for a discussion on square waves and sine waves.
- **Tools for Dynamic Analog Signals** for a discussion on how to use an automotive multimeter to detect sine waves and square waves.

Technical Notes

Note that the techniques described for determining the pinouts, checking for voltage, jumpering across the relay, wiring the fuel pump directly to the battery, and performing a voltage drop test apply not only to the fuel pump, but also to any electric motor, including window motors and cooling fans.

38

Testing Solenoids (Actuators)

Multimeter Test Set-Ups

⚠ WARNING —

Make personal safety your top priority. Watch what you are doing, stay alert. Do not work on your car if you are tired, not feeling well, or in a bad mood. Common sense, alertness, and good judgment are crucial to safe work on your vehicle.

Protect yourself from shock, trauma, and hazardous chemicals: Wear safety glasses, work boots, and protective gloves when working on your vehicle. Have hearing protection available for work that generates noise.

Be sure the work environment is well lit, properly ventilated, and that the vehicle is on a level surface. The floor, vehicle, and components inside the vehicle should be dry to minimize the possibility of shocks.

This book is designed as background technical training and as a general overview of how to do electrical diagnosis and service. It is designed to be used in conjunction with the appropriate service information for your vehicle, not to replace factory service information.

If you lack the skills, tools and equipment, or a suitable workshop for any procedure described in this book, have the job done by a qualified shop or an authorized dealer.

For additional warnings and cautions, see the **Safety**, *chapter and the warnings, cautions, and notices in your vehicle manufacturer's service information.*

Solenoid Voltage

Test Set-Up 38-A

See page 38-6 for step-by-step procedure

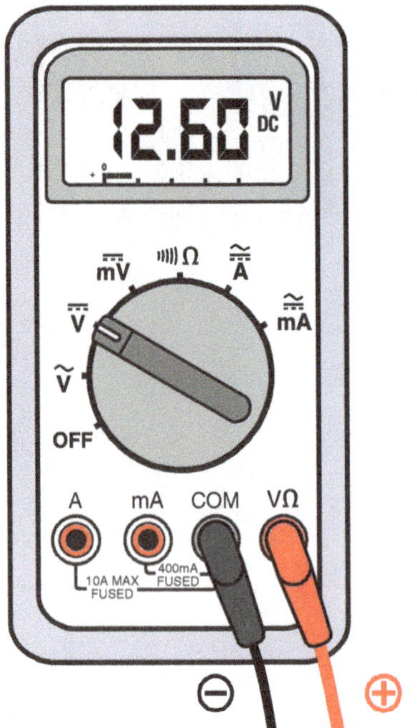

Testing for Solenoid Supply Voltage

Test Conditions

Solenoid

Dial set to: V DC Volts

Red input terminal: VΩ

Black input terminal: COM

Red lead: power wire (+)

Black lead: chassis ground (–)

Connector connected, or disconnected and probed at power wire if necessary

Engine off, solenoid powered

Typical Normal Values

Voltage: Battery voltage (approximately 12.6 volts)

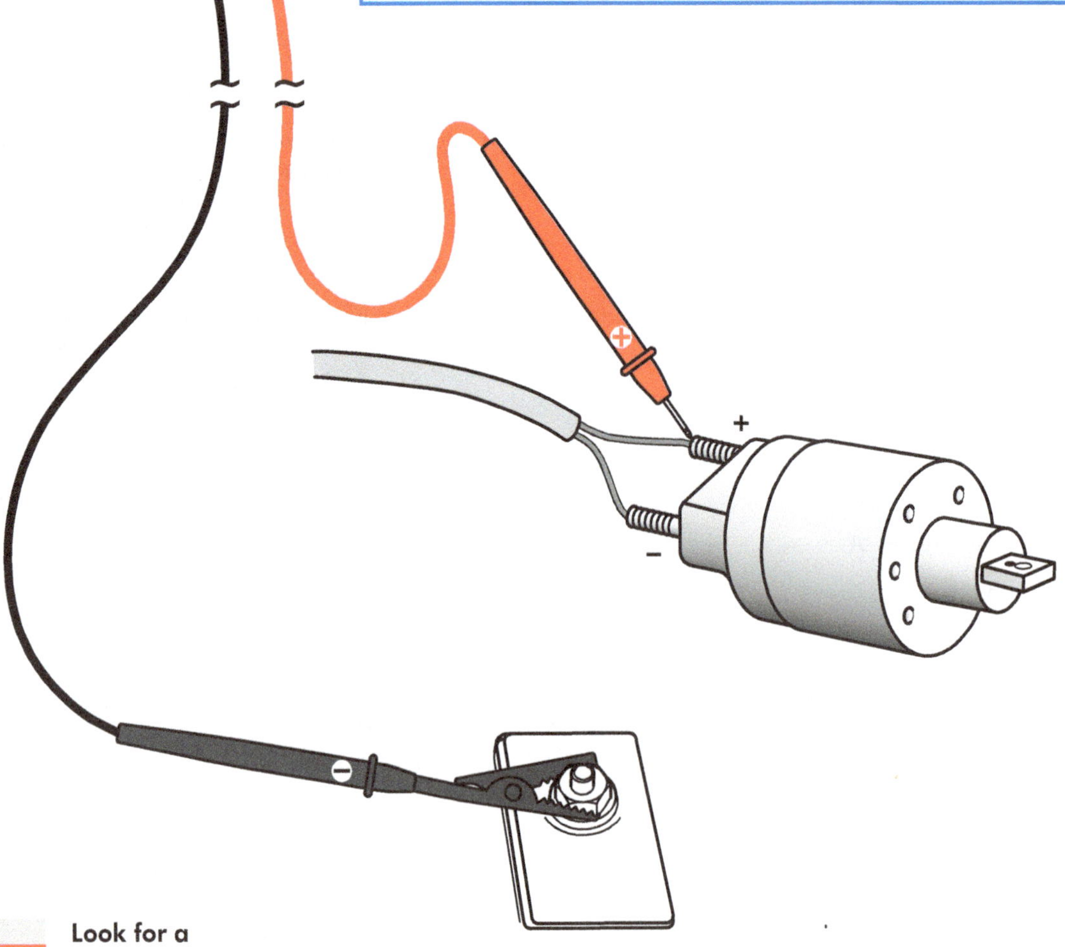

Look for a Static Value

Solenoid Resistance

See page 38-6 for step-by-step procedure

Test Set-Up **38-B**

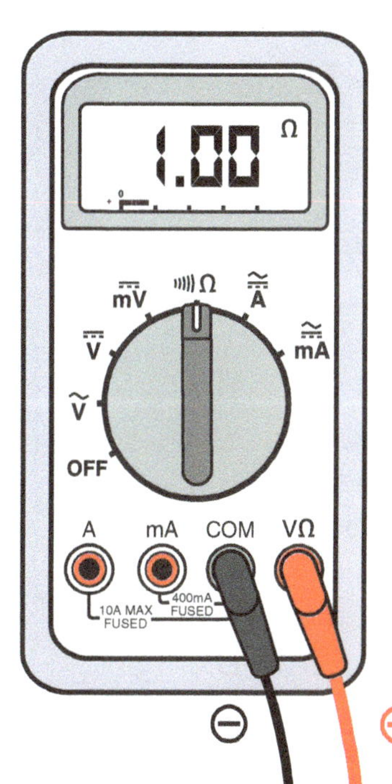

Testing Resistance Across Solenoid

Test Conditions

Solenoid

Dial set to:	Ω resistance
Red input terminal:	VΩ
Black input terminal:	COM
Red lead:	solenoid power pin (+)
Black lead:	solenoid ground pin (–)

Connector disconnected, testing at solenoid

Engine off

Typical Normal Values

Resistance: Approximately 1Ω

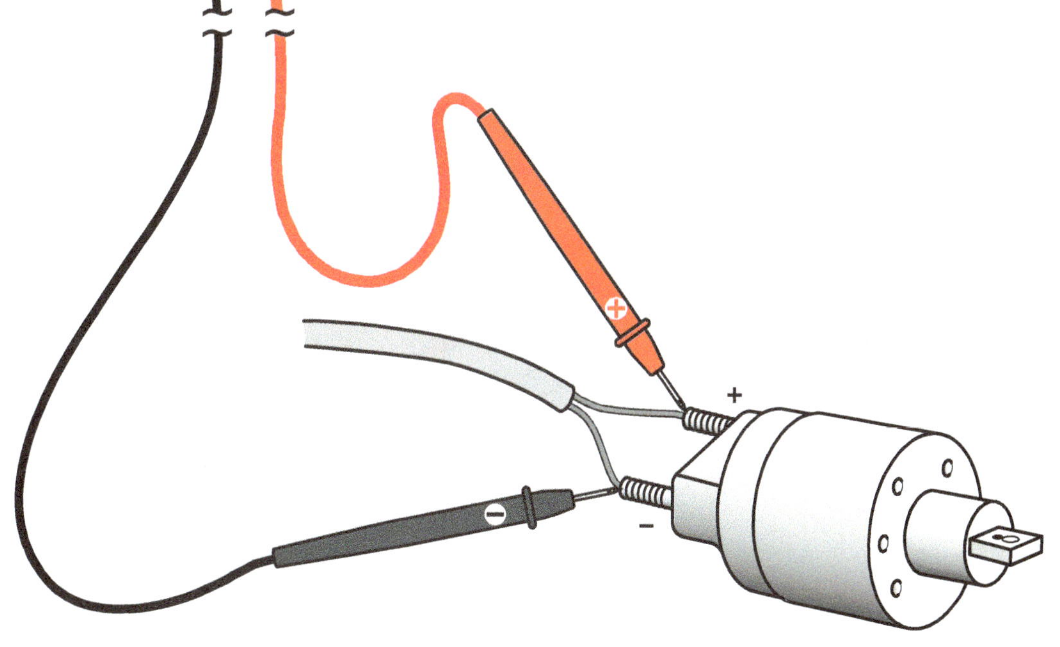

Look for a Static Value

PWM Solenoid Trigger Signal

Test Set-Up 38-C

See page 38-7 for step-by-step procedure

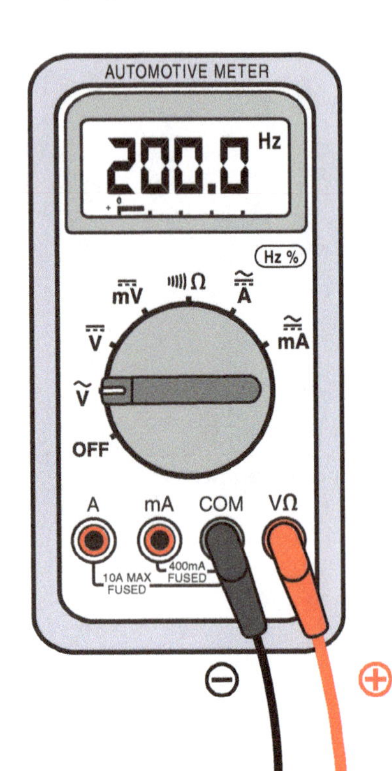

Testing for Varying Duty Cycle of Square Wave

Test Conditions

Solenoid

Automotive meter that reports duty cycle of a square wave

Dial set to:	Ṽ AC Volts Hz %
Red input terminal:	VΩ
Black input terminal:	COM
Red lead:	solenoid signal wire (+)
Black lead:	chassis ground (–)

Connector remains connected, back-probed at signal wire

Solenoid exercised

Typical Normal Values

Signal Duty Cycle: Varies with solenoid control

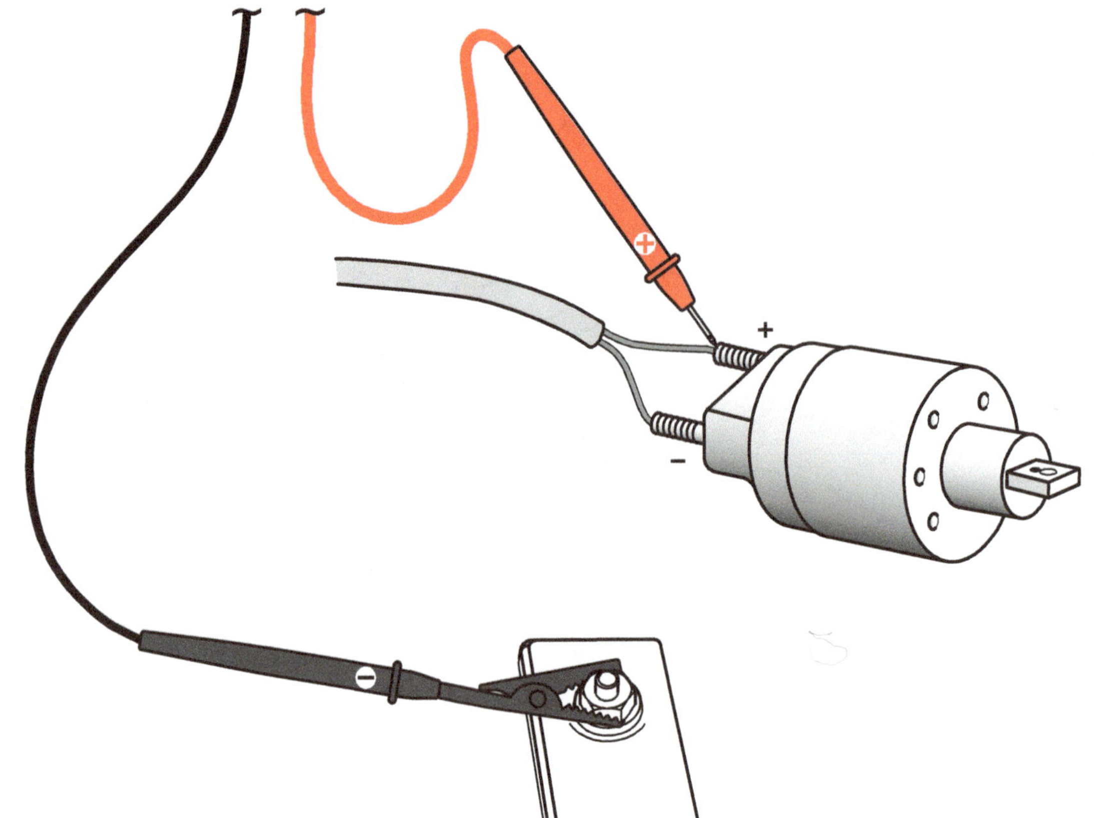

Look for a Square Wave

Solenoids and Other Actuators

Solenoids are devices that use electrical current to perform linear physical motion (moving something forward and backward, or in and out). They are used in electronically controlled valves, and in certain types of electrical relays such as those found on starter motors. See **Figure 1**.

Figure 1 VANOS solenoid on a BMW M52TU motor.

Operation

The primary element of a solenoid is an electromagnet, but instead of having a coil of wire wound around a fixed iron core like a regular electromagnet, the core is instead free to move back and forth inside the coil of wire. When employed in this configuration, the core is often referred to as a plunger. When the coil of wire is energized with current, the induced magnetic field drives the plunger out of the coil. A spring usually presses the plunger back into place when the current is removed.

The forward-back action makes solenoids useful as electronically actuated valves to control the flow of fluids, air, or heat. Applying voltage to the valve will cause the plunger to extend, opening or closing the valve. The small valves that modulate the vacuum lines on mid-1970s cars, or open and close ports in carburetors, are solenoids. Fuel injectors are solenoids as well.

In addition to being used as simple valves, solenoids are integrated into a variety of mechanical applications from electronic door locks to ABS brakes to camshaft advance mechanisms. In this context, they're sometimes referred to as "actuators," though that term is also applied to many devices that move when supplied with electricity.

Signal

Simple solenoids fully extend the plunger when battery voltage is applied and fully retract it when voltage is absent. In addition, there are Pulse Width Modulated (PWM) solenoids which accept a PWM square wave as a control signal and extend the plunger a proportional amount in response to the square wave duty cycle (e.g., the valve is extended halfway in response to a square wave with a 50% duty cycle).

Circuit

Solenoid wiring varies with the application. A simple solenoid such as a valve has two wires – power and ground. Starter solenoids, because they are grounded directly to the engine, have no ground wire, but they have additional wires because they are also functioning as relays.

A PWM solenoid generally has two wires but may also have a shielded wire. In the case of ABS control modules and variable valve timing modules, solenoids may be integrated into larger assemblies (that is, they may be soldered directly onto a printed circuit board inside an enclosure).

Testing

The voltage to, the current drawn by, and the resistance across a solenoid can be verified with any multimeter, but if the solenoid is a PWM solenoid receiving a square wave pulse train from the ECM, an automotive multimeter that can detect a square wave and measure its duty cycle, such as the Fluke 88V, should be employed to detect the signal.

Equipment Needed for Solenoid Testing

Test	Tool
Voltage and resistance	Multimeter
Square wave signal	Automotive multimeter that detects a square wave and reports its duty cycle, or oscilloscope

Locate the solenoid. Physically inspect both sides of the connector. Make sure no wires are broken and no pins are pushed out. Inspect the solenoid itself to make sure it isn't cracked.

Determining Pinouts

Pinouts are best determined using factory service information (a wiring diagram). However, we have found that the following procedure is useful for confirming that the factory service information is correct, or determining pinouts if a wiring diagram is not available.

- Locate the connector to the solenoid and disconnect it.
- Set the meter to measure DC voltage.
- Connect the black lead to any chassis ground.
- Do whatever needs to be done to power the solenoid. This could simply be turning the key to ignition, or it may require pressing a switch or jumpering across a relay.
- On the wiring harness side of the connector, touch the red lead to each of the pins. The pin that reads battery voltage (about 12.6 volts) should be the supply voltage pin. The other pin should be ground.
- Turn the power off.

Solenoid Voltage Test

See Test Set-Up on page 38-2

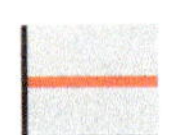

Look for a Static Value

The voltage input to a simple solenoid is usually the car's charging voltage, whereas a PWM solenoid will require a square wave as input. The voltage input to a PWM solenoid is may be 12V or 5V; consult a repair manual for your car to know for certain.

These are summarized in the **Typical Normal Solenoid Voltage Values Table**.

Typical Normal Solenoid Voltage Values

	Simple Solenoid	PWM Solenoid
Voltage	Battery voltage (12.6V) or charging voltage (14.2V)	5V or 12V square wave

- Look up the location of the solenoid fuse in your owner's manual. Locate the fuse and verify it is not blown. If it's blown, replace it and see if it blows again. If it does, the solenoid is likely internally shorted. If the new fuse doesn't blow, see if the solenoid now works. If the fuse is not blown, continue.
- Do whatever is necessary to power the solenoid. This may simply involve turning the key to ignition, or may require pushing a control button or jumpering across a relay.
- Touch the solenoid with your finger. You should be able to hear and feel it click when it is powered.
- If you don't feel the solenoid click, use a multimeter set to measure voltage to verify the presence of battery voltage (about 12.6V) across the power and ground terminals of the connector to the solenoid.
- If there is not voltage across the terminals, use the techniques described in the **Troubleshooting** chapter to determine whether the fault is on the power side or on the ground side. Find the cause of the fault and fix it.
- If battery voltage and ground are present at the solenoid but the solenoid still does not actuate, it is easy to test the solenoid directly if it is out in the open and if it is not a PWM solenoid. As described in the **Troubleshooting** chapter, you can connect solenoid's two terminals directly to the car's battery using a long pair of wires with alligator clips on the ends.
    - If you do this, care must be taken not to short the clips against the body of the car or each other.
    - Slide a piece of heat shrink tubing over each alligator clip to insulate them from touching each other.
    - Connect the two clips to the solenoid. Verify that they are secure and do not touch.
    - Connect the clips to the battery, positive to positive and negative to negative.
    - With the solenoid directly connected to the battery, you should hear and feel it click. If it does not click when wired directly to the battery in this fashion, the solenoid is likely defective and should be replaced.

Solenoid Resistance Test

See Test Set-Up on page 38-3

Look for a Static Value

The resistance across solenoid terminals should be about an ohm. If the resistance between the two terminals is infinite (shows no continuity), the winding is likely broken somewhere. Note that if the resistance is correct, it doesn't guarantee the solenoid is good; it simply means nothing inside is broken or shorted. In contrast, because a PWM solenoid almost certainly has integrated circuitry to process the square wave signal, a resistance measurement across the terminals may not be useful.

Typical Normal Simple Solenoid Resistance
Approximately 1 ohm

- Use a multimeter set to measure resistance.
- Unplug the connector from the solenoid.
- Touch one multimeter lead to one of the solenoid's terminals.

- Touch the other multimeter lead to the solenoid's other terminal.
- Verify that the meter shows approximately 1 ohm.
- If it instead shows "OL" (open circuit) or 0 ohms (short circuit), the solenoid is likely defective and should be replaced.

PWM Solenoid Trigger Signal Test

See Test Set-Up on page 38-4

Look for a Square Wave

If the solenoid is a pulse width modulated (PWM) solenoid, there should be a square wave signal on the solenoid's signal wire.

Typical Normal PWM Solenoid Signal
PWM square wave, pulse width (duty cycle) varies with solenoid input

- Use an automotive meter set to detect a square wave and measure duty cycle, or an oscilloscope.
- Use the red lead to back-probe the signal pin on the solenoid's connector.
- Connect the black lead to chassis ground.
- Power the solenoid. This may require starting the car and turning on whatever control exercises the solenoid (e.g., if it is a heater valve solenoid, turn on the heat).
- As the solenoid is exercised, you should see the pulse width and duty cycle vary.

Supplemental Testing

If you have direct confirmation via an automotive multimeter or a scope that the ECM is outputting a signal but the solenoid is not responding, there may be a broken wire or bad connection between the solenoid and the ECM.

- Use a multimeter set to measure resistance to verify continuity between the ground wire on the connector and chassis ground.
- Use a multimeter set to measure resistance to verify continuity between the signal wire on the connector and the ECM. This will require a wiring diagram for your car, knowledge of the location of the ECM, and knowledge of which pin on the ECM connector carries the signal from the solenoid.

Is a Scan Tool Useful?

If a solenoid is part of the emission control system (the EGR valve, for example), incorrect operation should generate a code, and the code should be displayed on an OBD-II scan tool. A model-specific scan tool may provide additional information, such as whether the fault is on the power or ground side.

Is a Noid Light Useful?

Maybe. A "Noid light" ("noid" for "soleNOID") is a small incandescent light, usually mounted in a connector with pins that are designed to be plugged directly into a solenoid's connector. If the Noid light flashes, the solenoid is receiving a signal. If the Noid light stays on solid, the signal is not being toggled. If the Noid light does not come on, the solenoid is not receiving a signal. Noid lights are usually available in a set with different lights for different applications. You would need to verify the set has a Noid light for the specific connector on your solenoid.

Is an Oscilloscope Useful?

Only for a PWM solenoid. A scope can be used to verify the presence of the square wave at the solenoid. As the solenoid is extended, its duty cycle (pulse width) should increase.

Additional Information

Related Chapters

- **Common Multimeter Tests** or a discussion on how to use a multimeter to measure voltage and resistance.
- **Dynamic Analog Signals** for a discussion on square waves and sine waves.
- **Tools for Dynamic Analog Signals** for a discussion on how to use an automotive multimeter to detect sine waves and square waves.

Technical Notes

Most people probably know the solenoid best where it is employed as part of a starter motor. As we discussed in the **Starter** chapter, the electromagnet in a solenoid can be used in a very similar manner to the one in a relay to make and break a pair of electrical contacts, creating a remote-controlled high-current switch with a slightly different configuration from a relay. The solenoid's linear plunger movement is perfect for a starter motor because that linear action is what engages and retracts the starter pinion gear from the flywheel. So, in the case of the starter solenoid, it's being used *both* as a relay *and* as a linear translation device.

Note that, although fuel injectors are solenoids that are a controlled by a square wave, they are not, in fact, PWM solenoids. They do not interpret the square wave as a partial extension command. Instead, they are regulator solenoids that open and close fully and very rapidly in response to the timing of the square wave.

> **WARNING**
>
> *Your common sense, good judgment and general alertness are crucial to safe and successful electrical work. Before attempting any electrical service work on your vehicle, be sure to read chapter* **1 Safety** *and the copyright page at the front of the book. Review these warnings and cautions each time you prepare to work on your car. Please also read any warnings and cautions that accompany the procedures in the book.*

A

B

C

D

E

WARNING

Your common sense, good judgment and general alertness are crucial to safe and successful electrical work. Before attempting any electrical service work on your vehicle, be sure to read chapter **1 Safety** *and the copyright page at the front of the book. Review these warnings and cautions each time you prepare to work on your car. Please also read any warnings and cautions that accompany the procedures in the book.*

F

G

H

I

J

K

L

M

N

O

> **WARNING**
>
> *Your common sense, good judgment and general alertness are crucial to safe and successful electrical work. Before attempting any electrical service work on your vehicle, be sure to read chapter* **1 Safety** *and the copyright page at the front of the book. Review these warnings and cautions each time you prepare to work on your car. Please also read any warnings and cautions that accompany the procedures in the book.*

P

Q

R

S

T

U

V

W

Z

Selected Books and Repair Information From Bentley Publishers

Motorsports

Alex Zanardi: My Sweetest Victory
Alex Zanardi and Gianluca Gasparini
ISBN 978-0-8376-1249-2

The Unfair Advantage *Mark Donohue and Paul van Valkenburgh*
ISBN 978-0-8376-0069-7

Equations of Motion - Adventure, Risk and Innovation
William F. Milliken
ISBN 978-0-8376-1570-7

Engineering

Bosch Automotive Handbook
Robert Bosch GmbH
ISBN 978-0-8376-1732-9

Bosch Fuel Injection and Engine Management *Charles O. Probst, SAE*
ISBN 978-0-8376-0300-1

Maximum Boost: Designing, Testing, and Installing Turbocharger Systems
Corky Bell ISBN 978-0-8376-0160-1

Supercharged! Design, Testing and Installation of Supercharger Systems
Corky Bell ISBN 978-0-8376-0168-7

Scientific Design of Exhaust and Intake Systems *Phillip H. Smith & John C. Morrison* ISBN 978-0-8376-0309-4

Physics for Gearheads
Randy Beikmann
ISBN 978-0-8376-1615-5

Audi

Audi A4 Service Manual: 2002-2008, 1.8L Turbo, 2.0L Turbo, 3.0L, 3.2L
Bentley Publishers
ISBN 978-0-8376-1574-5

Audi A4 Service Manual: 1996-2001, 1.8L Turbo, 2.8L *Bentley Publishers*
ISBN 978-0-8376-1675-9

Audi TT Service Manual: 2000-2006, 1.8L turbo, 3.2 L *Bentley Publishers*
ISBN 978-0-8376-1625-4

Audi A6 (C5 platform) Service Manual: 1998-2004 *Bentley Publishers*
ISBN 978-0-8376-1670-4

BMW

Memoirs of a Hack Mechanic
Rob Siegel ISBN 978-0-8376-1720-6

BMW 3 Series (F30, F31, F34) Service Manual: 2012-2015 *Bentley Publishers*
ISBN 978-0-8376-1752-7

BMW X3 (E83) Service Manual: 2004-2010 *Bentley Publishers*
ISBN 978-0-8376-1731-2

BMW X5 (53) Service Manual: 2000-2006 *Bentley Publishers*
ISBN 978-0-8376-1643-8

BMW 3 Series (E90, E91, E92, E93) Service Manual: 2006-2011
Bentley Publishers ISBN 978-0-8376-1723-7

BMW 3 Series (E46) Service Manual: 1999-2005 *Bentley Publishers*
ISBN 978-0-8376-1657-5

BMW 3 Series (E36) Service Manual: 1992-1998 *Bentley Publishers*
ISBN 978-0-8376-1709-1

BMW Z3 (E36/7) Service Manual: 1996-2002 *Bentley Publishers*
ISBN 978-0-8376-1617-9

BMW 5 Series (E60, E61) Service Manual: 2004-2010 *Bentley Publishers*
ISBN 978-0-8376-1689-6

BMW 5 Series (E39) Service Manual: 1997-2003 *Bentley Publishers*
ISBN 978-0-8376-1672-8

Corvette

Corvette: America's Star-Spangled Sports Car *Karl Ludvigsen*
ISBN 978-0-8376-1659-9

Corvette by the Numbers: The Essential Corvette Parts Reference 1955-1982 *Alan Covin*
ISBN 978-0-8376-0228-2

Zora Arkus-Duntov: The Legend Behind Corvette *Jerry Burton*
ISBN 978-08376-0858-7

Porsche

Porsche 911 (996) Service Manual: 1999-2005 *Bentley Publishers*
ISBN 978-0-8376-1710-7

Porsche 911 (993) Service Manual: 1995-1998 *Bentley Publishers*
ISBN 978-0-8376-1719-0

Porsche Boxster Service Manual: 1997-2004 *Bentley Publishers*
ISBN 978-0-8376-1645-2

Porsche 911 Carrera Service Manual: 1984-1989 *Bentley Publishers*
ISBN 978-0-8376-1696-4

Porsche 911 SC Service Manual: 1978-1983 *Bentley Publishers*
ISBN 978-0-8376-1705-3

Porsche: Excellence Was Expected
Karl Ludvigsen
ISBN 978-0-8376-0235-6

Porsche — Origin of the Species
Karl Ludvigsen
ISBN 978-0-8376-1331-4

Volkswagen

Volkswagen Rabbit, GTI Service Manual: 2006-2009 *Bentley Publishers*
ISBN 978-0-8376-1664-3

Volkswagen Jetta Service Manual: 2005-2010 *Bentley Publishers*
ISBN 978-0-8376-1616-2

Volkswagen Jetta, Golf, GTI Service Manual: 1999-2005 *Bentley Publishers*
ISBN 978-0-8376-1678-0

Volkswagen Jetta, Golf, GTI: 1993-1999, Cabrio: 1995-2002 Service Manual *Bentley Publishers*
ISBN 978-0-8376-1660-5

Volkswagen GTI, Golf, Jetta Service Manual: 1985-1992 *Bentley Publishers*
ISBN 978-0-8376-1637-7

Volkswagen Corrado Repair Manual: 1990-1994 *Bentley Publishers*
ISBN 978-0-8376-1699-5

Volkswagen Passat, Passat Wagon Service Manual: 1998-2005
Bentley Publishers
ISBN 978-0-8376-1669-8

MINI Repair Manuals

MINI Cooper Service Manual: 2007-2013 *Bentley Publishers*
ISBN 978-0-8376-1730-5

MINI Cooper Service Manual: 2002-2006 *Bentley Publishers*
ISBN 978-0-8376-1639-1

Mercedes-Benz

Mercedes-Benz C-Class (W202) Service Manual 1994-2000
Bentley Publishers
ISBN 978-0-8376-1692-6

Mercedes Benz E-Class (W124) Owner's Bible: 1986-1995
Bentley Publishers
ISBN 978-0-8376-0230-1

Mercedes-Benz Technical Companion
Staff of The Star and members of Mercedes-Benz Club of America
ISBN 978-0-8376-1033-7

Bentley Publishers has published service manuals and automobile books since 1950. For more information, please contact Bentley Publishers at 1734 Massachusetts Avenue, Cambridge, MA 02138 USA, or visit our web site at
BentleyPublishers.com